Mechanical Behavior of Concrete Materials and Structures: Experimental Evidence and Analytical Models

Mechanical Behavior of Concrete Materials and Structures: Experimental Evidence and Analytical Models

Editors

Dario De Domenico
Luís Filipe Almeida Bernardo

MDPI • Basel • Beijing • Wuhan • Barcelona • Belgrade • Manchester • Tokyo • Cluj • Tianjin

Editors

Dario De Domenico
Department of Engineering
University of Messina
Messina
Italy

Luís Filipe Almeida Bernardo
Centre of Materials and
Building Technologies
(C-MADE), Department of
Civil Engineering and
Architecture
University of Beira Interior
Covilhã
Portugal

Editorial Office
MDPI
St. Alban-Anlage 66
4052 Basel, Switzerland

This is a reprint of articles from the Special Issue published online in the open access journal *Materials* (ISSN 1996-1944) (available at: www.mdpi.com/journal/materials/special_issues/mech_concrete_structure).

For citation purposes, cite each article independently as indicated on the article page online and as indicated below:

LastName, A.A.; LastName, B.B.; LastName, C.C. Article Title. *Journal Name* **Year**, *Volume Number*, Page Range.

ISBN 978-3-0365-4912-5 (Hbk)
ISBN 978-3-0365-4911-8 (PDF)

Contents

About the Editors

Dario De Domenico

Dario De Domenico is an Assistant Professor of Structural Engineering at the Department of Engineering of the University of Messina, Italy, and Research Fellow at C.E.R.I.S.I. (Centre of Excellence in Research and Innovation of Large Dimension Structures and Infrastructures). After graduating in Civil Engineering, he achieved his PhD in Materials and Structural Engineering at the University of Reggio Calabria. He has also been a Visiting Researcher at both the Department of Civil and Structural Engineering, University of Sheffield, UK, and at the Laboratory of Mechanics and Materials, Aristotle University of Thessaloniki, Greece. He has more than 10 years of teaching experience in academic courses, workshops, and specialized seminars in the subject area of civil engineering, particularly structural engineering and structural seismic designs of structures. His Ph.D. thesis, dealing with a numerical limit analysis procedure for composite materials and reinforced concrete structures, has been awarded a special mention from the ACI (American Concrete Institute) Italy Chapter, for the best PhD theses defended in 2012–2014 in Italy in the field of concrete structures. His research activity, since the PhD thesis, has been directed towards the development of numerical models for simulations of the mechanical behavior of concrete elements at ultimate limit states. He has authored/co-authored over 80 articles in international scientific journals. He has been the co-advisor of four PhD students and is currently the co-advisor of one PhD candidate, as well as more than twenty Master's students. He has produced editorial and scientific review work for numerous international journals, having reviewed more than 200 articles.

Luís Filipe Almeida Bernardo

Luís F. A. Bernardo achieved his PhD in Civil Engineering (expertise in Mechanics of Structures and Materials) in 2004 at the University of Coimbra, Portugal. Currently, he is an Associate Professor with Aggregation and a Research Member at C-MADE (Centre of Materials and Building Technologies) in the Department of Civil Engineering and Architecture at the University of Beira Interior in Covilhã, Portugal. His research interests include structural analysis and design, numerical modelling and the optimization of steel/concrete structures, and new structural materials and building systems. He has been a researcher in several national research projects and he collaborates with researchers from Portugal, Brazil, and Italy. He has authored/coauthored about 70 articles in international peer-reviewed journals. He has been working as a reviewer for several international scientific journals and has revised about 270 articles. He has supervised 21 MSc theses and 6 PhD theses in his research field. He has also co-supervised some MSc and PhD Theses, including in Brazil. Additionally, he has developed activities as a consulter and civil engineer for the structural design of buildings and special structures in Portugal.

materials

MDPI

Editorial

Editorial for "Mechanical Behavior of Concrete Materials and Structures: Experimental Evidence and Analytical Models"

Dario De Domenico [1,*] and Luís F. A. Bernardo [2]

1 Department of Engineering, University of Messina, 98166 Messina, Italy
2 Centre of Materials and Building Technologies (C-MADE), Department of Civil Engineering and Architecture, University of Beira Interior, 6201-001 Covilhã, Portugal; lfb@ubi.pt
* Correspondence: dario.dedomenico@unime.it; Tel.: +39-0906765921

Citation: De Domenico, D.; Bernardo, L.F.A. Editorial for "Mechanical Behavior of Concrete Materials and Structures: Experimental Evidence and Analytical Models". *Materials* **2022**, *15*, 4921. https://doi.org/10.3390/ma15144921

Received: 7 July 2022
Accepted: 13 July 2022
Published: 15 July 2022

Publisher's Note: MDPI stays neutral with regard to jurisdictional claims in published maps and institutional affiliations.

Concrete is one of the most widespread materials in the civil engineering field due to its versatility for both structural and non-structural applications depending on the density range, competitiveness in terms of durability and manufacturing costs, as well as ease in finding raw constituent elements. For this reason, the mechanical behavior of concrete and, even more, reinforced concrete (RC) has been a research theme tackled by many researchers through different approaches for years. Although the relevant literature is full of papers on this topic, ranging from experimental works to theoretical contributions, an accurate and comprehensive description of the actual mechanical behavior exhibited by concrete and reinforced concrete at service and ultimate conditions still remains a challenge in the field of structural engineering. This is due to several intricate and interconnected phenomena involved, such as tensile cracking, compression crushing, strain softening, interaction between aggregates and matrix, interaction between concrete and reinforcement, stiffness degradation, energy dissipation, and ductility exhibited under cycling loading.

Following these research motivations, this Special Issue collects 15 papers focused on the mechanical behavior of concrete materials and structures, including both experimental findings and numerical analyses using both conventional and advanced methodologies. In the Editors' opinion, each article contains clear scientific novelty from various standpoints (analytical, numerical, experimental, conceptual), thus representing a major contribution to the understanding of the mechanical behavior of concrete materials and structures. The Editors hope that this article collection can somehow contribute, even if modestly, to the continuous research for a more thorough and reliable understanding of the mechanical behavior of ordinary and prestressed concretes, as well as special concretes, including high-strength, recycled, and fiber-reinforced concretes, for both structural and non-structural applications, and for the development of related numerical/analytical predictive models.

Among the experimental contributions, Ni et al. [1] presented static and dynamic shear tests on four types of transverse connections used in adjacent precast concrete box-beam bridges to evaluate their shear transfer performance before and after cracking. In addition to experimental tests, a finite element model was also developed to calibrate and validate the interfacial material parameters. This contribution provided quantitative information on the effects of shear key cracking on vertical stiffness, and on the relationship between shear transfer and relative displacement across the shear key.

Three contributions were focused on recycled concretes, from either an experimental or numerical perspective. Garbaya et al. [2] incorporated phosphogypsum, which is a by-product of the production of phosphoric acid, into a new construction material. Based on an experimental campaign comprising physicochemical, mechanical, and thermal analyses, it was demonstrated that the different degrees of hydration that this material possesses facilitate the exchange of water with the external environment by creating a water pump that helps to condition the ambient air. Jahandari et al. [3] studied, through experimental tests, the compression behavior of concretes prepared with recycled coarse aggregates at replacement levels of 50% and 100% of natural coarse aggregates, resulting from both low- and high-strength concretes. The experimental campaign, including 60 tests, was aimed to analyze the effect of hooked-end steel fibers and silica fume, introduced as a partial replacement of cement, in such recycled concrete specimens. It was shown that the addition of steel fibers and silica fume considerably increased the strength (especially for recycled aggregates resulting from high-strength concretes), the elastic modulus, and the post-peak ductility of concretes. The third paper pertinent to the field of recycled concrete was authored by Peng et al. [4], who proposed a numerical analysis method for recycled concrete called the parallel homogenization method. An equivalent meso-damage model was developed through the base force element method, based on the complementary energy principle. The recycled concrete was treated as a five-phase medium, including the aggregate, the old mortar, the new mortar, the old interface, and the new interface. Based on the simulation of experimental uniaxial tensile tests of recycled concrete, it was demonstrated that the proposed method shows significant computational advantages over alternative mesoscopic damage models.

The mechanical behavior of concrete was analyzed not only under static loading. The dynamic characteristics of concrete were experimentally investigated by Sun et al. [5] through large-diameter split Hopkinson pressure bar (SHPB) tests performed on concretes and mortars comparatively. The experimental tests made it possible to analyze the influence of strain rate on the actual dynamic strength of concrete materials and the influence of strain acceleration on inertial effects. It was also shown that the strain rate effect of concrete is more sensitive than that of mortar, but the inertia effect of mortar is more sensitive than that of concrete.

Another aspect playing a key role in structural serviceability analyses of RC structures is the concrete shrinkage. Dey et al. [6] presented a comprehensive experimental program aimed to investigate the long-term (estimated at five years from casting) shrinkage effects of concrete on the deformative and tension stiffening response of RC members. Experimental tests on 14 RC ties with various geometry and mechanical characteristics demonstrated that the long-term shrinkage of concrete remarkably lowered the cracking load of the RC members and caused an apparent tension stiffening reduction, especially for members with higher reinforcement ratios.

The influence of multi-walled and single-walled carbon nanotubes (CNTs) on the carbonation, compressive and flexural strength, electrical resistance, and porosity of special mortars was experimentally analyzed by Lee et al. [7]. Based on the experimental outcomes, the introduction of CNTs led to a decrease in compressive and flexural strengths compared to plain mortars, due to an increase in the internal pore volume. However, the mortars prepared with CNTs exhibited a much lower electrical resistance (of around 10–20% of the plain specimens) and a higher acceleration rate of conductive cement mortar (the carbonation rate of conductive cement mortar increased by 1.5 times as the dosage of CNT was doubled in the mixture).

An increasing number of researchers have recently advanced the use of machine learning methodologies for predicting the mechanical characteristics of concrete based on training sets of data. For estimating the compressive strength of ultra-high-strength concrete (UHSC), Shen et al. [8] used soft computing techniques by considering 372 different mix proportions with 10 input variables, namely, cement content, fly ash, silica fume and silicate content, sand and water content, superplasticizer content, steel fiber, steel fiber aspect ratio, and curing time. The effect of these ten input parameters on the output parameter (compressive strength) and their interaction was evaluated using SHapley Additive exPlanations. It was demonstrated that the curing time has the highest impact on UHSC compressive strength estimation, followed by silica fume, sand, and superplasticizer content. Along a similar research line, Ahmad et al. [9] used supervised machine learning techniques, in particular, for comparing individual and ensemble algorithms to predict the compressive strength of fly-ash-based concretes, trained by 270 experimental data collected from the literature. In this study, the input parameters included cement content, aggregates, water, binder-to-water ratio, fly ash, and superplasticizer.

Bernardo et al. [10] addressed the problem of predicting the torsional strength of RC members through an analytical approach, inspired by the space truss analogy, with some empirical coefficients obtained by regression analysis. A wide database containing 202 tests of RC beams tested under pure torsion was first compiled, including under- and over-reinforced beams, plain and hollow beams, as well as normal- and high-strength concrete beams. Based on this database, correlation studies between the torsional strength and geometrical and mechanical parameters of the RC beams (compressive concrete strength, concrete area enclosed within the outer perimeter of the cross section, and amount of reinforcement) were carried out, following which refined predicting equations were elaborated to predict the torsional strength of RC beams. It was also demonstrated that the accuracy of the proposed equations is superior to that of alternative code-based formulations. The torsional behavior of RC beams, with particular attention to the transition from the uncracked to the cracked stage, was also investigated by Teixeira and Bernardo [11] through the generalized softened variable angle truss-model (GSVATM). The GSVATM was used to check the accuracy of some smeared constitutive laws for tensile concrete proposed in the literature. Among five different smeared constitutive laws analyzed in this paper, the formulation proposed by Belarbi and Hsu in 1994 exhibited the best accuracy and reliability against a wide database of experimental results including 103 RC beams with plain and hollow rectangular cross sections tested under pure torsion.

The deflection behavior of horizontal structural members was the object of the investigation of D'Antino and Pisani [12]. The recommendations of the various Eurocodes on the maximum deflection limits were critically analyzed by focusing on the integrity of the superstructures. Different types of horizontal members, namely, rib and clay pot (or hollow block), composite steel–concrete, and timber beam slabs, were designed to respect the deflection limit enforced by the Eurocodes. The authors proposed a curvature control method in place of the deflection control method adopted by the Eurocodes; this approach would allow for defining a general limit curvature value for floorings that could be adopted as the minimum performance level of the standards and would be able to guarantee the absence of flooring cracking.

One of the most complex phenomena to simulate in concrete structures is fracture behavior, due to the heterogeneous microstructure and interaction between aggregates and matrix. In this context, Tawfik et al. [13] employed state-of-the-art numerical techniques for simulating the fracture behavior of concrete. In particular, the authors proposed different crack simulation techniques, namely, the contour integral technique, the extended finite element method, and the virtual crack closure technique, implemented within the commercial finite element software ABAQUS, to investigate the flexural response and the fracture behavior of notched plain and reinforced concrete beams under three-point bending and four-point bending tests. The comparison of numerical outcomes with experimental findings demonstrated that the extended finite element method exhibited the best fracture energy estimation and solution-dependent crack path, and the most reliable results among the analyzed numerical techniques.

It is well known that durability of existing concrete structures was not always regarded as a crucial performance requirement in the past, and periodical maintenance plans have not been performed over the years. As a consequence, many damage phenomena have been recently observed in existing concrete structures dated from around 50 years ago; in some extreme cases, these damage mechanisms have led to complete structural collapse, as recently observed for bridges. One of the most serious factors negatively affecting the durability of concrete structures is the corrosion of steel reinforcement. In this context, De Domenico et al. [14] developed a systematic numerical-experimental approach for the load-bearing capacity assessment of existing prestressed concrete bridge decks in which the corrosion of prestressing strands was explicitly considered. The developed procedure can represent a convenient assessment tool to rapidly identify critical portions of a large infrastructure network prior to performing detailed analyses to establish a list of intervention priorities in a timely and reasonable way.

Finally, the review paper by Dong et al. [15] illustrated the various methods proposed in the literature to simulate the coupled hygro-thermo-mechanical behavior of concrete pavement. Among the analyzed methods, it emerged that an area of research that was not sufficiently explored by previous researchers is the deformation and failure mechanism of pavement concrete under the coupling action of moisture, temperature, and wheel load. In this context, COMSOL software was identified as a promising numerical tool for performing such coupled hygro-thermal-mechanical analyses of concrete pavement.

Author Contributions: Writing—original draft preparation, D.D.D.; writing—review and editing, D.D.D., L.F.A.B. All authors have read and agreed to the published version of the manuscript.

Funding: This research received no external funding.

Acknowledgments: We would like to thank all authors for their valuable contributions, the many dedicated referees for their useful guidance to improve the papers, the Editorial Team of *Materials* and, above all, Daisy Liu for her professional assistance and kindness during this Special Issue.

Conflicts of Interest: The authors declare no conflict of interest.

References

1. Ni, X.; Anselme, A.O.; Wang, G.; Xing, Y.; Xu, R. Experimental Investigation of Shear Keys for Adjacent Precast Concrete Box Beam Bridges. *Materials* **2022**, *15*, 1459. [CrossRef] [PubMed]
2. Garbaya, H.; Jraba, A.; Khadimallah, M.A.; Elaloui, E. The Development of a New Phosphogypsum-Based Construction Material: A Study of the Physicochemical, Mechanical and Thermal Characteristics. *Materials* **2021**, *14*, 7369. [CrossRef] [PubMed]
3. Jahandari, S.; Mohammadi, M.; Rahmani, A.; Abolhasani, M.; Miraki, H.; Mohammadifar, L.; Rashidi, M. Mechanical properties of recycled aggregate concretes containing silica fume and steel fibres. *Materials* **2021**, *14*, 7065. [CrossRef] [PubMed]
4. Peng, Y.; Zakaria, S.; Sun, Y.; Chen, Y.; Zhang, L. Analysis of Tensile Strength and Failure Mechanism Based on Parallel Homogenization Model for Recycled Concrete. *Materials* **2021**, *15*, 145. [CrossRef] [PubMed]
5. Sun, B.; Chen, R.; Ping, Y.; Zhu, Z.; Wu, N.; Shi, Z. Research on Dynamic Strength and Inertia Effect of Concrete Materials Based on Large-Diameter Split Hopkinson Pressure Bar Test. *Materials* **2022**, *15*, 2995. [CrossRef] [PubMed]
6. Dey, A.; Vastrad, A.V.; Bado, M.F.; Sokolov, A.; Kaklauskas, G. Long-term concrete shrinkage influence on the performance of reinforced concrete structures. *Materials* **2021**, *14*, 254. [CrossRef] [PubMed]
7. Lee, G.C.; Kim, Y.; Seo, S.Y.; Yun, H.D.; Hong, S. Influence of CNT Incorporation on the Carbonation of Conductive Cement Mortar. *Materials* **2021**, *14*, 6721. [CrossRef] [PubMed]
8. Shen, Z.; Deifalla, A.F.; Kamiński, P.; Dyczko, A. Compressive Strength Evaluation of Ultra-High-Strength Concrete by Machine Learning. *Materials* **2022**, *15*, 3523. [CrossRef] [PubMed]
9. Ahmad, A.; Farooq, F.; Niewiadomski, P.; Ostrowski, K.; Akbar, A.; Aslam, F.; Alyousef, R. Prediction of compressive strength of fly ash based concrete using individual and ensemble algorithm. *Materials* **2021**, *14*, 794. [CrossRef] [PubMed]
10. Bernardo, L.F.; Teixeira, M.M.; De Domenico, D.; Gama, J.M. Improved Equations for the Torsional Strength of Reinforced Concrete Beams for Codes of Practice Based on the Space Truss Analogy. *Materials* **2022**, *15*, 3827. [CrossRef] [PubMed]
11. Teixeira, M.; Bernardo, L. Evaluation of Smeared Constitutive Laws for Tensile Concrete to Predict the Cracking of RC Beams under Torsion with Smeared Truss Model. *Materials* **2021**, *14*, 1260. [CrossRef] [PubMed]
12. D'Antino, T.; Pisani, M.A. A Proposal to Improve the Effectiveness of the Deflection Control Method Provided by Eurocodes for Concrete, Timber, and Composite Slabs. *Materials* **2021**, *14*, 7627. [CrossRef] [PubMed]
13. Tawfik, A.B.; Mahfouz, S.Y.; Taher, S.E.D.F. Nonlinear ABAQUS Simulations for Notched Concrete Beams. *Materials* **2021**, *14*, 7349. [CrossRef] [PubMed]
14. De Domenico, D.; Messina, D.; Recupero, A. A Combined Experimental-Numerical Framework for Assessing the Load-Bearing Capacity of Existing PC Bridge Decks Accounting for Corrosion of Prestressing Strands. *Materials* **2021**, *14*, 4914. [CrossRef] [PubMed]
15. Dong, W.; Liu, C.; Bao, X.; Xiang, T.; Chen, D. Advances in the Deformation and Failure of Concrete Pavement under Coupling Action of Moisture, Temperature, and Wheel Load. *Materials* **2020**, *13*, 5530. [CrossRef] [PubMed]

materials

MDPI

Article

Improved Equations for the Torsional Strength of Reinforced Concrete Beams for Codes of Practice Based on the Space Truss Analogy

Luís F. A. Bernardo [1,*], Mafalda M. Teixeira [1], Dario De Domenico [2] and Jorge M. R. Gama [3]

1 Centre of Materials and Building Technologies (C-MADE), Department of Civil Engineering and Architecture, University of Beira Interior, 6201-001 Covilhã, Portugal; mafalda.m.teixeira@ubi.pt
2 Department of Engineering, University of Messina, 98166 Messina, Italy; dario.dedomenico@unime.it
3 Center of Mathematics and Applications (CMA), Department of Mathematics, University of Beira Interior, 6201-001 Covilhã, Portugal; jgama@ubi.pt
* Correspondence: lfb@ubi.pt

Abstract: Design codes provide the necessary tools to check the torsional strength of reinforced concrete (RC) members. However, some researchers have pointed out that code equations still need improvement. This study presents a review and a comparative analysis of the calculation procedures to predict the torsional strength of RC beams from some reference design codes, namely the Russian, American, European, and Canadian codes for RC structures. The reliability and accuracy of the normative torsional strengths are checked against experimental results from a broad database incorporating 202 RC rectangular beams tested under pure torsion and collected from the literature. The results show that both the readability and accuracy of the codes' equations should be improved. Based on a correlation study between the experimental torsional strengths, and geometrical and mechanical properties of the beams, refined yet simple equations are proposed to predict torsional strength. It is demonstrated that the proposed formulation is characterized by a significant improvement over the reference design codes. The efficiency of the proposed formulae is also assessed against another equation earlier proposed in the literature, and an improvement is noted as well. From the results, it can be concluded that the proposed equations in this study can contribute to a more accurate and economical design for practice.

Keywords: reinforced concrete; beams; torsional strength; correlation study; codes of practice; space truss model; thin-walled tube analogy

Citation: Bernardo, L.F.A.; Teixeira, M.M.; De Domenico, D.; Gama, J.M.R. Improved Equations for the Torsional Strength of Reinforced Concrete Beams for Codes of Practice Based on the Space Truss Analogy. *Materials* **2022**, *15*, 3827. https://doi.org/10.3390/ma15113827

Academic Editor: Angelo Marcello Tarantino

Received: 26 April 2022
Accepted: 25 May 2022
Published: 27 May 2022

1. Introduction

In engineering practice, structural members under pure torsion are not a common situation. Usually, torsional effects are combined with other internal forces in the critical sections of the member. However, there are several practical cases in structures in which structural members must sustain primary torsional effects in their critical sections. Typical examples are encountered in bridge structures or geometrically complex building structures, in which reinforced concrete (RC) girders and columns can be subjected to primary torsion due to the high eccentricity of static loads or complex geometry of the members (e.g., curvature in plan). For such members, an accurate calculation of the torsional strength is essential for the design or assessment of the torsional capacity, namely to guarantee or check the safety at the ultimate limit states. For such purposes, current RC members and structural engineers usually base their calculations on the rules from design codes. Nowadays, each country has its own set of design codes that govern structural design, namely for RC structures, and some of them have a high impact in other countries and also constitute reference codes for the international community [1–7].

Despite all the research effort made in recent decades by the scientific community, several current design codes for concrete structures are still somewhat scarce in providing detailed and specific design rules for torsion. These include basic reinforcement detailing rules and limits for important design variables to ensure a good performance of RC members under torsion for both the ultimate and serviceability limit states. For example, some design codes do not provide any specific rule regarding the minimum amount of torsional reinforcement, which is considered a basic requirement to avoid a sudden failure after concrete cracking. The same can be stated for the maximum amount of torsional reinforcement to ensure ductility at failure (torsional reinforcement should yield before concrete crushing). Although such a maximum amount can be indirectly computed from the maximum compressive stress allowed for the concrete struts, this upper-stress limit can vary substantially among design codes. As there is a lack of specific rules for torsion, some codes refer to the rules related to the reinforcement requirements for other internal forces, such as for bending (for the longitudinal reinforcement) and for shear (for the transverse reinforcement). In addition to the aforementioned missing aspects, when the rules from several design codes are used to predict the torsional strength of RC beams, small accuracy and high dispersion of the results, including unsafe predictions, are still observed when they are compared with experimental data [8,9]. This observation shows that research work on the torsion of RC beams still needs to be carried out to propose more accurate design rules to be incorporated in future revisions of design codes.

The first reference design codes, which incorporated specific design rules for torsion for RC members, were based on the so-called skew-bending theory. This model was proposed by Hsu in 1968 [10] and was established from empirical observations based on the failure pattern observed in several experiments with RC rectangular beams under pure torsion. This model showed to provide accurate predictions for the torsional strength of RC beams with common rectangular cross-sections, such as the ones used in building structures. However, when applied to cross-sections with large aspect ratios or to cross-sections with more complex geometries, such as the ones used in bridge girders, this model produces more complex formulations and shows to be much less accurate. In spite of this, the skew-bending theory was developed over more than two decades [11–13] and had a considerable influence on the design rules for torsion in some reference design codes, such as the ACI code (American code) up to 1995. Presently, the design rules for torsion of the Eurasian code SiNiP 2018 [1] are still based on the skew-bending theory.

Nowadays, the torsion design rules from most design codes for concrete structures are based on the space truss analogy. This model, first proposed by Raush in 1929 [14], was later combined with the classical thin tube theory from Bredt [15] and further developed in the second half of the last century [16–19]. The space truss analogy allows for a better physical understanding of how an RC beam behaves under torsion in the cracked stage and provides simple equations to compute the torsional strength, even for geometrically complex cross-sections. However, due to different hypotheses incorporated in the model, to allow for a simple torsional design, the calculation procedures for torsion can be somewhat different among the design codes. Because of this, noticeable differences can be observed in the results when different design codes are used to predict the torsional strength of RC beams, although all these codes are formally based on the same space truss resisting mechanism. These observations justify the need for additional improvements to be incorporated in future revisions of the codes.

Since the 1980s, refined versions of models based on the space truss analogy have been proposed that allow one to compute with accuracy the strength of RC beams under pure torsion [18,20–26], RC beams under torsion combined with other internal forces [27–30]. More advanced analytical models have also been proposed in the literature and applied to beams under torsion and combined loadings [31–36]. Although these models have been shown to be very reliable when compared with experimental results, they are not easy to be used by practitioners as they require advanced calculation procedures to be implemented on the computer. Hence, simple and reliable equations would be preferable for practice.

Based on these motivations and research needs, this study first presents a critical overview and a comparative analysis of the calculation procedures from design codes to predict the torsional strength of RC rectangular beams. For this, some reference design codes considered important due to their territorial scope were chosen. Such design codes are the following ones: the Eurasian code, SiNiP 2018 [1], two different versions of American codes, namely ACI 318R-89 [2] and ACI 318R-19 [3], the European codes MC90 [4], MC10 [5] and Eurocode 2 [6], and the Canadian code CSA A23.3-14 [7]. For comparison purposes, this list includes codes based on different mechanical models to establish the design rules for torsion, namely the skew-bending theory and the space truss analogy, and also codes with different application scopes (laws or recommendation documents). The calculation procedures from the codes are summarized and checked against a broad database incorporating 202 RC rectangular beams tested under pure torsion collected from the literature. This database includes under- and over-reinforced beams, plain and hollow beams, as well as normal- and high-strength concrete beams. Then, based on correlations studies, improved and simple equations are proposed to compute the torsional strength of RC beams. The proposed model correlates the torsional strength and three main properties of the beams: the compressive concrete strength, the concrete area enclosed within the outer perimeter of the cross-section, and the amount of torsional reinforcement. The accuracy and reliability of the proposed equations are checked against the results from the reference design codes. They are also checked against simple equations proposed by Rahal in 2013 [8], which have a similar form to the ones proposed here and were based on a similar approach to that used in this study (by fitting experimental results). For these reasons, the research from Rahal [8] was considered a benchmark. The results show that the proposed equations significantly improve the accuracy and reliability of the torsional strength of RC beams when compared with the same ones from the reference design codes. They also give slightly better results when compared with the ones from the equations previously proposed by Rahal [8].

When compared to previous research, namely the one from Rahal [8], which constitutes a reference study, the main novelty in this study is related to the higher number of reference design codes studied, the higher number of reference beams considered in the database, and the somewhat different methodology used to fit the experiment results in obtaining simple and improved equations for torsional strength.

It should be noted that this article addresses only the particular case of RC beams under pure torsion. It is well known that in real concrete structures, the critical section of members usually carries combined loadings, for instance, torsion combined with other internal forces (bending, shear, and axial forces). However, for some concrete members, such as girders curved in plan and girders with eccentric loadings, torsion could be the primary internal acting force. Furthermore, the design provisions to check the interaction between internal forces requires the calculation of the torsional strength of the cross-section, considering only torsion as loading. Hence, the ultimate strength of the cross-section needs to be well known. This justifies the importance of this study.

2. Equations from the Reference Design Codes

From the studied design codes, the SiNiP 2018 [1] and ACI 318R-89 [2] codes are the only ones whose equations for torsional design, i.e., to compute the torsional strength of RC beams, are based on the skew-bending theory. Although ACI 318R-89 is no longer in use, it was included in this study for comparison with SiNiP 2018 (which is still in use) and to better understand the influence of the underlying mechanical model in the code formulation. The calculation procedure for torsion for all the other reference design codes (ACI 318R-19 [3], MC90 [4], MC10 [5], Eurocode 2 [6], and CSA A23.3-14 [7]) are based on the space truss analogy. As far as the European design codes are concerned, it should be stated that MC10 substituted MC90. However, as the design rules for torsion were simplified in MC10, MC90 was also considered for comparison.

Table A1 in Appendix A summarizes the equations incorporated in each reference design code for torsional design and to compute the torsional strength of RC beams. In Table A1, all codes' equations were rewritten to uniformize the symbology for better clarity and to facilitate comparison. The meaning of the used symbology can be found in the Nomenclature.

American and Canadian codes incorporate specific rules to design both the longitudinal and transverse torsional reinforcement. In general, the European codes incorporate specific equations to design longitudinal torsional reinforcement. However, they refer to the rules for shear reinforcement to design the transverse torsional reinforcement.

From the presented equations in Table A1, one can highlight five main parameters that can be defined by somewhat different rules but strongly influence the magnitude of the calculated torsional strength:

- The limit of the wall's thickness of the equivalent hollow beam, which determines and limits the area enclosed within the flow of shear stress acting on the beam's cross-section;
- The flow of shear stress, which is induced by the external torque, and the corresponding shear resultant forces in each wall;
- The criteria to compute the torsional strength, which, depending on the underlying mechanical model and design code, consider separately the strength contributed by the torsional tensile reinforcement and the strength contributed by the compressive concrete;
- The maximum limit allowed for the compressive stress in the concrete struts to avoid a brittle failure of the beam due to concrete crushing;
- The angle of the concrete struts to the longitudinal axis of the beam.

A more detailed analysis of the summarized equations in Table A1 can be found in some of the reference design codes and also in a previous study from the two first authors [9].

3. Database with Reference Beams

For this study, an extensive literature review was performed to compile the main properties and experimental results of RC rectangular beams tested under pure torsion until failure. A total of 202 beams were compiled from several studies [10,12,19,37–49] to build the database (with 158 plain beams and 44 hollow beams). The number of beams found in the literature was higher than 202; however, some of them were disregarded based on the following criteria:

1. The main properties of the tested RC beams needed to compute the normative torsional strength should be given;
2. The experimental torsional strength should be given and the RC beams should have failed in pure torsion in their ultimate stage as expected;
3. The beams should comply with reinforcement requirements from ACI 318R-19 [3]. For instance, and among other requirements, the spacing of the hoops should be less than the upper limit set by the code to avoid untypical behavior (for instance, premature failure) during testing. ACI 318R-19 [3] was the chosen code because it was found to be the one that incorporates a higher number of specific detailing rules for RC beams under torsion.

Table A2 in Appendix A summarizes the main geometric and mechanical properties of the reference RC beams that are necessary to compute the torsional strength from the design codes. The meaning of each parameter can be found in the Nomenclature.

Figure 1 presents graphs with the distribution of some key parameters for the 202 reference RC beams from the database. In the abscissa of the graphs, the parameters are: f_{cm} is the average compressive concrete strength, $\rho_{tot} = \rho_l + \rho_t$ is the total ratio of torsional reinforcement, f_{ly} and f_{ty} are the average yielding stresses for longitudinal and transverse reinforcement, respectively.

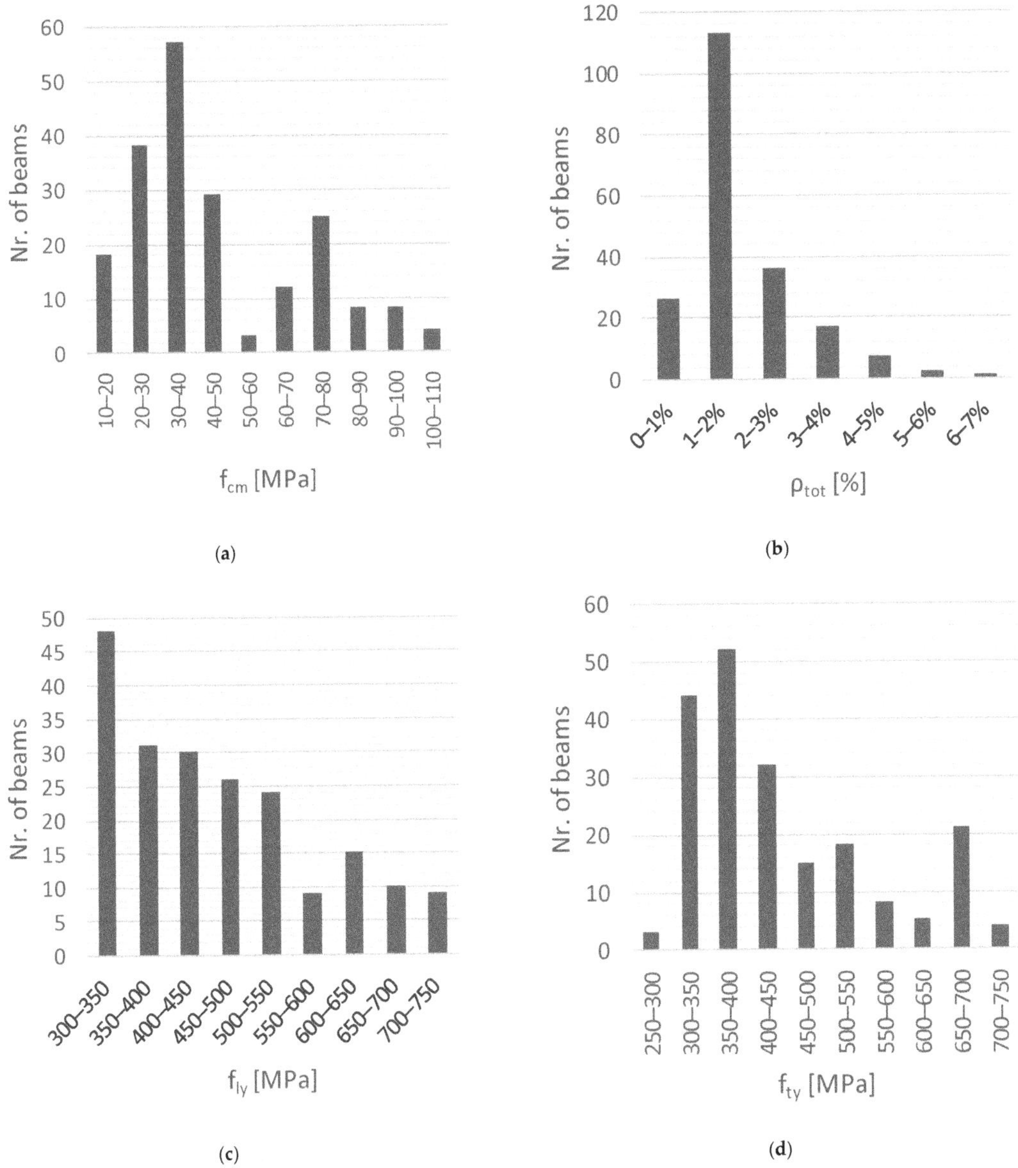

Figure 1. Distribution of key parameters for the reference RC beams. (**a**) Concrete strength. (**b**) Total reinforcement ratio. (**c**) Yielding stress of longitudinal reinforcement. (**d**) Yielding stress of transverse reinforcement.

Figure 1 shows that 142 and 60 beams are built with normal- (up to 50 MPa) and high-strength concrete (over 50 MPa, according to [5]), respectively. The average concrete compressive strength ranges between 14 MPa and 110 MPa. The total reinforcement ratio ranges between a minimum of 0.37% and a maximum of 6.36%, being for most of the beams in the range of 1 to 2%. The yielding stress ranges between 308.8 MPa and 723.9 MPa for the longitudinal reinforcement and between 285 MPa and 714.8 MPa for the transverse reinforcement. For most of the beams, it ranges between 300 MPa and 500 MPa.

The database used in this study is wider than the ones used in previous studies on the torsion of RC beams. For instance, the database used by Rahal [8], which is an important reference for this study, included 50 beams less than the database used in this study.

4. Evaluation of Design Codes

For each reference beam from the database (see Table A2), the theoretical torsional strength, $T_{R,th}$, was computed according to the calculation procedures from each reference design code (see Table A1). The obtained values are presented in Table A3 in Appendix A, which also presents the corresponding experimental values, $T_{R,exp}$. The calculated ratios $T_{R,exp}/T_{R,th}$ are presented in a Table A4 in Appendix A. Table 1 summarizes, for each design code, the average value, $\overline{x}$, and the coefficient of variation, cv, computed for the ratios, $T_{R,exp}/T_{R,th}$, from all reference beams. The results are presented separately for plain beams (P), hollow beams (H), and also for all beams together (P + H). This separation can be justified because some design codes include corrections to the equations for hollow beams, while others do not.

Table 1. Comparative analysis between design codes.

Cross Section		P $T_{R,exp}/T_{R,th}$	H $T_{R,exp}/T_{R,th}$	P + H $T_{R,exp}/T_{R,th}$
SiNiP 2018 [1]	$\overline{x}$ =	1.21	1.47	1.25
	cv =	34%	20%	32%
ACI 318R-89 [2]	$\overline{x}$ =	1.12	1.27	1.15
	cv =	23%	26%	24%
ACI 318R-19 [3]	$\overline{x}$ =	1.40	1.38	1.40
	cv =	31%	18%	28%
MC90 [4]	$\overline{x}$ =	1.28	1.61	1.36
	cv =	24%	29%	28%
MC10 [5]	$\overline{x}$ =	1.41	1.07	1.33
	cv =	44%	33%	44%
Eurocode 2 [6]	$\overline{x}$ =	1.07	1.29	1.12
	cv =	24%	31%	28%
CSA A23.3-14 [7]	$\overline{x}$ =	0.98	1.13	1.01
	cv =	22%	27%	24%

From Table 1, it can be stated that all design codes show a relatively high dispersion for the ratio $T_{R,exp}/T_{R,th}$ (in general $cv > 20\%$), which represents a motivation for developing more accurate and reliable torsional strength equations.

Design codes based on the skew-bending theory, namely Si-NiP 2018 [1] and ACI 318R-89 [2] codes, present similar results. In general, they both tend to underestimate the torsional strength, ($\overline{x} > 1$), with $\overline{x} = 1.25$ for Si-NiP 2018 code and $\overline{x} = 1.15$ for ACI 318R-89 code. Among these two design codes, the ACI 318R-89 code shows the best results, with $\overline{x}$ closer to 1 and less dispersion of the results disregarding the cross-section type (with $cv = 24\%$, against $cv = 32\%$ for SiNiP 2018 code). Regarding the cross-section type, it can be observed that the accuracy of the ACI 318R-89 code seems to be better for plain beams (with $\overline{x} = 1.12$ and $cv = 23\%$ for plain beams and $\overline{x} = 1.27$ and $cv = 26\%$ for hollow beams); similar trends are observed for Si-NiP 2018 code (with $\overline{x} = 1.21$ and $cv = 34\%$ for plain beams and $\overline{x} = 1.47$ and $cv = 20\%$ for hollow beams). This observation can be explained because the model based on the skew-bending theory was calibrated for plain beams [10].

The other design codes based on the space truss analogy (ACI 318R-19 [3], MC90 [4], MC10 [5], Eurocode 2 [6], and CSA A23.3-14 [7]) show results with some differences among them. Among those design codes, and disregarding the cross-section type, the CSA A23.3-14 code seems to be the most accurate (with $\overline{x} = 1.01$ and $cv = 24\%$), while the MC10 code seems to be one of the worst (with $\overline{x} = 1.33$ and $cv = 44\%$). ACI 318R-19, MC90, and Eurocode 2 codes show the same level of dispersion (with $cv = 28\%$ for all of them), and all tend to underestimate the torsional strength (with $\overline{x}$ ranging from 1.33 and 1.40). Regarding the cross-section type, it can be observed that the CSA A23.3-14 code is the only one that tends to slightly overestimate the torsional strength of hollow beams (with $\overline{x}$ = 0.98 and cv = 22 %). Eurocode 2 seems to be one that provides the best results for plain sections (with $\overline{x}$ = 1.07 and cv = 24%). For hollow beams, it is not so clear because of the higher dispersion of the results.

It is also worth noting that the ACI 318R-89 code (currently not in use) provides, in general, more accurate results and with less dispersion when compared with the ACI 318R-19 code. This is because the majority of the reference tested beams have small rectangular cross-sections, for which the skew-bending theory provides better results (as previously referred).

Based on the above considerations and balancing the accuracy with the degree of safety, it can be concluded that, among the codes currently in use, Eurocode 2 seems to be the one that presents the most satisfactory results. However, some caution is required with this conclusion (and other ones previously stated) as the dispersion of the results is high for all codes.

The predictive accuracy of each code formulation can be assessed in Figure 2, which presents scatter plots relating to the experimental torsional strengths (in ordinate) with the theoretical ones (in abscissa) for each of the reference codes. In the graphs, different markers were used to distinguish the results for plain beams ("■") and for hollow beams ("□").

In each graph, an inclined line with a 45° angle is drawn, which represents the location of the points in case both the experimental and theoretical torsional strengths are equal, i.e., the code predicts exactly the torsional strength of the beam. Points located on the left side of the referred line correspond to the case in which the code underestimates the torsional strength of the beams. If the points are located on the right side of the line, then the code overestimates the torsional strength.

From Figure 2, it can be seen that some of the design codes can overestimate the torsional strength of several reference beams noticeably, in particular, for hollow beams. This is the case for the MC10 and CSA A23.3-14 codes.

The results from Table 1 and Figure 2 show clearly that the level of accuracy of all analyzed codes, as well as the level of safety for some of them, should be improved.

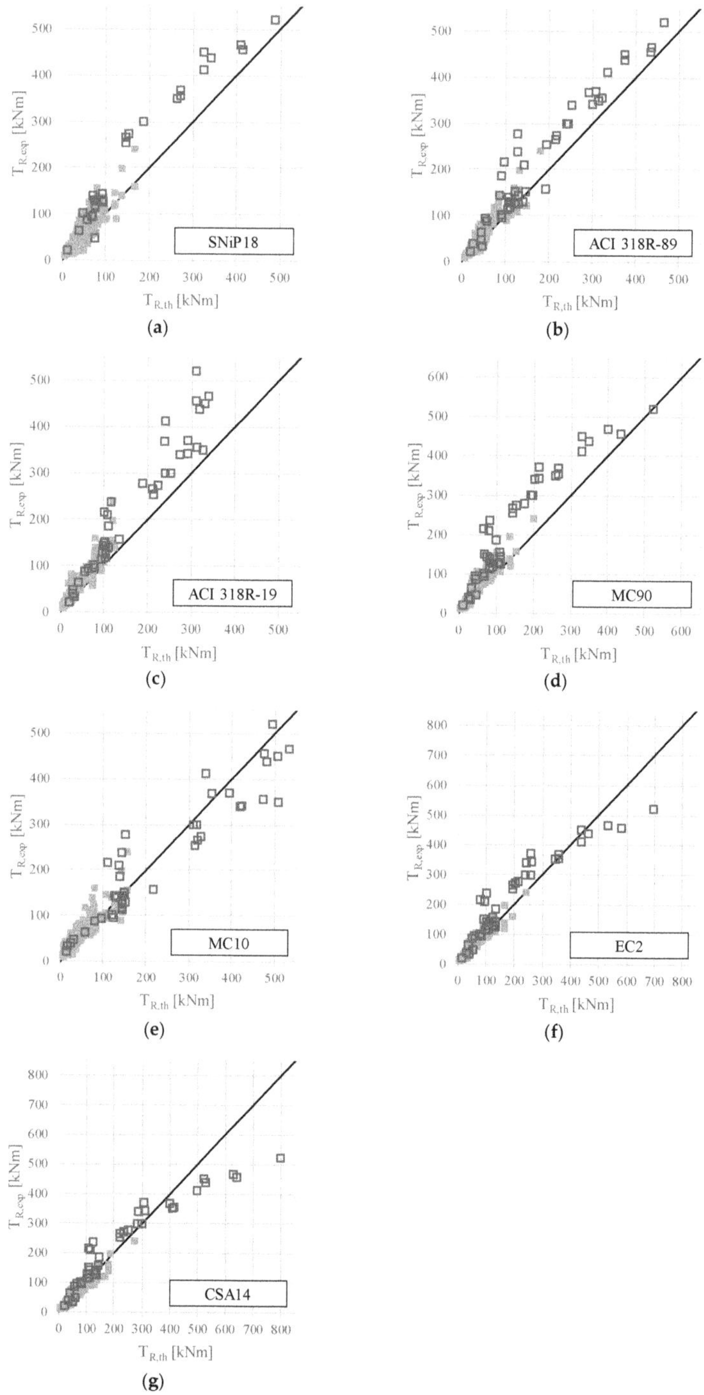

Figure 2. Experimental vs. theoretical torsional strengths (design codes). (**a**) SNiP18. (**b**) ACI 318R_89. (**c**) ACI 318R_19. (**d**) MC90. (**e**) MC10. (**f**) EC2. (**g**) CSA14.

5. Equations Proposed by Rahal

Rahal, in 2013 [8], showed that for both the ACI and CSA codes, after some basic algebraic manipulations, a general and simple torsional strength equation can be written in the form of Equation (1). This equation is written here with some parameters standardized according to the nomenclature used in the code equations previously presented. Equation (1) was used in some previous models [18,23].

$$T_R = 2A_k\sqrt{\frac{A_l f_{yl}}{u_k}\frac{A_t f_{yt}}{s}} \tag{1}$$

It can be shown that the general form of Equation (1) can be obtained from all the code equations, which are based on the space truss analogy. This is the case of the reference codes considered in this study (ACI 318R-19 [3], MC90 [4], MC10 [5], Eurocode 2 [6], and CSA A23.3-14 [7])).

Rahal [8] pointed out the following drawbacks for Equation (1) based on experimental evidence:

1. In the hollow tube model used in the space truss analogy, the shear flow is constant around the perimeter of the tube walls. Design code formulations consider a constant effective thickness for all the walls. As a consequence, the model assumes the same shear stress and shear. As pointed out by Rahal [8], this is not consistent with the experimental results on RC rectangular beams under torsion that show different conditions on the different faces of the cross-section [10,12,37]. Experiments show that larger tensile strains are observed in the longer legs of the hoops and larger diagonal strains are observed in the longer faces of the cross-section. In this regard and based on these observations, a refined variable-angle space truss model incorporating different strut inclination angles in the different faces of the cross-section was recently proposed by De Domenico [50];
2. Rahal [8] also pointed out that Equation (1) disregards the effect of the concrete compressive strength, while experiments [10,12] show that this parameter has a significant influence on the torsional strength;
3. In addition, Rahal [8] also noted that in most experiments [17,40,51], the concrete of the beams did not spall at the maximum torque or was limited to the corners of the cross-section [37,42]. However, in Equation (1), the torsional strength is related to the spalled concrete dimensions through area A_k (area enclosed within the shear flow path).

To solve the first drawback, Rahal [8] proposed to reduce the power by 0.5 for the reinforcement term in Equation (1) to compensate for the relatively smaller contribution of the hoops and concrete on the shorter side of the cross-section. To solve the second drawback, the author suggested incorporating an additional term to consider the effect of the concrete compressive strength in Equation (1). For the third drawback, the author simply suggested correcting Equation (1) to relate the torsional strength with the unspalled concrete dimensions, substituting the reduced area, A_k, with the concrete area, A_c (area enclosed within the outer perimeter of the cross-section), and the corresponding perimeter is denoted as p_c (in place of p_h).

Based on the experimental results collected from the literature (which included 152 RC beams tested under torsion) and based on separated nonlinear correlations (using appropriate subsets of the reference beams) for each of the previously referred terms/parameters to correct Equation (1), Rahal [8] proposed the improved Equations (2) and (3) to compute the torsional strength of RC beams. These equations are written here according to the nomenclature and metric units for the parameters used in this study (as referred to in Section 6):

$$T_R = 0.33(f_c)^{0.16}A_c\left(A_l f_{yl}\frac{A_t f_{yt}}{s}\right)^{0.35} \tag{2}$$

$$\leq 2500(f_c)^{0.3}\frac{A_c{}^2}{p_c} \quad (3)$$

It should be noted that, in Equation (2), the power for the concrete strength term was empirically selected by Rahal [8] to provide good results.

Equation (2) governs the torsional strength for under-reinforced sections (the failure is governed by the yielding of the torsional reinforcement) and includes the "reinforcement term" $(A_l f_{yl} A_t f_{yt}/s)$ and the "concrete strength term" (a term related to f_c). The upper limit stated in Equation (3) governs the torsional strength for over-reinforced sections (the failure is governed by concrete crushing before reinforcement yielding) and includes the "concrete strength term." Equations (2) and (3) are not limited to rectangular cross-sections and can be applied to arbitrary cross-section shapes.

Rahal [8] checked the results from Equations (2) and (3) against the experimental results from the 152 test specimens and very good agreement was observed. In addition, a comparison with the ACI and CSA codes showed that the proposed equations provide better results, with higher accuracy and much less dispersion. The author also showed that such good results were observed for both normal- and high-strength concrete beams, as well as for under- and over-reinforced beams.

For this study, the torsional strengths computed from Equations (2) and (3) are rechecked against the experimental results of all 202 RC beams included in the wider database built for this research. The obtained results are presented for each reference beam in Table A3 and the respective ratios $T_{R,exp}/T_{theo}^{Rahal}$ are presented in Table A4. The results are summarized in Table 2 and Figure 3, in the same way as previously presented in Table 1 and Figure 2. The obtained results still confirm the conclusions from Rahal [8], namely that Equations (2) and (3) provide accurate results (with $\overline{x}$ = 1.06) with a very acceptable dispersion (with cv = 15%). Table 2 also shows that the results for both plain and hollow sections are very similar.

Table 2. Comparative analysis for the torsional strength from improved equations.

Cross-Section		P $T_{R,exp}/T_{R,th}$	H $T_{R,exp}/T_{R,th}$	P + H $T_{R,exp}/T_{R,th}$
Rahal [8]	$\overline{x}$ =	1.05	1.08	1.06
	cv =	14%	17%	15%
Equations (6) and (7)	$\overline{x}$ =	1.01	1.01	1.01
	cv =	14%	9%	13%
Equations (8) and (9)	$\overline{x}$ =	0.96	0.96	0.96
	cv =	15%	9%	14%

6. Alternative Improvement of the Equations from Rahal

In this section, Equation (1) to compute the torsional strength of RC beams is improved based on the wider database built for this study and also on a somewhat different correlation methodology than the one used by Rahal [8], namely for under-reinforced beams. The performed studies are presented separately for under- and over-reinforced beams.

6.1. Upper Limit to Control Concrete Crushing (Over-Reinforced Beams)

Following the same methodology from Rahal [8] to refine the upper limit stated in Equation (3) to control concrete crushing, a subset of 70 beams (62 plain beams and 8 hollow beams) was created from the database. The failure of such beams was governed by concrete crushing in the struts without yielding the torsional reinforcement. These beams are marked with an asterisk in Table A3 and represent beams with fragile failure. For these beams, a scatter plot is presented in Figure 4, with the average concrete compressive strength (f_{cm}) in abscissa and the factor $T_{R,exp}p_c/A_c{}^2$ in ordinate (with the following units: $T_{R,exp}$ [kNm], p_c [m], and A_c [m^2]). From the scatterplot, a power trendline was computed to fit the data ($1494 f_c^{0.4}$). In the same graph, the power trendline computed by Rahal [8] and based on less reference beams is also plotted for comparison ($2500 f_c^{0.3}$, see upper limit stated in Equation (3)). This power trendline is slightly shifted up when compared to the power trendline computed from the scatter plot in Figure 4. After computing, the torsional strengths for the over-reinforced beams from the database using an equation based on $1494 f_c^{0.4}$ and after a comparative analysis with the experimental strengths, it was observed that more unsafe values were obtained for the reference beams (the predicted torsional strength is higher than the real one for more beams, i.e., more points are located above the trend curve). This observation can be explained due to the high dispersion observed for the points in the scatter plot in Figure 4. For practical design, this situation is not acceptable and a correction of the power trendline was studied. The results suggested that the power trendline should be slightly shifted up to minimize the referred unsafe predictions. After some attempts, it was concluded that the power trendline suggested by Rahal [8] was quite appropriate. For this reason, the power trendline $2500 f_c^{0.3}$ was also adopted in this study and the upper limit stated in Equation (3) remained unchanged to define the upper limit to control concrete crushing (see upper limit stated in Equation (7)).

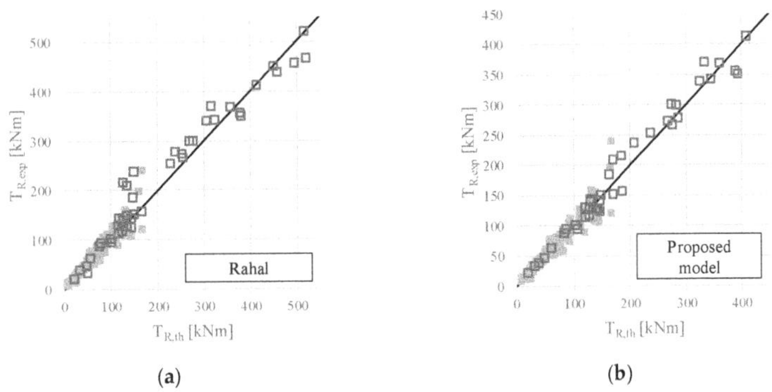

Figure 3. Experimental vs. theoretical torsional strengths (improved equations). (**a**) Model from Rahal. (**b**) Proposed model.

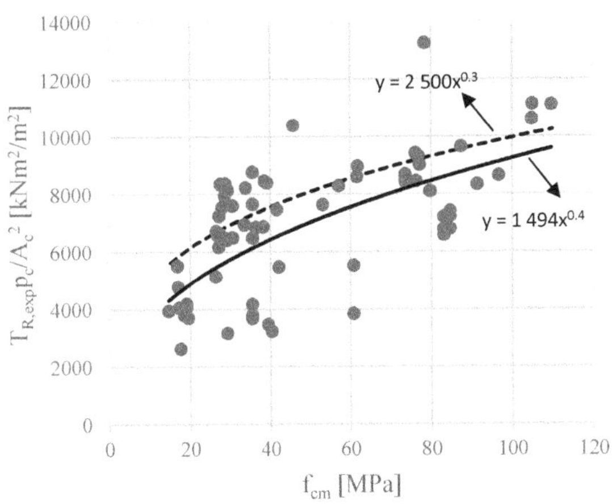

Figure 4. Torsional strength in over-reinforced beams.

6.2. Reinforcement, Concrete Strength, and Concrete Area Terms (Under-Reinforced Beams)

For under-reinforced beams, Rahal [8] studied separated nonlinear correlations using appropriate subsets of reference beams from its database to study the influence of both the "reinforcement term" ($A_l f_{yl} A_t f_{yt}/s$) and the "concrete strength term" (related to f_c).

Additional explanations of the approach followed by the referred author can be found in [8]. In this study, a different correlation methodology was used. Considering 132 reference under-reinforced beams from the database, a correlation was studied between the torsional strength and three terms. Two terms are the ones referred to previously, namely the "reinforcement term" and the "concrete strength term". In addition, a third term was added and related to the concrete area enclosed within the external perimeter of the cross-section, the so-called "concrete area term" (related to A_c). From preliminary analysis, it was found that by adding this term the correlation is noticeably improved.

As Equation (1) is linearizable with a logarithmic transformation, a multiple linear regression was performed. Applying a logarithmic transformation and adding the additional "concrete area term," Equation (1) can be rewritten in the following general linear form:

$$ln(T_R) = A + B\,ln(f_c) + C\,ln(A_c) + D\,ln\left(A_l f_{yl}\frac{A_t f_{yt}}{s}\right), \tag{4}$$

where A, B, C, and D are numerical coefficients to be determined.

To perform the multiple linear regression, the data and terms for each reference beam from the database (T_R, f_c, A_c, and $A_l f_{yl}\frac{A_t f_{yt}}{s}$) were previously log-transformed. Then, a multiple linear model was fitted, obtaining a high R^2 equal to 0.984, and a mean squared error equal to 0.013552. For this analysis, IBM SPSS software version 28 was used. The obtained linear fitted model is the following:

$$ln(T_R) = 0.087 + 0.218\,ln(f_c) + 1.013\,ln(A_c) + 0.318\,ln\left(A_l f_{yl}\frac{A_t f_{yt}}{s}\right) \tag{5}$$

From the previous equation, the equivalent Equation (6) can be written to compute the torsional strength for under-reinforced beams.

6.3. Proposed Equations

From the results obtained in the previous subsections, Equations (6) and (7) are proposed as an improvement for Equation (1) to compute the torsional strength of RC beams. Equation (6) constitutes an alternative to Equation (2) proposed by Rahal in 2013 [8]. In Equations (6) and (7), the units of parameters are: f_c [MPa], A_c [m^2], A_l [cm^2], f_{yl} [MPa], A_t/s [cm^2/m], f_{yl} [MPa], A_c [m^2], and p_c [m]. The torsional strength T_R is computed in units [kNm].

As for Equations (2) and (3), Equation (6) governs the torsional strength for under-reinforced beams, while the upper limit stated in Equation (7) governs the torsional strength for over-reinforced beams, respectively. Equations (6) and (7) can also be applied to arbitrary cross-section shapes.

$$T_R = 1.091(f_c)^{0.218}(A_c)^{1.013}\left(A_l f_{yl}\frac{A_t f_{yt}}{s}\right)^{0.318} \tag{6}$$

$$\leq 2500(f_c)^{0.3}\frac{A_c{}^2}{p_c}, \tag{7}$$

The results from Equations (6) and (7) are checked against the results from all 202 tested beams from the database. The torsional strengths computed from the previous equations for each reference beam ($T_{T,th}^{Prop}$) are presented in Table A3, and the ratios $T_{R,exp}/T_{T,th}^{Prop}$ are presented in Table A4. The results are summarized in Table 2 and Figure 3. They show that Equations (6) and (7) provide accurate results (with $\overline{x}$ = 1.01) with a very acceptable dispersion (with cv = 13%). The results are also good for both plain and hollow RC beams. A comparison with the results from the reference codes used in this study (Table 1 and Figure 2) shows that the proposed equations provide much better results, with higher accuracy and much less dispersion. When compared with the results from Equations (2) and (3) from Rahal [8], it can be concluded that they are quite similar, although the results from the equations proposed in this study are slightly better. It should also be noted that these good results were observed for both normal- and high-strength concrete beams, as well as for under- and over-reinforced beams.

Finally, Table 2 also summarizes the obtained results substituting Equation (6) with a simplified version, Equation (8). Table 2 shows that, despite the very small changes in the powers and the numerical factors, the results show that the computed torsional strengths tend to be slightly unsafe (with $\overline{x}$ = 0.96 < 1.00). This shows that the model is highly sensitive to the precision of the numerical values (numerical factor and powers).

$$T_R = 1.09(f_c)^{0.22} A_c \left(A_l f_{yl} \frac{A_t f_{yt}}{s} \right)^{0.32} \quad (8)$$

$$\leq 2500(f_c)^{0.3} \frac{{A_c}^2}{p_c}, \quad (9)$$

Although it is not discussed in this paper, it is worth noting that the calibration of appropriate safety factors for material parameters as well as of a model uncertainty factor γ_{Rd} for Equations (6) and (7) would produce a code-formatted design capacity equation compliant with a predefined reliability level [52], which could be used in the design of RC beams failing in torsion.

7. Conclusions

In this study, a review and comparative analysis of the calculation procedures to compute the torsional strength of RC beams from some reference design codes was performed. For this, a wide database was built, incorporating the experimental torsional strengths of 202 RC rectangular beams tested until failure and found in the literature. In addition, based on the reference RC beams from the database and on correlation studies between the torsional strength and some properties (amount of reinforcement, concrete strength, and concrete area enclosed within the external perimeter of the cross-section), simple equations to compute the torsional strength were proposed and checked.

From the obtained results, the following main conclusions can be drawn:

- In general, equations from the studied reference codes still need improvements to increase the accuracy and reduce the observed statistical dispersion;
- Some reference design codes overestimate the torsional strength of several reference RC beams from the database noticeably, which is not acceptable for design and justifies further improvements;
- The proposed equations to compute the torsional strength of RC beams showed to be much more reliable and accurate in comparison with code's equations;
- The proposed equations are simple and can easily be used for practice to assess with accuracy the torsional strength of RC members, including plain and hollow beams, normal- and high-strength concrete beams, as well as under- and over-reinforced beams;
- When compared with similar equations from a previous study [8], the proposed equations were shown to be slightly better at predicting torsional strength;
- This study confirms that simple and reliable design equations can be obtained by simply fitting the results with experimental data existing in the literature and related to the torsional strength of the RC beams.

Author Contributions: L.F.A.B.—conceptualization, supervision, validation, writing—review and editing. M.M.T.—methodology, investigation, formal analysis, and writing—original draft preparation. D.D.D.—validation and writing—review and editing. J.M.R.G.—data curation, software, and writing—review and editing. All authors have read and agreed to the published version of the manuscript.

Funding: This research received no external funding.

Institutional Review Board Statement: Not applicable.

Informed Consent Statement: Not applicable.

Data Availability Statement: Not applicable.

Conflicts of Interest: The authors declare no conflict of interest.

Nomenclature

A_c	area enclosed within the outer perimeter of the cross-section
A_k	area within the centerline of wall's thickness (assumed to coincide with the shear flow path)
A_l	total area of longitudinal reinforcement
A_{l1}	area of bottom longitudinal reinforcement
A_{l2}	area of tensile longitudinal steel near the vertical face
A_o	area enclosed within circular shear flow
A_{oh}	area enclosed within centerline of hoops
A_t	area for one bar of the hoop
f_c	average compressive strength of concrete
f_{ck}	characteristic value of the compressive strength of concrete at 28 days
f_{yl}	average yield strength of longitudinal reinforcement
f_{yt}	average yield strength of tranerse reinforcement
k_c	concrete reduction factor
p_c	gross perimeter of the cross-section
p_h	perimeter of the centerline of hoop
q	shear stress flow due to torsion
s	longitudinal spacing between hoops
T	applied torsional moment
t	Wall's thickness of the equivalent hollow section
T_c	torsional moment resisted by concrete
T_{max}	maximum torsional mome
T_R	torsional moment resistance
$T_{R,exp}$	experimental torsional moment resistance
$T_{R,th}$	theoretical torsional moment resistance
T_s	torsional moment resisted by the reinforcement
T_u	ultimate torsional moment
u_k	perimeter of the area A_k
V	alied shear force
V_R	resistance shear force
V_{Rc}	shear force due to the compressive stresses resisted by concrete
V_{Rl}	shear force due to the axial stresses resisted by the longitudinal reinforcement
$V_{R,max}$	maximum resistance shear force
V_{Rt}	shear force due to shear stresses resisted by the transverse reinforcement
x	smaller dimension of the cross-section
x_1	smaller dimension of hoops
y	larger dimension of the cross-section
y_1	larger dimension of hoops
z_i	length of wall i, equal to the distance between the centerline intersection of adjacent walls
α	angle of the hoops to the longitudinal axis
α_{cw}	coefficient to account for the stress state of the compressed chord
α_t	efficiency coefficient
γ_c	partial safety factor for concrete properties
δ	numerical coefficient to account for imperfections
θ	angle of the diagonal compressive stresses in concrete struts to the longitudinal axis
ν	strength reduction factor for cracked concrete
ρ_l	ratio of longitudinal reinforcement: $\rho_l = A_l/xy$
ρ_t	ratio of transverse reinforcement: $\rho_t = A_t u/As$
τ	shear stress due to torsion
ϕ	reduction coefficient
ϕ_c	concrete resistance factor
ϕ_s	reinforcement resistance factor

Appendix A

Table A1. Equations from the reference design codes.

	SinNiP 2018	ACI 318R-89
Limit of the section wall thickness	Not available	Wall thickness of the hollow section: If $t \geq x/4$ the section is considered as a plain section; If $x/10t \leq t < x/4$ the section is considered as an equivalent plain section.
Shear stresses and shear forces due to torsion	Not available	Not available
Safety condition	Condition: $T_R \leq T_{max}$ where $T_R = min\{T_{R1}; T_{R2}\}$ $T_{R1} = 0.5A_{l1}f_{yl}y + A_t f_{yt}\left(\frac{x^2y}{s(2y+x)}\right)$ $T_{R2} = 0.5A_{l2}f_{yl}x + A_t f_{yt}\left(\frac{y^2x}{s(2x+y)}\right)$	Condition: $T_u \leq T_R$ where $T_R = T_c + T_s$ For plain section: $T_c = 0.8(f_c)^{1/2}x^2y$ For hollow section: $T_c = 0.8(f_c)^{1/2}x^2y\left(\frac{4t}{x}\right)$ $T_s = \frac{A_t}{s}\alpha_t x_1 y_1 f_{ty}$ where $\alpha_t = 0.66 + 0.33(y_1/x_1) \leq 1.5$ and $x_1 \leq y_1$

Maximum limit by the tension in concrete struts	Maximum limit:
	$T_{max} = 0.1f_c x^2 y$
		Maximum limit:
		If $T_s \leq 4T_c$, then
		$T_R = T_c + T_s$
		If $T_s > 4T_c$, then
		$T_R = 5T_c$

*	Not available	Not available

* Angle between the concrete struts and the longitudinal axis of the beams.

	ACI 318R-19	MC 90
Limit of the section wall thickness	Wall thickness of the hollow section: If $t \geq A_{oh}/p_h$ the section is considered as a plain section; If $t < A_{oh}/p_h$ the section is considered as an equivalent plain section.	Wall thickness of the hollow section: If $t = A/u \leq t_{real}$ the section is considered as a hollow section; If $t = A/u > t_{real}$ the section is considered as a plain section (equivalent hollow section).Wall thickness of the plain section (equivalent hollow section): $t = A/u$ and $t >$ 2x distance between the face of the section and the axis of the longitudinal reinforcement.
Shear stresses and shear forces due to torsion	Shear flow for a thin-walled tube: $q = \tau t$ Shear stress along the perimeter of the section: $\tau = \frac{T}{2A_o t}$ with $A_o = 0.85A_{oh}$	Shear flow at a wall i: $\tau_i t_i = \frac{T}{2A_o\delta}$ Shear stress at a wall i: $V_i = \frac{Tz_i}{2A_o\delta}$ where $\delta = 1.0 - 0.25(x/y)$, with $y \geq x$.
Safety condition	Condition: $T_u \leq \phi T_R$ where $\phi = 1$ (for the present study). $T_R = min\{(a);(b)\}$ $(a)\ T_R = 2A_o \frac{A_t}{s} f_{yt} \cos\theta$ $(b)\ T_R = 2A_l f_{yl} \frac{\tan\theta}{p_h}$	Condition: $V_R = min\{(a);(b);(c)\}$ (a) Shear force due to the force on the longitudinal reinforcement: $V_{Rl} \leq A_l f_{yl} / \cot\theta$ (c) Shear force due to the force on the transverse reinforcement: $V_{Rt} = A_t f_{yt} u_k \cot\theta / s$
Maximum limit by the tension in concrete struts	Maximum limit to plain section: $T_u \leq \phi 8(f_c)^{1/2} 1.7 \frac{A_{oh}^2}{p_h}$ Maximum limit to hollow section: $T_u \leq \phi 8(f_c)^{1/2} 1.7 A_{oh} t$ where $\phi = 1$ (for the present study).	Maximum limit: Shear force due to the force on the concrete struts: $V_{Rc} \leq f_{cd2} t u_k \cos\theta \sin\theta$ where $f_{cd2} = 0.60(1 - f_{ck}/250) f_c$
*	$\cot^2\theta = \frac{A_l f_{yl} s}{A_t f_{yt} p_h}$	$\cot^2\theta = \frac{A_l f_{yl} s}{A_t f_{yt} u_k}$

* Angle between the concrete struts and the longitudinal axis of the beams.

	MC 10	Eurocode 2
Limit of the section wall thickness	Wall thickness of the hollow section: $t = t_{real}$ Wall thickness of the plain section (equivalent hollow section): t = 2x distance between the face of the section and the axis of the longitudinal reinforcement and $t \leq x/8$.	Wall thickness of the hollow section: If $t = A/u \leq t_{real}$ the section is considered as a hollow section; If $t = A/u > t_{real}$ the section is considered as a plain section (equivalent hollow section).Wall thickness of the plain section (equivalent hollow section): $t = A/u$ e $t >$ 2x distance between the face of the section and the axis of the longitudinal reinforcement.
Shear stresses and shear forces due to torsion	Shear stress at a wall i: $V_i = \frac{Tz_i}{2A_k}$	Shear flow at a wall i: $\tau_i t_i = \frac{T}{2A_k}$ Shear stress at a wall i: $V_i = \tau_i t_i z_i$
Safety condition	Condition: $V_R = V_{Rc} + V_{Rt} \leq V_{R,max}$ *Approximation level II:* $V_{Rc} = 0$ $V_{Rt} = \frac{A_t}{s} u_k f_{yt} \cot\theta$	Condition: $\frac{T}{T_{R,max}} \leq 1.0$ $V_R = \frac{A_t}{s} u_k f_{yt} \cot\theta$
Maximum limit by the tension in concrete struts	Maximum limit: $V_{R,max} = k_c\left(\frac{f_{ck}}{\gamma_c}\right) t u_k \left(\frac{\cot\theta - \cot\alpha}{1+\cot^2\theta}\right)$ Approximation level II: $k_c = 0.55(30/f_{ck})^{1/3} \leq 0.55$ and $\alpha = 90$	Maximum limit: $T_{R,max} = 2\nu\alpha_{cw} f_c A_k t \sin\theta\cos\theta$ where $\nu = 0.6[1 - (f_{ck}/250)]$ and $\alpha_{cw} = 1$, for non-prestressed beams.
*	$\cot^2\theta = \frac{2A_l f_{yl} s}{A_t f_{yt} u_k}$	$\cot^2\theta = \frac{A_l f_{yl} s}{A_t f_{yt} u_k}$

* Angle between the concrete struts and the longitudinal axis of the beams.

	CSA A23.3-14
Limit of the section wall thickness	Wall thickness of the hollow section: If $t \geq A_{oh}/p_h$ the section is considered as a plain section; If $t < A_{oh}/p_h$ the section is considered as an equivalent plains section.
Shear stresses and shear forces due to torsion	Not available
Safety condition	Condition: $T_u \leq T_R$ where $T_R = 2A_o\phi_s \frac{A_t}{s} f_{yt} \cos\theta$ with $A_o = 0,85A_{oh}$ and $\phi_s = 1$ (for the present study).
Maximum limit by the tension in concrete struts	Maximum limit to plain section: $T_u \leq 0.25\phi_c f_c 1.7\frac{A_{oh}^2}{p_h}$ Maximum limit to hollow section: $T_u \leq 0.25\phi_c f_c 1.7A_{oh}t$ where $\phi_c = 1$ (for the present study).
*	$\cot^2\theta = \frac{A_l f_{yl} s}{0.45A_t f_{yt} p_h}$

* Angle between the concrete struts and the longitudinal axis of the beams.

Table A2. Geometric and mechanical properties of the reference RC beams.

Ref.	Beam	**	x (m)	y (m)	t (m)	x_1 (m)	y_1 (m)	A_{l1} (cm^2)	A_{l2} (cm^2)	A_l (cm^2)	A_t/s (cm^2/m)	f_c (MPa)	f_{yl} (MPa)	f_{yt} (MPa)
Hsu (1968)	B1	P	0.254	0.381	-	0.216	0.343	2.53	2.53	5.07	4.68	27.6	314.0	341.0
	B3	P	0.254	0.381	-	0.216	0.343	5.73	5.73	11.36	10.16	28.1	327.6	320.0
	B4	P	0.254	0.381	-	0.216	0.343	7.74	7.74	15.48	14.01	29.2	320.0	323.4
	B5	P	0.254	0.381	-	0.216	0.343	10.20	10.20	20.39	18.47	30.6	332.4	321.4
	B6	P	0.254	0.381	-	0.216	0.343	12.90	12.90	25.81	22.58	28.8	331.7	322.8
	B7	P	0.254	0.381	-	0.216	0.343	2.53	2.53	5.16	10.16	26.0	320.0	318.6
	B8	P	0.254	0.381	-	0.216	0.343	2.53	2.53	5.16	22.58	26.8	322.1	320.0
	B9	P	0.254	0.381	-	0.216	0.343	5.73	5.73	11.36	4.66	28.8	319.3	342.8
	B10	P	0.254	0.381	-	0.216	0.343	12.90	12.90	25.8	4.66	26.5	334.0	342.0
	C2	P	0.254	0.254	-	0.216	0.216	2.53	2.53	5.07	6.07	26.5	334.0	345.0
	C4	P	0.254	0.254	-	0.216	0.216	5.73	5.73	11.36	13.11	27.2	336.6	327.6
	C5	P	0.254	0.254	-	0.216	0.216	7.74	7.74	15.48	17.67	27.2	328.3	329.0
	C6	P	0.254	0.254	-	0.216	0.216	10.20	10.20	20.39	23.91	27.6	315.9	327.6
	G2	P	0.254	0.508	-	0.216	0.470	3.97	3.97	7.94	5.91	30.9	323.0	334.0
	G3	P	0.254	0.508	-	0.216	0.470	5.73	5.73	11.36	8.29	26.8	338.6	327.6
	G4	P	0.254	0.508	-	0.216	0.470	7.74	7.74	15.48	11.29	28.3	325.5	321.4
	G5	P	0.254	0.508	-	0.216	0.470	10.20	10.20	20.39	15.05	26.9	331.0	327.6
	G6	P	0.254	0.508	-	0.216	0.470	2.53	3.80	7.60	5.61	29.9	334.0	350.0
	G7	P	0.254	0.508	-	0.216	0.470	3.97	5.96	12.00	8.84	31.0	319.3	322.8
	G8	P	0.254	0.508	-	0.216	0.470	5.73	8.60	17.03	12.32	28.3	322.1	329.0
	I2	P	0.254	0.381	-	0.216	0.343	3.97	3.97	7.94	7.25	45.2	325	349
	I3	P	0.254	0.381	-	0.216	0.343	5.73	5.73	11.36	10.16	44.8	343.4	333.8
	I4	P	0.254	0.381	-	0.216	0.343	7.74	7.74	15.48	14.01	45.0	315.2	326.2
	I5	P	0.254	0.381	-	0.216	0.343	10.20	10.20	20.39	18.47	45.0	310.3	325.5
	I6	P	0.254	0.381	-	0.216	0.343	12.90	12.90	25.81	22.58	45.8	325.5	329.0
	J1	P	0.254	0.381	-	0.216	0.343	2.53	2.53	5.16	4.66	14.3	327.6	346.2
	J2	P	0.254	0.381	-	0.216	0.343	3.97	3.97	8.00	7.21	14.6	320.0	340.7
	J3	P	0.254	0.381	-	0.216	0.343	5.73	5.73	11.36	10.16	16.9	388.6	337.2
	J4	P	0.254	0.381	-	0.216	0.343	7.74	7.74	15.48	14.01	16.8	324.1	331.7
	K2	P	0.152	0.495	-	0.114	0.457	2.53	3.80	7.74	6.77	30.6	335.9	337.9
	K3	P	0.152	0.495	-	0.114	0.457	3.97	5.96	12.00	10.42	29.0	315.9	320.7
	K4	P	0.152	0.495	-	0.114	0.457	5.73	8.60	17.03	15.05	28.6	344.1	340.0
	M1	P	0.254	0.381	-	0.216	0.343	3.97	3.97	8.00	4.76	29.9	326.2	353.1
	M2	P	0.254	0.381	-	0.216	0.343	5.73	5.73	11.36	6.77	30.6	329.0	357.2
	M3	P	0.254	0.381	-	0.216	0.343	7.74	7.74	15.48	9.24	26.8	322.1	326.2
	M4	P	0.254	0.381	-	0.216	0.343	10.20	10.20	20.39	12.33	26.6	318.6	326.9
	M5	P	0.254	0.381	-	0.216	0.343	12.90	12.90	25.81	15.63	28.0	335.2	331.0
	M6	P	0.254	0.381	-	0.216	0.343	10.20	15.30	30.58	15.63	29.4	317.9	340.7
	N1	P	0.152	0.305	-	0.130	0.283	1.43	1.43	2.84	3.50	29.5	352.4	341.4
	N1a	P	0.152	0.305	-	0.130	0.283	1.43	1.43	2.84	3.50	28.7	346.2	344.8

** P—Plain section; H—Hollow section.

Ref.	Beam	**	x (m)	y (m)	t (m)	x_1 (m)	y_1 (m)	A_{l1} (cm^2)	A_{l2} (cm^2)	A_l (cm^2)	A_t/s (cm^2/m)	f_c (MPa)	f_{yl} (MPa)	f_{yt} (MPa)
	N2	P	0.152	0.305	-	0.130	0.283	2.53	2.53	5.16	6.35	30.4	331.0	337.9
	N2a	P	0.152	0.305	-	0.130	0.283	2.53	2.53	1.61	6.21	28.4	333.1	360.7
	N3	P	0.152	0.305	-	0.130	0.283	1.43	2.14	4.26	5.08	27.3	351.7	351.7
	N4	P	0.152	0.305	-	0.130	0.283	2.53	3.25	6.58	7.98	27.3	340.9	355.9
[17]	T4	P	0.500	0.500	-	0.454	0.454	5.65	5.65	18.10	10.28	35.3	356.7	356.7
Leonhardt, Schelling (1974)	VB2	P	0.440	0.240	-	0.420	0.220	2.63	1.46	7.01	5.84	26.4	541.4	541.4
	VB3	P	0.440	0.240	-	0.420	0.220	2.63	1.46	7.01	5.84	39.1	541.4	541.4
	VB4	P	0.440	0.240	-	0.420	0.220	2.63	1.46	7.01	5.84	49.8	541.4	541.4
	VM1	P	0.294	0.160	-	0.280	0.146	1.50	1.00	3.00	3.63	39.1	442.4	568.9
	VM2	P	0.440	0.240	-	0.420	0.220	3.30	2.20	6.60	5.32	36.1	431.6	436.5
	VM3	P	0.587	0.320	-	0.561	0.294	6.42	4.28	12.84	7.14	40.0	461.0	442.4
	VQ1	P	0.324	0.324	-	0.304	0.304	1.15	1.15	3.46	2.88	19.0	557.1	557.1
	VQ3	P	0.580	0.186	-	0.560	0.166	1.83	0.61	4.27	3.05	17.6	432.6	432.6
	VQ9	P	0.806	0.140	-	0.786	0.120	2.54	0.56	5.08	2.82	19.5	441.4	441.4
	VS2-VQ2	P	0.440	0.240	-	0.420	0.220	1.53	0.92	3.66	3.05	19.0	432.6	432.6
	VS3	P	0.440	0.240	-	0.420	0.220	2.14	1.22	5.49	4.55	19.5	432.6	432.6
	VS4-VQ5	P	0.440	0.240	-	0.420	0.220	2.75	1.53	7.32	6.10	19.0	432.6	432.6
	VS9	P	0.440	0.240	-	0.420	0.220	1.16	1.16	3.48	2.90	17.6	570.9	570.9
	VS10-VB1	P	0.440	0.240	-	0.420	0.220	2.61	1.45	6.96	5.80	19.0	570.9	570.9
	VU1	P	0.440	0.240	-	0.420	0.220	1.40	0.84	3.36	5.60	19.5	441.4	441.4
	VU2	P	0.440	0.240	-	0.420	0.220	1.96	1.12	5.04	5.60	19.5	441.4	441.4
	VU3	P	0.440	0.240	-	0.420	0.220	2.52	1.40	6.72	4.18	18.5	441.4	441.4
	VU4	P	0.440	0.240	-	0.420	0.220	2.52	1.40	6.72	2.80	18.5	441.4	441.4
McMullen, Rangan (1978)	A2	P	0.254	0.254	-	0.222	0.222	2.53	2.53	5.16	7.82	38.2	380.0	285.0
	A3	P	0.254	0.254	-	0.219	0.219	3.97	3.97	8.00	8.94	39.4	352.4	360.0
	A4	P	0.254	0.254	-	0.219	0.219	5.73	5.73	11.36	12.42	39.2	351.0	360.0
	B1r	P	0.178	0.356	-	0.146	0.324	1.43	1.43	2.85	3.87	36.3	360.0	285.0
	B2	P	0.178	0.356	-	0.146	0.324	2.53	2.53	5.07	7.19	39.6	380.0	285.0
	B3	P	0.178	0.356	-	0.143	0.321	3.97	3.97	8.00	8.60	38.6	352.4	360.0
	B4	P	0.178	0.356	-	0.143	0.321	5.73	5.73	11.36	11.76	38.5	351.0	360.0
Rasmussen, Baker (1995)	B30.1	P	0.160	0.275	-	0.120	0.235	5.15	7.72	15.44	8.73	41.7	620.0	665.0
	B30.2	P	0.160	0.275	-	0.120	0.235	5.15	7.72	15.44	8.73	38.2	638.0	669.0
	B30.3	P	0.160	0.275	-	0.120	0.235	5.15	7.72	15.44	8.73	36.3	605.0	672.0
	B50.1	P	0.160	0.275	-	0.120	0.235	5.15	7.72	15.44	8.73	61.8	612.0	665.0
	B50.2	P	0.160	0.275	-	0.120	0.235	5.15	7.72	15.44	8.73	57.1	614.0	665.0
	B50.3	P	0.160	0.275	-	0.120	0.235	5.15	7.72	15.44	8.73	61.7	612.0	665.0
	B70.1	P	0.160	0.275	-	0.120	0.235	5.15	7.72	15.44	8.73	77.3	617.0	658.0
	B70.2	P	0.160	0.275	-	0.120	0.235	5.15	7.72	15.44	8.73	76.9	614.0	656.0
	B70.3	P	0.160	0.275	-	0.120	0.235	5.15	7.72	15.44	8.73	76.2	617.0	663.0
	B110.1	P	0.160	0.275	-	0.120	0.235	5.09	7.63	15.44	8.73	109.8	618.0	655.0
	B110.2	P	0.160	0.275	-	0.120	0.235	5.09	7.63	15.44	8.73	105.0	634.0	660.0

** P—Plain section; H—Hollow section.

Ref.	Beam	**	x	y	t	x_1	y_1	A_{l1}	A_{l2}	A_l	A_t/s	f_c	f_{yl}	f_{yt}
			(m)	(m)	(m)	(m)	(m)	(cm^2)	(cm^2)	(cm^2)	(cm^2/m)	(MPa)	(MPa)	(MPa)
	B110.3	P	0.160	0.275	-	0.120	0.235	5.15	7.72	15.44	8.73	105.1	629.0	655.0
Koutchoukali, Belarbi (2001)	B5UR1	P	0.203	0.305	-	0.165	0.267	2.53	2.53	5.16	6.56	39.6	386.0	373.0
	B7UR1	P	0.203	0.305	-	0.165	0.267	2.53	2.53	5.16	6.56	64.6	386.0	399.0
	B9UR1	P	0.203	0.305	-	0.165	0.267	2.53	2.53	5.16	6.56	75.0	386.0	373.0
	B12UR1	P	0.203	0.305	-	0.165	0.267	2.53	2.53	5.16	6.56	80.6	386.0	399.0
	B12UR2	P	0.203	0.305	-	0.165	0.267	2.53	2.53	5.16	6.95	76.2	386.0	386.0
	B12UR3	P	0.203	0.305	-	0.165	0.267	2.53	3.25	6.58	7.46	72.9	379.5	386.0
	B12UR4	P	0.203	0.305	-	0.165	0.267	2.53	3.80	7.74	7.88	75.9	373.0	386.0
	B12UR5	P	0.203	0.305	-	0.165	0.267	3.97	3.97	8.00	10.13	76.7	380.0	386.0
	B14UR1	P	0.203	0.305	-	0.165	0.267	2.53	2.53	5.16	6.56	93.9	386.0	386.0
Fang, Shiau (2004)	H-06-06	P	0.350	0.500	-	0.300	0.450	3.97	9.93	11.92	7.13	78.5	440.0	440.0
	H-06-12	P	0.350	0.500	-	0.300	0.450	5.07	12.67	20.65	7.10	78.5	410.0	440.0
	H-07-10	P	0.350	0.500	-	0.300	0.450	5.73	14.33	17.03	7.89	68.4	500.0	420.0
	H-07-16	P	0.350	0.500	-	0.300	0.450	8.60	22.92	28.39	7.89	68.4	500.0	420.0
	H-12-12	P	0.350	0.500	-	0.300	0.450	5.07	12.67	20.65	14.19	78.5	410.0	440.0
	H-12-16	P	0.350	0.500	-	0.300	0.450	8.60	18.62	28.39	14.19	78.5	520.0	440.0
	H-14-10	P	0.350	0.500	-	0.300	0.450	5.73	14.33	17.03	16.13	68.4	500.0	360.0
	H-20-20	P	0.350	0.500	-	0.300	0.450	8.60	22.92	34.06	23.46	78.5	560.0	440.0
	N-06-06	P	0.350	0.500	-	0.300	0.450	3.97	9.93	12.00	7.10	35.5	440.0	440.0
	N-06-12	P	0.350	0.500	-	0.300	0.450	5.07	12.67	20.65	7.10	35.5	410.0	440.0
	N-07-10	P	0.350	0.500	-	0.300	0.450	5.73	14.33	17.03	7.89	33.5	500.0	420.0
	N-07-16	P	0.350	0.500	-	0.300	0.450	8.60	20.06	28.39	7.89	33.5	500.0	420.0
	N-12-12	P	0.350	0.500	-	0.300	0.450	5.07	12.67	20.65	14.19	35.5	410.0	440.0
	N-12-16	P	0.350	0.500	-	0.300	0.450	8.60	20.06	28.39	14.19	35.5	520.0	440.0
	N-14-10	P	0.350	0.500	-	0.300	0.450	5.73	14.33	17.03	16.13	33.5	500.0	360.0
	N-20-20	P	0.350	0.500	-	0.300	0.450	8.60	22.92	34.06	23.46	35.5	560.0	440.0
Chiu et al. (2007)	HAS-51-50	P	0.420	0.420	-	0.370	0.370	3.80	3.25	9.03	5.94	76.0	396.0	385.0
	NAS-61-35	P	0.420	0.420	-	0.370	0.370	4.68	4.69	10.80	4.19	48.0	394.0	385.0
	HAS-90-50	P	0.420	0.420	-	0.370	0.370	5.96	5.96	15.89	5.94	78.0	400.0	385.0
	NBS-43-44	P	0.350	0.500	-	0.300	0.450	2.53	3.80	7.60	5.09	35.0	400.0	385.0
	HBS-74-17	P	0.350	0.500	-	0.300	0.450	5.73	6.44	12.89	2.02	67.0	505.0	600.0
	HBS-82-13	P	0.350	0.500	-	0.300	0.450	5.73	7.16	14.31	1.49	67.0	493.0	600.0
	NBS-82-13	P	0.350	0.500	-	0.300	0.450	5.73	7.16	14.31	1.49	35.0	493.0	600.0
	HBS-60-61	P	0.350	0.500	-	0.300	0.450	3.97	5.24	10.48	7.13	67.0	402.0	385.0
	HCS-52-50	P	0.250	0.700	-	0.200	0.650	2.53	4.51	9.03	5.09	76.0	396.0	385.0
	HCS-91-50	P	0.250	0.700	-	0.200	0.650	3.97	7.94	15.89	5.09	78.0	400.0	385.0
Lee and Kim (2010)	T1-1	P	0.300	0.350	-	0.260	0.310	2.53	2.53	5.07	5.48	43.2	410.0	370.0
	T1-2	P	0.300	0.350	-	0.260	0.310	2.53	3.80	7.60	8.39	44.0	410.0	370.0
	T1-3	P	0.300	0.350	-	0.260	0.310	2.53	5.07	10.14	10.97	41.7	410.0	370.0
	T1-4	P	0.300	0.350	-	0.260	0.310	3.97	5.96	11.92	16.89	42.6	510.0	355.0
	T2-2	P	0.300	0.350	-	0.260	0.310	3.97	5.96	7.94	5.48	41.7	510.0	370.0

** P—Plain section; H—Hollow section.

Ref.	Beam	**	x (m)	y (m)	t (m)	x_1 (m)	y_1 (m)	A_{l1} (cm^2)	A_{l2} (cm^2)	A_l (cm^2)	A_t/s (cm^2/m)	f_c (MPa)	f_{yl} (MPa)	f_{yt} (MPa)
	T2-3	P	0.300	0.350	-	0.260	0.310	3.97	5.96	11.92	8.10	42.7	510.0	370.0
	T2-4	P	0.300	0.350	-	0.260	0.310	5.73	7.00	14.00	9.51	42.6	512.4	370.0
Peng, Wong (2011)	SW12-1	P	0.150	1.200	-	0.100	1.150	2.26	11.31	22.62	3.93	44.2	480.0	459.0
	SW10-1	P	0.150	1.000	-	0.100	0.950	2.26	9.05	18.10	3.93	29.5	499.0	459.0
	SW10-2	P	0.150	1.000	-	0.098	0.948	2.26	9.05	18.10	7.54	44.2	480.0	480.0
	SW10-3	P	0.150	1.000	-	0.098	0.948	2.26	9.05	18.10	11.31	29.5	499.0	499.0
	SW10-4	P	0.150	1.000	-	0.094	0.944	4.02	16.08	32.16	16.08	33.8	497.0	497.0
	SW8-1	P	0.150	0.800	-	0.102	0.752	1.57	7.07	14.14	4.02	29.5	459.0	433.0
	SW8-2	P	0.150	0.800	-	0.098	0.748	1.57	7.07	14.14	11.31	29.5	459.0	499.0
Joh et al. (2019)	RA-SD4-3.2-0.3-3.28	P	0.300	0.400	-	0.270	0.370	11.91	11.91	30.97	7.13	73.7	452.0	484.0
	RA-SD5-3.2-0.3-3.21	P	0.300	0.400	-	0.270	0.370	11.61	11.61	30.97	6.48	73.7	499.0	538.0
	RA-SD6-3.2-0.2-3.21	P	0.300	0.400	-	0.270	0.370	11.61	11.61	30.97	6.48	73.7	630.0	538.0
	RA-SD4-3.2-0.5-2.13	P	0.300	0.400	-	0.270	0.370	5.73	8.60	17.19	7.13	84.7	456.0	484.0
	RA-SD5-3.2-0.7-1.63	P	0.300	0.400	-	0.270	0.370	3.97	5.96	11.92	6.48	84.7	529.0	538.0
	RA-SD6-3.2-0.6-1.63	P	0.300	0.400	-	0.270	0.370	3.97	5.96	11.92	6.48	84.7	627.0	538.0
	RA-SD4-3.2-1.1-1.33	P	0.300	0.400	-	0.270	0.370	2.53	3.80	7.60	7.13	83.1	474.0	484.0
	RA-SD5-3.2-1.0-1.26	P	0.300	0.400	-	0.270	0.370	2.53	3.80	7.60	6.48	83.1	522.0	538.0
	RA-SD6-3.2-0.8-1.26	P	0.300	0.400	-	0.270	0.370	3.97	3.97	7.94	6.48	83.1	627.0	538.0
Ju et al. (2019)	MR30-0.77	P	0.350	0.500	-	0.300	0.450	2.53	3.80	7.60	3.96	29.3	489.8	467.5
	MT30-1.32	P	0.350	0.500	-	0.300	0.450	5.73	8.60	17.19	3.96	29.3	500.4	467.5
	MT40-1.32	P	0.350	0.500	-	0.300	0.450	5.73	8.60	17.19	3.96	40.3	500.4	467.5
	MT40-1.89	P	0.350	0.500	-	0.300	0.450	5.73	8.60	17.19	10.57	40.3	489.8	489.8
Ibrahim et al. (2020)	NSC-S1-C30	P	0.200	0.300	-	0.168	0.268	2.26	2.26	4.52	6.28	42.1	689.7	534.1
	NSC-S1-C45	P	0.200	0.300	-	0.138	0.238	2.26	2.26	4.52	6.28	39.4	689.7	534.1
	HSC-C30	P	0.200	0.300	-	0.168	0.268	2.26	2.26	4.52	6.28	60.8	689.7	534.1
	HSC-C45	P	0.200	0.300	-	0.138	0.238	2.26	2.26	4.52	6.28	60.8	689.7	534.1
Kim et al. (2020)	S08-3-65	P	0.400	0.600	-	0.310	0.510	5.24	7.77	18.08	10.97	35.4	313.3	334.9
	S08-4-90	P	0.400	0.600	-	0.310	0.510	2.85	5.23	13.30	7.92	35.4	474.6	485.8
	S08-5-122.5	P	0.400	0.600	-	0.310	0.510	3.80	3.96	10.45	5.82	35.4	569.6	595.9
	S10-3-52.5	P	0.400	0.600	-	0.310	0.510	5.24	9.93	22.39	13.58	35.4	320.5	334.0
	S10-4-72.5	P	0.400	0.600	-	0.310	0.510	5.07	5.78	16.63	9.83	35.4	467.6	485.1
	S10-5-100	P	0.400	0.600	-	0.310	0.510	3.80	5.23	12.98	7.13	35.4	567.6	595.2
	S06-3-90	P	0.400	0.600	-	0.310	0.510	2.85	5.23	13.30	7.92	35.4	319.4	334.3
	S10-5-90	P	0.400	0.600	-	0.310	0.510	3.96	5.78	14.41	7.92	35.4	569.6	594.8
	S08-3-72.5	P	0.400	0.600	-	0.310	0.510	5.07	5.78	16.63	9.83	35.4	308.8	334.8
	S12-5-72.5	P	0.400	0.600	-	0.310	0.510	5.24	7.77	18.05	9.83	35.4	565.1	595.2
Hsu (1968)	D3	H	0.254	0.381	0.064	0.216	0.343	*	*	11.36	10.16	28.4	341.4	333.1
	D4	H	0.254	0.381	0.064	0.216	0.343	30.97	30.97	15.48	14.01	30.6	330.3	333.1
[17,38]	T0	H	0.500	0.500	0.080	0.430	0.430	*	*	32.16	10.28	45.1	345.2	357.0
	T1	H	0.500	0.500	0.080	0.454	0.454	4.52	4.52	18.10	10.28	35.3	356.7	356.7
	T2	H	0.500	0.500	0.080	0.430	0.430	*	*	18.10	10.28	35.3	357.0	357.0

* No sufficient data; ** P—Plain section; H—Hollow section.

Ref.	Beam	**	x (m)	y (m)	t (m)	x_1 (m)	y_1 (m)	A_{l1} (cm^2)	A_{l2} (cm^2)	A_l (cm^2)	A_t/s (cm^2/m)	f_c (MPa)	f_{yl} (MPa)	f_{yt} (MPa)
	T5	H	0.800	0.400	0.080	0.730	0.330	*	*	10.00	10.28	47.1	528.6	512.9
[19]	VH1	H	0.324	0.324	0.080	0.304	0.304	1.15	1.15	3.46	2.88	17.2	447.3	447.3
	VH2	H	0.324	0.324	0.080	0.304	0.304	*	*	6.91	5.76	17.2	447.3	447.3
Bernardo, Lopes (2009)	A1	H	0.600	0.600	0.098	0.537	0.547	*	*	6.53	3.14	48.4	695.9	636.7
	A2	H	0.600	0.600	0.107	0.538	0.531	4.62	4.62	13.95	6.28	47.3	672.4	695.9
	A3	H	0.600	0.600	0.109	0.535	0.535	5.65	5.65	18.10	8.27	46.2	672.4	714.8
	A4	H	0.600	0.600	0.104	0.520	0.525	7.95	7.95	23.75	11.22	54.8	723.9	714.8
	A5	H	0.600	0.600	0.104	0.528	0.528	9.68	9.68	30.66	14.14	53.1	723.9	672.4
	B2	H	0.600	0.600	0.108	0.533	0.534	4.78	4.78	14.58	6.70	69.8	672.4	695.9
	B3	H	0.600	0.600	0.109	0.535	0.537	7.95	7.95	23.75	11.22	77.8	723.9	714.8
	B4	H	0.600	0.600	0.112	0.523	0.536	10.05	10.05	32.17	15.08	79.8	723.9	672.4
	B5	H	0.600	0.600	0.117	0.518	0.518	12.06	12.06	40.21	18.85	76.4	723.9	672.4
	C1	H	0.600	0.600	0.097	0.540	0.549	*	*	6.53	3.14	91.7	695.9	636.7
	C2	H	0.600	0.600	0.100	0.532	0.533	4.62	4.62	13.95	6.28	94.8	672.4	695.9
	C3	H	0.600	0.600	0.103	0.545	0.540	7.95	7.95	23.75	10.47	91.6	723.9	714.8
	C4	H	0.600	0.600	0.103	0.546	0.545	9.68	9.68	30.66	14.14	91.4	723.9	672.4
	C5	H	0.600	0.600	0.104	0.540	0.543	12.32	12.32	36.69	17.40	96.7	723.9	672.4
	C6	H	0.600	0.600	0.104	0.533	0.529	14.07	14.07	48.25	22.62	87.5	723.9	672.4
Chiu et al. (2007)	HAH-81-35	H	0.420	0.420	0.075	0.370	0.370	5.73	6.44	14.31	4.19	78.0	493.0	385.0
	NCH-62-33	H	0.250	0.700	0.075	0.200	0.650	3.97	5.40	10.80	3.40	48.0	394.0	385.0
	HCH-91-42	H	0.250	0.700	0.075	0.200	0.650	3.97	7.94	15.89	4.32	78.0	400.0	385.0
Jeng (2014)	A095c	H	0.497	0.711	0.145	0.437	0.651	*	*	13.16	9.93	35.1	371.0	381.0
	A120a	H	0.502	0.719	0.184	0.442	0.659	*	*	20.00	7.59	27.6	464.0	380.0
	B065b	H	0.503	0.710	0.092	0.443	0.650	*	*	50.97	9.93	39.2	452.0	380.0
	B080a	H	0.500	0.721	0.112	0.440	0.661	*	*	28.39	12.90	46.5	454.0	392.0
	B110a	H	0.498	0.710	0.155	0.438	0.650	*	*	20.00	8.60	48.1	453.0	369.0
	C100a	H	0.499	0.723	0.127	0.439	0.663	*	*	28.39	12.90	90.6	466.0	447.0
	D075a	H	0.498	0.734	0.087	0.438	0.674	*	*	28.39	12.90	94.9	469.0	381.0
	D090a	H	0.501	0.722	0.105	0.441	0.662	*	*	28.39	12.90	105.7	466.0	447.0
Kim et al. (2020)	H08-3-65	H	0.400	0.600	0.100	0.310	0.510	5.24	7.77	18.08	10.97	36.5	361.1	352.2
	H08-4-90	H	0.400	0.600	0.100	0.310	0.510	3.96	5.23	13.30	7.92	36.5	445.7	448.9
	H08-5-100	H	0.400	0.600	0.100	0.310	0.510	3.25	5.23	11.88	7.13	36.5	545.5	539.8
	H10-3-52.5	H	0.400	0.600	0.100	0.310	0.510	5.96	9.21	22.39	13.58	36.5	356.9	352.4
	H10-4-72.5	H	0.400	0.600	0.100	0.310	0.510	5.07	5.78	16.63	9.83	36.5	444.5	447.9
	H10-5-80	H	0.400	0.600	0.100	0.310	0.510	3.96	5.78	14.41	8.91	36.5	546.3	538.6
	H06-3-90	H	0.400	0.600	0.100	0.310	0.510	2.85	5.23	13.30	7.92	36.5	359.1	351.0
	H10-5-135	H	0.400	0.600	0.100	0.310	0.510	3.25	3.8	9.02	5.28	36.5	548.1	540.3
	H08-3-72.5	H	0.400	0.600	0.100	0.310	0.510	5.07	5.78	16.63	9.83	36.5	359.4	352.7
	H12-5-72.5	H	0.400	0.600	0.100	0.310	0.510	5.07	5.78	16.63	9.83	36.5	404.1	538.7

* No sufficient data; ** P—Plain section; H—Hollow section.

Table A3. Theoretical torsional strengths.

Ref.	Beam	**	$T_{R,exp}$ (kNm)	$T_{R,th}^{SiNiP18}$ (kNm)	$T_{R,th}^{ACI89}$ (kNm)	$T_{R,th}^{ACI19}$ (kNm)	$T_{R,th}^{MC90}$ (kNm)	$T_{R,th}^{MC10}$ (kNm)	$T_{R,th}^{EC2}$ (kNm)	$T_{R,th}^{CSA14}$ (kNm)	$T_{R,th}^{Rahal}$ (kNm)	$T_{R,th}^{Prop}$ (kNm)
Hsu (1968)	B1	P	22.30	16.71	22.58	19.0	20.4	12.7	24.9	20.0	21.3	22.97
	B3	P	37.48	37.32	37.16	29.4	32.0	27.0	38.4	43.6	36.8	37.87
	B4 *	P	47.30	50.25	44.11	29.9	43.6	29.2	52.3	59.5	46.0	46.51
	B5 *	P	56.11	67.68	45.14	30.6	57.2	31.2	70.0	63.7	51.5	51.42
	B6 *	P	61.64	70.86	43.85	29.7	54.9	28.9	68.2	60.1	50.6	50.54
	B7	P	26.87	23.25	36.71	27.5	21.3	25.3	25.5	29.0	27.3	28.73
	B8	P	32.51	33.00	42.25	28.7	31.9	24.5	38.2	43.5	36.4	37.40
	B9	P	29.80	29.86	22.77	28.7	22.1	23.1	26.5	30.2	28.6	30.14
	B10	P	34.40	61.33	22.39	28.6	34.1	15.4	40.9	46.5	38.1	38.94
	C2	P	15.30	15.23	15.28	14.7	15.3	10.5	22.4	16.9	15.8	16.79
	C4 *	P	25.29	33.72	25.50	14.8	24.0	15.2	32.0	29.1	27.2	27.48
	C5 *	P	29.69	44.64	28.42	14.8	26.9	15.4	37.2	29.1	27.6	27.60
	C6 *	P	34.21	45.21	28.60	14.9	27.5	15.9	37.9	29.5	27.7	27.71
	G2	P	40.30	29.01	39.72	33.2	40.3	36.1	42.6	34.9	36.7	39.24
	G3	P	49.56	42.17	49.28	43.9	37.2	35.8	56.3	50.2	46.3	47.91
	G4	P	64.80	55.40	57.91	45.1	49.3	39.4	69.0	66.4	56.8	57.91
	G5 *	P	71.91	74.68	56.48	44.0	66.5	37.4	76.0	85.8	69.4	69.29
	G6	P	39.10	26.53	39.41	32.9	25.7	39.5	29.4	34.7	36.3	38.78
	G7	P	52.61	39.56	52.00	47.2	38.1	43.3	43.5	51.3	48.1	50.14
	G8	P	73.38	57.33	57.98	45.1	54.3	39.1	62.1	73.2	60.8	61.64
	I2	P	36.00	26.87	33.17	30.4	30.0	37.4	31.3	32.1	32.0	34.54
	I3	P	45.61	39.05	40.66	37.1	33.4	46.2	40.1	45.6	40.9	43.14
	I4	P	58.02	49.93	51.03	37.2	43.4	47.8	52.1	59.3	49.2	51.00
	I5	P	70.67	65.14	54.81	37.2	56.7	48.5	68.0	77.4	57.8	57.77
	I6 *	P	76.65	84.13	55.27	37.5	71.1	49.0	85.3	95.4	58.1	58.06
	J1	P	21.45	17.22	20.33	19.7	15.2	8.2	18.2	20.7	19.6	20.37
	J2 *	P	29.13	26.33	27.78	21.1	23.1	8.7	27.7	30.3	26.4	26.66
	J3 *	P	35.22	41.53	33.57	22.8	33.3	11.7	41.3	35.2	36.7	36.40
	J4 *	P	40.64	41.19	33.44	22.7	33.3	12.1	41.3	34.9	42.6	41.72
	K2	P	23.71	20.43	21.15	14.9	17.6	18.6	19.0	21.4	22.7	23.95
	K3 *	P	28.45	29.96	20.60	14.5	23.8	17.5	25.8	29.5	29.3	30.11
	K4 *	P	35.00	32.92	20.45	14.4	35.2	17.1	38.2	29.1	30.1	30.08
	M1	P	30.37	23.41	23.66	24.9	19.2	26.5	23.1	26.3	26.1	27.80
	M2	P	40.53	33.98	30.26	30.6	27.6	27.9	33.2	37.7	33.7	35.17
	M3	P	43.80	44.15	34.87	28.7	35.6	23.5	42.8	48.6	39.4	40.16
	M4 *	P	49.56	57.97	42.09	28.5	47.0	23.6	56.4	55.3	47.8	47.83
	M5 *	P	55.65	68.83	43.22	29.3	51.8	25.5	64.3	58.4	50.1	50.10
	M6 *	P	60.06	72.22	44.27	30.0	52.8	26.3	65.5	61.2	50.9	50.82
	N1	P	9.09	6.62	8.63	7.5	5.9	8.5	6.7	7.9	7.9	8.72
	N1a	P	8.99	6.58	8.65	7.5	5.8	8.2	6.7	7.9	7.9	8.65

* Fragile failure; ** P—Plain section; H—Hollow section.

Ref.	Beam	**	$T_{R,exp}$ (kNm)	$T_{R,th}^{SiNiP18}$ (kNm)	$T_{R,th}^{ACI89}$ (kNm)	$T_{R,th}^{ACI19}$ (kNm)	$T_{R,th}^{MC90}$ (kNm)	$T_{R,th}^{MC10}$ (kNm)	$T_{R,th}^{EC2}$ (kNm)	$T_{R,th}^{CSA14}$ (kNm)	$T_{R,th}^{Rahal}$ (kNm)	$T_{R,th}^{Prop}$ (kNm)
	N2	P	14.45	11.37	12.97	10.2	10.3	9.3	11.7	13.9	11.8	12.54
	N2a	P	13.21	11.63	12.54	7.6	5.9	9.1	6.7	8.0	7.9	8.66
	N3	P	12.19	9.33	11.53	9.7	8.8	7.6	10.0	11.9	10.4	11.08
	N4 *	P	15.69	15.04	12.29	9.7	13.5	8.2	15.5	18.3	14.1	14.60
[17]	T4	P	138.61	80.95	124.21	126.6	83.8	121.6	111.7	133.4	124.7	128.92
Leonhardt, Schelling (1974)	VB2 *	P	42.11	24.55	53.57	38.7	34.4	17.3	39.8	50.7	39.6	40.73
	VB3	P	46.40	24.55	57.01	47.1	34.4	29.0	39.8	50.7	42.2	44.37
	VB4	P	48.54	24.55	59.50	48.1	34.4	35.3	39.8	50.7	43.9	46.77
	VM1	P	13.89	8.58	16.66	12.5	9.0	10.2	10.4	13.1	11.2	12.23
	VM2	P	39.17	26.14	46.25	35.7	25.5	29.9	29.5	37.6	33.8	36.08
	VM3	P	100.80	70.62	113.58	92.8	66.1	76.4	76.5	97.7	88.0	92.02
	VQ1	P	21.11	15.99	24.54	25.1	15.8	12.3	21.1	26.4	23.3	24.48
	VQ3	P	19.98	9.62	35.85	20.5	15.1	9.0	16.4	21.6	21.7	22.95
	VQ9	P	21.90	11.08	44.31	19.9	14.9	10.6	15.6	21.0	24.2	25.64
	VS2-VQ2	P	19.53	11.74	29.19	20.1	14.3	10.4	16.6	21.2	20.4	21.74
	VS3	P	28.56	16.07	37.12	30.0	21.5	10.8	24.8	31.6	27.2	28.25
	VS4-VQ5 *	P	34.32	20.53	44.92	32.8	28.7	10.4	33.2	42.3	33.1	33.79
	VS9 *	P	21.56	16.31	32.69	25.2	18.0	9.1	20.8	26.5	23.6	24.70
	VS10-VB1 *	P	33.30	25.70	52.94	32.8	36.0	10.4	41.7	53.1	38.8	39.03
	VU1	P	23.93	13.75	43.11	26.6	19.0	11.4	22.0	28.0	24.9	26.15
	VU2 *	P	30.37	16.47	43.11	32.6	23.3	11.2	26.9	34.3	28.7	29.75
	VU3 *	P	31.04	17.77	35.28	32.4	23.2	9.4	26.9	34.2	28.5	29.36
	VU4	P	25.96	16.39	28.02	26.6	19.0	8.4	22.0	28.0	24.7	25.85
McMullen, Rangan (1978)	A2	P	22.58	17.01	17.62	18.6	13.0	21.6	17.4	19.6	18.1	19.43
	A3	P	27.77	24.69	22.16	18.7	18.8	23.6	25.1	27.7	23.4	24.68
	A4 *	P	34.43	35.16	28.10	18.6	26.3	23.8	35.1	38.8	29.6	30.55
	B1r	P	12.30	8.08	11.79	8.8	12.3	9.9	10.8	9.3	11.0	12.29
	B2	P	20.80	15.05	18.22	16.5	20.8	16.5	17.7	17.4	17.2	18.62
	B3	P	25.29	22.23	23.21	15.9	19.8	21.6	22.6	25.2	22.6	23.78
	B4 *	P	31.72	31.27	23.17	15.9	26.7	21.6	30.5	35.1	28.0	28.00
Rasmussen, Baker (1995)	B30.1 *	P	16.62	29.36	15.11	8.2	21.2	11.1	25.7	19.9	17.0	17.04
	B30.2 *	P	15.29	26.89	14.46	7.8	19.6	10.3	23.8	18.2	16.6	16.59
	B30.3 *	P	15.25	25.56	14.09	7.6	19.1	9.8	23.1	17.3	16.4	16.34
	B50.1 *	P	19.95	43.51	18.39	9.9	28.6	15.3	34.8	29.4	19.2	19.17
	B50.2 *	P	18.46	40.20	17.68	9.6	27.0	14.3	32.9	27.2	18.7	18.72
	B50.3 *	P	19.13	43.44	18.37	9.9	28.6	15.2	34.8	29.4	19.2	19.16
	B70.1 *	P	20.06	49.42	20.57	11.1	31.2	17.9	36.5	36.8	20.5	20.50
	B70.2 *	P	20.74	49.19	20.51	11.1	31.1	17.9	36.4	36.6	20.5	20.47
	B70.3 *	P	20.96	49.46	20.42	11.0	31.3	17.8	36.7	36.3	20.4	20.41
	B110.1 *	P	24.72	48.95	24.51	13.3	31.2	23.1	36.5	44.4	22.8	22.78
	B110.2 *	P	23.62	50.12	23.97	13.0	31.7	22.3	37.1	45.2	22.5	22.47

* Fragile failure; ** P—Plain section; H—Hollow section.

Ref.	Beam	**	$T_{R,exp}$ (kNm)	$T_{R,th}^{SiNiP18}$ (kNm)	$T_{R,th}^{ACI89}$ (kNm)	$T_{R,th}^{ACI19}$ (kNm)	$T_{R,th}^{MC90}$ (kNm)	$T_{R,th}^{MC10}$ (kNm)	$T_{R,th}^{EC2}$ (kNm)	$T_{R,th}^{CSA14}$ (kNm)	$T_{R,th}^{Rahal}$ (kNm)	$T_{R,th}^{Prop}$ (kNm)
	B110.3 *	P	24.77	50.24	23.98	13.0	31.5	22.3	36.8	44.8	22.5	22.48
Koutchoukali, Belarbi (2001)	B5UR1	P	19.40	16.42	18.14	16.0	14.5	22.4	17.4	18.8	18.1	19.46
	B7UR1	P	18.90	16.87	20.50	18.4	15.0	25.7	18.0	19.4	20.0	22.11
	B9UR1	P	21.10	16.42	20.12	17.8	14.5	24.8	17.4	18.8	20.0	22.36
	B12UR1	P	19.40	16.87	21.28	18.4	15.0	25.7	18.0	19.4	20.8	23.21
	B12UR2	P	18.40	17.04	21.41	18.6	15.2	26.0	18.2	19.6	20.8	23.10
	B12UR3	P	22.50	19.10	22.29	21.6	17.6	30.1	21.2	22.8	22.9	25.15
	B12UR4	P	23.70	19.09	23.28	22.1	19.5	33.3	23.4	25.2	24.7	27.03
	B12UR5	P	24.00	25.70	27.89	22.2	22.2	37.7	26.6	29.3	27.5	29.83
	B14UR1	P	21.00	16.64	21.43	18.1	14.8	25.2	17.7	19.1	21.0	23.74
Fang, Shiau (2004)	H-06-06	P	92.00	57.90	85.00	76.0	57.8	113.1	70.1	80.1	87.4	95.25
	H-06-12	P	115.10	66.14	84.79	96.4	73.3	135.1	88.9	101.6	103.1	110.75
	H-07-10	P	126.70	86.66	85.35	99.6	75.8	128.9	91.8	104.9	103.2	109.71
	H-07-16	P	144.50	122.53	85.35	113.5	97.8	109.1	118.5	135.5	123.4	129.06
	H-12-12	P	155.30	80.31	133.52	121.6	103.7	158.3	125.7	143.6	131.5	138.06
	H-12-16	P	196.00	140.15	133.52	121.6	137.0	146.3	166.0	189.7	159.7	164.77
	H-14-10	P	135.20	97.96	124.21	113.5	100.3	149.9	121.5	138.9	125.6	131.16
	H-20-20 *	P	239.00	167.22	180.32	121.6	200.1	158.5	242.5	277.1	166.9	166.74
	N-06-06	P	79.70	57.84	72.98	76.1	57.9	80.6	70.2	80.2	77.0	80.17
	N-06-12	P	95.20	66.14	72.98	81.8	73.3	70.1	88.9	101.6	90.8	93.16
	N-07-10	P	111.70	86.66	75.24	79.4	75.8	68.7	91.8	104.9	92.1	93.90
	N-07-16	P	117.30	122.53	75.24	79.4	97.8	58.2	118.5	135.5	110.1	110.46
	N-12-12 *	P	116.80	80.31	121.26	81.8	103.7	82.1	125.7	143.6	115.8	116.13
	N-12-16 *	P	138.00	140.15	121.26	81.8	137.0	75.9	166.0	183.3	131.5	131.41
	N-14-10 *	P	125.00	97.96	114.11	79.4	100.3	79.9	121.5	138.9	112.0	112.26
	N-20-20 *	P	158.00	167.22	121.26	81.8	155.4	82.2	195.1	183.3	131.5	131.41
Chiu et al. (2007)	HAS-51-50	P	84.90	40.48	73.94	54.7	37.9	82.9	50.6	57.7	68.6	76.34
	NAS-61-35	P	74.70	48.22	56.00	50.2	34.8	75.0	46.3	52.8	60.0	65.34
	HAS-90-50	P	104.23	63.51	74.50	73.0	50.6	109.1	67.4	76.9	84.2	92.19
	NBS-43-44	P	60.60	34.20	54.67	45.8	34.8	69.1	42.2	48.2	53.8	57.82
	HBS-74-17	P	62.20	65.76	52.24	52.7	40.1	78.5	48.6	55.5	65.9	72.85
	HBS-82-13	P	56.30	68.29	47.26	47.1	35.8	70.1	43.4	49.6	60.9	67.83
	NBS-82-13	P	52.90	68.29	38.02	47.1	35.8	47.5	43.4	49.6	54.9	58.88
	HBS-60-61	P	93.70	52.35	76.14	63.7	48.5	94.8	58.8	67.2	75.3	82.25
	HCS-52-50	P	73.54	40.26	63.60	44.9	37.4	65.8	41.1	47.3	64.5	72.11
	HCS-91-50	P	95.86	59.72	63.93	59.9	49.9	86.1	54.8	63.1	79.2	87.07
Lee and Kim (2010)	T1-1	P	32.90	23.41	30.99	26.4	19.3	38.6	24.5	27.8	29.6	32.32
	T1-2	P	45.90	27.93	40.26	39.9	29.2	50.6	37.1	42.1	39.7	42.27
	T1-3	P	54.10	30.94	47.99	41.6	38.5	48.2	49.1	55.6	47.8	49.85
	T1-4	P	62.40	54.32	64.60	42.0	56.7	51.1	72.1	81.7	62.8	63.97
	T2-2	P	38.10	41.82	30.76	36.8	26.9	42.2	34.2	38.8	37.2	39.67

* Fragile failure; ** P—Plain section; H—Hollow section.

Ref.	Beam	**	$T_{R,exp}$ (kNm)	$T_{R,th}^{SiNiP18}$ (kNm)	$T_{R,th}^{ACI89}$ (kNm)	$T_{R,th}^{ACI19}$ (kNm)	$T_{R,th}^{MC90}$ (kNm)	$T_{R,th}^{MC10}$ (kNm)	$T_{R,th}^{EC2}$ (kNm)	$T_{R,th}^{CSA14}$ (kNm)	$T_{R,th}^{Rahal}$ (kNm)	$T_{R,th}^{Prop}$ (kNm)
	T2-3	P	50.20	44.88	39.14	42.1	40.1	42.8	51.0	57.7	49.3	51.36
	T2-4	P	56.40	62.46	43.55	42.0	47.1	43.4	60.0	68.0	55.2	56.94
Peng, Wong (2011)	SW12-1	P	74.00	66.82	43.22	39.7	48.7	59.0	50.3	57.8	87.3	91.40
	SW10-1	P	50.70	54.79	33.97	26.4	39.8	30.7	41.3	47.5	63.9	65.62
	SW10-2	P	68.00	58.02	49.70	31.0	52.2	59.0	54.3	64.5	76.3	76.21
	SW10-3	P	74.30	62.29	40.61	25.3	66.5	39.8	69.1	51.7	67.6	67.51
	SW10-4 *	P	80.50	76.05	43.46	24.9	87.1	44.2	91.6	54.5	70.4	70.32
	SW8-1	P	40.40	30.61	26.43	21.1	30.8	24.6	32.3	35.3	44.8	46.59
	SW8-2	P	60.10	34.63	32.48	19.5	50.2	31.7	52.7	39.8	52.3	52.30
Joh et al. (2019)	RA-SD4-3.2-0.3-3.28 *	P	86.80	97.32	58.91	75.6	74.0	60.9	91.1	110.0	86.5	90.30
	RA-SD5-3.2-0.3-3.21 *	P	88.00	103.65	59.32	75.6	78.2	58.9	96.2	116.2	89.8	93.41
	RA-SD6-3.2-0.2-3.21 *	P	89.40	126.46	59.32	75.6	87.8	53.7	108.1	130.5	93.5	93.41
	RA-SD4-3.2-0.5-2.13 *	P	76.30	63.56	60.39	78.1	55.4	80.1	68.1	82.3	72.2	77.41
	RA-SD5-3.2-0.7-1.63 *	P	74.50	53.42	60.80	70.4	49.9	82.7	61.4	74.2	67.1	72.47
	RA-SD6-3.2-0.6-1.63 *	P	70.00	61.20	60.80	76.7	54.3	79.0	66.9	80.8	71.2	76.49
	RA-SD4-3.2-1.1-1.33 *	P	74.20	35.28	60.18	53.0	37.5	78.8	46.2	55.8	54.8	60.21
	RA-SD5-3.2-1.0-1.26 *	P	67.70	37.83	60.59	55.9	39.6	83.1	48.7	58.9	56.9	62.29
	RA-SD6-3.2-0.8-1.26 *	P	69.90	54.08	60.59	62.6	44.4	86.2	54.6	66.0	61.6	66.96
Ju et al. (2019)	MR30-0.77	P	55.00	39.38	50.92	49.2	37.4	55.3	45.4	51.9	55.0	58.27
	MT30-1.32 *	P	57.00	80.08	50.92	74.3	56.9	43.7	69.0	78.8	73.8	76.05
	MT40-1.32 *	P	58.30	80.08	54.73	74.8	56.9	64.7	69.0	78.8	77.6	81.52
	MT40-1.89	P	98.40	93.65	106.57	87.1	94.1	89.9	114.1	130.4	110.4	112.27
Ibrahim et al. (2020)	NSC-S1-C30 *	P	19.70	24.21	23.08	17.0	20.8	20.9	24.9	27.9	23.1	24.34
	NSC-S1-C45 *	P	12.50	24.21	18.54	10.2	16.9	21.1	20.3	21.9	22.9	23.99
	HSC-C30 *	P	19.90	24.21	24.13	20.5	20.8	28.0	24.9	27.9	24.5	26.37
	HSC-C45 *	P	13.80	24.21	19.75	12.6	16.9	28.7	20.3	21.9	24.5	26.37
Kim et al. (2020)	S08-3-65	P	123.00	71.29	107.84	95.8	82.9	143.1	99.4	100.9	114.5	118.81
	S08-4-90	P	124.00	63.67	111.18	102.4	89.6	149.3	107.5	109.0	120.9	124.81
	S08-5-122.5	P	89.00	80.78	103.94	95.4	82.5	142.5	99.1	100.5	114.2	118.51
	S10-3-52.5	P	126.00	77.60	124.27	102.4	103.7	150.3	124.4	126.2	133.9	136.98
	S10-4-72.5 *	P	109.00	99.74	128.72	102.4	110.7	149.5	132.8	134.7	140.2	142.79
	S10-5-100 *	P	108.00	90.17	118.70	102.4	101.6	147.7	121.9	123.7	132.1	135.24
	S06-3-90	P	101.00	43.20	88.35	70.4	60.9	105.2	73.1	74.2	92.3	97.71
	S10-5-90 *	P	106.00	95.94	127.61	102.4	113.0	147.6	135.6	137.6	142.3	144.70
	S08-3-72.5	P	106.00	66.73	100.60	86.3	74.7	129.0	89.7	91.0	106.5	111.23
	S12-5-72.5 *	P	120.00	123.96	149.33	102.4	140.4	147.6	168.5	170.9	165.6	166.13
Hsu (1968)	D3	H	39.11	66.5	27.11	28.3	26.2	23.6	31.5	35.8	32.1	38.97
	D4 *	H	47.93	75.27	45.20	29.4	44.9	29.6	53.9	61.2	47.3	47.92
[17,38]	T0	H	185.50	116.4	91.25	112.2	99.9	139.4	133.3	146.6	146.8	161.65
	T1	H	140.01	70.87	106.45	110.6	83.8	132.0	111.7	133.4	124.7	128.92
	T2	H	143.10	74.0	87.13	99.3	76.2	127.2	101.6	111.8	116.7	128.99

* Fragile failure. ** P—Plain section; H—Hollow section.

Ref.	Beam	**	$T_{R,exp}$ (kNm)	$T_{R,th}^{SiNiP18}$ (kNm)	$T_{R,th}^{ACI89}$ (kNm)	$T_{R,th}^{ACI19}$ (kNm)	$T_{R,th}^{MC90}$ (kNm)	$T_{R,th}^{MC10}$ (kNm)	$T_{R,th}^{EC2}$ (kNm)	$T_{R,th}^{CSA14}$ (kNm)	$T_{R,th}^{Rahal}$ (kNm)	$T_{R,th}^{Prop}$ (kNm)
	T5	H	156.88	87.7	192.68	135.1	109.3	216.9	124.9	142.3	165.6	185.69
[19]	VH1	H	21.79	12.84	21.04	20.1	12.7	14.5	16.9	21.2	19.6	20.82
	VH2 *	H	34.50	25.6	46.24	32.9	30.2	15.5	41.6	51.2	49.9	32.35
Bernardo, Lopes (2009)	A1	H	150.79	108.5	122.50	101.7	67.9	150.9	90.6	107.1	134.7	147.43
	A2	H	254.08	145.39	193.85	212.2	144.6	313.0	192.8	223.6	227.8	236.84
	A3	H	299.92	185.31	239.37	239.4	192.6	317.6	256.7	298.2	277.2	281.58
	A4	H	368.22	269.46	291.30	238.1	266.6	352.9	355.5	398.5	357.8	359.54
	A5 *	H	412.24	324.19	334.84	238.9	328.7	338.0	438.3	499.1	412.2	408.84
	B2	H	273.28	152.10	216.87	223.3	152.7	329.1	203.6	235.3	251.7	266.80
	B3	H	355.85	269.46	321.28	312.1	266.6	471.8	355.5	414.1	378.5	388.14
	B4 *	H	437.85	339.83	374.47	317.1	347.8	480.9	463.7	529.4	457.6	463.09
	B5 *	H	456.19	413.87	434.23	309.5	434.7	474.4	579.6	641.3	496.1	495.73
	C1	H	151.76	109.4	146.65	102.5	67.9	151.5	90.6	108.0	149.2	169.46
	C2	H	266.14	145.39	215.62	211.0	144.6	319.0	192.8	222.4	254.5	275.56
	C3	H	351.17	263.00	314.00	326.1	257.6	508.0	343.5	407.3	379.2	393.43
	C4 *	H	450.31	324.19	374.22	330.7	328.7	506.0	438.3	524.1	449.6	460.21
	C5 *	H	467.26	407.83	436.52	338.1	398.9	532.7	531.9	629.1	519.5	526.94
	C6 *	H	521.33	487.91	464.52	309.2	521.6	494.1	695.5	798.7	516.7	516.31
Chiu et al. (2007)	HAH-81-35	H	94.31	68.82	52.95	64.6	44.8	96.7	59.7	68.1	77.4	85.31
	NCH-62-33	H	64.14	39.94	45.64	40.0	33.3	57.7	36.6	42.1	55.2	60.60
	HCH-91-42	H	87.51	56.61	58.00	55.0	45.8	79.4	50.3	58.0	74.6	82.64
Jeng (2014)	A095c	H	209.98	184.5	142.52	108.4	79.9	137.5	96.8	114.3	134.4	168.97
	A120a	H	215.25	189.8	96.76	101.6	65.4	112.1	79.3	107.0	124.7	184.33
	B065b	H	278.00	252.2	128.27	187.4	174.4	151.5	211.9	250.3	237.8	286.31
	B080a	H	300.66	275.2	243.23	250.9	198.2	309.1	239.8	283.5	267.9	273.75
	B110a	H	237.48	212.0	127.34	117.3	83.7	143.6	101.6	123.6	149.4	208.45
	C100a	H	370.15	344.7	308.37	291.2	214.7	393.6	259.4	306.8	315.3	333.10
	D075a	H	339.48	314.0	251.86	272.3	201.1	418.3	242.2	287.0	305.0	324.75
	D090a	H	343.08	317.6	298.37	292.0	214.9	424.1	260.1	307.6	324.0	345.40
Kim et al. (2020)	H08-3-65	H	128.00	79.94	112.05	104.0	91.2	146.4	109.5	111.1	123.1	127.15
	H08-4-90	H	130.00	74.29	106.20	96.4	83.4	146.8	100.1	101.6	115.6	120.10
	H08-5-100	H	117.00	76.27	111.77	104.0	90.7	146.5	108.9	110.5	122.6	126.70
	H10-3-52.5	H	143.00	92.53	129.59	104.0	112.4	146.7	134.8	136.8	142.4	145.15
	H10-4-72.5	H	127.00	94.04	122.35	104.0	103.7	146.6	124.4	126.2	134.6	137.91
	H10-5-80	H	125.00	93.70	129.86	104.0	111.7	147.2	134.0	136.0	141.8	144.59
	H06-3-90	H	102.00	47.39	91.44	76.5	66.2	122.2	79.5	80.6	98.3	103.69
	H10-5-135	H	95.00	70.56	92.83	78.9	68.2	126.0	81.9	83.1	100.4	105.70
	H08-3-72.5	H	114.00	75.47	104.53	95.6	82.7	145.9	99.3	100.7	114.9	119.46
	H12-5-72.5	H	129.00	93.25	139.33	104.0	108.4	152.7	130.1	132.0	138.9	141.88

* Fragile failure. ** P—Plain section; H—Hollow section.

Table A4. Ratios of experimental to theoretical torsional strengths.

Ref.	Beam	**	$\frac{T_{R,exp}}{T_{R,th}^{SiNiP18}}$	$\frac{T_{R,exp}}{T_{R,th}^{ACI89}}$	$\frac{T_{R,exp}}{T_{R,th}^{ACI19}}$	$\frac{T_{R,exp}}{T_{R,th}^{MC90}}$	$\frac{T_{R,exp}}{T_{R,th}^{MC10}}$	$\frac{T_{R,exp}}{T_{R,th}^{EC2}}$	$\frac{T_{R,exp}}{T_{R,th}^{CSA14}}$	$\frac{T_{R,exp}}{T_{R,th}^{Rahal}}$	$\frac{T_{R,exp}}{T_{R,th}^{Prop}}$
Hsu (1968)	B1	P	1.33	0.99	1.17	1.09	1.76	0.90	1.11	1.05	0.97
	B3	P	1.00	1.01	1.28	1.17	1.39	0.98	0.86	1.02	0.99
	B4	P	0.94	1.07	1.58	1.09	1.62	0.90	0.80	1.03	1.02
	B5	P	0.83	1.24	1.83	0.98	1.80	0.80	0.88	1.09	1.09
	B6	P	0.87	1.41	2.07	1.12	2.13	0.90	1.03	1.22	1.22
	B7	P	1.16	0.73	0.98	1.26	1.06	1.05	0.93	0.98	0.94
	B8	P	0.99	0.77	1.13	1.02	1.33	0.85	0.75	0.89	0.87
	B9	P	1.00	1.31	1.04	1.35	1.29	1.12	0.99	1.04	0.99
	B10	P	0.56	1.54	1.20	1.01	2.24	0.84	0.74	0.90	0.88
	C2	P	1.00	1.00	1.04	1.00	1.46	0.68	0.90	0.97	0.91
	C4	P	0.75	0.99	1.71	1.05	1.67	0.79	0.87	0.93	0.92
	C5	P	0.67	1.04	2.00	1.10	1.93	0.80	1.02	1.07	1.08
	C6	P	0.76	1.20	2.29	1.24	2.15	0.90	1.16	1.23	1.23
	G2	P	1.39	1.01	1.22	1.00	1.12	1.16	1.15	1.10	1.03
	G3	P	1.18	1.01	1.13	1.33	1.38	1.15	0.99	1.07	1.03
	G4	P	1.17	1.12	1.44	1.32	1.64	0.58	0.98	1.14	1.12
	G5	P	0.96	1.27	1.64	1.08	1.92	0.95	0.84	1.04	1.04
	G6	P	1.47	0.99	1.19	1.52	0.99	1.33	1.13	1.08	1.01
	G7	P	1.33	1.01	1.12	1.38	1.22	1.21	1.03	1.09	1.05
	G8	P	1.28	1.27	1.63	1.35	1.88	1.18	1.00	1.21	1.19
	I2	P	1.34	1.09	1.18	1.20	0.96	1.15	1.12	1.13	1.04
	I3	P	1.17	1.12	1.23	1.36	0.99	1.14	1.00	1.12	1.06
	I4	P	1.16	1.14	1.56	1.34	1.21	1.11	0.98	1.18	1.14
	I5	P	1.08	1.29	1.90	1.25	1.46	1.04	0.91	1.22	1.22
	I6	P	0.91	1.39	2.04	1.08	1.57	0.90	0.80	1.32	1.32
	J1	P	1.25	1.06	1.09	1.41	2.62	1.18	1.04	1.09	1.05
	J2	P	1.11	1.05	1.38	1.26	3.36	1.05	0.96	1.11	1.09
	J3	P	0.85	1.05	1.55	1.06	3.01	0.85	1.00	0.96	0.97
	J4	P	0.99	1.22	1.79	1.22	3.36	0.98	1.16	0.95	0.97
	K2	P	1.16	1.12	1.59	1.35	1.27	1.25	1.11	1.04	0.99
	K3	P	0.95	1.38	1.96	1.20	1.63	1.10	0.97	0.97	0.94
	K4	P	1.06	1.71	2.42	0.99	2.05	0.92	1.20	1.16	1.16
	M1	P	1.30	1.28	1.22	1.58	1.14	1.31	1.16	1.17	1.09
	M2	P	1.19	1.34	1.32	1.47	1.45	1.22	1.07	1.20	1.15
	M3	P	0.99	1.26	1.53	1.23	1.86	1.02	0.90	1.11	1.09
	M4	P	0.85	1.18	1.74	1.05	2.10	0.88	0.90	1.04	1.04
	M5	P	0.81	1.29	1.90	1.07	2.19	0.87	0.95	1.11	1.11
	M6	P	0.83	1.36	2.00	1.14	2.28	0.92	0.98	1.18	1.18
	N1	P	1.37	1.05	1.20	1.55	1.07	1.36	1.14	1.15	1.04
	N1a	P	1.37	1.04	1.20	1.54	1.10	1.35	1.13	1.14	1.04
	N2	P	1.27	1.11	1.41	1.41	1.56	1.23	1.04	1.23	1.15

** P—Plain section; H—Hollow section.

Ref.	Beam	**	$\frac{T_{R,exp}}{T_{R,th}^{SINiP18}}$	$\frac{T_{R,exp}}{T_{R,th}^{ACI89}}$	$\frac{T_{R,exp}}{T_{R,th}^{ACI19}}$	$\frac{T_{R,exp}}{T_{R,th}^{MC90}}$	$\frac{T_{R,exp}}{T_{R,th}^{MC10}}$	$\frac{T_{R,exp}}{T_{R,th}^{EC2}}$	$\frac{T_{R,exp}}{T_{R,th}^{CSA14}}$	$\frac{T_{R,exp}}{T_{R,th}^{Rahal}}$	$\frac{T_{R,exp}}{T_{R,th}^{Prop}}$
	N2a	P	1.14	1.05	1.75	2.25	1.45	1.97	1.66	1.67	1.52
	N3	P	1.31	1.06	1.26	1.39	1.60	1.22	1.03	1.17	1.10
	N4	P	1.04	1.28	1.62	1.16	1.91	1.02	0.86	1.12	1.07
[17]	T4	P	1.71	1.12	1.10	1.65	1.14	1.24	1.04	1.11	1.08
Leonhardt, Schelling (1974)	VB2	P	1.72	0.79	1.09	1.23	2.43	1.06	0.83	1.06	1.03
	VB3	P	1.89	0.81	0.98	1.35	1.60	1.17	0.92	1.10	1.05
	VB4	P	1.98	0.82	1.01	1.41	1.38	1.22	0.96	1.11	1.04
	VM1	P	1.62	0.83	1.11	1.55	1.36	1.34	1.06	1.24	1.14
	VM2	P	1.50	0.85	1.10	1.54	1.31	1.33	1.04	1.16	1.09
	VM3	P	1.43	0.89	1.09	1.53	1.32	1.32	1.03	1.14	1.10
	VQ1	P	1.32	0.86	0.84	1.34	1.72	1.00	0.80	0.91	0.86
	VQ3	P	2.08	0.56	0.98	1.32	2.23	1.21	0.93	0.92	0.87
	VQ9	P	1.98	0.49	1.10	1.47	2.07	1.41	1.04	0.90	0.85
	VS2-VQ2	P	1.66	0.67	0.97	1.36	1.88	1.18	0.92	0.96	0.90
	VS3	P	1.78	0.77	0.95	1.33	2.64	1.15	0.90	1.05	1.01
	VS4-VQ5	P	1.67	0.76	1.04	1.20	3.31	1.03	0.81	1.04	1.02
	VS9	P	1.32	0.66	0.86	1.20	2.38	1.04	0.81	0.91	0.87
	VS10-VB1	P	1.30	0.63	1.01	0.93	3.21	0.80	0.63	0.86	0.85
	VU1	P	1.74	0.56	0.90	1.26	2.09	1.09	0.85	0.96	0.92
	VU2	P	1.84	0.70	0.93	1.31	2.70	1.13	0.88	1.06	1.02
	VU3	P	1.75	0.88	0.96	1.34	3.32	1.16	0.91	1.09	1.06
	VU4	P	1.58	0.93	0.98	1.37	3.08	1.18	0.93	1.05	1.00
McMullen, Rangan (1978)	A2	P	1.33	1.28	1.21	1.73	1.05	1.30	1.15	1.25	1.16
	A3	P	1.12	1.25	1.49	1.48	1.17	1.11	1.00	1.19	1.13
	A4	P	0.98	1.23	1.85	1.31	1.45	0.98	0.89	1.16	1.13
	B1r	P	1.52	1.04	1.39	1.00	1.24	1.24	1.32	1.12	1.00
	B2	P	1.38	1.14	1.26	1.00	1.26	1.26	1.20	1.21	1.12
	B3	P	1.14	1.09	1.59	1.28	1.17	1.12	1.00	1.12	1.06
	B4	P	1.01	1.37	2.00	1.19	1.47	1.04	0.90	1.13	1.13
Rasmussen, Baker (1995)	B30.1	P	0.57	1.10	2.03	0.78	1.49	0.65	0.84	0.97	0.98
	B30.2	P	0.57	1.06	1.96	0.78	1.49	0.64	0.84	0.92	0.92
	B30.3	P	0.60	1.08	2.00	0.80	1.55	0.66	0.88	0.93	0.93
	B50.1	P	0.46	1.08	2.01	0.70	1.31	0.57	0.68	1.04	1.04
	B50.2	P	0.46	1.04	1.93	0.68	1.29	0.56	0.68	0.99	0.99
	B50.3	P	0.44	1.04	1.92	0.67	1.26	0.55	0.65	1.00	1.00
	B70.1	P	0.41	0.98	1.80	0.64	1.12	0.55	0.55	0.98	0.98
	B70.2	P	0.42	1.01	1.87	0.67	1.16	0.57	0.57	1.01	1.01
	B70.3	P	0.42	1.03	1.90	0.67	1.18	0.57	0.58	1.03	1.03
	B110.1	P	0.50	1.01	1.86	0.79	1.07	0.68	0.56	1.08	1.09
	B110.2	P	0.47	0.99	1.82	0.75	1.06	0.64	0.52	1.05	1.05
	B110.3	P	0.49	1.03	1.91	0.79	1.11	0.67	0.55	1.10	1.10
	B5UR1	P	1.18	1.07	1.21	1.34	0.87	1.11	1.03	1.07	1.00

** P—Plain section; H—Hollow section.

Ref.	Beam	**	$\frac{T_{R,exp}}{T_{R,th}^{SiNiP18}}$	$\frac{T_{R,exp}}{T_{R,th}^{ACI89}}$	$\frac{T_{R,exp}}{T_{R,th}^{ACI19}}$	$\frac{T_{R,exp}}{T_{R,th}^{MC90}}$	$\frac{T_{R,exp}}{T_{R,th}^{MC10}}$	$\frac{T_{R,exp}}{T_{R,th}^{EC2}}$	$\frac{T_{R,exp}}{T_{R,th}^{CSA14}}$	$\frac{T_{R,exp}}{T_{R,th}^{Rahal}}$	$\frac{T_{R,exp}}{T_{R,th}^{Prop}}$
Koutchoukali, Belarbi (2001)	B7UR1	P	1.12	0.92	1.03	1.26	0.74	1.05	0.97	0.94	0.85
	B9UR1	P	1.29	1.05	1.19	1.45	0.85	1.21	1.12	1.05	0.94
	B12UR1	P	1.15	0.91	1.05	1.29	0.76	1.08	1.00	0.93	0.84
	B12UR2	P	1.08	0.86	0.99	1.21	0.71	1.01	0.94	0.89	0.80
	B12UR3	P	1.18	1.01	1.04	1.28	0.75	1.06	0.99	0.98	0.89
	B12UR4	P	1.24	1.02	1.07	1.22	0.71	1.01	0.94	0.96	0.88
	B12UR5	P	0.93	0.86	1.08	1.08	0.64	0.90	0.82	0.87	0.80
	B14UR1	P	1.26	0.98	1.16	1.42	0.83	1.18	1.10	1.00	0.88
Fang, Shiau (2004)	H-06-06	P	1.59	1.08	1.21	1.59	0.81	1.31	1.15	1.05	0.97
	H-06-12	P	1.74	1.36	1.19	1.57	0.85	1.29	1.13	1.12	1.04
	H-07-10	P	1.46	1.48	1.27	1.67	0.98	1.38	1.21	1.23	1.15
	H-07-16	P	1.18	1.69	1.27	1.48	1.32	1.22	1.07	1.17	1.12
	H-12-12	P	1.93	1.16	1.28	1.50	0.98	1.24	1.08	1.18	1.12
	H-12-16	P	1.40	1.47	1.61	1.43	1.34	1.18	1.03	1.23	1.19
	H-14-10	P	1.38	1.09	1.19	1.35	0.90	1.11	0.97	1.08	1.03
	H-20-20	P	1.43	1.33	1.97	1.19	1.51	0.99	0.86	1.43	1.43
	N-06-06	P	1.38	1.09	1.05	1.38	0.99	1.14	0.99	1.03	0.99
	N-06-12	P	1.44	1.30	1.16	1.30	1.36	1.07	0.94	1.05	1.02
	N-07-10	P	1.29	1.48	1.41	1.47	1.63	1.22	1.06	1.21	1.19
	N-07-16	P	0.96	1.56	1.48	1.20	2.02	0.99	0.87	1.07	1.06
	N-12-12	P	1.45	0.96	1.43	1.13	1.42	0.93	0.81	1.01	1.01
	N-12-16	P	0.98	1.14	1.69	1.01	1.82	0.83	0.75	1.05	1.05
	N-14-10	P	1.28	1.10	1.57	1.25	1.56	1.03	0.90	1.12	1.11
	N-20-20	P	0.94	1.30	1.93	1.02	1.92	0.81	0.86	1.20	1.20
Chiu et al. (2007)	HAS-51-50	P	2.10	1.15	1.55	2.24	1.02	1.68	1.47	1.24	1.11
	NAS-61-35	P	1.55	1.33	1.49	2.15	1.00	1.61	1.41	1.25	1.14
	HAS-90-50	P	1.64	1.40	1.43	2.06	0.96	1.55	1.36	1.24	1.13
	NBS-43-44	P	1.77	1.11	1.32	1.74	0.88	1.44	1.26	1.13	1.05
	HBS-74-17	P	0.95	1.19	1.18	1.55	0.79	1.28	1.12	0.94	0.85
	HBS-82-13	P	0.82	1.19	1.20	1.57	0.80	1.30	1.14	0.92	0.83
	NBS-82-13	P	0.77	1.39	1.12	1.48	1.11	1.22	1.07	0.96	0.90
	HBS-60-61	P	1.79	1.23	1.47	1.93	0.99	1.59	1.40	1.24	1.14
	HCS-52-50	P	1.83	1.16	1.64	1.97	1.12	1.79	1.55	1.14	1.02
	HCS-91-50	P	1.61	1.50	1.60	1.92	1.11	1.75	1.52	1.21	1.10
Lee and Kim (2010)	T1-1	P	1.41	1.06	1.25	1.71	0.85	1.34	1.18	1.11	1.02
	T1-2	P	1.64	1.14	1.15	1.57	0.91	1.24	1.09	1.16	1.09
	T1-3	P	1.75	1.13	1.30	1.40	1.12	1.10	0.97	1.13	1.09
	T1-4	P	1.15	0.97	1.49	1.10	1.22	0.87	0.76	0.99	0.98
	T2-2	P	0.91	1.24	1.04	1.42	0.90	1.11	0.98	1.03	0.96
	T2-3	P	1.12	1.28	1.19	1.25	1.17	0.98	0.87	1.02	0.98
	T2-4	P	0.90	1.30	1.34	1.20	1.30	0.94	0.83	1.02	0.99
	SW12-1	P	1.11	1.71	1.86	1.52	1.25	1.47	1.28	0.85	0.81

** P—Plain section; H—Hollow section.

Ref.	Beam	**	$\frac{T_{R,exp}}{T_{R,th}^{SiNiP18}}$	$\frac{T_{R,exp}}{T_{R,th}^{ACI89}}$	$\frac{T_{R,exp}}{T_{R,th}^{ACI19}}$	$\frac{T_{R,exp}}{T_{R,th}^{MC90}}$	$\frac{T_{R,exp}}{T_{R,th}^{MC10}}$	$\frac{T_{R,exp}}{T_{R,th}^{EC2}}$	$\frac{T_{R,exp}}{T_{R,th}^{CSA14}}$	$\frac{T_{R,exp}}{T_{R,th}^{Rahal}}$	$\frac{T_{R,exp}}{T_{R,th}^{Prop}}$
Peng, Wong (2011)	SW10-1	P	0.93	1.49	1.92	1.28	1.65	1.23	1.07	0.79	0.77
	SW10-2	P	1.17	1.37	2.19	1.30	1.15	1.25	1.05	0.89	0.89
	SW10-3	P	1.19	1.83	2.93	1.12	1.86	1.08	1.44	1.10	1.10
	SW10-4	P	1.06	1.85	3.23	0.92	1.82	0.88	1.48	1.14	1.14
	SW8-1	P	1.32	1.53	1.91	1.31	1.64	1.25	1.15	0.90	0.87
	SW8-2	P	1.74	1.85	3.08	1.20	1.90	1.14	1.51	1.15	1.15
Joh et al. (2019)	RA-SD4-3.2-0.3-3.28	P	0.89	1.47	1.15	1.17	1.42	0.95	0.79	1.00	0.96
	RA-SD5-3.2-0.3-3.21	P	0.85	1.48	1.16	1.13	1.49	0.91	0.76	0.98	0.94
	RA-SD6-3.2-0.2-3.21	P	0.71	1.51	1.18	1.02	1.66	0.83	0.68	0.96	0.96
	RA-SD4-3.2-0.5-2.13	P	1.20	1.26	0.98	1.38	0.95	1.12	0.93	1.06	0.99
	RA-SD5-3.2-0.7-1.63	P	1.39	1.23	1.06	1.49	0.90	1.21	1.00	1.11	1.03
	RA-SD6-3.2-0.6-1.63	P	1.14	1.15	0.91	1.29	0.89	1.05	0.87	0.98	0.92
	RA-SD4-3.2-1.1-1.33	P	2.10	1.23	1.40	1.98	0.94	1.61	1.33	1.35	1.23
	RA-SD5-3.2-1.0-1.26	P	1.79	1.12	1.21	1.71	0.81	1.39	1.15	1.19	1.09
	RA-SD6-3.2-0.8-1.26	P	1.29	1.15	1.12	1.58	0.81	1.28	1.06	1.13	1.04
Ju et al. (2019)	MR30-0.77	P	1.40	1.08	1.12	1.47	0.99	1.21	1.06	1.00	0.94
	MT30-1.32	P	0.71	1.12	0.77	1.00	1.30	0.83	0.72	0.77	0.75
	MT40-1.32	P	0.73	1.07	0.78	1.02	0.90	0.85	0.74	0.75	0.72
	MT40-1.89	P	1.05	0.92	1.13	1.05	1.09	0.86	0.75	0.89	0.88
Ibrahim et al. (2020)	NSC-S1-C30	P	0.81	0.85	1.16	0.95	0.94	0.79	0.71	0.85	0.81
	NSC-S1-C45	P	0.52	0.67	1.23	0.74	0.59	0.62	0.57	0.55	0.52
	HSC-C30	P	0.82	0.82	0.97	0.96	0.71	0.80	0.71	0.81	0.75
	HSC-C45	P	0.57	0.70	1.09	0.82	0.48	0.68	0.63	0.56	0.52
Kim et al. (2020)	S08-3-65	P	1.73	1.14	1.28	1.48	0.86	1.24	1.22	1.07	1.04
	S08-4-90	P	1.95	1.12	1.21	1.38	0.83	1.15	1.14	1.03	0.99
	S08-5-122.5	P	1.10	0.86	0.93	1.08	0.62	0.90	0.89	0.78	0.75
	S10-3-52.5	P	1.62	1.01	1.23	1.22	0.84	1.01	1.00	0.94	0.92
	S10-4-72.5	P	1.09	0.85	1.06	0.98	0.73	0.82	0.81	0.78	0.76
	S10-5-100	P	1.20	0.91	1.05	1.06	0.73	0.89	0.87	0.82	0.80
	S06-3-90	P	2.34	1.14	1.43	1.66	0.96	1.38	1.36	1.09	1.03
	S10-5-90	P	1.10	0.83	1.03	0.94	0.72	0.78	0.77	0.75	0.73
	S08-3-72.5	P	1.59	1.05	1.23	1.42	0.82	1.18	1.17	1.00	0.95
	S12-5-72.5	P	0.97	0.80	1.17	0.85	0.81	0.71	0.70	0.72	0.72
Hsu (1968)	D3	H	0.59	1.44	1.38	1.49	1.65	1.24	1.09	1.22	1.00
	D4	H	0.64	1.06	1.63	1.07	1.62	0.89	0.78	1.01	1.00
[17,38]	T0	H	1.59	2.03	1.65	1.86	1.33	1.39	1.27	1.26	1.15
	T1	H	1.98	1.32	1.27	1.67	1.06	1.25	1.05	1.12	1.09
	T2	H	1.93	1.64	1.44	1.88	1.12	1.41	1.28	1.23	1.11
	T5	H	1.79	0.81	1.16	1.44	0.72	1.26	1.10	0.95	0.84
[19]	VH1	H	1.70	1.04	1.08	1.72	1.50	1.29	1.03	1.11	1.05
	VH2	H	1.35	0.75	1.05	1.14	2.22	0.83	0.67	0.69	1.07
	A1	H	1.39	1.23	1.48	2.22	1.00	1.66	1.41	1.12	1.02

** P—Plain section; H—Hollow section.

Ref.	Beam	**	$\frac{T_{R,exp}}{T_{R,th}^{SiNiP18}}$	$\frac{T_{R,exp}}{T_{R,th}^{ACI89}}$	$\frac{T_{R,exp}}{T_{R,th}^{ACI19}}$	$\frac{T_{R,exp}}{T_{R,th}^{MC90}}$	$\frac{T_{R,exp}}{T_{R,th}^{MC10}}$	$\frac{T_{R,exp}}{T_{R,th}^{EC2}}$	$\frac{T_{R,exp}}{T_{R,th}^{CSA14}}$	$\frac{T_{R,exp}}{T_{R,th}^{Rahal}}$	$\frac{T_{R,exp}}{T_{R,th}^{Prop}}$
Bernardo, Lopes (2009)	A2	H	1.75	1.31	1.20	1.76	0.81	1.32	1.14	1.12	1.07
	A3	H	1.62	1.25	1.25	1.56	0.94	1.17	1.01	1.08	1.07
	A4	H	1.37	1.26	1.55	1.38	1.04	1.04	0.92	1.03	1.02
	A5	H	1.27	1.23	1.73	1.25	1.22	0.94	0.83	1.00	1.01
	B2	H	1.80	1.26	1.22	1.79	0.83	1.34	1.16	1.09	1.02
	B3	H	1.32	1.11	1.14	1.33	0.75	1.00	0.86	0.94	0.92
	B4	H	1.29	1.17	1.38	1.26	0.91	0.94	0.83	0.96	0.95
	B5	H	1.10	1.05	1.47	1.05	0.96	0.79	0.71	0.92	0.92
	C1	H	1.39	1.03	1.48	2.23	1.00	1.68	1.40	1.02	0.90
	C2	H	1.83	1.23	1.26	1.84	0.83	1.38	1.20	1.05	0.97
	C3	H	1.34	1.12	1.08	1.36	0.69	1.02	0.86	0.93	0.89
	C4	H	1.39	1.20	1.36	1.37	0.89	1.03	0.86	1.00	0.98
	C5	H	1.15	1.07	1.38	1.17	0.88	0.88	0.74	0.90	0.89
	C6	H	1.07	1.12	1.69	1.00	1.06	0.75	0.65	1.01	1.01
Chiu et al. (2007)	HAH-81-35	H	1.37	1.78	1.46	2.11	0.97	1.58	1.39	1.22	1.11
	NCH-62-33	H	1.61	1.41	1.60	1.93	1.11	1.75	1.52	1.16	1.06
	HCH-91-42	H	1.55	1.51	1.59	1.91	1.10	1.74	1.51	1.17	1.06
Jeng (2014)	A095c	H	1.13	1.47	1.94	2.63	1.53	2.17	1.84	1.56	1.24
	A120a	H	1.08	2.22	2.12	3.29	1.92	2.72	2.01	1.73	1.17
	B065b	H	1.06	2.17	1.48	1.59	1.83	1.31	1.11	1.17	0.97
	B080a	H	1.06	1.24	1.20	1.52	0.97	1.25	1.06	1.12	1.10
	B110a	H	1.12	1.86	2.02	2.84	1.65	2.34	1.92	1.59	1.14
	C100a	H	1.06	1.20	1.27	1.72	0.94	1.43	1.21	1.17	1.11
	D075a	H	1.06	1.35	1.25	1.69	0.81	1.40	1.18	1.11	1.05
	D090a	H	1.05	1.15	1.18	1.60	0.81	1.32	1.12	1.06	0.99
Kim et al. (2020)	H08-3-65	H	1.60	1.14	1.23	1.40	0.87	1.17	1.15	1.04	1.01
	H08-4-90	H	1.75	1.22	1.35	1.56	0.89	1.30	1.28	1.12	1.08
	H08-5-100	H	1.53	1.05	1.12	1.29	0.80	1.07	1.06	0.95	0.92
	H10-3-52.5	H	1.55	1.10	1.37	1.27	0.97	1.06	1.05	1.00	0.99
	H10-4-72.5	H	1.35	1.04	1.22	1.22	0.87	1.02	1.01	0.94	0.92
	H10-5-80	H	1.33	0.96	1.20	1.12	0.85	0.93	0.92	0.88	0.86
	H06-3-90	H	2.15	1.12	1.33	1.54	0.83	1.28	1.27	1.04	0.98
	H10-5-135	H	1.35	1.02	1.20	1.39	0.75	1.16	1.14	0.95	0.90
	H08-3-72.5	H	1.51	1.09	1.19	1.38	0.78	1.15	1.13	0.99	0.95
	H12-5-72.5	H	1.38	0.93	1.24	1.19	0.84	0.99	0.98	0.93	0.91

** P—Plain section; H—Hollow section.

References

1. *SP 63.13330.2018*; Concrete and Reinforced Cocrete Structures. Ministry of Regional Development of the Russian Federation: Moscow, Russia, 2018.
2. ACI Committe 318. *Building Code Requirements for Structural Concrete (ACI 318-89) and Commentary (ACI 318R-89)*; American Concrete Institute: Detroi, MI, USA, 1989.
3. ACI Committe 318. *Building Code Requirements for Structural Concrete (ACI 318-19) and Commentary (ACI 318R-19)*; American Concrete Institute: Detroi, MI, USA, 2019.
4. *CEB-FIP MC 90*; Design of Concrete Structures. CEB-FIP Model Code 1990. Thomas Telford: London, UK, 1993.
5. CEB-FIP Model Code 2010. Comité Euro-International du Béton Secretariat Permanent. Case Postale 88, CH-1015 Lausanne, Switzerland, 2013.
6. *EN 1992-1-1*; Eurocode 2: Design of Concrete Structures—Part 1-1: General Rules and Rules for Buildings. European Committee for Standardization: Brussels, Belgium, 2005; 225p.
7. *CSA Committee A23.3-14*; Design of Concrete Strutures (CAN/CSA-A23.3-14). Canadian Standards Association: Toronto, ON, Canada, 2014.
8. Rahal, K.N. Torsional strength of normal and high strength reinforced concrete beams. *Eng. Struct.* **2013**, *56*, 2206–2216. [CrossRef]
9. Teixeira, M.M.; Bernardo, L.F.A. Torsional strength of reinforced concrete beams—Evaluation of some codes of practice. Concrete structures: New trends for eco-efficiency and performance. In Proceedings of the Fib Symposium 2021, Lisbon, Portugal, 14–16 June 2021; pp. 2035–2046.

10. Hsu, T.T.C. *Torsion of Structural Concrete—Behavior of Reinforced Concrete Rectangular Members*; Torsion of Structural Concrete, SP-18; American Concrete Institute: Detroit, MI, USA, 1968; Volume 18, pp. 261–306.
11. McMullen, A.E.; Warwaruk, J. Concrete beams in bending, torsion and shear. *J. Struct. Div.* **1970**, *96*, 885–903. [CrossRef]
12. McMullen, A.E.; Rangan, B.V. Pure Torsion in Rectangular Sections—A Re-examination. *ACI J. Proc.* **1978**, *75*, 511–519.
13. Elfgren, L.; Karlsson, I.; Losberg, A. Torsion-bending-shear interaction for concrete beams. *J. Struct. Div.* **1974**, *100*, 1657–1676. [CrossRef]
14. Rausch, E. Berechnung des Eisenbetons Gegen Verdrehung (Design of Reinforcerd Concrete in Torsion). Ph.D. Thesis, Technische Hoechschule, Berlin, Germany, 1929; 53p. (In German)
15. Bredt, R. Kritische Bemerkungen zur Drehungselastizitat (Critical remarks on torsional elasticity). *Z. Ver. Dtsch. Ing.* **1896**, *40*, 785–790. (In German)
16. Cowan, H.J. Elastic Theory for Torsional Strength of Rectangular Reinforced Concrete Beams. *Mag. Concr. Res.* **1950**, *2*, 3–8. [CrossRef]
17. Lampert, P.; Thurlimann, B. *Torsionsversuche an Stahlbetonbalken (Torsion Tests of Reinforced Concrete Beams)*; Bericht, Nr. 6506-2; Institut fur Baustatik, ETH, Zurich: Graz, Austria, 1969; 101p. (In German)
18. Mitchell, D.; Collins, M.P. Diagonal Compression Field Theory—A Rational Model for Structural Concrete in Pure Torsion. *ACI Struct. J.* **1974**, *71*, 396–408.
19. Leonhardt, F.; Schelling, G. *Torsionsversuche an Stahl Betonbalken*; Bulletin, No. 239; Deutscher Ausschuss fur Stahlbeton: Berlin, Germany, 1974; 122p. (In German)
20. Hsu, T.T.C.; Mo, Y.L. Softening of Concrete in Torsional Members—Theory and Tests. *J. Am. Concr. Inst.* **1985**, *82*, 290–303.
21. Jeng, C.-H.; Hsu, T.T.C. A Softened Membrane Model for Torsion in Reinforced Concrete Members. *Eng. Struct.* **2009**, *31*, 1944–1954. [CrossRef]
22. Bernardo, L.F.A.; Andrade, J.M.A. A Unified Softened Truss Model for RC and PC Concrete Beams under Torsion. *J. Build. Eng.* **2020**, *32*, 101467. [CrossRef]
23. Rahal, K.N.; Collins, M.P. Simple model for predicting torsional strength of reinforced and prestressed concrete sections. *J. Am. Concr. Inst.* **1996**, *93*, 658–666.
24. Bhatti, M.A.; Almughrabi, A. Refined model to estimate torsional strength of reinforced concrete beams. *J. Am. Concr. Inst.* **1996**, *93*, 614–622.
25. Wang, W.; Hsu, T.T.C. Limit analysis of reinforced concrete beams subjected to pure torsion. *J. Struct. Eng.* **1997**, *123*, 86–94. [CrossRef]
26. Bernardo, L.F.A.; Taborda, C.S.B.; Andrade, J.M.A. Generalized Softened Variable Angle Truss Model for PC Beams under Torsion. *Int. J. Concr. Struct. Mater.* **2018**, *12*, 62. [CrossRef]
27. Silva, J.R.B.; Horowitz, B.; Bernardo, L.F.A. Efficient procedure to analyze RC beam sections using the softened truss model. *ACI Struct. J.* **2017**, *114*, 765–774.
28. Greene, G.G.; Belarbi, A. Model for RC members under torsion, bending, and shear. I: Theory. *J. Eng. Mech.* **2009**, *135*, 961–969. [CrossRef]
29. Rahal, K.N.; Collins, M.P. Combined torsion and bending in reinforced and prestressed concrete beams. *ACI Struct. J.* **2003**, *100*, 157–165.
30. Bernardo, L.; Taborda, C. Softened Truss Model for Reinforced Concrete Beams under Torsion Combined with Axial Force. *Appl. Mech.* **2020**, *1*, 79–96. [CrossRef]
31. Alnauimi, A.S.; Bhatt, P. Direct design of hollow reinforced concrete beams. Part I: Design procedure. *Struct. Concr.* **2004**, *5*, 139–146. [CrossRef]
32. Bairan Garcia, J.M.; Mari Bernat, A.R. Shear-bending-torsion interaction in structural concrete members: A nonlinear coupled sectional approach. *Arch. Comput. Methods Eng.* **2007**, *14*, 249–278. [CrossRef]
33. Karayannis, C.G. Smeared crack analysis for plain concrete in torsion. *J. Struct. Eng.* **2000**, *126*, 638–645. [CrossRef]
34. Karayannis, C.G.; Chalioris, C.E. Experimental validation of smeared analysis for plain concrete in torsion. *J. Struct. Eng.* **2000**, *126*, 646–653. [CrossRef]
35. Deifalla, A.; Salem, N.M. A Machine Learning Model for Torsion Strength of Externally Bonded FRP-Reinforced Concrete Beams. *Polymers* **2022**, *14*, 1824. [CrossRef]
36. Lyu, Z.; Yu, Y.; Samali, B.; Rashidi, M.; Mohammadi, M.; Nguyen, T.N.; Nguyen, A. Cross-Validation for Prediction of Torsional Strength of Reinforced Concrete Beam. *Materials* **2022**, *15*, 1477. [CrossRef]
37. Peng, X.-N.; Wong, Y.-L. Behaviour of reinforced concrete walls subjected to monotonic pure torsion-An experimental study. *Eng. Struct.* **2011**, *33*, 2495–2508. [CrossRef]
38. Lampert, P.; Thürlimann, B. Essais de Poutre en Béton Armé sous Torsion Simple et Flexion Combinées (Torsionsversuche und Torsions-Biege-Versuche an Stahlbetonbalken). *Com. Eur. Béton BI* **1969**, *71*, 177–207. (In German)
39. Rasmussen, L.J.; Baker, G. Torsion in reinforced normal and high-strength concrete beams—Part 1: Experimental test series. *ACI Struct. J.* **1995**, *92*, 56–62.
40. Koutchoukali, N.-E.; Belarbi, A. Torsion of high-strength reinforced concrete beams and minimum reinforcement requirement. *ACI Struct. J.* **2001**, *98*, 462–469.
41. Fang, I.-K.; Shiau, J.-K. Torsional behavior of normal- and high-strength concrete beams. *ACI Struct. J.* **2004**, *101*, 304–313.

42. Chiu, H.-J.; Fang, I.-K.; Young, W.-T.; Shiau, J.-K. Behavior of reinforced concrete beams with minimum torsional reinforcement. *Eng. Struct.* **2007**, *29*, 2193–2205. [CrossRef]
43. Lee, J.-Y.; Kim, S.-W. Torsional strength of RC beams considering tension stiffening effect. *J. Struct. Eng.* **2010**, *136*, 1367–1378. [CrossRef]
44. Joh, C.; Kwahk, I.; Lee, J.; Yang, I.-H.; Kim, B.-S. Torsional behavior of high-strngth concrete beams with minimum reinforcemente ratio. *Adv. Civ. Eng.* **2019**, *2019*, 1432697.
45. Ju, H.; Han, S.-J.; Zhang, D.; Kim, J.; Wu, W.; Kim, K.S. Estimation of minimum torsional reinforcement of reinforced concrete and steel fiber-reforced concrete members. *Adv. Mater. Sci. Eng.* **2019**, *2019*, 4595363. [CrossRef]
46. Ibrahim, M.S.; Gebreyouhannes, E.; Muhdin, A.; Gebre, A. Effect of concrete cover on the pure torsional behavior of reinforced contrete beams. *Eng. Struct.* **2020**, *216*, 110790. [CrossRef]
47. Kim, M.-J.; Kim, H.-G.; Lee, Y.-J.; Kim, D.-H.; Lee, J.-Y.; Kim, K.-H. Pure torsional behavior of RC beams in relation to the amount of torsional reinforcment and cross-sectional properties. *Constr. Build. Mater.* **2020**, *260*, 119801. [CrossRef]
48. Bernardo, L.F.A.; Lopes, S.M.R. Torsion in HSC hollow beams: Strength and ductility analysis. *ACI Struct. J.* **2009**, *106*, 39–48.
49. Jeng, C.-H. Unified softened membrane model for torsion in hollow and solid reinforced concrete members modeling the entire pre- and post-cracking behavior. *J. Struct. Eng.* **2015**, *141*, 1–20. [CrossRef]
50. De Domenico, D. Torsional strength of RC members using a plasticity-based variable-angle space truss model accounting for non-uniform longitudinal reinforcement. *Eng. Struct.* **2021**, *228*, 111540. [CrossRef]
51. Lopes, S.M.R.; Bernardo, L.F.A. Twist behavior of high-strength concrete hollow beams—Formation of plastic hinges along the length. *Eng. Struct.* **2009**, *31*, 138–149. [CrossRef]
52. *EN 1990: 2002*; Eurocode 0: Basis of Structural Design. Part 1-General Rules and Rules for Buildings. European Committee for Standardization: Brussels, Belgium, 2002.

materials

MDPI

Article

Compressive Strength Evaluation of Ultra-High-Strength Concrete by Machine Learning

Zhongjie Shen [1,*], Ahmed Farouk Deifalla [2,*], Paweł Kamiński [3] and Artur Dyczko [4]

1. Xijing University, Xi'an 710123, China
2. Structural Engineering and Construction Management Department, Faculty of Engineering and Technology, Future University in Egypt, Cairo 11835, Egypt
3. Faculty of Civil Engineering and Resource Management, AGH University of Science and Technology, Mickiewicza 30, 30-059 Kraków, Poland; pkamin@agh.edu.pl
4. Mineral and Energy Economy Research Institute of the Polish Academy of Sciences, J. Wybickiego 7a, 31-261 Kraków, Poland; arturdyczko@min-pan.krakow.pl

* Correspondence: zhongjieshen@outlook.com (Z.S.); ahmed.deifalla@fue.edu.eg (A.F.D.)

Abstract: In civil engineering, ultra-high-strength concrete (UHSC) is a useful and efficient building material. To save money and time in the construction sector, soft computing approaches have been used to estimate concrete properties. As a result, the current work used sophisticated soft computing techniques to estimate the compressive strength of UHSC. In this study, XGBoost, AdaBoost, and Bagging were the employed soft computing techniques. The variables taken into account included cement content, fly ash, silica fume and silicate content, sand and water content, superplasticizer content, steel fiber, steel fiber aspect ratio, and curing time. The algorithm performance was evaluated using statistical metrics, such as the mean absolute error (MAE), root mean square error (RMSE), and coefficient of determination (R^2). The model's performance was then evaluated statistically. The XGBoost soft computing technique, with a higher R^2 (0.90) and low errors, was more accurate than the other algorithms, which had a lower R^2. The compressive strength of UHSC can be predicted using the XGBoost soft computing technique. The SHapley Additive exPlanations (SHAP) analysis showed that curing time had the highest positive influence on UHSC compressive strength. Thus, scholars will be able to quickly and effectively determine the compressive strength of UHSC using this study's findings.

Keywords: UHSC; building material; compressive strength; soft computing technique; concrete

Citation: Shen, Z.; Deifalla, A.F.; Kamiński, P.; Dyczko, A. Compressive Strength Evaluation of Ultra-High-Strength Concrete by Machine Learning. *Materials* **2022**, *15*, 3523. https://doi.org/10.3390/ma15103523

Academic Editors: Dario De Domenico and Luís Filipe Almeida Bernardo

Received: 13 April 2022
Accepted: 7 May 2022
Published: 13 May 2022

1. Introduction

Ultra-high-strength concrete (UHSC) is becoming increasingly popular because of its superior mechanical qualities, increased ductility, and longer life expectancy [1]. If UHSC is cured for 28 days, it will have a compressive strength of more than 120 MPA, even after cracking. To attain excellent characteristics, UHSC's maximum density was carefully designed. Particle packing density in UHSC results in low permeability and dense microstructures [2,3]. Due to the inclusion of distributed discrete fibers, UHSC has increased mechanical strength and crack resistance. There are a variety of civil engineering applications using UHSC, which range from building to rehabilitation to repair. The mechanical characteristics of UHSC are now being evaluated in current practice, by conducting complete experimental examinations. Experimental approaches can be used to determine the precise link between material qualities and mix design, but this requires a significant investment, in terms of both time and money [4]. Variables for UHSC include the cement content, the water content, the additive material content, the fiber content (e.g., steel fibers), the content and type of admixtures, and aggregates content and type (e.g., superplasticizer) [5–7]. The addition of dispersed short-discrete fibers to concrete increased crack resistance and improved mechanical characteristics [8–16]. Steel fibers are employed to increase the toughness and post-cracking behavior of the cementitious material [17–20]. Despite several experimental studies in the literature, it is still difficult to predict the characteristics of UHSCs containing various mixtures of components using computational methodologies. Thus, in this work, an attempt was made to anticipate the compressive property of UHSC using soft computing techniques.

Complex issues in a variety of engineering domains can be effectively solved using soft computing approaches. Machine learning (ML) approaches may be used to predict the final output after being provided an input data set. In order to forecast the characteristics of concrete, two ML strategies were used, i.e., a standalone approach (based on a single model) and an ensemble approach (such as AdaBoost and bagging). Ensemble models beat individual ML models in terms of performance, according to studies. However, there are examples of ML models that may be used to predict cement composites characteristics. There has been a detailed evaluation of the use of ML approaches to anticipate concrete mechanical characteristics [21]. In addition, a number of studies have been done to predict the mechanical characteristics of different types of concretes, such as high-performance concrete (HPC) [22], self-healing concrete [23], recycled aggregate concrete (RCA) [24], phase change material-integrated concrete [25], etc. Han, et al. [26] employed a machine learning technique to forecast HPC compressive strength. Cement, fine aggregates, FA, GGBFS, coarse aggregates, age, water, and five other combination variables were included in the dataset's input parameters. The compressive strength of HPC was accurately predicted by the established model. This article forecasts the compressive strength of UHSC using soft computing techniques and will serve as a baseline to save time and money for future researchers.

The previous studies were related to high performance concrete with a compressive strength around 10–80 MPa [27]. However, this study is related to ultra-high strength concrete (UHSC) with a compressive strength of 100–160 MPa, where the particle packing theory is important. Additionally, the effect of raw ingredients on compressive strength was not investigated by previous studies, which remains a research gap. Therefore, the effect of input parameters (raw materials) on the output parameter (compressive strength) was evaluated using SHapley Additive exPlanations and their interaction was explained. The compressive strength of UHSC may also be predicted using machine learning methods in an alternate approach, to save experimental time and money. In this paper, a variety of ensembled machine learning approaches were used to estimate the compressive of UHSC. XGBoost, AdaBoost, and Bagging are included as ensemble machine learning models. In addition, all models were tested using statistical criteria, and a comparison was made between several machine learning models. A better model was then proposed based on the performance of several statistical parameters to predict UHSC outcomes. Furthermore, a post hoc model-agnostic technique, i.e., SHapley Additive exPlanations (SHAP), was also implemented to give ML model insight [28,29]. The integration of SHAP with ML algorithms was performed in the current research to provide a comprehensive understanding of the mix design of concrete, regarding its strength parameters through its non-linear complex behavior, and to describe the contribution of input parameters by assigning a weight factor to each input parameter. This will be highly beneficial for the development of durable and sustainable concrete mixes.

2. Soft Computing Techniques

In order to get the best results, ensemble learning trains numerous base learners to aggregate their findings according to a predetermined methodology [30]. The design and building of fundamental learners, as well as their integration, is critical to ensemble learning algorithms. Based on how base learners collaborate, ensemble learning may be divided into parallel and sequential forms. No substantial connections between individual learners can be found in the parallel ensemble, as demonstrated by the bagging technique. Learners in a sequential ensemble are highly interdependent and sequentially formed, as shown by boosting [31]. Here, the fundamentals of ensemble approaches are briefly discussed.

Iteratively updating the previous classifier's parameters reduces the gradient of the loss function and generates a new classifier. The regression tree group is assured to have the highest generalization ability by minimizing the error of prediction across numerous regression trees. The loss function of the model is enhanced by including the regular term. As part of this process, a Taylor expansion of the loss function is used to calculate the split node. The performance of generalization and computation has been enhanced by the use of the regularization approach and second-order derivative information [32]. Figure 1 shows the XGBoost algorithm's structure.

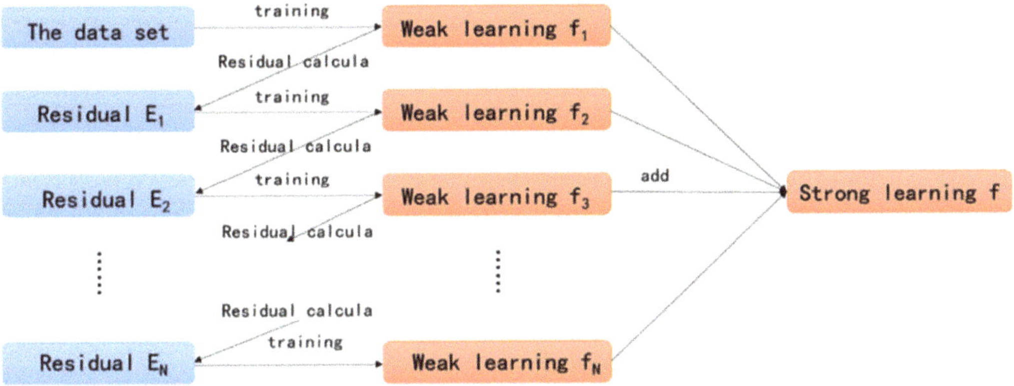

Figure 1. Structure of XGBoost algorithm [32].

A sequential ensemble may be built using the boosting approach. It creates a mediocre learner based on the first set of data. After that, a new weak learner is created to try to correct the mistakes of the previous weak learner. To approach the final prediction model, all weak learners must be included into it. All samples are given equal weight when AdaBoost is used to start the dataset. When a new learner makes a mistake, the samples that it gains weight on, obtain the weight that the first learner gets right. This process has a predetermined number of repetitions, before an error occurs. Updating the training sample weights in subsequent rounds takes the weaker learners' performance into account. Figure 2 depicts an ensembled support vector regressor technique with AdaBoost.

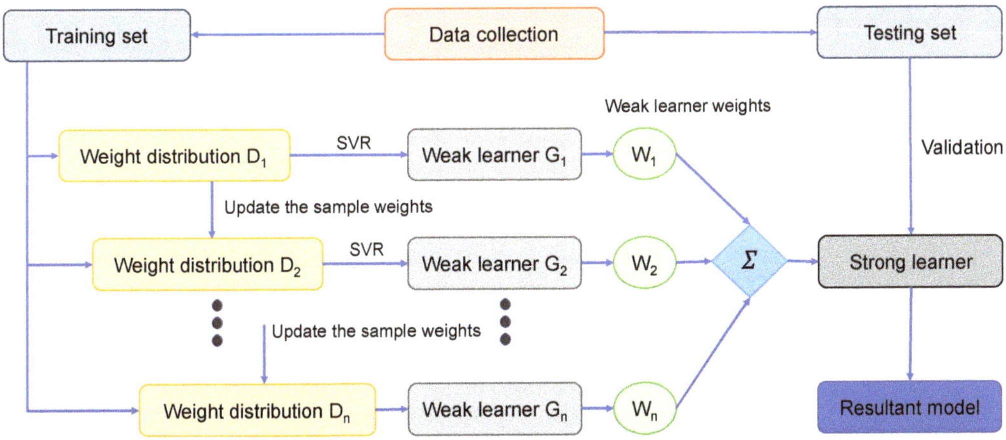

Figure 2. AdaBoost algorithm [33].

Bootstrapping and aggregation are two parts of the process of bagging. Training several models is made possible by regularly dividing the full dataset into smaller groups (base learners). The final forecast is the sum of the individual model results. These estimations are averaged together to obtain this forecast in the regression example. According to the categorization example, the voting process is used to make a final forecast. Algorithms such as support vector regressor, adaptive boosting, and bagging were used in this work to predict concrete properties, all of which have been demonstrated to perform well in previous studies for normal strength concrete. The process flow of the bagging algorithm is shown in Figure 3.

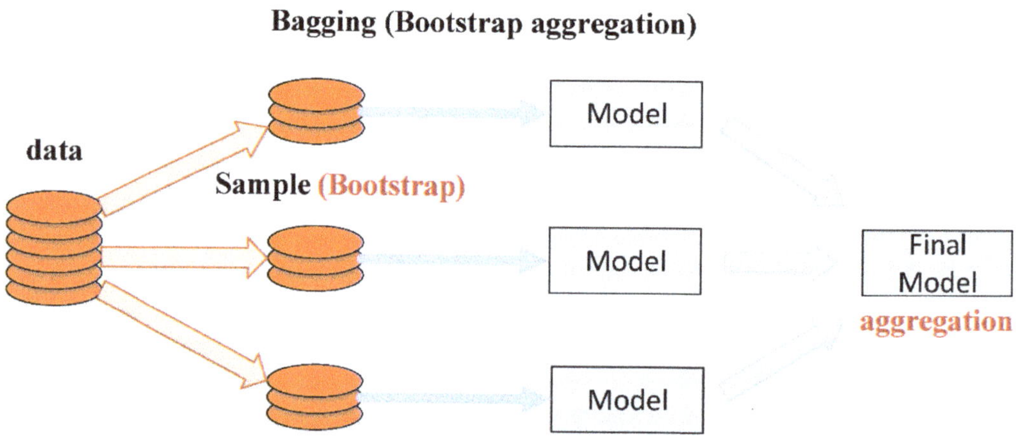

Figure 3. Process flow of bagging algorithm [34].

3. Interpretability of Model Using SHAP

The establishment of a robust prediction tool is gaining attention due to the ML models learning ability from recognized data and for prediction responses in unknown areas. However, lower interpretability and greater complexity is common in most machine learning modelling approaches [35]. SHAP is derived from game theory Shapley values [36]. Its employment is intended to determine the importance of different features within models [35,37]. In SHAP, the feature importance (j) for model outcome f; $\phi^j(f)$, is allotted weight for feature contribution summation towards output of model $f(x_i)$ for the overall potential combinations of features [38]. The expression for $\phi^j(f)$ is shown in Equation (1), as given below:

$$\phi^j(f) = \sum_{S \subseteq \{x^1,\ldots\ldots,\ x^p\}/\{x^j\}} \frac{|S|!(p-|S|-1)!}{p!} \Big(f(S \sqcup \{x^j\}) - f(S) \Big) \tag{1}$$

where; S = features subset, p = feature number in model, and x_j = feature j.

In the SHAP process, the importance of a feature is investigated by quantifying the prediction errors when disturbing a specified value of a feature. The prediction error sensitivity is considered for assigning a weight to feature importance, while perturbing its value. The trained ML model performance is also explained by using SHAP. SHAP uses an additional feature attribution method, i.e., linear input factor addition, to explain an interpretable model, is taken by the model output. As an illustration, a model having input factors x_i; where i ranges from 1 to k, and; k represents input factors number and h (x_s), shows an explanation model having x_s as a simplified input, whereas; Equation (2) is implemented to present an original model $f(x)$:

$$f(x) = h(x_s) = \varnothing_0 + \sum_{i=1}^{p} \varnothing_i x_s^i \tag{2}$$

where $\varnothing_0$= constant without any information (i.e., no input), and p = input feature number.

The mapping function, i.e., $x = m_x(x_s)$, has a correlation with both x and x_s inputs. Lundberg and Lee [35] explained Equation (9), in which the prediction value, i.e., (h ()) is improved by $\varnothing_0$, $\varnothing_1$, and $\varnothing_3$ terms and a decline of $\varnothing_4$ in h () value is also noted (Figure 4). There is a single value solution to Equation (9) that includes three preferred properties, i.e., missingness, consistency, and local accuracy. In missingness, it is ensured that no value for importance is assigned to the missing features, i.e., $\varnothing_i = 0$ is employed by $x_s^i = 0$. Consistency ensures no reduction in attribution, assigned to the respective features, as a change in feature with more impact. In local accuracy, it is ensured that the summation of feature attribution is taken as a function for the outcome, which includes a requirement for the model to match the outcome f for x_s as a simplified input. $x = m_x x_s$ represents the attainment of local accuracy.

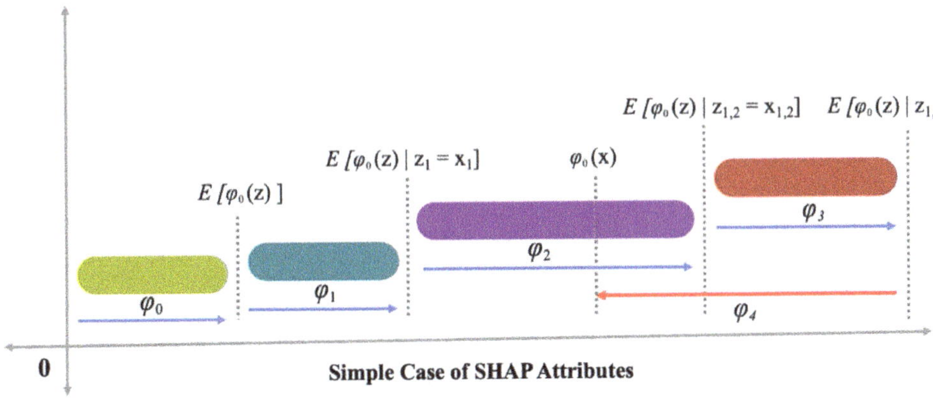

Figure 4. Attributes of SHAP.

4. Data Set

Figure 5 shows the data set utilized to forecast UHSC's compressive strength. The literature [39] provides a compressive database and there were 372 mix proportions with 10 input parameters selected from the data in the range of 100–160 MPa. These include cement content, fly ash, and silica fume content, as well as sand and water. Input parameters of steel fiber aspect ratio and curing time are also included. Predictor variables of the output parameter (compressive strength) are based on these input parameters. Each variable's range and lowest and maximum values are shown in Figure 5. There is also a figure that presents the mean and standard deviation for each variable. Compressive strength was predicted using Anaconda software's Spyder and Python scripting. The histogram of compressive strength used in this study is shown in Figure 6.

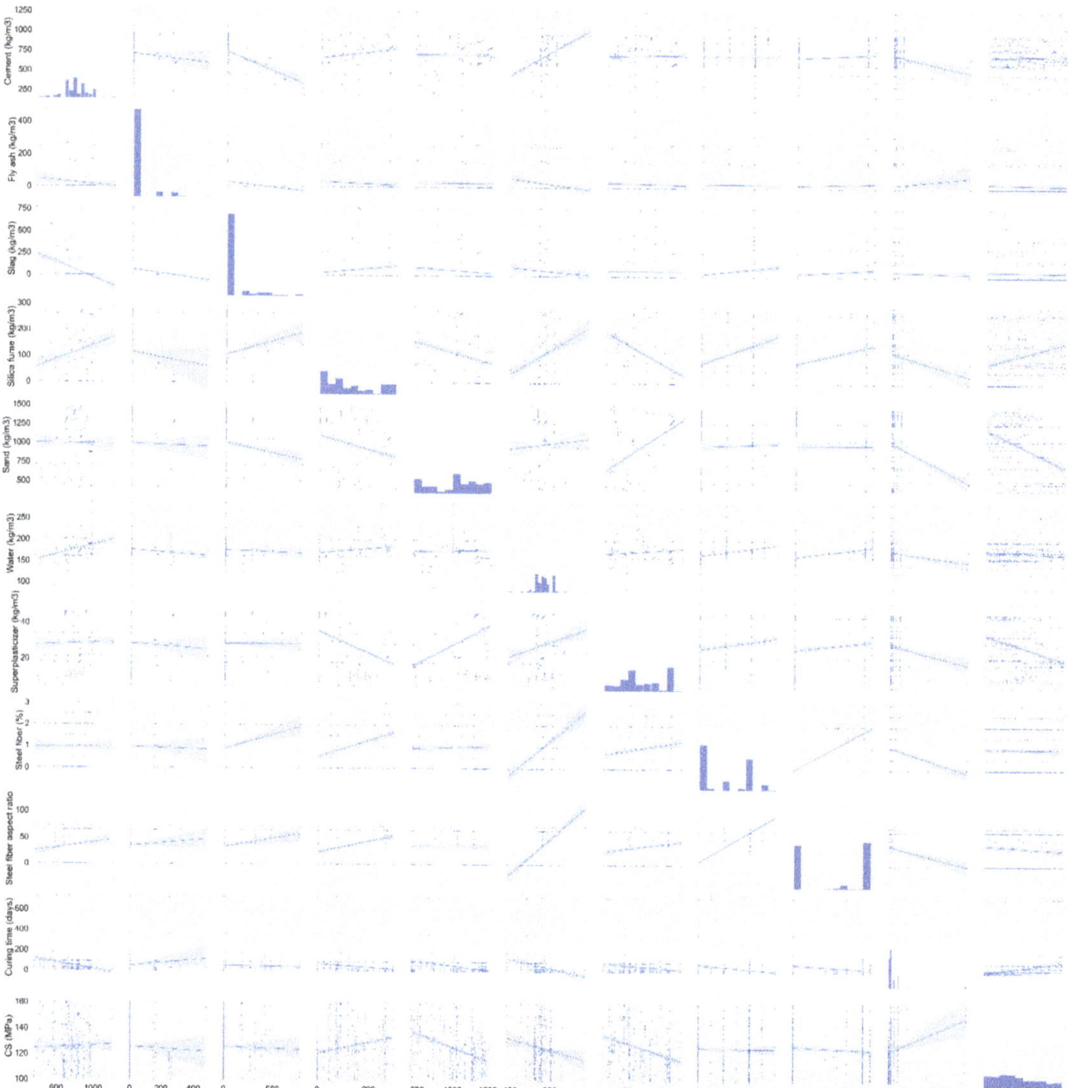

Figure 5. Data description of parameters.

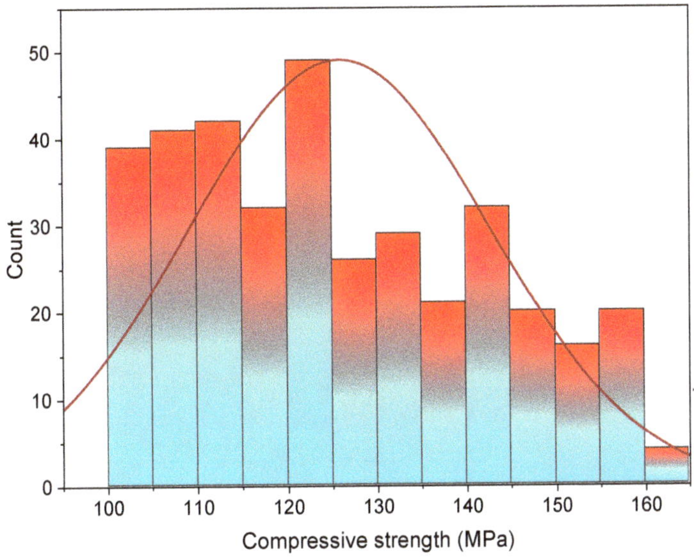

Figure 6. UHSC compressive strength distribution.

5. Results and Discussion

5.1. XGBoost

The comparison of experimental and predicted values with the XGBoost algorithm for compressive strength of UHSC is presented in Figure 7. The XGBoost exhibited reasonable predicted results, with low variation for the compressive strength of UHSC. An acceptable R^2 value of 0.89 shows the suitability of the XGBoost model. Figure 8 illustrates the error distribution of the experimental and XGBoost predicted values of compressive strength for UHSC. The average values of error for compressive strength are 6.42 MPa. Whereas 50% of error values are less than 5 MPa, 37% are from 5 to 10 MPa, and 24% are higher than 10 MPa.

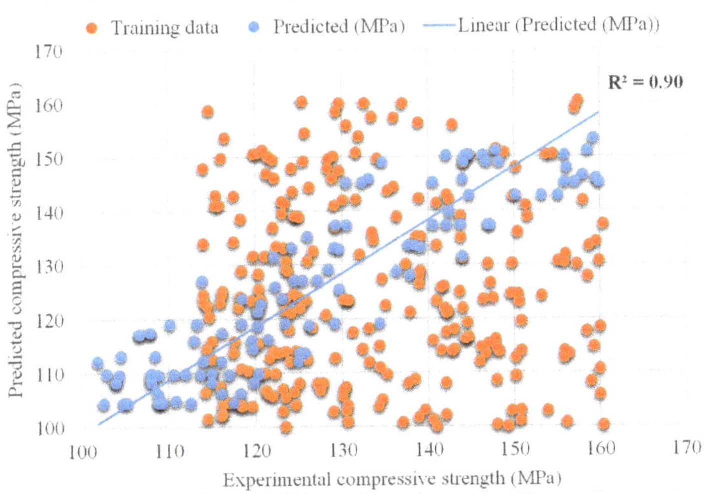

Figure 7. Experimental and XGBoost predicted results.

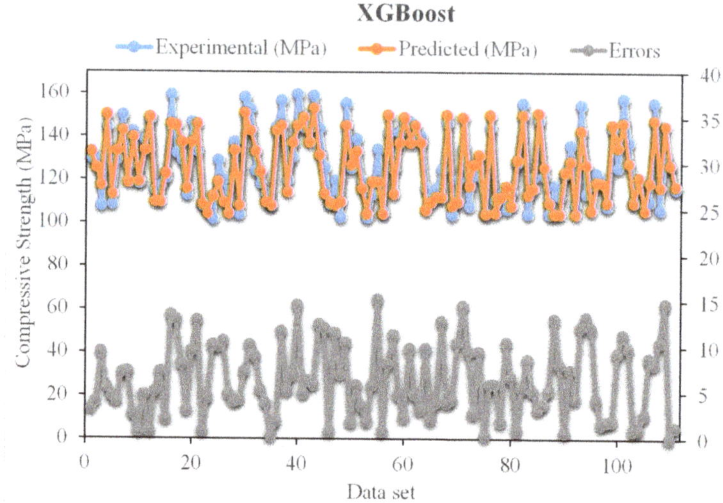

Figure 8. Experimental and SVR predicted values with errors.

5.2. *AdaBoost*

Figure 9 shows the experimental and predicted AdaBoost algorithm results for compressive strength of UHSC. The R^2 value for AdaBoost is 0.82 and represents less precise results than that of the XGBoost algorithm. The distribution of experimental and Adaboost predicted values with errors for compressive strength of UHSC is demonstrated in Figure 10. It is noted that 30% of error data is below 5 MPa, 29% is from 5 to 10 MPa, and 52% is higher than 10 MPa. The lower error and higher R^2 value indicated the better accuracy of XGBoost model compared to AdaBoost.

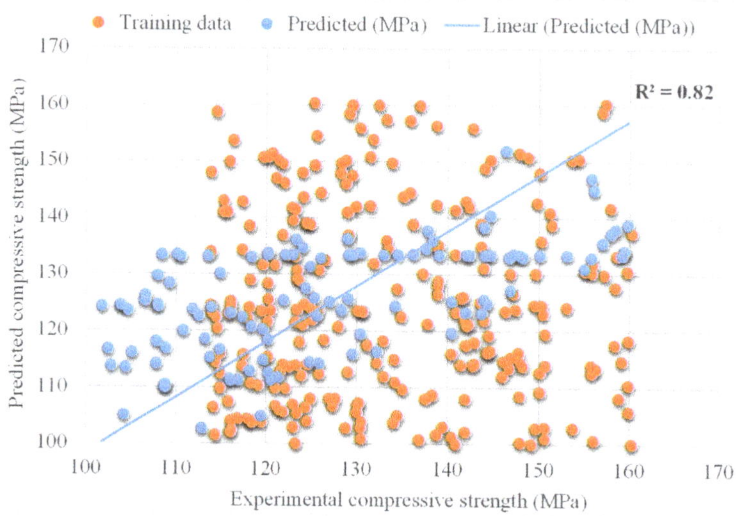

Figure 9. Experimental and AdaBoost predicted results.

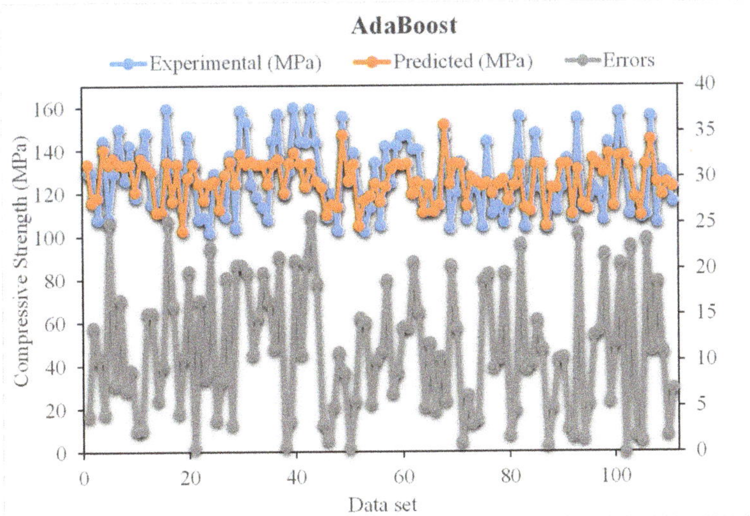

Figure 10. Experimental and AdaBoost predicted values with errors.

5.3. Bagging

The experimental and bagging predicted results of UHSC for compressive strength are shown in Figure 11. The R^2 for this model is 0.78, which shows less suitable results compared to the above two models. However, the predicted compressive strength results of UHSC for XGBoost are better than the other ensembled models. Figure 12 demonstrates the distribution of experimental and bagging predicted values with errors for compressive strength of UHSC. Whereas 30% of error values are below 5 MPa, 17% of values range from 5 to 10 MPa, and 62% of values are found above 10 MPa. The error and R^2 values for the compressive strength of UHSC for XGBoost are more accurate than the bagging model. Wang, et al. [33] reported that the AdaBoost machine learning approaches predicted the best compressive strength of geopolymer composites. Zhu, et al. [40] used machine learning to forecast the splitting tensile strength (STS) of concrete containing recycled aggregate (RA) and revealed that the precision level of the bagging model was better. Ahmad, et al. [41] studied the boosting and AdaBoost ML approaches to predict the compressive strength of a high-calcium fly-ash-based geopolymer. Bagging indicated the best results. However, the R^2 and error values obtained for the XGBoost ensemble machine learning models are acceptable. Thus, this finding implies that XGBoost could predict outcomes with a higher degree of accuracy than the other models.

5.4. Comparison of All Models

The validity of a model during execution is assessed by employing the K-fold cross-validation method. Statistical checks are used to evaluate the performance of models [42–45]. Usually, random dispersion is performed by splitting data into ten groups for k-fold cross-validation, and this process is repeated ten times to obtain acceptable results. Table 1 shows statistical checks for all models. The R^2 values for the XGBoost, AdaBoost, and Bagging models were 0.90, 0.82, and 0.78, respectively, as shown in Figure 13a–c. It was found that the R^2 of XGBoost was higher than that of all other models, with low error values for the compressive strength of UHSC.

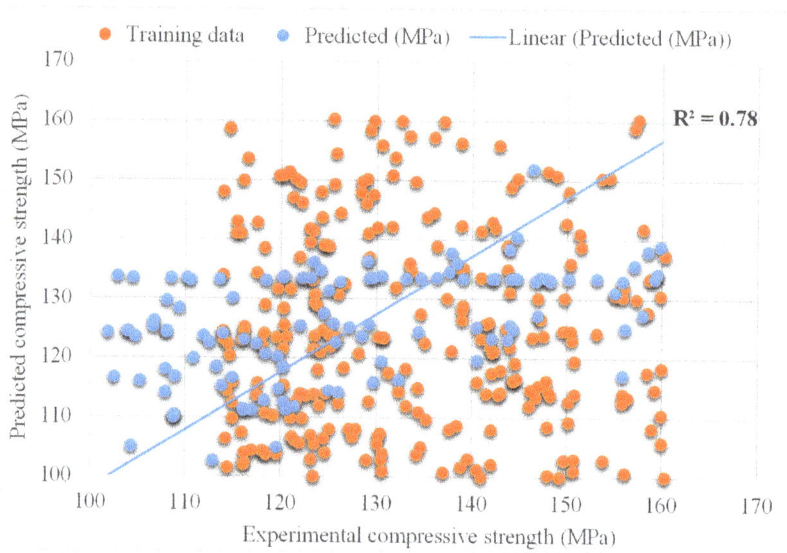

Figure 11. Experimental and BSA predicted results.

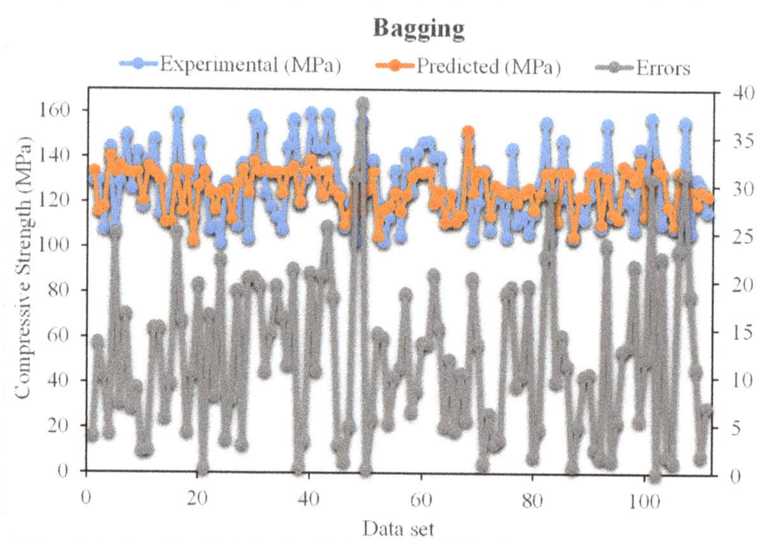

Figure 12. Distribution of experimental and BSA predicted values with errors.

Table 1. Statistical checks of the XGBoost, AdaBoost, and Bagging models.

Techniques	MAE (MPa)	RMSE (MPa)	R^2
XGBoost	6.4	7.6	0.90
Adaboost	11.0	13.1	0.82
Bagging	11.9	14.6	0.78

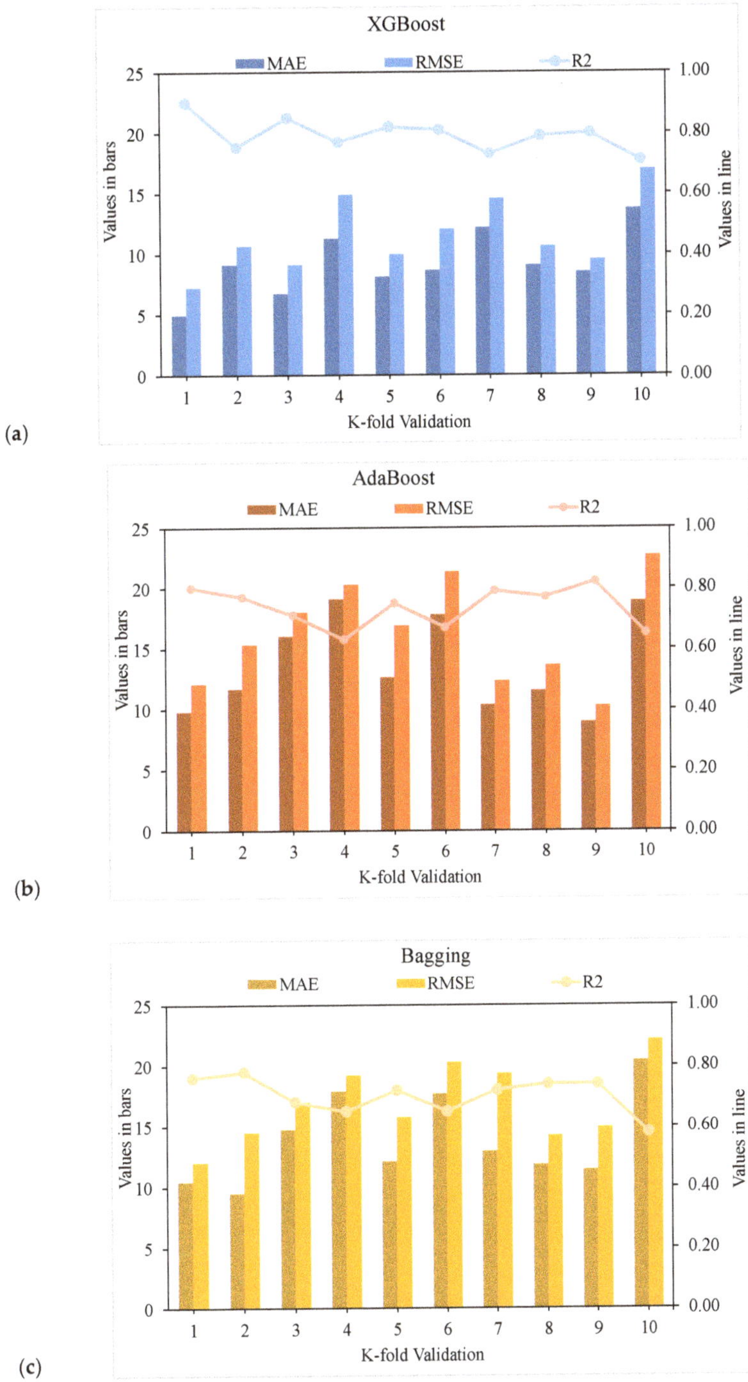

Figure 13. Statistical representation: (**a**) XGBoost; (**b**) AdaBoost; (**c**) Bagging.

The compressive strength of UHSC was predicted using ensembles of machine learning approaches in this work, which aimed to provide efficient and reliable findings. With an R^2 value of 0.90, XGBoost's output provided more exact predictions for UHSC compressive strength. Using an optimized model from the 20 sub-models shown in Figure 14a–c to predict compressive strength, the XGBoost ensemble machine learning models performed better. It was, thus, shown that, compared to the other models, the XGBoost ensembled models demonstrated an excellent accuracy and low error.

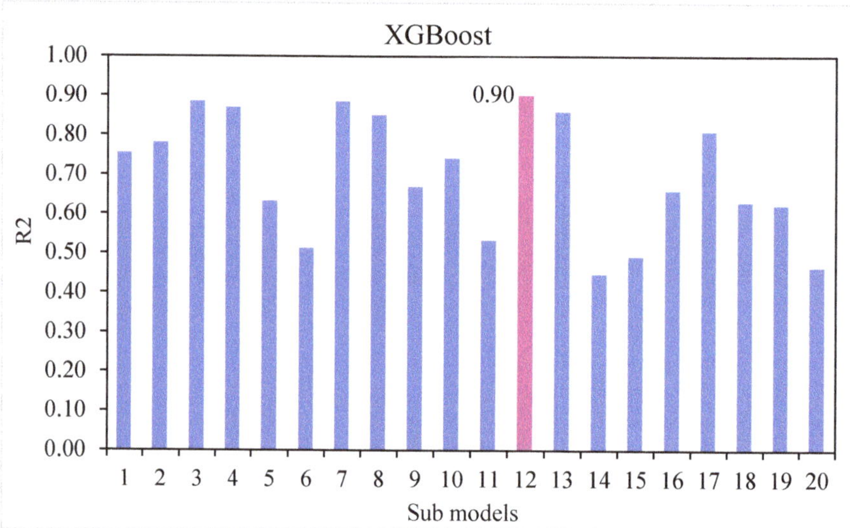

(a)

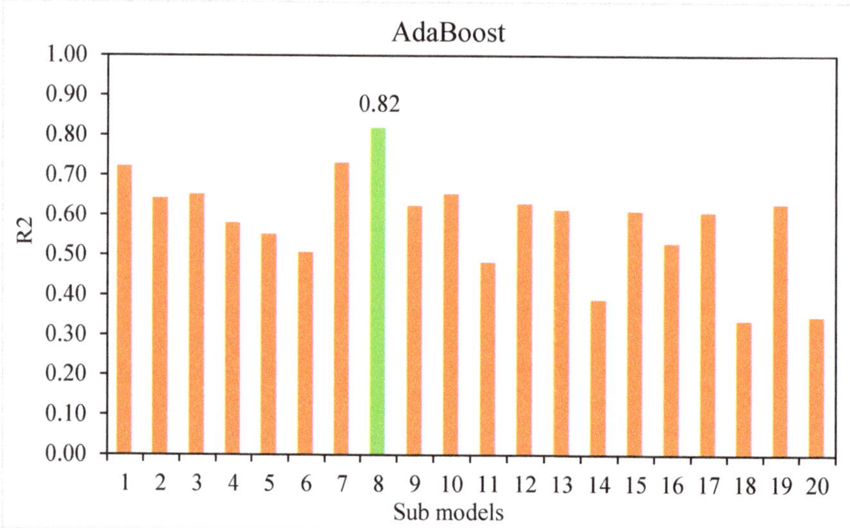

(b)

Figure 14. *Cont.*

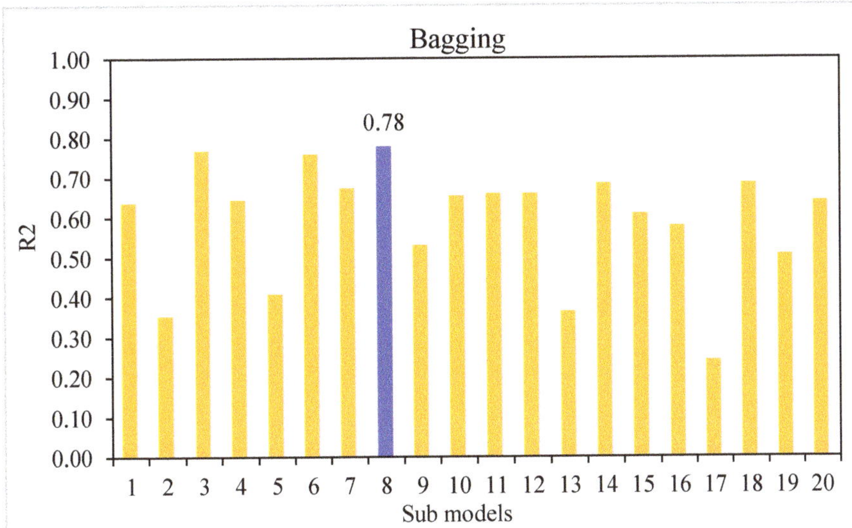

(c)

Figure 14. Sub-model results: (**a**) XGBoost; (**b**) AdaBoost; (**c**) Bagging.

Mahjoubi, et al. [46] constructed an auto-tune learning framework for ultra-high-performance concrete flowability, mechanical characteristics, and porosity prediction (UHPC). Other models were also considered by Mahjoubi et al. [47,48] in previous studies for multiple functions, and can be applied to similar types of studies in the future. This study evaluated compressive strength in the range of 100–160 MPa, considering 372 mix proportions with 10 input parameters selected from the database of Mahjoubi et al. [39,46]. A much more relevant model could be obtained by increasing the number of datasheets and by importing a significantly higher number of mixtures, as well as by considering higher input parameters. Therefore, it is suggested that the number of data points and outcomes in future investigations be raised by experimental work, field tests, and numerical analysis, using a range of approaches (e.g., Monte Carlo simulation, among others). Environmental factors (such as high temperatures and humidity) could be included in the input parameters, along with a detailed explanation of the raw materials, to improve the models' responses.

6. Enhanced Explainability of ML Models

In the current research, an in-depth description of the ML model and dependencies/interactions of all the considered features is provided. Initially, by implementing the SHAP tree explainer for the entire dataset, an enhanced global representation of feature influences, by merging local explanations from SHAP, is provided. A tree-like SHAP approximation technique, named TreeExplainer, was employed [49]. In this technique, the internal structure of tree-based models was evaluated; that is the sum of calculations set having a linkage with the leaf node of a tree model that led to low-order complexity [49]. The XGBoost model denotes the performance forecasting with higher precision for compressive strength of ultra-high strength concrete (UHSC), so in the current section, the model's interpretation is done for compressive strength of UHSC using SHAP. The correlation of various features with SHAP values for compressive strength of UHSC (as obtained from the XGBoost ensemble modelling) is presented in Figure 15.

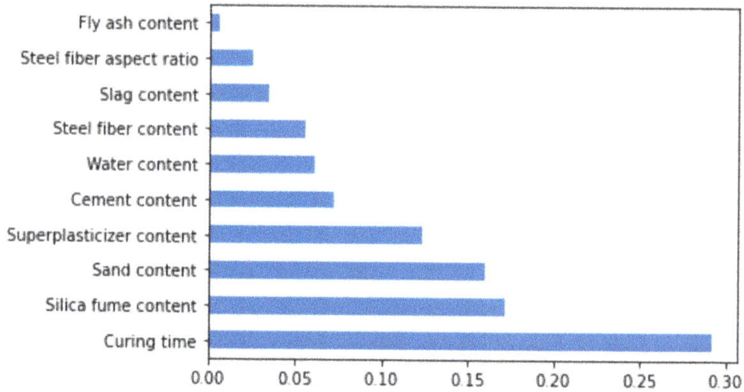

Figure 15. Feature importance.

It can be noted here that the curing time has highest SHAP value in the case of compressive strength prediction for UHSC. Increasing curing time would result in greater compressive strength, as UHSC has a high quantity of binders, i.e., silica fume, slag, fly-ash etc., so the hydration process requires more curing time, ultimately resulting in enhanced compressive strength. The silica fume content feature, i.e., a key parameter of UHSC and directly influencing the compressive strength, has the second highest SHAP value. Subsequently, sand is the third most influential feature, as shown in Figure 15. In UHSC, particle packing density would be difficult to achieve in the case of higher sand contents. Super-plasticizer is fourth in the row, due to its higher SHAP value. More super-plasticizer and a lesser water content positively influences the compressive strength of UHSC. Similarly, the influence of cement is next in terms of SHAP value, followed by the water, steel fiber, and fly-ash features. All these features have their unique roles in the compressive strength of UHSC. Fly ash has little effect on compressive strength and influences the workability of UHSC more.

Figure 16 depicts the violin plot SHAP values for all the corresponding features that were considered to predict the compressive strength of UHSC. In the said plot, a unique color represents every feature's value and the corresponding SHAP value at the x-axis represents the outcome contribution. For instance, for curing time and silica fume content as input features, a positive influence can be observed from the right side of the axis, showing a direct relationship for both the features with the compressive strength of UHSC. A SHAP value of almost 14, in the form of red points at the rightmost, shows that a higher curing time enhances the UHSC compressive strength. However, in case of the super-plasticizer feature, a positive influence is seen, but only up to the optimal content. Beyond this content, it has a negative influence, in the form of a blue color (i.e., lower values). It is usually observed that upon enhancing the water-binder ratio, the compressive strength tends to increase up to a certain limit, and then further enhancement of the water-binder ratio decreases the compressive strength. In the same manner, a higher quantity of sand negatively influences the compressive strength of UHSC, as its particle packing is disturbed. Furthermore, a weaker bond would be observed in the case of a higher sand content with respect to binder. Steel fiber and cement content also show a positive influence. Last, water has both positive and negative influences and is directly related to the binder content. A higher water content would result in a reduced UHSC compressive strength. Fly ash and slag, although they do not have a considerable impact on compressive strength of UHSC, still display more or less similar feature influences. These observations are dependent on the database used in this study, and results with greater accuracy may be acquired in the case of more data points.

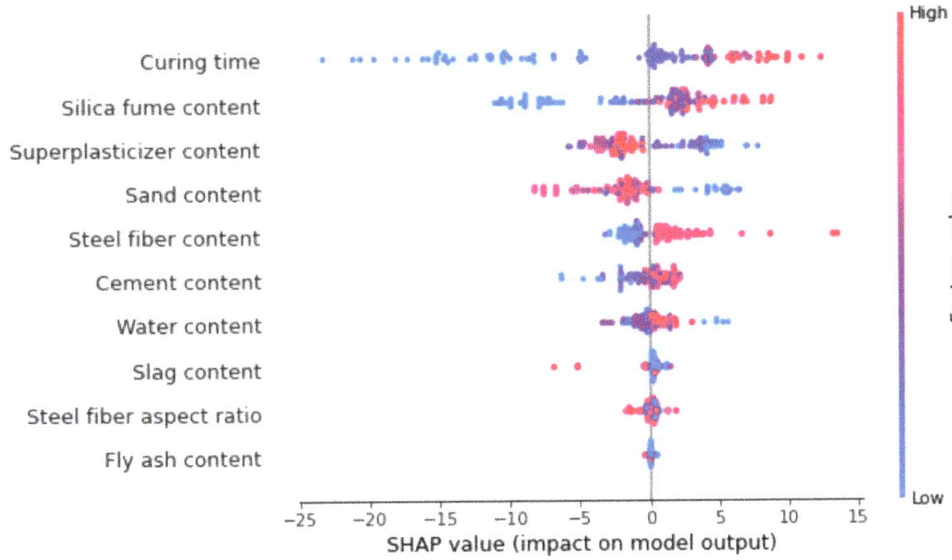

Figure 16. SHAP plot.

The interaction of the various considered features with the compressive strength of UHSC is presented in Figure 17. The curing feature interaction is shown in Figure 17a. It may be observed from the plot that curing time is a major influence of the compressive strength of UHSC and is in a positive/direct relationship. In this scenario, the maximum interaction of curing is with silica fume, hence, aiding in the enhancement of UHSC strength. In Figure 17b, a positive influence of silica on the compressive strength of UHSC is observed. A greater interaction of silica is found with curing time and it is negatively influential. The fine aggregate/sand feature interaction is plot in Figure 17c. The sand content has a negative influence, due to its effect on silica fume. Therefore, the effect of sand on silica fume results in decreased compressive strength. Then, in a row, super-plasticizer shows both positive and negative interactions, depending upon the content (Figure 17d). A lesser content, up to the optimum content, would result in a positive interaction and vice versa. In the same manner, cement content positively interacts and greatly influences the water content, as the w/c ratio has a major role in the development of strength, due to multiple factors, including the hydration process (Figure 17e). In Figure 17f, the interaction of silica fume with water content is shown. The lesser surface area of silica fume demands a higher water content. Furthermore, during pozzolanic activity in the hydration reaction, silica fume needs more water; therefore, a higher interaction of silica fume with water content is observed.

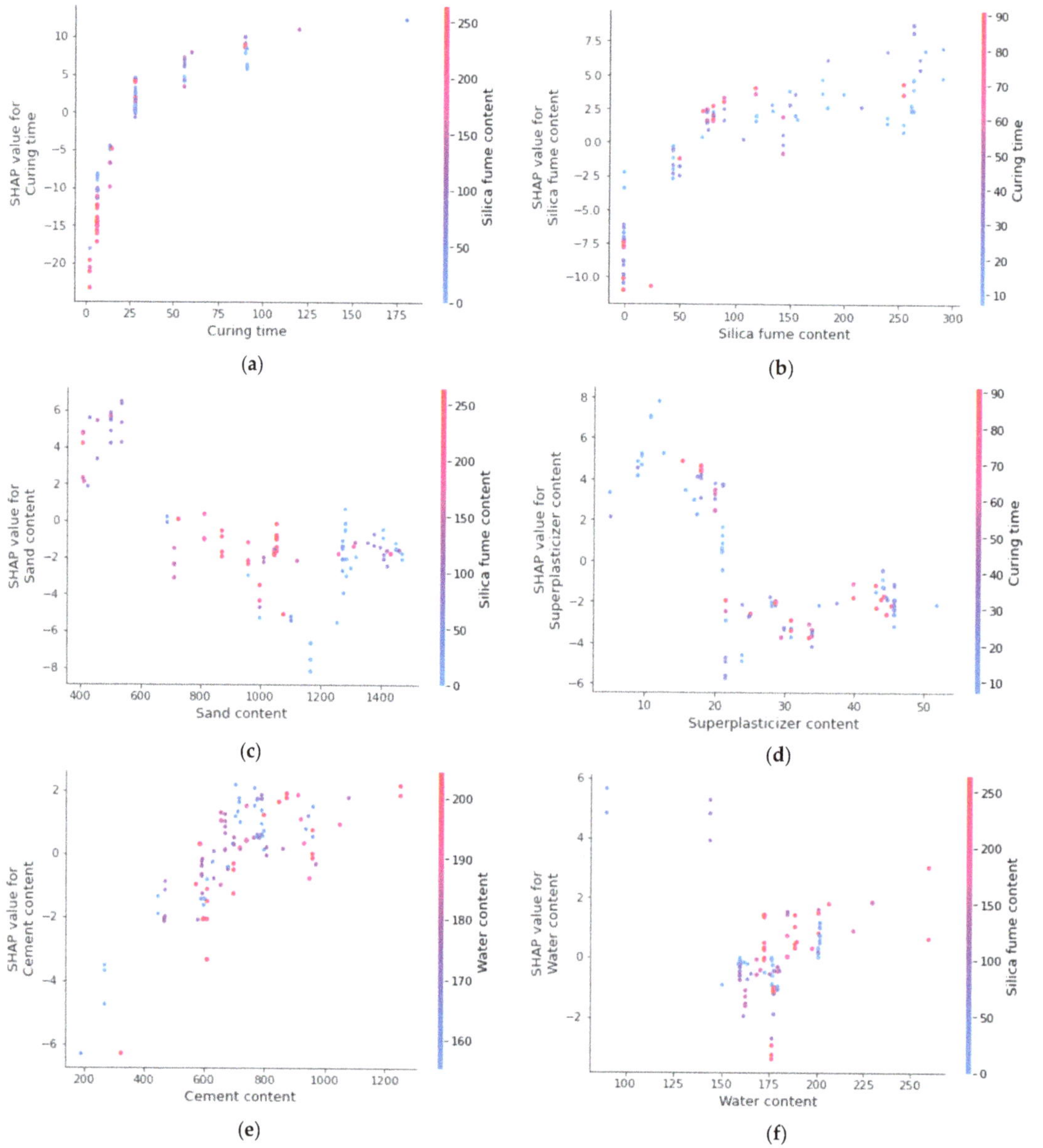

Figure 17. Interaction plot of various parameters: (**a**) Curing time; (**b**) Silica fume content; (**c**) Sand content; (**d**) Superplasticizer; (**e**) Cement content; (**f**) Water content.

7. Conclusions

Soft computing has recently been employed in the construction sector to forecast the mechanical characteristics of concrete, which has gained the attention of the industry. It was the goal of this study to evaluate the accuracy of soft computing approaches for predicting the compressive strength of UHSC. Ten input variables were used for estimation: i.e., cement content, fly ash, silica fume and silicate content, sand and water content, superplasticizer content, steel fiber, steel fiber aspect ratio, and curing time. As a result of our research, we have come to the following conclusions:

- As evidenced by the R^2 value of 0.90, the XGBoost method was able to accurately estimate the compressive strength of UHSC from its actual data. However, the ensembled machine learning models, i.e., AdaBoost and Bagging with R^2 values of 0.82 and 0.78, predicted unacceptable findings for the compressive strength of UHSC.
- A total of twenty sub-models, ranging from 10 to 200 estimators, were utilized to optimize the anticipated compressive strength of UHSC. An ensembled model XGBoost was able to accurately forecast the compressive strength more effectively than the other models.
- XGBoost models demonstrated lower MAE and RMSE, with a higher R^2 value for compressive strength of UHSC, compared to the other model in the k-fold validation results. XGBoost was proven to have the best compressive strength prediction accuracy for UHSC.
- The model's performance was evaluated using statistical measures such as MAE and RMSE. However, XGBoost projected superior results, with less error and a higher coefficient of determination for evaluating the compressive strength of UHSC.
- The XGBoost is the best method for predicting the compressive strength of UHSC utilizing soft computing approaches.
- Curing time has highest impact on UHSC compressive strength estimation, followed by silica fume, sand and super-plasticizer content, as depicted by SHAP analysis. Whereas, the compressive strength of UHSC with fly ash content is the least influential.
- The feature interaction plot showed that curing time, cement content, and silica fume positively influence UHSC compressive strength.

Author Contributions: Z.S.: Conceptualization, Investigation, Methodology, Formal analysis, Writing-original draft. A.F.D.: Conceptualization, Investigation, Methodology, Formal analysis, Funding acquisition, Writing—Review & Editing, Supervision. P.K.: Software, Visualization, Resources, Funding acquisition, Writing—Review & Editing. A.D.: Validation, Data Curation, Writing—Review & Editing, Project administration. All authors have read and agreed to the published version of the manuscript.

Funding: This research received no external funding.

Institutional Review Board Statement: Not applicable.

Informed Consent Statement: Not applicable.

Data Availability Statement: Data will be available on request from corresponding author.

Conflicts of Interest: The authors declare no conflict of interest.

References

1. Akhnoukh, A.K.; Buckhalter, C. Ultra-high-performance concrete: Constituents, mechanical properties, applications and current challenges. *Case Stud. Constr. Mater.* **2021**, *15*, e00559. [CrossRef]
2. Fan, L.; Meng, W.; Teng, L.; Khayat, K.H. Effect of steel fibers with galvanized coatings on corrosion of steel bars embedded in UHPC. *Compos. Part B Eng.* **2019**, *177*, 107445. [CrossRef]
3. Pyo, S.; Tafesse, M.; Kim, H.; Kim, H.-K. Effect of chloride content on mechanical properties of ultra high performance concrete. *Cem. Concr. Compos.* **2017**, *84*, 175–187. [CrossRef]
4. Xu, M.; Bao, Y.; Wu, K.; Xia, T.; Clack, H.L.; Shi, H.; Li, V.C. Influence of TiO_2 incorporation methods on NOx abatement in Engineered Cementitious Composites. *Constr. Build. Mater.* **2019**, *221*, 375–383. [CrossRef]
5. Li, P.; Yu, Q.; Brouwers, H. Effect of coarse basalt aggregates on the properties of Ultra-high Performance Concrete (UHPC). *Constr. Build. Mater.* **2018**, *170*, 649–659. [CrossRef]
6. Meng, W.; Khayat, K.H. Effect of graphite nanoplatelets and carbon nanofibers on rheology, hydration, shrinkage, mechanical properties, and microstructure of UHPC. *Cem. Concr. Res.* **2018**, *105*, 64–71. [CrossRef]
7. Meng, W.; Yao, Y.; Mobasher, B.; Khayat, K.H. Effects of loading rate and notch-to-depth ratio of notched beams on flexural performance of ultra-high-performance concrete. *Cem. Concr. Compos.* **2017**, *83*, 349–359. [CrossRef]
8. Cao, M.; Mao, Y.; Khan, M.; Si, W.; Shen, S. Different testing methods for assessing the synthetic fiber distribution in cement-based composites. *Constr. Build. Mater.* **2018**, *184*, 128–142. [CrossRef]

9. Khan, M.; Cao, M.; Hussain, A.; Chu, S. Effect of silica-fume content on performance of $CaCO_3$ whisker and basalt fiber at matrix interface in cement-based composites. *Constr. Build. Mater.* **2021**, *300*, 124046. [CrossRef]
10. Arshad, S.; Sharif, M.B.; Irfan-ul-Hassan, M.; Khan, M.; Zhang, J.-L. Efficiency of supplementary cementitious materials and natural fiber on mechanical performance of concrete. *Arab. J. Sci. Eng.* **2020**, *45*, 8577–8589. [CrossRef]
11. Xie, C.; Cao, M.; Guan, J.; Liu, Z.; Khan, M. Improvement of boundary effect model in multi-scale hybrid fibers reinforced cementitious composite and prediction of its structural failure behavior. *Compos. Part B Eng.* **2021**, *224*, 109219. [CrossRef]
12. Cao, M.; Khan, M. Effectiveness of multiscale hybrid fiber reinforced cementitious composites under single degree of freedom hydraulic shaking table. *Struct. Concr.* **2021**, *22*, 535–549. [CrossRef]
13. Khan, U.A.; Jahanzaib, H.M.; Khan, M.; Ali, M. Improving the tensile energy absorption of high strength natural fiber reinforced concrete with fly-ash for bridge girders. In *Key Engineering Materials*; Trans Tech Publication: Stafa-Zurich, Switzerland, 2018; pp. 335–342.
14. Khan, M.; Cao, M.; Ai, H.; Hussain, A. Basalt Fibers in Modified Whisker Reinforced Cementitious Composites. *Period. Polytech. Civil Eng.* **2022**, *66*, 344–354. [CrossRef]
15. Zhang, N.; Yan, C.; Li, L.; Khan, M. Assessment of fiber factor for the fracture toughness of polyethylene fiber reinforced geopolymer. *Constr. Build. Mater.* **2022**, *319*, 126130. [CrossRef]
16. Khan, M.; Ali, M. Improvement in concrete behavior with fly ash, silica-fume and coconut fibres. *Constr. Build. Mater.* **2019**, *203*, 174–187. [CrossRef]
17. Khan, M.; Cao, M.; Chu, S.; Ali, M. Properties of hybrid steel-basalt fiber reinforced concrete exposed to different surrounding conditions. *Constr. Build. Mater.* **2022**, *322*, 126340. [CrossRef]
18. Li, L.; Khan, M.; Bai, C.; Shi, K. Uniaxial tensile behavior, flexural properties, empirical calculation and microstructure of multi-scale fiber reinforced cement-based material at elevated temperature. *Materials* **2021**, *14*, 1827. [CrossRef]
19. Khan, M.; Cao, M.; Xie, C.; Ali, M. Hybrid fiber concrete with different basalt fiber length and content. *Struct. Concr.* **2022**, *23*, 346–364. [CrossRef]
20. Khan, M.; Cao, M.; Xie, C.; Ali, M. Effectiveness of hybrid steel-basalt fiber reinforced concrete under compression. *Case Stud. Constr. Mater.* **2022**, *16*, e00941. [CrossRef]
21. Chaabene, W.B.; Flah, M.; Nehdi, M.L. Machine learning prediction of mechanical properties of concrete: Critical review. *Constr. Build. Mater.* **2020**, *260*, 119889. [CrossRef]
22. Castelli, M.; Vanneschi, L.; Silva, S. Prediction of high performance concrete strength using genetic programming with geometric semantic genetic operators. *Expert Syst. Appl.* **2013**, *40*, 6856–6862. [CrossRef]
23. Ramadan Suleiman, A.; Nehdi, M.L. Modeling self-healing of concrete using hybrid genetic algorithm–artificial neural network. *Materials* **2017**, *10*, 135. [CrossRef] [PubMed]
24. Zhang, J.; Huang, Y.; Aslani, F.; Ma, G.; Nener, B. A hybrid intelligent system for designing optimal proportions of recycled aggregate concrete. *J. Clean. Prod.* **2020**, *273*, 122922. [CrossRef]
25. Marani, A.; Nehdi, M.L. Machine learning prediction of compressive strength for phase change materials integrated cementitious composites. *Constr. Build. Mater.* **2020**, *265*, 120286. [CrossRef]
26. Han, Q.; Gui, C.; Xu, J.; Lacidogna, G. A generalized method to predict the compressive strength of high-performance concrete by improved random forest algorithm. *Constr. Build. Mater.* **2019**, *226*, 734–742. [CrossRef]
27. Xu, Y.; Ahmad, W.; Ahmad, A.; Ostrowski, K.A.; Dudek, M.; Aslam, F.; Joyklad, P. Computation of High-Performance Concrete Compressive Strength Using Standalone and Ensembled Machine Learning Techniques. *Materials* **2021**, *14*, 7034. [CrossRef]
28. Lauritsen, S.M.; Kristensen, M.; Olsen, M.V.; Larsen, M.S.; Lauritsen, K.M.; Jørgensen, M.J.; Lange, J.; Thiesson, B. Explainable artificial intelligence model to predict acute critical illness from electronic health records. *Nat. Commun.* **2020**, *11*, 3852. [CrossRef]
29. Johnsen, P.V.; Riemer-Sørensen, S.; DeWan, A.T.; Cahill, M.E.; Langaas, M. A new method for exploring gene–gene and gene–environment interactions in GWAS with tree ensemble methods and SHAP values. *BMC Bioinform.* **2021**, *22*, 230. [CrossRef]
30. Salami, B.A.; Rahman, S.M.; Oyehan, T.A.; Maslehuddin, M.; Al Dulaijan, S.U. Ensemble machine learning model for corrosion initiation time estimation of embedded steel reinforced self-compacting concrete. *Measurement* **2020**, *165*, 108141. [CrossRef]
31. Liu, K.; Dai, Z.; Zhang, R.; Zheng, J.; Zhu, J.; Yang, X. Prediction of the sulfate resistance for recycled aggregate concrete based on ensemble learning algorithms. *Constr. Build. Mater.* **2022**, *317*, 125917. [CrossRef]
32. Zhang, M.; Hao, S.; Hou, A. Study on the Intelligent Modeling of the Blade Aerodynamic Force in Compressors Based on Machine Learning. *Mathematics* **2021**, *9*, 476. [CrossRef]
33. Wang, Q.; Ahmad, W.; Ahmad, A.; Aslam, F.; Mohamed, A.; Vatin, N.I. Application of Soft Computing Techniques to Predict the Strength of Geopolymer Composites. *Polymers* **2022**, *14*, 1074. [CrossRef] [PubMed]
34. Wu, L.-Y.; Weng, S.-S. Ensemble Learning Models for Food Safety Risk Prediction. *Sustainability* **2021**, *13*, 12291. [CrossRef]
35. Lundberg, S.M.; Lee, S.-I. A unified approach to interpreting model predictions. *Adv. Neural Inf. Process. Syst.* **2017**, *30*. [CrossRef]
36. Shapley, L.S.; Roth, A.E. *The Shapley Value: Essays in Honor of Lloyd S. Shapley*; Cambridge University Press: Cambridge, UK, 1988.
37. Cohen, S.; Ruppin, E.; Dror, G. Feature selection based on the shapley value. *Other Words* **2005**, *1*, 98Eqr.
38. Molnar, C. *Interpretable Machine Learning*; Lulu. Com: Morrisville, NC, USA, 2020.
39. Mahjoubi, S.; Bao, Y. The key material properties of ultra-high-performance concrete (UHPC). *Mendeley Data V1* **2021**, *1*. [CrossRef]
40. Zhu, Y.; Ahmad, A.; Ahmad, W.; Vatin, N.I.; Mohamed, A.M.; Fathi, D. Predicting the Splitting Tensile Strength of Recycled Aggregate Concrete Using Individual and Ensemble Machine Learning Approaches. *Crystals* **2022**, *12*, 569. [CrossRef]

41. Ahmad, A.; Ahmad, W.; Chaiyasarn, K.; Ostrowski, K.A.; Aslam, F.; Zajdel, P.; Joyklad, P. Prediction of geopolymer concrete compressive strength using novel machine learning algorithms. *Polymers* **2021**, *13*, 3389. [CrossRef]
42. Ahmad, W.; Ahmad, A.; Ostrowski, K.A.; Aslam, F.; Joyklad, P.; Zajdel, P. Application of advanced machine learning approaches to predict the compressive strength of concrete containing supplementary cementitious materials. *Materials* **2021**, *14*, 5762. [CrossRef]
43. Ahmad, A.; Ahmad, W.; Aslam, F.; Joyklad, P. Compressive strength prediction of fly ash-based geopolymer concrete via advanced machine learning techniques. *Case Stud. Constr. Mater.* **2021**, *16*, e00840. [CrossRef]
44. Yuan, X.; Tian, Y.; Ahmad, W.; Ahmad, A.; Usanova, K.I.; Mohamed, A.M.; Khallaf, R. Machine Learning Prediction Models to Evaluate the Strength of Recycled Aggregate Concrete. *Materials* **2022**, *15*, 2823. [CrossRef] [PubMed]
45. Shang, M.; Li, H.; Ahmad, A.; Ahmad, W.; Ostrowski, K.A.; Aslam, F.; Joyklad, P.; Majka, T.M. Predicting the Mechanical Properties of RCA-Based Concrete Using Supervised Machine Learning Algorithms. *Materials* **2022**, *15*, 647. [CrossRef] [PubMed]
46. Mahjoubi, S.; Meng, W.; Bao, Y. Auto-tune learning framework for prediction of flowability, mechanical properties, and porosity of ultra-high-performance concrete (UHPC). *Appl. Soft Comput.* **2022**, *115*, 108182. [CrossRef]
47. Mahjoubi, S.; Meng, W.; Bao, Y. Logic-guided neural network for predicting steel-concrete interfacial behaviors. *Expert Syst. Appl.* **2022**, *198*, 116820. [CrossRef]
48. Mahjoubi, S.; Barhemat, R.; Guo, P.; Meng, W.; Bao, Y. Prediction and multi-objective optimization of mechanical, economical, and environmental properties for strain-hardening cementitious composites (SHCC) based on automated machine learning and metaheuristic algorithms. *J. Clean. Prod.* **2021**, *329*, 129665. [CrossRef]
49. Lundberg, S.M.; Erion, G.; Chen, H.; DeGrave, A.; Prutkin, J.M.; Nair, B.; Katz, R.; Himmelfarb, J.; Bansal, N.; Lee, S.-I. Explainable AI for trees: From local explanations to global understanding. *arXiv* **2019**, arXiv:1905.04610. [CrossRef]

materials

MDPI

Article

Research on Dynamic Strength and Inertia Effect of Concrete Materials Based on Large-Diameter Split Hopkinson Pressure Bar Test

Bi Sun [1,2], Rui Chen [1], Yang Ping [3], ZhenDe Zhu [4], Nan Wu [5] and Zhenyue Shi [6,*]

1 Harbin Institute of Technology (Shenzhen), Shenzhen 518055, China; sunbi58@126.com (B.S.); cechenrui@hit.edu.cn (R.C.)
2 Shenzhen Water Planning and Design Institute Co., Ltd., Shenzhen 518001, China
3 PowerChina Eco-Environment Group Co., Ltd., Shenzhen 518102, China; pingy@swpdi.com
4 Key Laboratory of Ministry of Education of Geomechanics and Embankment Engineering, Hohai University, Nanjing 210098, China; zhendezhunj@163.com
5 Guangzhou University-Tamkang University Joint Research Centre for Engineering Structure Disaster Prevention and Control, Guangzhou University, Guangzhou 510006, China; wunan@gzhu.edu.cn
6 College of Safety and Environmental Engineering (College of Safety and Emergency Management), Shandong University of Science and Technology, Qingdao 266590, China
* Correspondence: 201881010015@sdust.edu.cn

Abstract: The Split Hopkinson Pressure Bar (SHPB) test device is an important tool to study the dynamic characteristics of concrete materials. Inertial effect is one of the main factors that cause inaccurate results in SHPB tests of concrete materials. To solve this problem, Large-diameter SHPB tests on concrete and mortar were performed. A dynamic increase factor (DIF) model considering strain rate effect and inertia effect was established. This model provides a scientific reference for studying the dynamic mechanical properties of concrete materials. The experimental results indicate that the strain rate effect of concrete is more sensitive than that of mortar, but the inertia effect of mortar is more sensitive than that of concrete. Under the same strain rate, the energy utilization rate, average fragment size, and impact potentiality of mortar are higher than concrete.

Keywords: large-diameter SHPB; high strain rate; concrete material; strain rate effect; inertia effect; dynamic strength

Citation: Sun, B.; Chen, R.; Ping, Y.; Zhu, Z.; Wu, N.; Shi, Z. Research on Dynamic Strength and Inertia Effect of Concrete Materials Based on Large-Diameter Split Hopkinson Pressure Bar Test. *Materials* **2022**, *15*, 2995. https://doi.org/10.3390/ma15092995

Academic Editors: Dario De Domenico and Luís Filipe Almeida Bernardo

Received: 20 March 2022
Accepted: 18 April 2022
Published: 20 April 2022

1. Introduction

Concrete materials, the largest and most widely used engineering building materials, are affected by static loads and high strain rate dynamic loads such as earthquakes, impacts, and explosions. As a heterogeneous, anisotropic, and strain rate-sensitive multiphase composite, concrete will show more complex dynamic mechanical characteristics under dynamic load than under static load, which has been one of the hot topics in recent years. Split Hopkinson Pressure Bar (SHPB) experimental technology is the most commonly used test method for the dynamic performance of concrete [1–7]. For heterogeneous materials, such as concrete, a specimen of considerable size is required to ensure a certain degree of homogeneity because concrete has a large aggregate size and has many microscopic defects. Ensuring a certain homogeneity in the SHPB test requires a fairly large specimen size. The pressure bar diameter of the SHPB device should also be increased correspondingly [8–10]. In the large-diameter SHPB test, the dynamic strength of the specimen is affected by the strain rate effect [11–13] and the inertia effect [14].

The dynamic increase factor (DIF), defined as the ratio of dynamic-to-static strength, is conventionally considered a material property [15,16]. A bilinear DIF model proposed by the Comité Euro-International du Béton (CEB) and the Fédération International de la Précontrainte (FIP) (CEB-FIP) [17] standard is the most representative achievement.

Tedesco et al. [18] conducted SHPB tests on concrete specimens with a diameter and length of 51 mm, and presented a regression equation to describe the relationship between the DIF and log10 of strain rate. Based on the SHPB test of concrete and mortar, Grote et al. [19] suggested a nonlinear DIF model at a high strain rate with a 250~1700 s^{-1} range. Ngo et al. [20] established a new relationship model between DIF and strain rate. The model considers the effects of various factors on dynamic strength and is suitable for concrete with a strength range of 32–160 MPa. Katayama et al. [21] adopted a quadratic equation of logarithmic strain rate to express the DIF model by introducing it into Drucker–Prager's equation. Hartmann et al. [22] used a power function to describe the relationship between DIF and strain rate.

With the progress of technology and the deepening of research, some scholars found that under high strain rate [23,24], the inertial effect is not eliminated, but dominant [25,26]. Under the axial dynamic load, part of the work of the external load is to provide kinetic energy to the particles so that the particles obtain axial acceleration. Due to the Poisson effect, particles will interact with adjacent particles, which will obtain radial acceleration. The load that provides acceleration to the particles is the inertial effect, which is part of the macro-bearing capacity [27]. Gorham [28] provided a relatively perfect inertia effect correction model based on the energy conservation law. Forrestal [14] also proved the existence of the inertia effect in theory. Guo et al. [29] considered that the radial inertia effect is significant only when the strain rate exceeds 110 s^{-1}. Flores-Johnson et al. [30] believed that the lateral confinement effect of the SHPB specimen is the main reason for the structural effect of all concrete-like materials. Li et al. [31] considered that the lateral inertial confinement of a cylindrical specimen was higher than a cubic specimen at the same strain rate. Zhou et al. [32] expressed that the increase of material strength is due to the inertia effect rather than strain rate effect. A quadratic equation was used to describe the relationship between DIF and the log of the strain rate and quantitatively confirmed by Li et al. [33,34]. Hao et al. [35,36] proposed that the reason for the large discreteness of the experimental results is that the inertial effect is unaccounted for. The quadratic equation was used to express the relationship between DIF and log of the strain rate. Xu et al. [37] presented semi-empirical equations for the concrete material DIF considering the internal configuration effect, by adopting a hyperbolic tangent function. Al-Salloum [38] used a power function to express the DIF model. Lu et al. [27] established a nonlinear dynamic uniaxial strength criterion, called the S criterion, based on understanding the physical mechanisms. Lee et al. [39,40] described pure rate DIF with strain rate and inertial effect with strain acceleration. The sum of the two obtains apparent DIF.

Inertia effect is an important factor causing the inaccuracy of SHPB test results of concrete materials. In order to explore the influence of the inertia effect on the strength of concrete materials under the dynamic load, SHPB tests of concrete and mortar were carried out in this paper. A DIF model considering strain rate effect and inertia effect was established. The dynamic mechanical response characteristics of mortar and concrete were compared and analyzed, which provides a theoretical basis and scientific support for seismic design and safety evaluation of concrete engineering.

2. Research on Strain Rate and Inertia Effect

2.1. Strain Rate Effect Research

Some typical empirical formulas for DIF have been developed based on the SHPB test of mortar and concrete specimens. The formulae were either based on power-law variation or followed the logarithmic trends [38]. One of the most commonly used empirical formulas for DIF was given by the CEB [17]. The DIF of the strain rate-dependent behavior of mortar and concrete can be obtained by the following piecewise function:

$$\mathrm{DIF} = \frac{f_{cd}}{f_{cs}} = \begin{cases} (\dot{\varepsilon}/\dot{\varepsilon}_s)^{1.026\alpha_s} & |\dot{\varepsilon}| \le 30\,\mathrm{s}^{-1} \\ \gamma_s(\dot{\varepsilon}/\dot{\varepsilon}_s)^{1/3} & |\dot{\varepsilon}| > 30\,\mathrm{s}^{-1} \end{cases} \tag{1}$$

where $\dot{\varepsilon}$ is the strain rate, f_{cs} and f_{cd} are the unconfined compressive strength in quasi-static and dynamic loading, respectively, and $\gamma_s = 10^{(6.156\alpha_s-2.0)}$, $\alpha_s = 1/(5+9f_{cs}/f_{c0})$, $f_{c0} = 10\,\text{MPa}$, $\dot{\varepsilon}_s = 30\times10^{-6}\,/s$. Equation (1) shows a nonlinear relationship between the dynamic strength of mortar and concrete and the high strain rate.

Tedesco and Ross [18] conducted a series of SHPB tests where the DIF rapidly increases with the strain rate. A logarithmic function can describe the relationship between DIF and strain rate:

$$\text{DIF} = \begin{cases} 0.00965\log\dot{\varepsilon} + 1.058 \geq 1.0 & \dot{\varepsilon} \leq 63.1\,\text{s}^{-1} \\ 0.758\log\dot{\varepsilon} - 0.289 \leq 2.5 & \dot{\varepsilon} > 63.1\,\text{s}^{-1} \end{cases} \tag{2}$$

Through explosion resistance tests of the ultra-high-strength concrete panel, Ngo et al. [20] also believed that there is a logarithmic relationship between DIF and strain rate at high strain rates with the formula:

$$\text{DIF} = \frac{f_{cd}}{f_{cs}} = \begin{cases} \left(\frac{\dot{\varepsilon}}{\dot{\varepsilon}_s}\right)^{1.026\alpha} & \dot{\varepsilon} \leq \dot{\varepsilon}_1 \\ A_1\ln(\dot{\varepsilon}) - A_2 & \dot{\varepsilon} > \dot{\varepsilon}_1 \end{cases} \tag{3}$$

where $\dot{\varepsilon}_s = 3\times10^{-5}\text{s}^{-1}$, $\alpha = 1/(20+f_{cs}/2)$, $\dot{\varepsilon}_1 = 0.0022f_{cs}^2 - 0.1989f_{cs} + 46.137$, $A_1 = -0.0044f_{cs} + 0.9866$, $A_2 = -0.0128f_{cs} + 2.1396$.

Guo et al. [29] viewed that the CEB-FIP 2010 equation [41] is not suitable for high strength concrete through the SHPB test of concrete with different strength, and proposed the following formula:

$$\text{DIF} = \begin{cases} (\dot{\varepsilon}/\dot{\varepsilon}_s)^{0.014} & \dot{\varepsilon} \leq \dot{\varepsilon}_{TR} \\ A\log_{10}(\dot{\varepsilon}/\dot{\varepsilon}_s) + B & \dot{\varepsilon} > \dot{\varepsilon}_{TR} \end{cases} \tag{4}$$

where $\dot{\varepsilon}_{TR}$ is the transition strain rate, and A and B are constants.

2.2. Inertial Effect Research

Because the linear function cannot accurately describe the relationship between the DIF and the logarithmic strain rate under high a strain rate, some scholars used quadratic or cubic equations to describe it. Grote et al. [19] tested the cement mortar specimens on SHPB with strain rates ranging from 250 to 1700 s^{-1} and gave the following equations:

$$\text{DIF} = \begin{cases} 0.0235\log\dot{\varepsilon} + 1.07 \geq 1.0 & \dot{\varepsilon} \leq 266.0\,\text{s}^{-1} \\ 0.882(\log\dot{\varepsilon})^3 - 4.4(\log\dot{\varepsilon})^2 + 7.22(\log\dot{\varepsilon}) - 2.64 & \dot{\varepsilon} > 266.0\,\text{s}^{-1} \end{cases} \tag{5}$$

Li et al. [34] conducted experimental and numerical studies on mortar samples. Their research results confirmed quantitatively that the apparent dynamic strength enhancement of concrete-like materials in a SHPB test is caused by the lateral inertia confinement instead of the strain rate sensitivity of the tested material. The DIF model was proposed as:

$$\text{DIF} = \begin{cases} 0.03438(\log\dot{\varepsilon} + 3) + 1 & \dot{\varepsilon} \leq 100\,\text{s}^{-1} \\ 1.729(\log\dot{\varepsilon})^2 - 7.1372\log\dot{\varepsilon} + 8.5303 & \dot{\varepsilon} > 100\,\text{s}^{-1} \end{cases} \tag{6}$$

Katayama et al. [21] believed that if the mass is retained, the inertia conservation and the spatial continuity of inertia can be maintained and presented another DIF model as:

$$\text{DIF} = 0.2583(\log\dot{\varepsilon})^2 - 0.05076\log\dot{\varepsilon} + 1.021 \tag{7}$$

Hao et al. [35,36] regarded that the friction at the sample bar interface is an important factor affecting the lateral inertia effect of the specimen under high-speed impact. They proposed an empirical formula to remove the influence of end friction confinement on dynamic strength increment of concrete material as:

$$\mathrm{DIF}_{Mortar} = \begin{cases} 0.0419\log\dot{\varepsilon}+1.2165 & 1\,\mathrm{s}^{-1} < \dot{\varepsilon} \le 10\,\mathrm{s}^{-1} \\ 0.8988(\log\dot{\varepsilon})^2 - 2.8255(\log\dot{\varepsilon}) + 3.4907 & 30\,\mathrm{s}^{-1} < \dot{\varepsilon} \le 1000\,\mathrm{s}^{-1} \end{cases} \quad (8)$$

$$\mathrm{DIF}_{Aggregate} = \begin{cases} 0.0191\log\dot{\varepsilon}+1.2222 & 1\,\mathrm{s}^{-1} < \dot{\varepsilon} \le 220\,\mathrm{s}^{-1} \\ 1.6607(\log\dot{\varepsilon})^2 - 6.9122(\log\dot{\varepsilon}) + 8.346 & 220\,\mathrm{s}^{-1} < \dot{\varepsilon} \le 1000\,\mathrm{s}^{-1} \end{cases} \quad (9)$$

where $\mathrm{DIF}_{Aggregate}$ is the DIF of concretes with aggregates.

Under a high strain rate, the inertia effect cannot be eliminated [23], and it dominates [26,29]. However, there is no further study on the influence of the inertia effect in Equations (5)–(9). Quadratic or cubic equations were used to fit the experimental data to obtain a higher fitting degree, leading to a non-conservative prediction [39]. Lee et al. [39,40] proposed a new concrete DIF that excludes inertia effects by considering the strain acceleration and geometry of the specimens based on SHPB test results, described by the formula:

$$\begin{cases} \mathrm{DIF}_{apparent} = \mathrm{DIF}_{rate} + \Delta f_{inertia} \\ \mathrm{DIF}_{rate} = \left(\frac{\dot{\varepsilon}}{\dot{\varepsilon}_s}\right)^{k_1} \\ \Delta f_{inertia} = k_2\frac{\rho_s d_s^2}{f_{cs}}\ddot{\varepsilon} + k_3\frac{\rho_s l_s^2}{f_{cs}}\ddot{\varepsilon} \end{cases} \quad (10)$$

where $\mathrm{DIF}_{apparent}$, DIF_{rate}, and $\Delta f_{inertia}$ are apparent DIF, pure rate DIF, and strength enhancement caused by inertia effects, respectively. $\ddot{\varepsilon}$, ρ_s, d_s, and l_s denote axial strain acceleration, density, the diameter of the specimen, and the initial specimen length, respectively. k_1, k_2, and k_3 are the material parameters.

Under the high strain rate, the particles in the specimen will obtain axial acceleration, i.e., axial inertial force. In addition, lateral inertial force is also generated due to the influence of Poisson's ratio. The macroscopic resistance of concrete must balance the actual failure force, axial inertia force, and lateral inertia force [27], as shown in Figure 1 and Equation (11).

$$Q = f(\sigma_f) + I(ma, \mu_d) \quad (11)$$

where Q is the macroscopic resistance, $f(\sigma_f)$ is a function of the actual failure stress, $I(ma, \mu_d)$ is a function of the inertial force, m is the quality of the particle, a is the acceleration of the particle, and μ_d is the dynamic Poisson's ratio of the specimen.

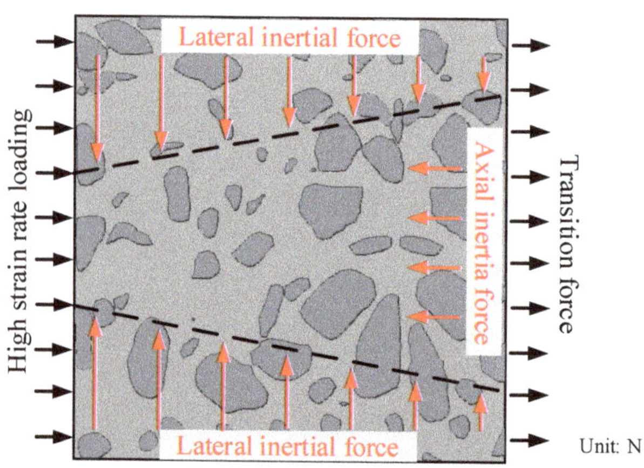

Figure 1. Inertia mechanism of specimen.

Based on the previous research results, we propose a DIF model for high strain rate, as reflected in the following formula:

$$\text{DIF} = K_1 \log_{10}(\dot{\varepsilon}/\varepsilon_s) + K_2\ddot{\varepsilon} + K_3 \tag{12}$$

where K_1, K_2, and K_3 are constants. The model considers the strain rate effect and the inertia effect. In order to verify the correctness of the model, SHPB experiments of mortar and concrete were carried out in this paper.

3. Experimental Research

3.1. Prepare for the Experiment

Ordinary Portland cement (OPC) and potable laboratory tap water were used for the experiments. Conventional crushed stone with particle sizes between 8 and 12 mm and natural river sand with particle sizes between 0.25 and 0.5 mm were employed as coarse and fine aggregates, respectively. The concrete was mixed at the proportion of 0.52:1:1.67:2.47 (water/cement/sand/aggregate) and subsequently set standing for 24 h. The mold of the specimen was removed, and the specimen was placed in a constant temperature (20 °C) and humidity (95%) curing box for 28 days. The specimens were drilled and polished to smooth the end face after curing. The mortar specimens have the same composition and preparation as the cement paste in the concrete. The diameter and height of the mortar and concrete specimens used for the SHPB test were 71 × 71 mm, respectively, as shown in Figure 2.

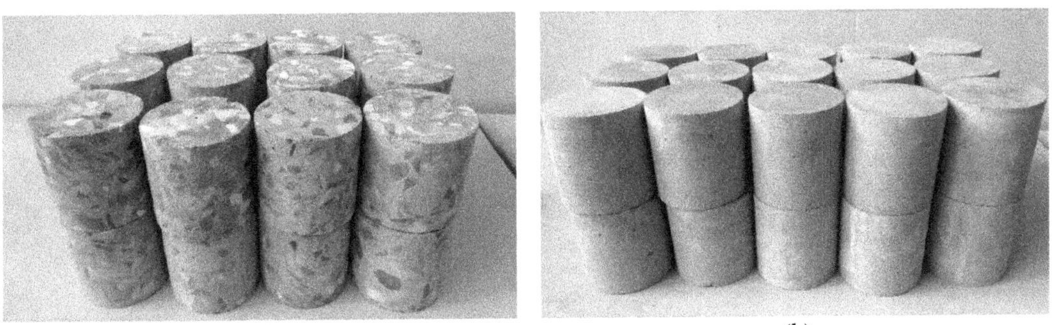

(a) (b)

Figure 2. Specimens for SHPB test: (**a**) Mortar, (**b**) Concrete.

A quasi-static test was conducted before the dynamic load test. An RMT-150B multi-functional full-automatic rigid rock servo testing machine was used for the static load test. The specimen radius and height were 50 and 100 mm, respectively. There were 3 test specimens of mortar and concrete, respectively. The average peak stresses of mortar and concrete specimens were 53.06 and 31.61 MPa, respectively, and their standard deviations were 2.31 and 1.91 MPa, respectively. In order to analyze the dynamic response characteristics of the two kinds of materials, the strain rates of mortar and concrete were extracted in the test. The average strain rates of mortar and concrete specimens were 1.03×10^{-5} and 1.12×10^{-5}/s, respectively, and their standard deviations were 4.71×10^{-8} and 2.16×10^{-7}, respectively. The stress–strain curves of mortar and concrete specimens, whose stress peak value is close to the average value, are shown in Figure 3.

From Figure 3, the compressive strength and elastic modulus of mortar are significantly greater than that of concrete under quasi-static load. This is because the aggregate of the concrete specimen has little effect under low strain rate, while the interfacial transition zone significantly reduces the bearing capacity. The compressive strength of mortar and concrete was 53.07 and 31.23 MPa, respectively. The strain rates of mortar and concrete were 1.03×10^{-5} and 1.12×10^{-5}/s, respectively.

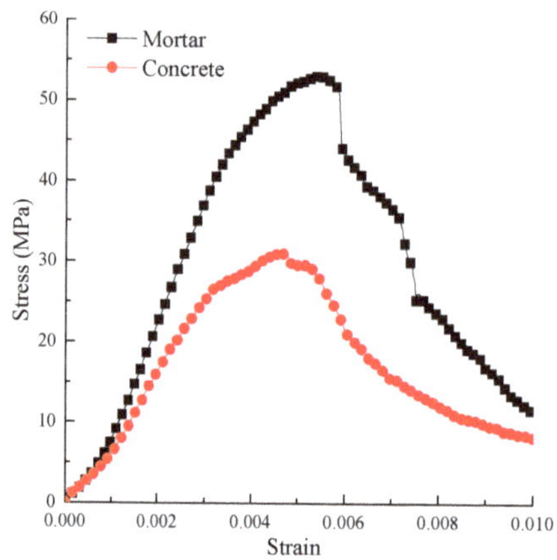

Figure 3. Stress–strain curve of the quasi-static test.

3.2. SHPB Experimental Instrument

Split Hopkinson Pressure Bar (SHPB) test technology is the most important and reliable test method to study the mechanical properties of materials under a high strain rate. The basic working principle of the SHPB test device [1] is that when the striker bar is pushed by air pressure to hit the incident pressure bar, an incident wave is produced in the bar. When the incident wave reaches the end, a portion is reflected to form a reflected wave. Another portion will penetrate the specimen and enter the transmission bar to become a transmission wave. The calculation formulas for strain $\varepsilon_S(t)$, strain rate $\dot{\varepsilon}_S(t)$, and stress $\sigma_S(t)$ of the specimen in the test are as follows:

$$\varepsilon_S(t) = -\frac{C_0}{l_0}\int_0^t [\varepsilon_I(t) - \varepsilon_R(t) - \varepsilon_T(t)]\mathrm{d}t \tag{13}$$

$$\sigma_S(t) = \frac{AE_0}{2A_S}[\varepsilon_I(t) + \varepsilon_R(t) + \varepsilon_T(t)] \tag{14}$$

$$\dot{\varepsilon}_S(t) = -\frac{C_0}{l_0}[\varepsilon_I(t) - \varepsilon_R(t) - \varepsilon_T(t)] \tag{15}$$

where: C_0 is the P-wave velocity of compression bar, l_0 is the length of the specimen, A, A_S are the cross-sectional areas of compression bar and specimen, respectively, E_0 is the elastic modulus of the compression bar, and $\varepsilon_I(t)$, $\varepsilon_R(t)$, and $\varepsilon_T(t)$ are the strain signals of the incident, reflected waves, and transmission waves, respectively.

To obtain accurate and reliable data, a tapered incident bar with a diameter of 74 mm was used in the SHPB system as shown in Figure 4. The steel bars had a Young's modulus E_0 = 200 GPa and Poisson's ratio υ = 0.3.

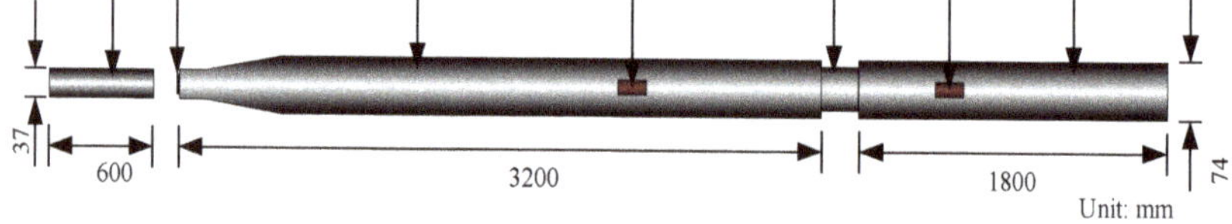

Figure 4. Schematic diagram of SHPB test.

There is severe waveform dispersion in the large diameter SHPB test. To prolong the rise time of the incident wave and filter its high-frequency oscillation, a pulse shaper was fixed on the end face of the incident bar in contact with the bullet [42,43]. The rectangular impact pulse was transformed into a triangular pulse to lengthen its rising edge by the pulse shaper. The wave dispersion can also be reduced by placing the strain gauge on the transmission bar as close to the specimen as possible. Vaseline was applied on both end faces of the sample to reduce the influence of the radial inertia effect by reducing friction. The brass and rubber shapers were tested with no specimen to examine the effect of different pulse shapers. Both shapers had a 20 mm diameter.

The SHPB test was carried out with 2 mm thick brass shaper and rubber shaper. The test results show that the shaping effect of rubber shaper is better. In order to obtain a better shaping effect, SHPB tests were carried out on rubber shapers with thicknesses of 1, 2, and 3 mm, respectively. The impact air pressure was 0.3 MPa, and the waveform test with no specimen is shown in Figure 5.

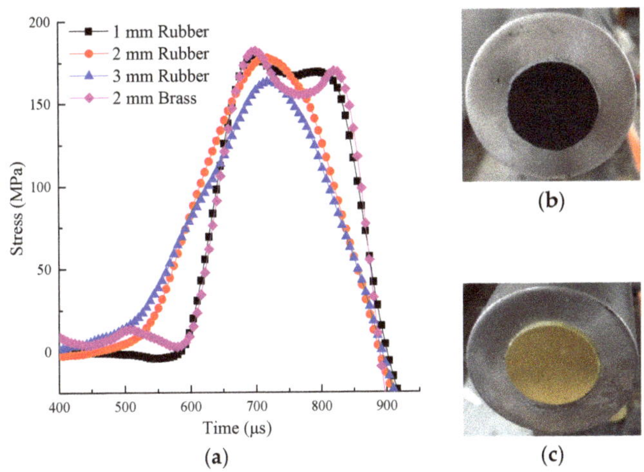

Figure 5. The waveforms of incident waves with different pulse shapers: (**a**) Stress time history curve, (**b**) Rubber sharper, (**c**) Brass sharper.

Figure 5 depicts the stress wave curves monitored in the incident bar. The positive and negative values of the curve indicate the direction of stress. It shows that the waveforms obtained by the brass shaper and 1 mm thick rubber shaper are relatively similar and both rectangular. The waveforms obtained by 2 and 3 mm thick rubber shapers are triangular waveforms, and the rise time is also long. When the incident waveform is a half-sine wave, the constant strain rate loading of the specimen is realized, and the inertia effect is greatly reduced [44]. At 0.25 MPa impact pressure, the 1 mm thickness rubber shaper was selected, and the 2mm thickness rubber shaper was selected in other cases.

3.3. SHPB Experimental

The striker bar can obtain different initial velocities by different impact air pressures. The selected five groups of impact pressures were 0.25, 0.3, 0.4, 0.5, and 0.6 MPa. When the impact air pressure is 0.25 MPa, a 1 mm thickness rubber shaper was selected because the current signal cannot be collected by a 2 mm thickness rubber shaper. A 2 mm thick rubber shaper was used in other cases. Each group was subjected to three impact tests. The data with a large error were removed. Then, a typical stress–strain curve was selected from each group to plot, as shown in Figure 6.

Both graphs have the same scale on the x and y axes for comparison in Figure 6. It shows that the strain rate increases the peak stress of mortar and concrete samples. It indicates that mortar and concrete specimens have a noticeable strain rate effect. The corresponding strain rate time history curve is shown in Figure 6.

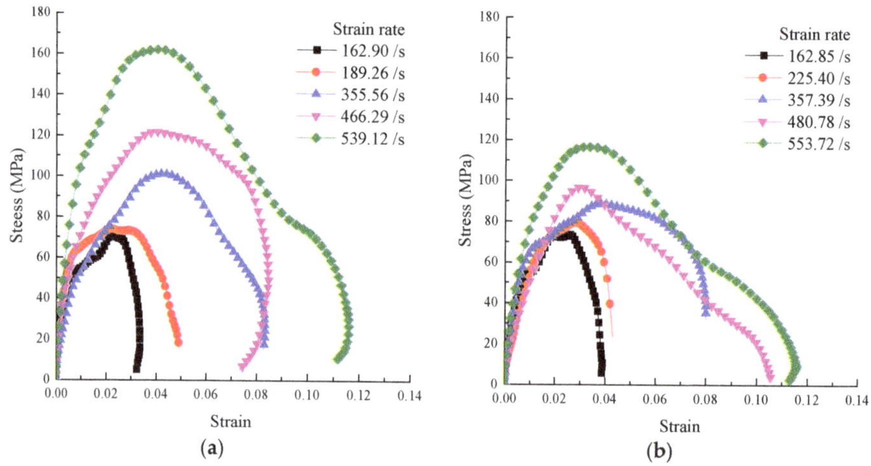

Figure 6. Stress–strain curve: (**a**) Mortar, (**b**) Concrete.

Figure 7 shows that high strain rates were obtained for the specimens by the SHPB test with a large-diameter bar. The maximum mortar and concrete strain reached 539.12 and 553.72/s, respectively. When the impact air pressure was 0.25 MPa, a 1 mm thickness rubber shaper was selected because a 2 mm thickness rubber shaper cannot collect the current signal. When the impact pressure was 0.25 MPa, the thickness of the rubber shaper was 1 mm, and when the slope of the rising edge of the strain rate time history curve was greater than at 0.30 MPa impact pressure, the thickness was 2 mm. Under a high strain rate, the duration of the constant strain rate is short, and the inertia effect is dominant.

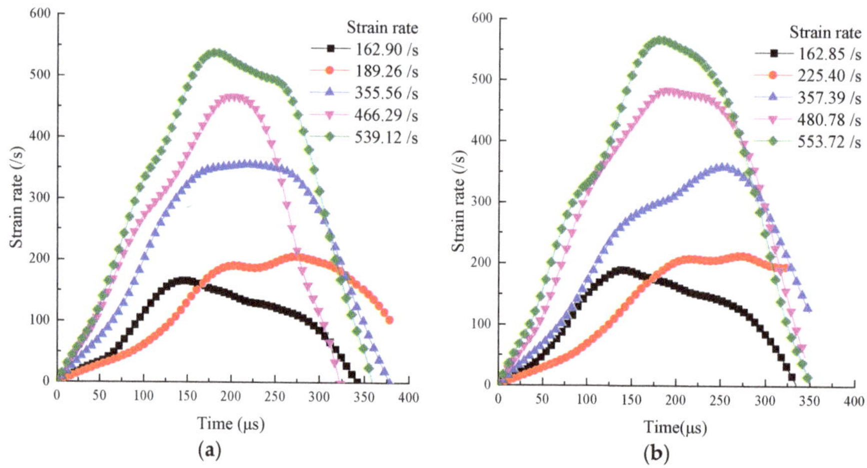

Figure 7. Strain rate time history curve: (**a**) Mortar, (**b**) Concrete.

When comparing the impact pressure of 0.25 and 0.30 MPa in Figures 4–6, the rubber shaper thickness was 1 mm and 2 mm, respectively. Although the impact pressure of 0.25 MPa is less than that of 0.30 MPa, the slope of the rising edge of the waveform obtained by 1 mm thick shaper is greater than that obtained by 2 mm. In the corresponding strain rate time history curve, the slope of the rising section with a 1 mm thick shaper is also greater than that obtained with a 2 mm. The rising slope of the strain rate time history curve is defined as the strain acceleration [40]. The strain acceleration is directly proportional to the inertial effect [27]. It shows that different strain accelerations can be obtained by the thickness of the pulse shaper. The peak stress of the two cases is close in Figure 5, indicating that the inertia effect increases the dynamic strength of the sample. When the thickness of rubber shaper is the same, the peak stress and strain acceleration increase significantly with the impact pressure. It can also be seen from Figure 6 that under high strain rate, the duration of the constant strain rate is short, and the inertia effect is dominant.

3.4. Dynamic Uniaxial Strength Criterion

According to the analysis in Section 2.1, it is common to describe the relationship between DIF and strain rate by logarithmic function under the high strain rate. Therefore, the logarithmic function was adopted to fit the relationship between DIF and strain rate. The fitting results are shown in Figure 8.

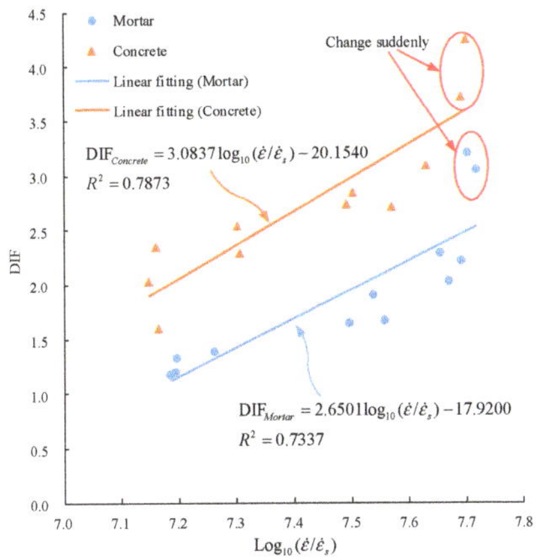

Figure 8. Variation of DIF with logarithm of strain rate.

In Figure 8, the relationship between DIF and strain rate of mortar and concrete is expressed by a logarithmic function as:

$$\begin{cases} \text{DIF}_{Mortar} = 2.6501\log_{10}(\dot{\varepsilon}/\dot{\varepsilon}_s) - 17.9200,\ R^2 = 0.7337 \\ \text{DIF}_{Concrete} = 3.0837\log_{10}(\dot{\varepsilon}/\dot{\varepsilon}_s) - 20.1540,\ R^2 = 0.7873 \end{cases} \tag{16}$$

where DIF_{Mortar}, $\text{DIF}_{Concrete}$ are dynamic increase factors of mortar and concrete, respectively. The logarithmic function can express the trend relationship between DIF and strain rate. The strain rate effect of concrete is more sensitive than that of mortar by comparing the coefficient of $\log_{10}(\dot{\varepsilon}/\dot{\varepsilon}_s)$. When $\log_{10}(\dot{\varepsilon}/\dot{\varepsilon}_s) > 7.7$, the DIF of mortar and concrete has a noticeable sudden change (see the mark in Figure 8). Therefore, the logarithmic strain rate cannot accurately describe DIF under a high strain rate.

Next, the inertia effect was considered. SHPB test data were fitted by Equation (12). The abscissa was set as $K_1\log_{10}(\dot{\varepsilon}/\varepsilon_s) + K_2\ddot{\varepsilon}$ to express the fitting relationship, and the fitting results are shown in Figure 9.

In Figure 9, the DIF fitting function of mortar and concrete considering strain rate and strain acceleration is as follows:

$$\begin{cases} \text{DIF}_{Mortar} = 1.0260\log_{10}(\dot{\varepsilon}/\varepsilon_s) + 0.6501\ddot{\varepsilon} - 7.2540,\ R^2 = 0.8606 \\ \text{DIF}_{Concrete} = 1.5410\log_{10}(\dot{\varepsilon}/\varepsilon_s) + 0.4580\ddot{\varepsilon} - 9.6820,\ R^2 = 0.8477 \end{cases} \tag{17}$$

When compared with Equation (16), the fitting degree is improved. It indicates that Equation (12) is feasible to fit the DIF of mortar and concrete under a high strain rate. When comparing the $\log_{10}(\dot{\varepsilon}/\varepsilon_s)$ coefficient in Equation (17), the strain rate effect of concrete is more sensitive than that of mortar, which is consistent with the above analysis. When compared with the $\ddot{\varepsilon}$ coefficient, the inertia effect of mortar is more sensitive than that of concrete. Therefore, the strain rate effect of the material is more sensitive, but the inertia effect is not necessarily more sensitive.

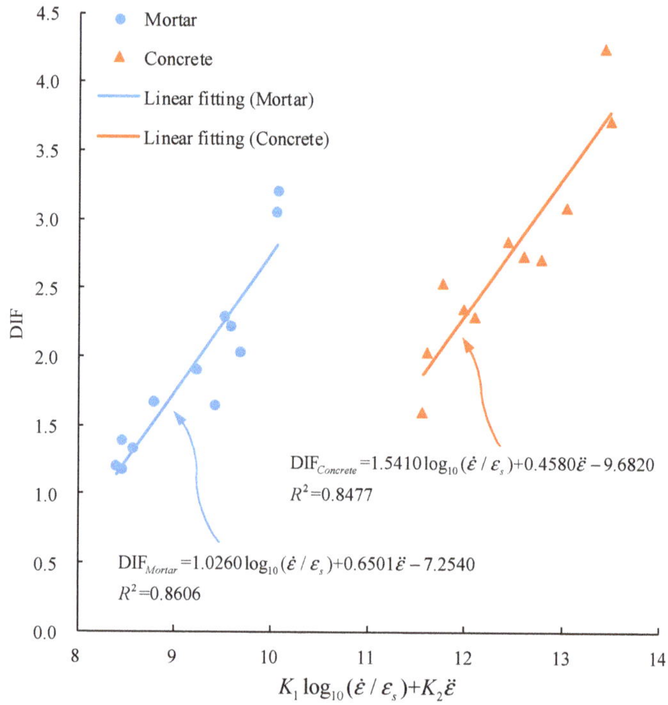

Figure 9. Variation of DIF with the logarithm of strain rate and strain acceleration.

4. Research on Energy Utilization and Fragmentation Morphology

4.1. Energy Utilization Research

The energy utilization was studied to compare the effect of strain rate on the energy utilization of mortar and concrete in large-diameter SHPB tests. The energy calculation formula of stress waves is:

$$\begin{cases} W_I(t) = E_0 C_0 A_S \int_0^t \varepsilon_I^2(t)\mathrm{d}t \\ W_T(t) = E_0 C_0 A_S \int_0^t \varepsilon_T^2(t)\mathrm{d}t \end{cases} \tag{18}$$

where W_I is incident energy and W_T is the transmission energy.

The calculation formula of energy utilization η is:

$$\eta = \frac{W_T}{W_I} \cdot 100\% \tag{19}$$

Figure 10 represents the energy utilization of mortar and concrete. The energy utilization of mortar and concrete increases with the strain rate, but the increase of mortar is faster. Under the same strain rate, the energy utilization of mortar is higher than that of concrete. Under the impact compression of large diameter SHPB, the energy utilization of mortar and concrete specimens is relatively low, and the highest is only 4.99%.

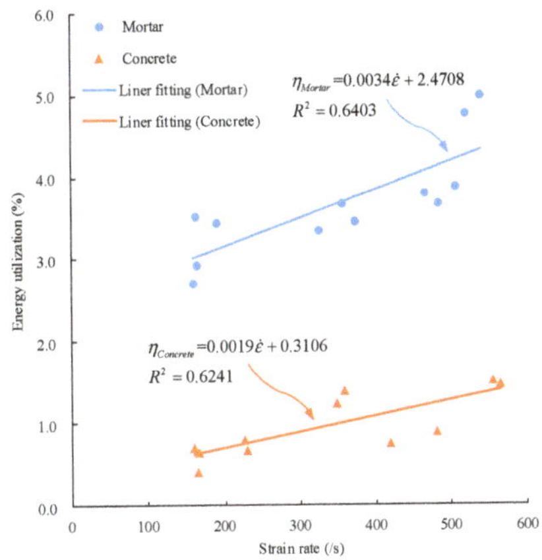

Figure 10. Relation between energy utilization and strain rate.

4.2. Fragmentation Morphology Research

The fracture morphology is an important aspect of evaluating the impact potentiality of concrete materials [45,46]. The pore sizes of classifying screens selected in this test were 2.0, 5.0, 10.0, 20, and 40 mm. The broken specimens were sieved into six groups with particle size ranges of 0.0–2.0, 2.0–5.0, 5.0–10.0, 10.0–20.0, 20.0–40.0, and 40–71 mm (71 mm is the diameter of the sample before fragmentation), respectively. The weighing instrument was a high-precision electronic scale with a measuring range of 1 kg and an accuracy of 0.1 g.

After impact, the broken specimens were collected, classified, and screened individually. First, the classifying screens were stacked from high to low according to the pore size. Then, the broken specimens were placed on the sieve with the largest mesh size on the uppermost layer, so that the specimens with different fragmentation degrees could be separated according to size. After screening, the fragmentation on each sieve was placed on the electronic scale for weighing, and the measurement results were recorded one by one. The screened fragmentations of mortar and concrete are shown in Figure 11.

Figure 11. Fragmentation morphologies of mortar and concrete: (**a**) Mortar (466.29/s), (**b**) Concrete (480.78/s).

Figure 11 shows the final fragmentation morphologies of mortar and concrete under the similar strain rate. It can be seen that the fragmentation morphologies of mortar and concrete specimens under impact load are different. The mortar sample was cracked along the axial direction. Although the mortar specimen was penetrated by cracks, the strip fragment still had high strength in the loading direction. The strips after impact splitting can still bear the impact load on the bar as a whole. However, the fragmentation degree of concrete specimen was very large, and the cracks mostly occurred in the interface transition zone (ITZ) between mortar and aggregate. From the fact that the strength of concrete was lower than that of mortar, the aggregate plays a small role in dynamic loading, so the ITZ reduces the strength of concrete.

To quantify the fragmentation degree of the specimen, the average fragment size of the broken specimen was adopted. The calculation formula is as follows:

$$\overline{X} = \frac{\sum n_i X_i}{\sum n_i} \tag{20}$$

where $\overline{X}$ is the average fragment size of the broken specimen in mm, X_i is the average size of specimen fragmentation retained on the classifying screen of class i, in mm, and n_i is the proportion of fragment mass with an average size X_i in %.

The median values of the average size of the fragmentation on each classifying screen were taken according to the sieve diameter, which are 1, 3.5, 7.5, 15, 30, and 55.5 mm. The relationship between the average fragment size of mortar and concrete and strain rate is shown in Figure 12.

In Figure 12, $\overline{X}_{Mortar}$ and $\overline{X}_{Concrete}$ represent the average fragment size of mortar and concrete, respectively. With the increase of strain rate, the average fragment size of mortar and concrete decreases, but that of the concrete decreases faster. Under the same strain rate, the average fragment size of mortar is larger than that of concrete. The average fragment size of mortar is 42.03–52.36 mm, and that of concrete is 20.89–40.21 mm. Therefore, mortar is better than concrete in the storage performance of elastic strain energy, indicating that the impact failure ability of mortar is stronger than that of concrete.

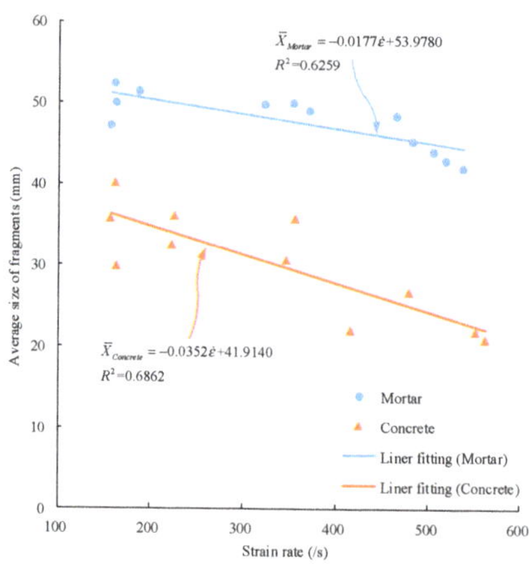

Figure 12. Relation between average block size and strain rate of mortar and concrete.

5. Discussion

Under high strain load, axial acceleration will be obtained by the grains in the specimen. The lateral acceleration will be obtained due to the Poisson effect. In the study of inertial effect, the lateral inertial force is often considered, while the axial inertial force is often ignored. Although the axial acceleration is consistent with the bearing capacity direction of the specimen, it is caused by the uneven stress of the sample. It is closely related to the slope of the rising edge of the loading waveform, and independent from the constant strain rate of the specimen. The axial acceleration is not a part of the real strength of the specimen. Therefore, the axial inertial force should also be taken into account in the study of inertial effect. The lateral strain or dynamic Poisson's ratio should be taken as the monitored object during the SHPB test.

The inertial effect was studied by theoretical analysis or numerical simulation, but the data extraction of strain acceleration was often ignored. In the SHPB test, it was found that the rising edge of the incident wave can be changed by rubber shapers with different thicknesses, and the inertia effect can be changed accordingly. Therefore, the control variable method can be used in the indoor test, that is, the strain rate and strain acceleration can be controlled respectively to study the inertial effect. The test can be repeated in the laboratory. One of the highlights of this paper is in establishing a DIF model considering inertial effect by considering strain rate and strain acceleration. In the future research, digital image correlation (DIC), CT scanning, and other technologies could be used to retrieve the strain, crack, and damage of the specimen under inertial effect [47,48]. It is helpful to deeply understand the inertial effect, establish an accurate dynamic model, and provide a theoretical basis for rock dynamics theory, disaster warnings, and safety assessments.

6. Conclusions

In this paper, the large-diameter SHPB tests on concrete and mortar were performed, the inertia effect was studied. The conclusions are as follows:

(1) The macroscopic resistance of concrete material is composed of the actual failure force, axial inertia force, and lateral inertia force. The dynamic growth factor (DIF) model was established. The DIF model comprehensively considers the influence of strain rate on the actual dynamic strength of concrete materials and the influence of strain acceleration on inertial effect.
(2) With the increase of bullet impact velocity, the influence of inertia effect becomes greater and greater. The strain rate effect of concrete is more sensitive than that of mortar, but the inertia effect of mortar is more sensitive than that of concrete.
(3) With the increase of strain rate, the energy utilization of mortar and concrete increases, while the average fragment size decreases. Under the same strain rate, the energy utilization rate, average fragment size, and impact potentiality of mortar are higher than that of concrete.

Author Contributions: Conceptualization, B.S., R.C., Z.Z. and N.W.; methodology, B.S. and Z.S.; software, B.S. and Z.S.; validation, B.S. and R.C.; formal Analysis, B.S., R.C. and N.W.; investigation, N.W. and Z.S.; resources, B.S., Y.P. and Z.Z.; data curation, B.S. and N.W.; writing—original draft preparation, B.S.; writing—review and editing, B.S., R.C., Z.Z., Y.P. and N.W.; visualization, Z.S.; supervision, R.C., Z.Z., Y.P. and N.W.; project administration, R.C.; funding acquisition, R.C., Z.Z., Y.P. and N.W. All authors have read and agreed to the published version of the manuscript.

Funding: This study was supported by the National Natural Science Foundation of China (Grant No. 52108448 and 51878249), and the Science and Technology Planning Project of Shenzhen Municipality (Grant No. JCYJ20190806144603586).

Institutional Review Board Statement: Not applicable.

Informed Consent Statement: Not applicable.

Data Availability Statement: Not applicable.

Conflicts of Interest: The authors declare no conflict of interest. The funders had no role in the design of the study; in the collection, analyses, or interpretation of data; in the writing of the manuscript, and in the decision to publish the results.

References

1. Kolsky, H. An Investigation of the Mechanical Properties of Materials at very High Rates of Loading. *Proc. Phys. Soc. Sect. B* **1949**, *62*, 676–700. [CrossRef]
2. Li, X.; Lok, T.; Zhao, J.; Zhao, P. Oscillation elimination in the Hopkinson bar apparatus and resultant complete dynamic stress–strain curves for rocks. *Int. J. Rock Mech. Min. Sci.* **2000**, *37*, 1055–1060. [CrossRef]
3. Yang, G.; Chen, X.; Xuan, W.; Chen, Y. Dynamic compressive and splitting tensile properties of concrete containing recycled tyre rubber under high strain rates. *Sādhanā* **2018**, *43*, 178. [CrossRef]

4. Wu, N.; Fu, J.; Zhu, Z.; Sun, B. Experimental study on the dynamic behavior of the Brazilian disc sample of rock material. *Int. J. Rock Mech. Min. Sci.* **2020**, *130*, 104326. [CrossRef]
5. Wu, N.; Zhu, Z.; Zhang, C.; Luo, Z. Dynamic Behavior of Rock Joint under Different Impact Loads. *KSCE J. Civ. Eng.* **2019**, *23*, 541–548. [CrossRef]
6. Sun, B.; Ping, Y.; Zhu, Z.; Jiang, Z.; Wu, N. Experimental Study on the Dynamic Mechanical Properties of Large-Diameter Mortar and Concrete Subjected to Cyclic Impact. *Shock Vib.* **2020**, *2020*, 8861197. [CrossRef]
7. Jankowiak, T.; Rusinek, A.; Voyiadjis, G.Z. Modeling and Design of SHPB to Characterize Brittle Materials under Compression for High Strain Rates. *Materials* **2020**, *13*, 2191. [CrossRef]
8. Merle, R.; Zhao, H. On the errors associated with the use of large diameter SHPB, correction for radially non-uniform distribution of stress and particle velocity in SHPB testing. *Int. J. Impact Eng.* **2006**, *32*, 1964–1980. [CrossRef]
9. Malvern, L.E.; Ross, C.A. *Dynamic Response of Concrete and Concrete Structures*; Deptment of Engineering Sciences, Florida University: Gainesville, FL, USA, 1986.
10. Huang, R.; Li, S.; Meng, L.; Jiang, D.; Li, P. Coupled Effect of Temperature and Strain Rate on Mechanical Properties of Steel Fiber-Reinforced Concrete. *Int. J. Concr. Struct. Mater.* **2020**, *14*, 48. [CrossRef]
11. Li, S.H.; Zhu, W.C.; Niu, L.L.; Yu, M.; Chen, C.F. Dynamic Characteristics of Green Sandstone Subjected to Repetitive Impact Loading: Phenomena and Mechanisms. *Rock Mech. Rock Eng.* **2018**, *51*, 1921–1936. [CrossRef]
12. Zhang, X.; Yang, Z.-J.; Huang, Y.-J.; Wang, Z.-Y.; Chen, X.-W. Micro CT Image-based Simulations of Concrete under High Strain Rate Impact using a Continuum-Discrete Coupled Model. *Int. J. Impact Eng.* **2021**, *149*, 103775. [CrossRef]
13. Zhou, J.-K.; Ge, L.-M. Effect of strain rate and water-to-cement ratio on compressive mechanical behavior of cement mortar. *J. Cent. South Univ.* **2015**, *22*, 1087–1095. [CrossRef]
14. Forrestal, M.; Wright, T.; Chen, W. The effect of radial inertia on brittle samples during the split Hopkinson pressure bar test. *Int. J. Impact Eng.* **2007**, *34*, 405–411. [CrossRef]
15. Huang, X.; Kong, X.; Chen, Z.; Fang, Q. A computational constitutive model for rock in hydrocode. *Int. J. Impact Eng.* **2020**, *145*, 103687. [CrossRef]
16. Hao, Y.; Hao, H.; Li, Z. Influence of end friction confinement on impact tests of concrete material at high strain rate. *Int. J. Impact Eng.* **2013**, *60*, 82–106. [CrossRef]
17. Comite Euro-International du Beton. *CEB-FIP Model Code 1990[S]*; Redwood Books: Trowbridge, UK, 1993.
18. Tedesco, J.W.; Ross, C.A. Strain-Rate Dependent Constitutive Equations for Concrete. *ASME J. Press. Vessel Technol.* **1998**, *120*, 398–405. [CrossRef]
19. Grote, D.L.; Park, S.W.; Zhou, M. Dynamic behavior of concrete at high strain rates and pressures: I. experimental characterization. *Int. J. Impact Eng.* **2001**, *25*, 869–886. [CrossRef]
20. Ngo, T.; Mendis, P.; Krauthammer, T. Behavior of Ultrahigh-Strength Prestressed Concrete Panels Subjected to Blast Loading. *J. Struct. Eng.* **2007**, *133*, 1582–1590. [CrossRef]
21. Katayama, M.; Itoh, M.; Tamura, S.; Beppu, M.; Ohno, T. Numerical analysis method for the RC and geological structures subjected to extreme loading by energetic materials. *Int. J. Impact Eng.* **2007**, *34*, 1546–1561. [CrossRef]
22. Hartmann, T.; Pietzsch, A.; Gebbeken, N. A Hydrocode Material Model for Concrete. *Int. J. Prot. Struct.* **2010**, *1*, 443–468. [CrossRef]
23. Xing, H.; Zhang, Q.; Ruan, D.; Dehkhoda, S.; Lu, G.; Zhao, J. Full-field measurement and fracture characterisations of rocks under dynamic loads using high-speed three-dimensional digital image correlation. *Int. J. Impact Eng.* **2018**, *113*, 61–72. [CrossRef]
24. Sun, X.; Wang, H.; Cheng, X.; Sheng, Y. Effect of pore liquid viscosity on the dynamic compressive properties of concrete. *Constr. Build. Mater.* **2020**, *231*, 117143. [CrossRef]
25. Xie, Y.-J.; Fu, Q.; Zheng, K.-R.; Yuan, Q.; Song, H. Dynamic mechanical properties of cement and asphalt mortar based on SHPB test. *Constr. Build. Mater.* **2014**, *70*, 217–225. [CrossRef]
26. Rossi, P.; Toutlemonde, F. Effect of loading rate on the tensile behaviour of concrete: Description of the physical mechanisms. *Mater. Struct.* **1996**, *29*, 116–118. [CrossRef]
27. Lu, D.; Wang, G.; Du, X.; Wang, Y. A nonlinear dynamic uniaxial strength criterion that considers the ultimate dynamic strength of concrete. *Int. J. Impact Eng.* **2017**, *103*, 124–137. [CrossRef]
28. Gorham, D.A. Specimen inertia in high strain-rate compression. *J. Phys. D Appl. Phys.* **1989**, *22*, 1888–1893. [CrossRef]
29. Guo, Y.; Gao, G.; Jing, L.; Shim, V. Response of high-strength concrete to dynamic compressive loading. *Int. J. Impact Eng.* **2017**, *108*, 114–135. [CrossRef]
30. Flores-Johnson, E.A.; Li, Q.M. Structural effects on compressive strength enhancement of concrete-like materials in a split Hopkinson pressure bar test. *Int. J. Impact Eng.* **2017**, *109*, 408–418. [CrossRef]
31. Li, M.; Hao, H.; Cui, J.; Hao, Y.-F. Numerical investigation of the failure mechanism of cubic concrete specimens in SHPB tests. *Def. Technol.* **2022**, *18*, 1–11. [CrossRef]
32. Zhou, X.; Hao, H. Modelling of compressive behaviour of concrete-like materials at high strain rate. *Int. J. Solids Struct.* **2008**, *45*, 4648–4661. [CrossRef]
33. Li, Q.; Lu, Y.; Meng, H. Further investigation on the dynamic compressive strength enhancement of concrete-like materials based on split Hopkinson pressure bar tests. Part II: Numerical simulations. *Int. J. Impact Eng.* **2009**, *36*, 1335–1345. [CrossRef]

34. Li, Q.M.; Meng, H. About the dynamic strength enhancement of concrete-like materials in a split Hopkinson pressure bar test. *Int. J. Solids Struct.* **2003**, *40*, 343–360. [CrossRef]
35. Hao, Y.F.; Hao, H. Numerical Evaluation of the Influence of Aggregates on Concrete Compressive Strength at High Strain Rate. *Int. J. Prot. Struct.* **2011**, *2*, 177–206. [CrossRef]
36. Hao, Y.F.; Hao, H.; Jiang, G.P.; Zhou, Y. Experimental confirmation of some factors influencing dynamic concrete compressive strengths in high-speed impact tests. *Cem. Concr. Res.* **2013**, *52*, 63–70. [CrossRef]
37. Xu, H.; Wen, H. Semi-empirical equations for the dynamic strength enhancement of concrete-like materials. *Int. J. Impact Eng.* **2013**, *60*, 76–81. [CrossRef]
38. Al-Salloum, Y.; Almusallam, T.; Ibrahim, S.; Abbas, H.; Alsayed, S. Rate dependent behavior and modeling of concrete based on SHPB experiments. *Cem. Concr. Compos.* **2015**, *55*, 34–44. [CrossRef]
39. Kim, K.-M.; Lee, S.; Cho, J.-Y. Effect of maximum coarse aggregate size on dynamic compressive strength of high-strength concrete. *Int. J. Impact Eng.* **2019**, *125*, 107–116. [CrossRef]
40. Lee, S.; Kim, K.-M.; Park, J.; Cho, J.-Y. Pure rate effect on the concrete compressive strength in the split Hopkinson pressure bar test. *Int. J. Impact Eng.* **2018**, *113*, 191–202. [CrossRef]
41. CEB; FIP. *FIB Model Code 2010*; Wiley: Hoboken, NJ, USA, 2011.
42. Zhang, H.; Liu, Y.; Sun, H.; Wu, S. Transient dynamic behavior of polypropylene fiber reinforced mortar under compressive impact loading. *Constr. Build. Mater.* **2016**, *111*, 30–42. [CrossRef]
43. Luo, G.; Wu, C.; Xu, K.; Liu, L.; Chen, W. Development of dynamic constitutive model of epoxy resin considering temperature and strain rate effects using experimental methods. *Mech. Mater.* **2021**, *159*, 103887. [CrossRef]
44. Jiang, T.; Xue, P.; Butt, H. Pulse shaper design for dynamic testing of viscoelastic materials using polymeric SHPB. *Int. J. Impact Eng.* **2015**, *79*, 45–52. [CrossRef]
45. Gong, F.; Wang, Y.; Wang, Z.; Pan, J.; Luo, S. A new criterion of coal burst proneness based on the residual elastic energy index. *Int. J. Min. Sci. Technol.* **2021**, *31*, 553–563. [CrossRef]
46. Gong, F.-Q.; Wang, Y.-L.; Luo, S. Rockburst proneness criteria for rock materials: Review and new insights. *J. Central South Univ.* **2020**, *27*, 2793–2821. [CrossRef]
47. Kucewicz, M.; Baranowski, P.; Małachowski, J. Dolomite fracture modeling using the Johnson-Holmquist concrete material model: Parameter determination and validation. *J. Rock Mech. Geotech.* **2021**, *13*, 335–350. [CrossRef]
48. Baranowski, P.; Kucewicz, M.; Gieleta, R.; Stankiewicz, M.; Konarzewski, M.; Bogusz, P.; Pytlik, M.; Małachowski, J. Fracture and fragmentation of dolomite rock using the JH-2 constitutive model: Parameter determination, experiments and simulations. *Int. J. Impact Eng.* **2020**, *140*, 103543. [CrossRef]

materials

Article

Experimental Investigation of Shear Keys for Adjacent Precast Concrete Box Beam Bridges

Xiaojing Ni [1,2], Ahehehinnou Ougbe Anselme [1,3], Guannan Wang [1,3], Yuan Xing [2] and Rongqiao Xu [1,3,*]

1 Department of Civil Engineering, Zhejiang University, Hangzhou 310058, China; nxj54837@126.com (X.N.); aheanselme@hotmail.com (A.O.A.); guannanwang@zju.edu.cn (G.W.)
2 Huahui Engineering Design Group Co., Ltd., Shaoxing 312000, China; xingyuan1393@163.com
3 Center for Balance Architecture, Zhejiang University, Hangzhou 310007, China
* Correspondence: xurongqiao@zju.edu.cn; Tel.: +86-571-88206843

Abstract: Longitudinal cracking in shear keys is one of the most frequently recurring problems in the adjacent precast concrete box beam bridges. The relative displacement across the shear key (RDSK) under loads has been used as a direct indicator for shear key cracking. Therefore, accurately simulating the interface between the shear key and beam or providing the correct relationship between shear transfer and RDSK is key to evaluating the damage of the shear key. In this study, the shear transfer properties of four types of composite specimens were studied by static displacement-controlled bi-shear (SDS), cyclic force-controlled bi-shear (CFS), and cyclic displacement-controlled bi-shear (CDS) tests. Two finite element models (FEMs) were established to calibrate and validate the interfacial material parameters. The results showed that adding reinforcement bars over the joints that connect the block and the overlay could improve the bearing capacity of the shear key. Formulae were proposed for the relation between shear force transfer and RDSK in engineering applications. The values of the interfacial material parameters used in the traction–separation model to simulate the interface between the shear key and beam were recommended.

Keywords: bridge; adjacent box beam; shear key; shear experiments; overlay

Citation: Ni, X.; Anselme, A.O.; Wang, G.; Xing, Y.; Xu, R. Experimental Investigation of Shear Keys for Adjacent Precast Concrete Box Beam Bridges. *Materials* **2022**, *15*, 1459. https://doi.org/10.3390/ma15041459

Academic Editor: Dario De Domenico

Received: 29 January 2022
Accepted: 11 February 2022
Published: 16 February 2022

1. Introduction

Adjacent precast concrete box beam bridges are widely used in short- to medium-span bridges. However, one of the most significant issues for this type of bridge is the longitudinal cracking of the shear keys. Generally, it is believed that cracked shear keys compromise the load transfer between beams [1,2]. In extreme cases, the load on a single beam exceeds its designed allowable load, leading to accidents [3]. However, field observation also found that the load transfer between beams could still be maintained for partially cracked shear keys [4–6]. Therefore, it is necessary to reasonably evaluate the damage and load transfer capacity of shear keys to predict the remaining service life of bridges and select appropriate maintenance and reinforcement strategies.

Some researchers conducted destructive tests on structures to evaluate the load transfer performance of cracked shear keys. Wang et al. [7] carried out a static load test on a structure composed of six beams connected by concrete shear keys. They found that at a load level of 70 kN, two shear keys cracked with relative displacement across the shear key (RDSK) of approximately 0.02 and 0.04 mm, respectively. As the load increased, the crack in the shear keys propagated and eventually failed at a load level of 140 kN, twice the cracking load. Yuan et al. [8,9] conducted four tests on two-beam structures connected by transverse post-tensioning (PT) and partially or fully cracked shear keys, which were cast with nonshrink grout. Over millions of cycles, the load levels increased from 80 to 400 kN, and the PT force dropped from 445 to 0 kN. The results showed that when the transverse PT force decreases from 445 to 45 kN, the load can still be transferred effectively, and the RDSK remains stable. Miller et al. [10] carried out three cyclic loading tests on four-beam structures connected by transverse tie rods and shear keys, cast with nonshrink grout. In the first two tests, there were initial cracks at shear keys in the middle caused by temperature, and in the third test, there were initial cracks near the beam end. During the cyclic loading, the cracks in the first two tests propagated, while the cracks in the third test did not. They found that the load was effectively transferred, and the load distribution changed by no more than 1% during all three tests. However, according to the test result of Leng et al. [11], it may be due to the position of the load and crack. Leng et al. tested an eight-beam structure connected only by concrete shear keys. They set different crack lengths on the first and the fourth shear keys to assess the influence of crack length and transverse position on load distribution. They found that the crack at the first shear key had a significant impact on the load distribution, but that at the fourth did not. These destructive tests indicate that the cracking of shear keys does not mean load transfer failure, and the ultimate bearing capacity may be much larger than the cracking load. However, destructive tests are unsuitable for bridges in service to evaluate residual capacity; the finite element method is more appropriate.

Since RDSK has been used as a direct indicator for shear key cracking [4,7–9,12–14], it is crucial to accurately simulate the interface of the beam and shear key in finite element models (FEMs) using solid elements. Three commonly used methods are the full bond [15–17], friction [13,18–21], and traction–separation model [13,21–23]. Full bond is suitable for the interface of concrete and grout materials with strong bonding ability, such as epoxy, MgNH4PO4, and UHPC, but not for commonly used nonshrink grout and concrete. Shear and flexural tests showed that the former is more prone to cracks in the concrete, while the latter is more prone to cracks in the interface [8,9,15,17,23–29]. Friction applies when the shear key has already cracked at the interface. The traction–separation model can be used for all of the above materials, uncracked and cracked. To sum up, the traction–separation model is more suitable for the interface of concrete and grout materials such as nonshrink grout and concrete. However, the material parameters used in this model were usually reversely determined by direct tensile or direct shear tests [17,20,29]. The problem is that the interface of the two materials in these tests was flat; factors such as joint configuration and cast direction, which affect the properties of concrete-like materials [30,31], were not considered. Murphy et al. [17] simulated the shear test of joint specimens by material parameters obtained from these tests, resulting in much larger cracking loads in the simulation than those in the experiment. Material parameters determined from tests on joint specimens have not been reported yet.

The grillage method is another common method to simulate adjacent box-beam bridges to analyze the load distribution between beams [32–35]. Two transverse connection types are usually used, namely shear transfer systems and shear–flexure transfer systems. For the shear transfer system, beams are simulated as longitudinal grillage members with transverse outriggers, and shear keys are represented by the pinned joints between the outriggers of adjacent beams. Cracks in shear keys are indicated by lowered vertical stiffness, resulting in larger RDSK [36,37]. For the shear–flexure transfer systems, the longitudinal properties of beams are simulated by longitudinal grillage members; the transverse properties of beams and shear keys are simulated by equivalent transverse grillage members. Cracks in shear keys are indicated by the lowered stiffness of the transverse grillage members [38]. The shear transfer system is recommended for bridges with partial depth shear keys [32]. Although there have been many studies on the shear performance of shear keys, the focus was on cracking loads and maximum shear loads [15,17,20,23–25,28,29]. There was little quantitative information about the effects of shear key cracking on vertical stiffness or the relation between shear transfer and RDSK.

2. Objectives

The main objectives of this study were to investigate the relationship between shear transfer and RDSK for transverse connection before and after shear key cracking and determine interfacial material parameters between beams and shear keys cast with concrete. To this end, the following studies were performed in this work:

- Static displacement-controlled bi-shear (SDS), cyclic force-controlled bi-shear (CFS), and cyclic displacement-controlled bi-shear (CDS) tests were conducted on four types of composite specimens to investigate the shear transfer performance;
- Based on the test results, curves and expressions for the relation between shear and RDSK of different types of transverse connections were proposed;
- Two FEMs were developed to calibrate and validate the interfacial material parameters.

3. Experimental Program

3.1. Configuration

Four connection details used to evaluate the shear transfer performance are shown in Figure 1. Type I and Type II specimens consisted of one concrete middle block and two concrete edge reaction blocks with grout joints between them. These two specimen types were 600 mm long, 270 mm in height, and 400 mm wide. Type III and Type IV specimens added a 70-mm-height concrete overlay and four 8-mm-diameter U-shaped steel bars (N1) based on Type I and Type II. Reinforcing steel bars in concrete blocks to prevent cracking are not drawn in Figure 1 for simplicity. Three-dimensional samples of N1 bars are shown in Figure 1e. N1 bars are spaced at 300–400 mm in the longitudinal direction in bridges; for safety reasons, 400 mm was used here as the width of specimens. Type II–IV connections had been widely used in Zhejiang Province in China before 2004 [39], and the bridges using these connection types now more or less experience longitudinal cracking problems. Type I was adopted here as a control for Type II and Type IV.

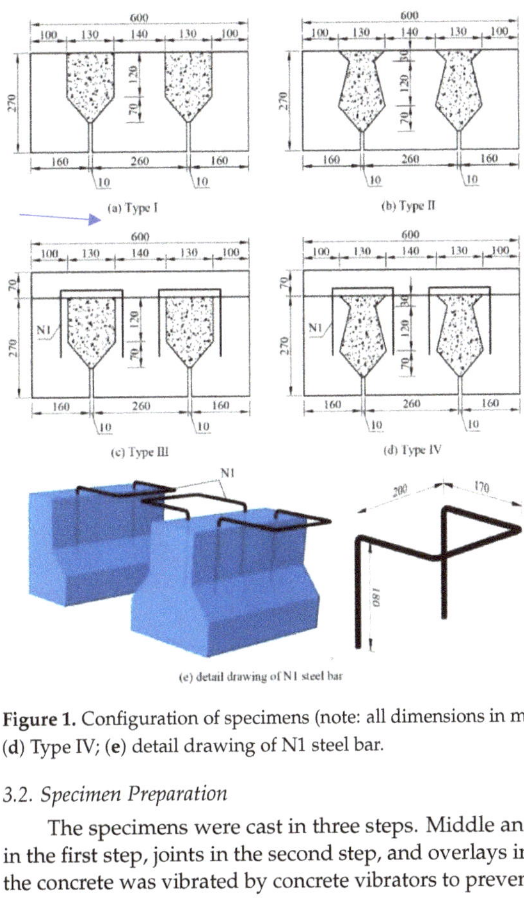

Figure 1. Configuration of specimens (note: all dimensions in mm): (**a**) Type I; (**b**) Type II; (**c**) Type III; (**d**) Type IV; (**e**) detail drawing of N1 steel bar.

3.2. Specimen Preparation

The specimens were cast in three steps. Middle and edge concrete blocks were cast in the first step, joints in the second step, and overlays in the third step. During each cast, the concrete was vibrated by concrete vibrators to prevent imperfect filling. All steps were

30 days apart and moist-cured for 28 days. Companion cubes were cast synchronously at each step to determine the actual compressive strength of the cast material. The cohesion and friction properties of the interface between blocks and joints are mainly affected by the surface treatment of the blocks. In this study, the surfaces of blocks were roughened to an amplitude of 6 mm by a concrete scrabbler [40]. The whole procedure is presented in Figure 2. For better data acquisition by the digital image correlation (DIC) instrument, the front surfaces of specimens were polished, painted white, and black speckles added before tests.

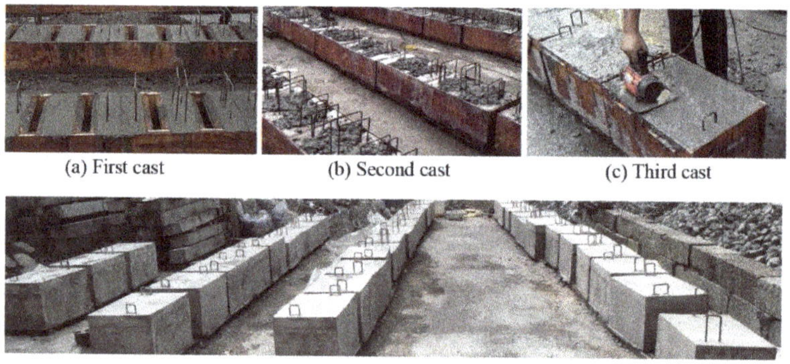

Figure 2. Specimen construction procedure.

Blocks and joints were constructed with commercial, ready-mixed concrete with a targeted 28-day compressive strength of 40 MPa (C40) and overlay of 30 MPa (C30) [41–43]. The constituents of the concrete are shown in Table 1. The average strength of the companion cubes was 46.1 and 45.2 MPa for C40 in the first and second cast and 34.6 MPa for C30 in the third cast [42]. N1 bars and stirrups in blocks had a nominal yield strength of 300 MPa (HPB 300) [44]. Other steel bars in blocks had a nominal yield strength of 400 MPa (HRB 400) [45].

Table 1. Mix design for C30.

Constituent	C30 (kg/m^3)	C40 (kg/m^3)
Aggregate	973	1000
Sand	845	791
Cement type 42.5/52.5 [1]	284	300
Water	92	89
Fly ash	35	20
Mineral powder	63	94
Polycarboxylates high-performance water-reducing admixture	8	8

[1] Cement type 42.5 was used for C30, and cement type 52.5 was used for C40.

3.3. Setup

The bi-shear test was performed using the setup depicted in Figure 3. A specimen was set on steel plates under edge blocks; the load was applied by a high-performance testing machine produced by INSTRON Company through a thin cushion and a steel plate placed on the top of the specimen. The machine could output displacement and load synchronously. The DIC instrument and dial gauges recorded displacement under load at the front and back surfaces, respectively (see Figure 4). DIC recorded the whole displacement field of the front surface during testing, and the data of specified points were extracted and analyzed for different purposes. Taking the front surface of the Type III specimen as an example (see Figure 5), Points 14 to 16 were used to calculate RDSK (Δ); Point 1 to compare with the displacement output by the loading machine; Points 2 to 37 to analyze displacement variation in the vertical direction; Points 38 to 41 to analyze boundary displacements, and Points 42 to 45 to calculate the cracking opening. On the back surface, dial gauges only recorded Points 14 to 16. Δ was calculated using Equation (1):

$$\Delta = (w_{14} + w_{16})/2 - w_{15}|_{\text{front or back}} \quad (1)$$

where w_i is the vertical displacement of Point i (i = 14–16).

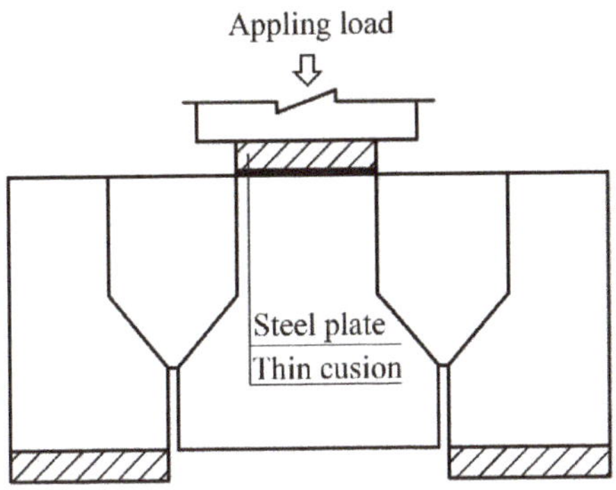

Figure 3. Test setup for bi-shear method.

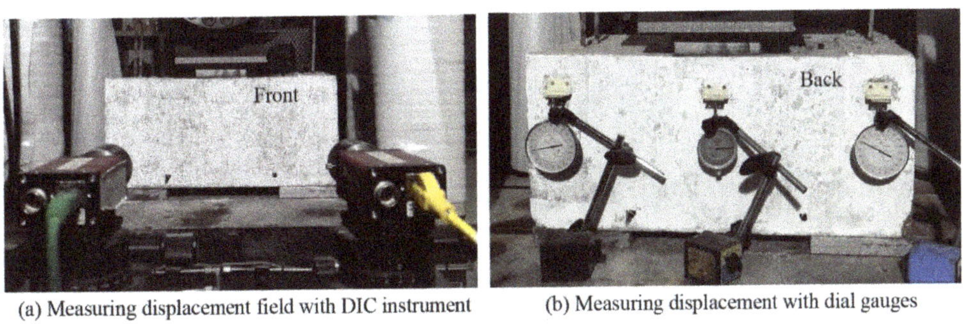

Figure 4. Layouts of instruments: (**a**) DIC instruments; (**b**) dial gauges.

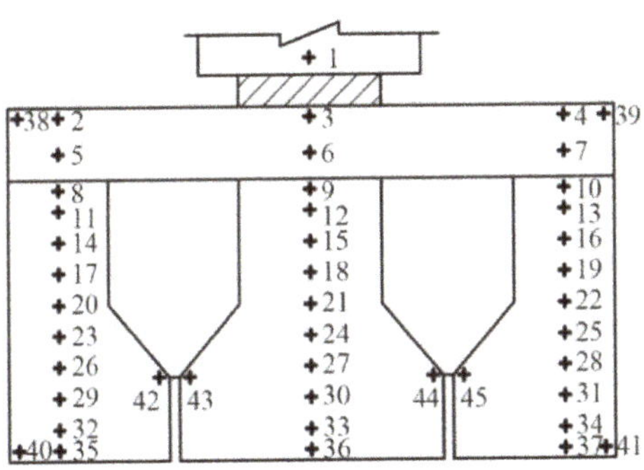

Figure 5. Data extracted positions.

3.4. Test Pocedure

Three different shear test types were conducted in sequence to evaluate the shear performance of four connection details before and after cracking, namely SDS, CFS, and CDS test.

In the SDS test, the displacement was applied at a rate of 0.02 mm/s until the specimen failed. The displacement field, cracks, associated cracking loads, and final failure mode were recorded during the test.

In the CFS test, the load varied linearly between 15 kN and the specified control force at a frequency of 1 Hz. In the first CFS test, the control force was set at approximately 60% of the cracking load obtained by the SDS test. This value was proposed on the assumption that the properties of the interface material are similar to concrete, for which 60% of the maximum tensile stress could be regarded as the elastic limit [46]. The control forces of the second and third CFS tests were determined based on the result of the last test. The controlled force increased if the specimen was uncracked in the previous test. Otherwise, it decreased. A total of 1400 cycles were applied in one test for two main reasons: (1) supposing that the structure was overloaded once a week over a 20-to-30-year period; and (2) too much data generated by DIC during the test, causing storage problems. An SDS test on the same specimen would follow if no cracks appeared after the CFS test. The displacement field and the failure mode were recorded during the test.

In the CDS test, the displacement varied linearly between 0.2 mm and the specified control displacement at a frequency of 1 Hz. The CDS test was proposed based on the phenomenon that RDSK could remain steady after shear key cracking during cyclic loading [8,9]. In addition, to compare dynamic load capacity with static load capacity under the same displacement, one static loading was conducted every 20 cycles. The control displacement was increased by 0.1 mm after each static loading. Figure 6 shows the loading procedure. The displacement field and the failure mode were recorded during the test.

3.5. Results and Discussion

The following notation of specimen names was used throughout this study: the first three letters stand for the test type; the number following the first three letters represents the type of specimen, and the last digit indicates the replicate number of the specimen. For example, Specimen SDS-II-3 indicates the third Type II specimen subjected to the SDS test.

The following method was used to determine the average value: (1) the arithmetic mean value of the measured values was used as an average value, and (2) data greater than or less than 15% of the arithmetic mean were excluded from the mean calculation [41,42].

The value of shear force V was set to half of the load F output by the machine.

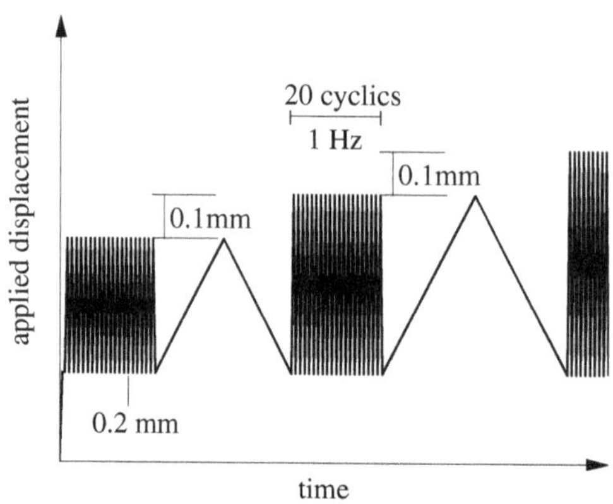

Figure 6. Loading procedure for CDS test.

3.5.1. Results of SDS Test

The typical V-Δ curves for four types of specimens are plotted in Figure 7. As shown, the V-Δ curves for Type I and Type II specimens are similar; the curves are approximately linear until a sudden failure occurs at the interface of block and joint. Figure 8 shows such a failure mode. The typical V-Δ curves for Type III and Type IV specimens are also similar, and the curves can be divided into two stages. In the first stage, the curves are similar to that of Type I and Type II; the interface of block and joint cracked during this stage, and the upper end of the crack extended to the bottom of the overlay (see Figure 9). In the second stage, V increases slowly and almost linearly with Δ until another sudden failure. During this stage, the upper end of the crack slowly reached up to the top of the overlay (see Figure 9).

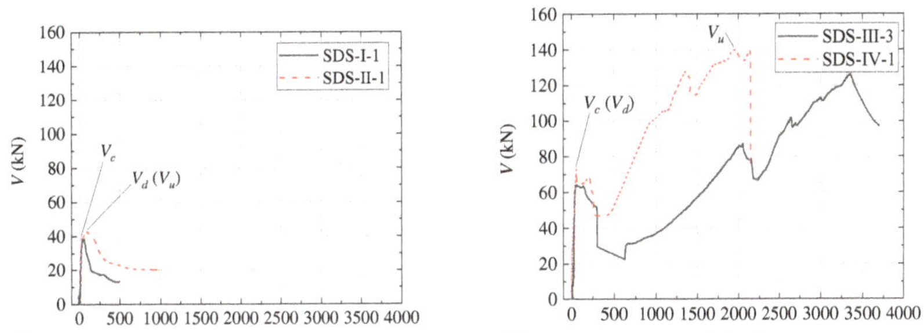

Figure 7. V-Δ plots of SDS test for typical specimens: (**a**) Specimen SDS-I-1 and SDS-II-1; (**b**) Specimen SDS-III-3 and SDS-IV-1.

(a) (b)

Figure 8. Failure modes of specimens: (**a**) Type I specimen; (**b**) Type II specimen.

The cracking shear force (V_c), the maximum shear force in the first ascending stage (V_p), the minimum shear force in the second ascending stage (V_d), the maximum shear force in the whole process (V_u), the corresponding slips across the joint (Δ_p and Δ_u), the stiffness before and after cracking (k_c and k_p), and crack opening (C_o) for each specimen are listed in Table 2. The value of k_c is the slope of the ascending curve in the first stage, calculated by Equation (2):

$$k_c = 0.6V_c/\Delta_{0.6} \quad (2)$$

where $\Delta_{0.6}$ is the value of Δ corresponding to $0.6V_c$ on the V-Δ curve. The value of k_p is the slope of the ascending curve in the second stage.

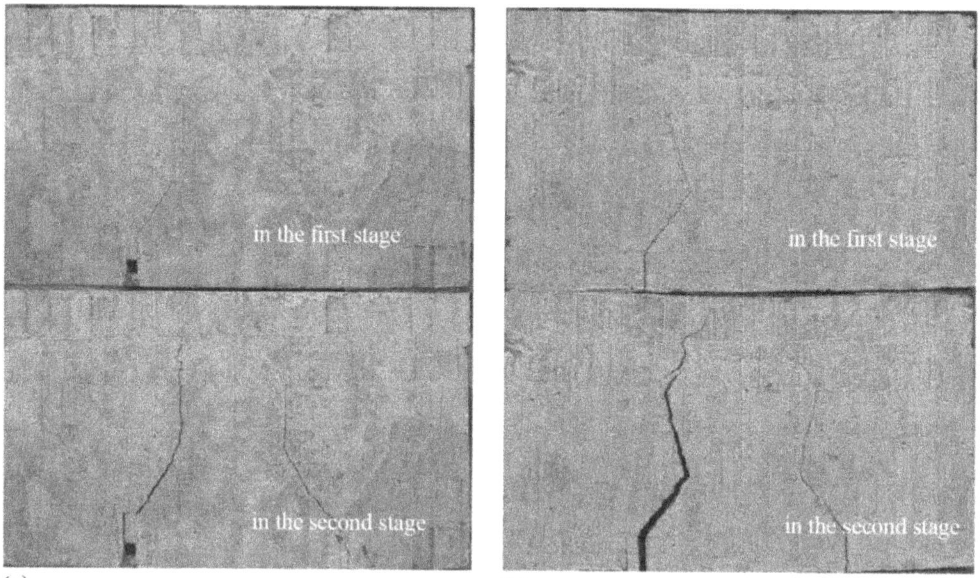

(a) (b)

Figure 9. Failure modes of specimens: (**a**) Type III specimen; (**b**) Type IV specimen.

Table 2. Results of SDS test.

Specimen	V_c (kN)	V_p (kN)	V_d (kN)	V_u (kN)	k_c (kN/μm)	k_p (kN/μm)	Δ_p (μm)	Δ_u (μm)	C_o (μm)
SDS-I-1	38.5	38.5		38.5	1.61		50	50	
SDS-I-2	49.0	49.0		49.0	0.87		72	72	
SDS-I-3	43.8	43.8		43.8	0.98		78	78	
CFS-I-1 [1]	45.4	45.4		45.4	1.15		55	55	
Average	44.2	44.2		44.2					
SDS-II-1	39.3	43.0		43.0	1.52		76	76	
SDS-II-2	53.2	55.5		55.5	0.43		226	226	
SDS-II-3	56.4	56.4		56.4	1.31		57	57	
CFS-II-2 [1]	72.2	72.2		72.2	1.60		45	45	
Average	54.8	56.0		56.0					
SDS-III-1 [2]	62.4	62.4	28.7	-		-	27	-	-
SDS-III-2	62.7	62.7	18.8	160.1	1.81	0.04	40	4020	4378
SDS-III-3	71.7	71.7	46.9	139.1	1.95	0.05	68	2132	3524
CFS-III-1 [1]	73.0	73.0	53.2	145.1	2.01	0.04	55	2864	4205
Average	67.5	67.5		148.1					
SDS-IV-1	64.0	64.0	22.4	126.1	1.93	0.04	63	3337	4743
SDS-IV-2	58.0	60.5	32.0	151.6	1.04	0.04	202	3287	4502
SDS-IV-3	55.3	55.3	27.3	164.1	2.80	0.03	55	5289	6389
CFS-IV-1 [1]	80.0	80.0	30.9	172.5	2.40	0.04	36	4092	7560
CFS-IV-2 [1]	78.7	78.7	29.2	155.3	2.21	0.04	45	4080	5020
Average	66.2	70.0		160.9					

[1] The specimen was subjected to SDS test after CFS test. [2] The test was terminated when the load dropped for the first time.

3.5.2. Results of CFS Test

The results of the CFS test are summarized in Table 3. The results agreed well with the previous assumption about the elastic limit (0.6 V_c). Figure 10 shows specimens' typical V-t and w-t curves during testing, where w is the displacement output by the machine and

t is the time. V-Δ curves are not presented here because the specimen response lagged significantly behind the added load during cyclic loading.

Table 3. Results of CFS test.

Specimen	V (kN)	V/V_c (%)	Cracked?	Failed?	Number of Cycles (Count)
CFS-I-1	30	68	No	No	1400
CFS-I-2	35	79	Yes	Yes	10
CFS-I-3	35	79	Yes	Yes	121
CFS-II-1	35	64	No	No	1400
CFS-II-2	40	73	Yes	Yes	1314
CFS-II-3	40	73	No	No	1400
CFS-III-1	40	59	No	No	1400
CFS-III-2	45	67	Yes	No	28
CFS-III-3	45	67	Yes	No	20
CFS-IV-1	40	60	No	No	1400
CFS-IV-2	45	68	No	No	1400
CFS-IV-3	50	76	Yes	No	25

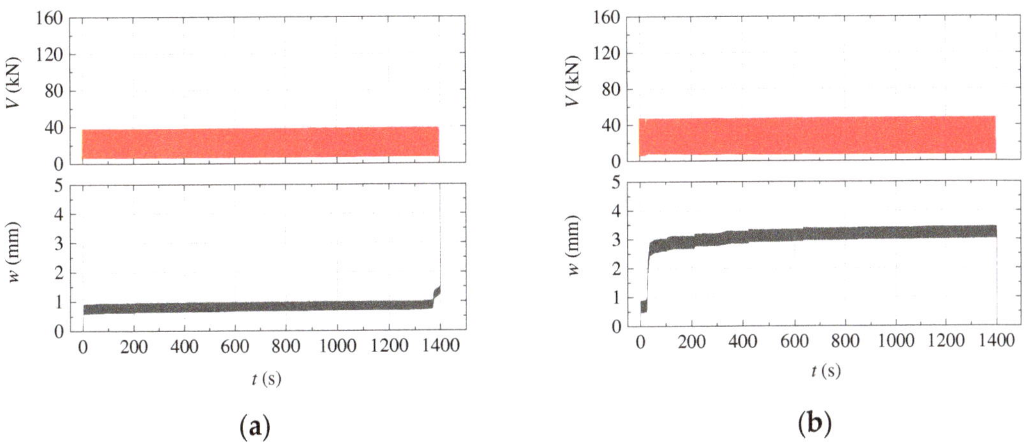

Figure 10. V-t and w-t plots of the CFS test for typical specimens: (**a**) Specimen CFS-II-2 and (**b**) Specimen CFS-IV-3.

For Type I and Type II specimens, once the specimen cracked, w increased quickly, and the specimen failed (see Figure 10a). In contrast, w found a new equilibrium position after several adjustment cycles for cracked Type III and Type IV specimens (see Figure 10b).

3.5.3. Results of CDS Test

The results of the CDS test are summarized in Table 4. The average cracking and maximum shear force under dynamic loading (V_c^d and V_u^d) of Type I to Type IV did not change much compared to those of the SDS test. The typical V-t and w-t curves are shown in Figure 11. Once the specimen of Type I and Type II cracked, increasing w did not result in increasing V synchronously, and the specimen failed quickly (see Figure 11a). In contrast, V increased with w for the Type III and Type IV specimens after cracking and could remain steady when w was relatively low (see Figure 11b).

Table 4. Results of CDS test.

Specimen	V_c^d (kN)	V_u^d (kN)	Failed?
CFS-I-1	38.4		Yes
CFS-I-2	48.5		Yes
CFS-I-3	44.9		Yes
Average	43.9		
CFS-II-1	55.3		Yes
CFS-II-2	44.2		Yes
CFS-II-3	55.8		Yes
Average	51.8		
CFS-III-1	61.7	130.6	Yes
CFS-III-2	66.4	147.9	Yes
CFS-III-3 [1]	-	-	
Average	64.1	139.3	
CFS-IV-1 [2]	62.4	>93.5	No
CFS-IV-2 [2]	73.4	>125.0	No
CFS-IV-3	72.0	127.1	Yes
Average	69.3	-	

[1] The specimen was damaged due to improper operation. [2] The preset displacement was small and the specimen did not fail.

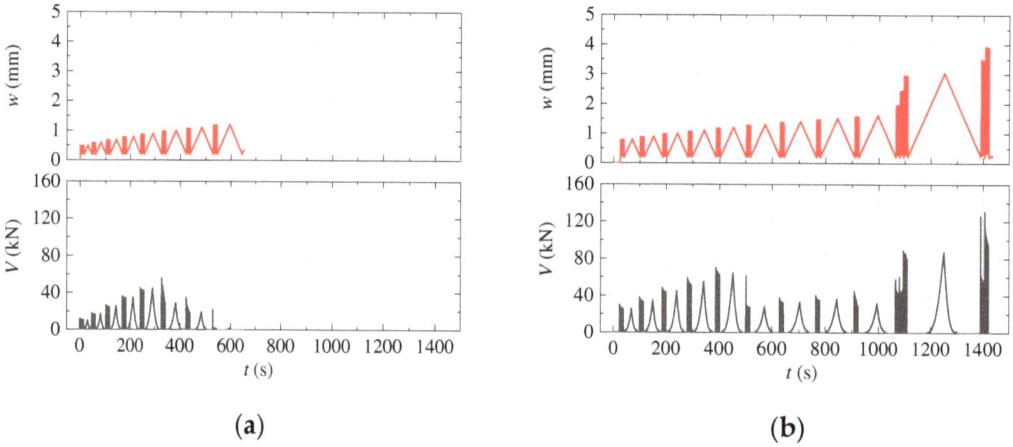

Figure 11. *w*-*t* and *V*-*t* plots of the CDS test for typical specimens: (**a**) Specimen CDS-II-1; (**b**) Specimen CDS-IV-2

3.5.4. Comparison of Different Connection Types

Using N1 bars and an overlay increased both the V_c and V_u of specimens. The V_c and V_u of the Type III connection were 1.5 and 3.4 times larger than those of the Type I connection, respectively. Moreover, the V_c and V_u of the Type IV connection were 1.2 and 2.9 times larger than those of the Type II connection, respectively. Type I and Type II connections failed quickly after cracking, while Type III and Type IV connections could still transfer shear force effectively. In addition, the shear transferring was steady when the force or displacement was not very high after cracking for Type III and Type IV connections.

3.5.5. Relationship between V and Δ

Little research has been done on the relationship of V and Δ for shear key connections used in adjacent box-beam bridges. Generally, when the shear key is intact, Δ is assumed to be zero, and when the shear key is damaged, V is set to the product of Δ and the stiffness of the shear key or just assumed as zero in the calculation [32,36,47].

Ye et al. [28] investigated the shear performance of shear keys by monolithically increasing force-controlled bi-shear tests. They presented two typical τ-Δ curves, where τ was the shear stress obtained from dividing V by the interface area (see Figure 12). Both the blocks and joints were cast with concrete. Curve 1 and Curve 2 presented shear keys without and with reinforcing steel bars connecting blocks and joints, respectively. For both curves, stress τ increased with Δ slowly before Δ around 130 μm; then, an almost linear relationship between τ and Δ was obtained until cracking. After the cracking, a minor increase in τ resulted in a large increase in Δ. The specimen without reinforcing steel bars failed at this time. Then, the specimen with reinforcing bars came into another almost linear relationship between τ and Δ until it failed. Because the load was monolithically increasingly applied in their study, the load dropping after cracking for specimens with reinforcing bars was not captured. Uneven surfaces of specimens may explain why τ increased slowly with Δ at the very beginning.

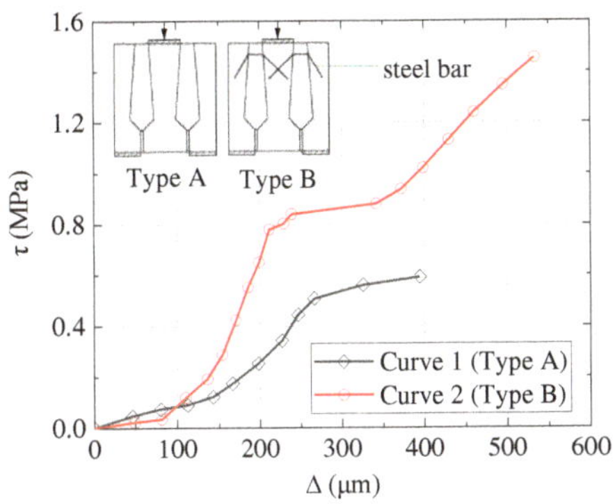

Figure 12. τ-Δ plots (data from Ye et al. [28]).

Rizkalla et al. [48] investigated the performance of flat and keyed joints used in shear wall panels with a compressive preload pressure of 2 MPa and 4 MPa normal to the shear-resistant surface under a direct shear test. Both outside blocks and the joint were cast with concrete. At first, the load was subjected to force control. After the maximum load was attained, the test continued with stroke control. The load–slip curves obtained from the test with different key configurations were similar. They presented a typical load–slip curve for multiple shear key connections, as shown in Figure 13. The curve is linear before cracking and has a load drop after the maximum load. For compressive stress used normal to the connection, the load remains steady even at large slips.

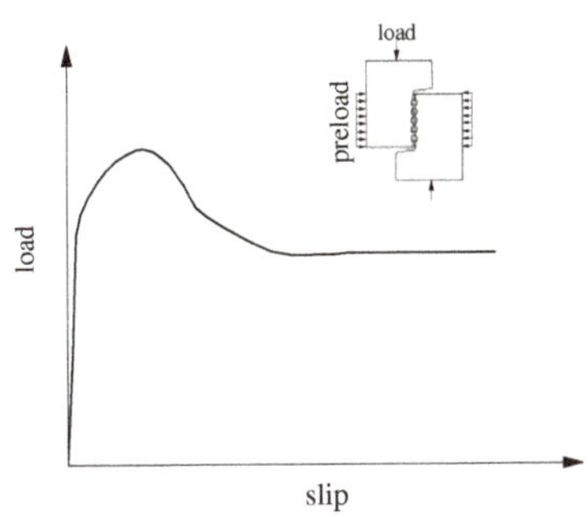

Figure 13. Load–slip curve (data from Rizkalla et al. [48]).

Instead of keyed joints, some researchers studied the relationship between the shear force and slip at two concrete interfaces cast at different times by bi-shear or direct shear tests [49–51]. The shapes of the shear–slip curves for specimens without and with steel bars at the interface are similar to those depicted in Figure 7a,b, respectively.

Based on the results of all these tests, the V-Δ curves can be simplified into two types, one without reinforcing bars as Type I and Type II connection (NRB connection), and the other with reinforcing bars as Type III and Type IV connection (RB connection), as shown in Figure 14. As the results of the CFS and CDS tests showed that the elastic limit is around 0.6 V_c, and when Δ is in the descending part, the NRB connection could not provide steady shear transfer; the relationship between V and Δ can be simplified as:

$$V = \begin{cases} k_c\Delta & \left(0 \leq \Delta \leq \Delta_c^d\right) \\ 0 & \left(\Delta > \Delta_c^d\right) \end{cases} \tag{3}$$

where 0.6 Δ_c is recommended for the value of Δ_c^d. Similarly, the shear transfer in the second descending part for the RB connection can also be set to zero. In addition, the relationship between V and Δ for the RB connection can be simplified as:

$$V = \begin{cases} k_c\Delta & \left(0 \leq \Delta \leq \Delta_c^d\right) \\ V_d & \left(\Delta_c^d < \Delta < \Delta_d\right) \\ V_d + k_p(\Delta - \Delta_d) & \left(\Delta_d \leq \Delta \leq \Delta_u^d\right) \\ 0 & \left(\Delta > \Delta_u^d\right) \end{cases} \tag{4}$$

where Δ_u^d is the correction value of Δ_u based on the difference in static and dynamic motion.

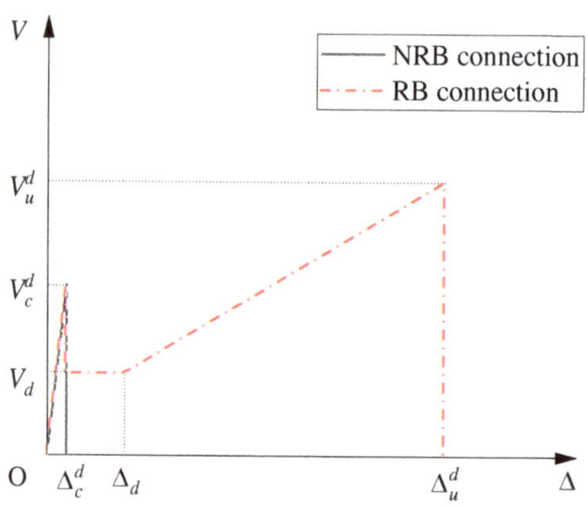

Figure 14. Typical V-Δ plots for NRB and RB connection.

4. FE Analysis

In the present study, FE analyses were performed using the software Abaqus 2018. Two FEMs were established, one based on the SDS test of the Type I specimen to calibrate the interface parameters and the other based on the Type III specimen to validate the interface parameters.

4.1. FEMs

The concrete blocks, joints, overlays, and steel plates were modeled with 8-node brick elements. Steel bars were modeled with 2-node trusses. The concrete damaged plasticity (CDP) model was used to model concrete behavior. The classical metal plasticity model with isotropic hardening was used to model steel behavior. Hard contact and friction were used to model the interface between the specimen and steel plate. The traction–separation constitutive model was used to model the interface of the block and joint. Full bond was used at the overlay–block and overlay–joint interfaces. Steel bars were embedded in the whole model. The two FEMs are shown in Figure 15.

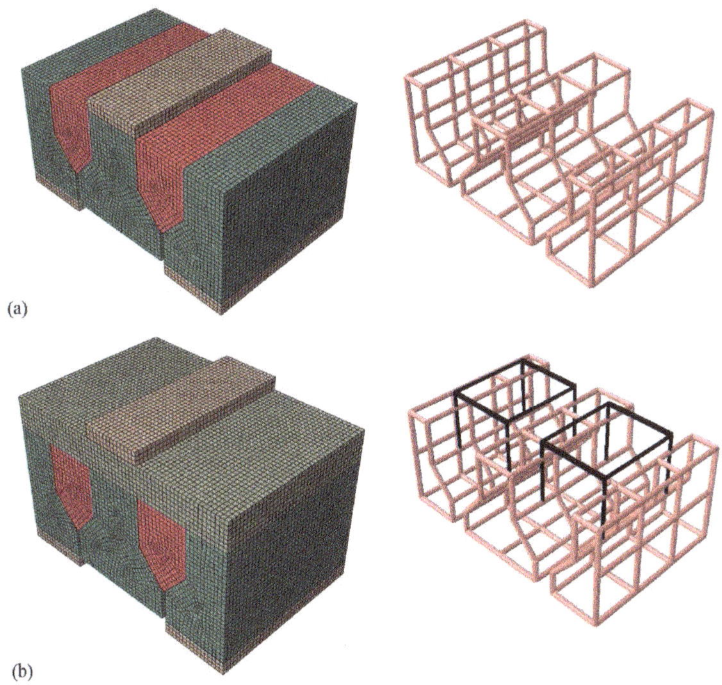

Figure 15. FEMs: (**a**) the SDS test for Type I specimen; (**b**) the SDS test for Type III specimen.

4.2. *Traction–Separation Constitutive Model*

The traction–separation constitutive model offers a method to model thin bonded interfaces whose geometric thickness may be considered to be zero for all practical purposes [52]. The constitutive thickness of interfaces is 1 unit by default and can be specified by users. Note that other input parameters in the model are based on the defined thickness value. The default number 1 was adopted in this study, and the length unit is mm.

The whole traction–separation model contains linear elastic traction–separation, damage initiation criteria, and a damage evolution model. The linear elastic traction–separation model contains stiffness parameters E_n, E_s, and E_t, representing normal and tangential stiffness components. The quadratic nominal stress criterion was used in this study as a damage initiation criterion and can be represented as

$$\left(\frac{\langle t_n \rangle}{t_n^0}\right)^2 + \left(\frac{t_s}{t_s^0}\right)^2 + \left(\frac{t_t}{t_t^0}\right)^2 = 1 \quad (5)$$

where t_n, t_s, and t_t represent the normal and tangential stress components; t_n^0, t_s^0, and t_t^0 represent peak values of the nominal stress when the deformation is either purely normal to the interface or purely in the first or the second shear direction. The symbol <> used in Equation (5) represents the Macaulay bracket with the usual interpretation. Damage is initiated only when the left part of Equation (5) equals 1. Once damage initiation has occurred, damage evaluation is determined on the fracture energy. In this study, the stiffness, peak stress, and fracture energy components in different directions were assumed to be the same [13], referred to as E, t_0, and G below.

In the FEM, the initial values of stiffness components were set to 380 MPa, deduced from the direct shear test on specimens composed of concretes cast at different times conducted by Harries et al. [50]. They used concretes with 28-day compressive strengths of 41.5 MPa and 29.1 MPa for old and new parts, respectively, and the interfaces were roughened to at least 6.4 mm amplitude before casting the new part. The initial values of peak stress components were determined by V_c divided by the connection area projected to the vertical plane. The initial values of fracture energy components were set to 0.1 N/mm based on that of concrete [53]. Then, these material parameters were calibrated during simulation so that the V-Δ curves of the model could match the experimental results.

4.3. FE Results

A comparison of the V-Δ curves resulting from the calibrated FEM and the experimental result (EXP) for Specimen SDS-I-1 is shown in Figure 16a. E and t_0 were calibrated to 450 MPa and 0.5 MPa, respectively. Values of fracture energy components ranging from 0.01 to 1.00 N/mm had been tried during the simulation, but little changed in the shape of V-Δ curves. The curves matched well in the ascending part but not in the descending part. However, the descending part is of minor importance; thus, the result is acceptable. The scalar stiffness degradation (SDEG), indicating the damage degree of model elements, is shown in Figure 16b, where SDEG = 0 indicates intact status and SDEG = 1 failure status. The damage status is similar to that shown in Figure 8a. For all Type I specimens, values of stiffness components with a range of 120 to 450 MPa and peak stress components with a range of 0.5 to 0.6 MPa are recommended.

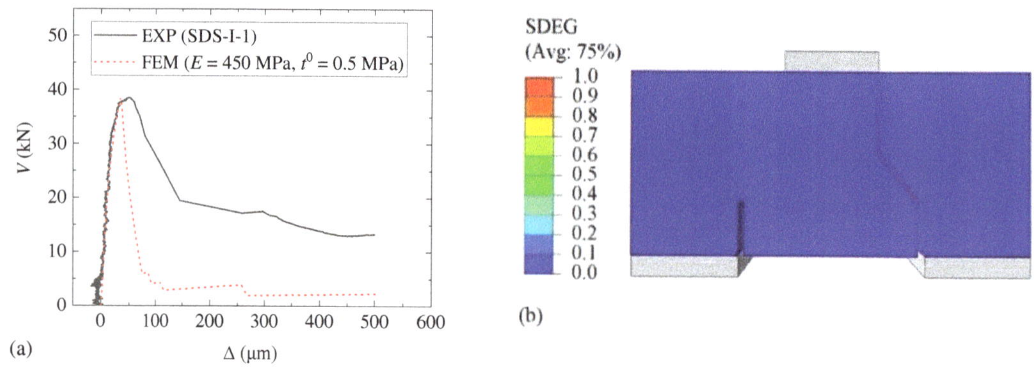

Figure 16. Results for Specimen SDS-I-1: (**a**) V-Δ curves; (**b**) damage status.

A comparison of the V-Δ curves for the second FEM result using material parameters obtained from the first FEM and the experimental result of SDS-III-1 is shown in Figure 17a. Values of stiffness components and peak stress components were 450 MPa and 0.5 MPa, respectively, the same as those for Specimen SDS-I-1. Both ascending parts before and after cracking showed good agreement, and the damage status shown in Figure 17b is similar to that in Figure 9a.

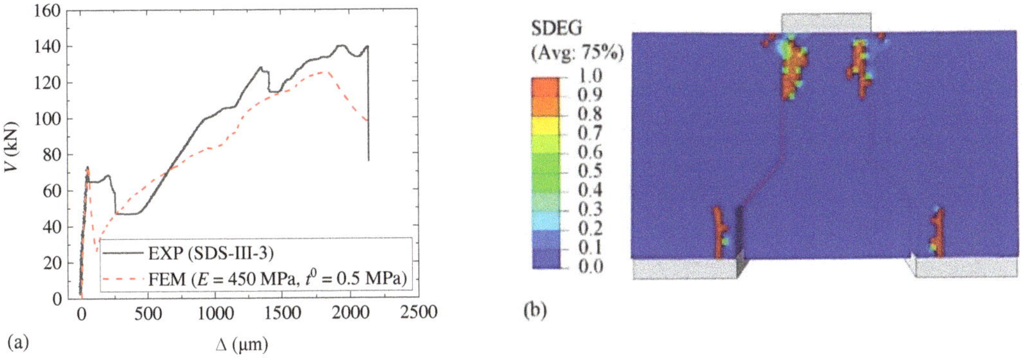

Figure 17. Results for Specimen SDS-III-3: (**a**) V-Δ curves; (**b**) damage status.

5. Conclusions

Static and dynamic bi-shear tests were conducted on four types of transverse connections used in adjacent box-beam bridges to evaluate their shear transfer performance before and after cracking. FEMs were developed to calibrate and validate the interfacial material parameters. Based on the results obtained from this study, the following conclusions can be drawn:

1. Adding overlays and reinforcing bars increased V_c and V_u by 53% and 235%, respectively, for the Type I specimen, and by 21% and 187%, respectively, for the Type II specimen.
2. All four types of connection could remain intact under the dynamic loading under approximately $0.6V_c$. When the Type I and Type II connections cracked, the load transfer failed quickly under cyclic loading. Although the Type III and Type IV connections cracked, the load transfer could still be maintained under a relatively low force or displacement cyclic loading.
3. The V-Δ curves for Type I and Type II could be simplified as a bilinear curve; the V-Δ curves for Type III and Type IV could be simplified as a combination of two bilinear curves corresponding to before and after cracking performance, respectively. The corresponding formulas, Equations (3) and (4), were proposed for engineering applications.
4. FEM results agreed well with EXP results. Values ranging from 120 MPa to 450 MPa for stiffness components and values ranging from 0.5 MPa to 0.6 MPa for peak stress components were recommended for interface materials with a unit thickness (1 mm) when using the traction–separation model.

Author Contributions: Conceptualization, X.N., Y.X. and R.X.; methodology, X.N., Y.X. and R.X.; investigation, X.N.; formal analysis, X.N. and R.X.; writing—original draft preparation, X.N., A.O.A. and G.W.; writing—review and editing, X.N., G.W. and R.X.; supervision, Y.X. and R.X. All authors have read and agreed to the published version of the manuscript.

Funding: This research was supported by the National Natural Science Foundation of China, grant number 51478422.

Institutional Review Board Statement: Not applicable.

Informed Consent Statement: Not applicable.

Data Availability Statement: Data are contained within the article.

Conflicts of Interest: The authors declare no conflict of interest.

References

1. Lall, J.; Alampalli, S.; DiCocco, E.F. Performance of full-depth shear keys in adjacent prestressed box beam bridges. *PCI J.* **1998**, *43*, 72–79. [CrossRef]
2. Russell, H.G. *Adjacent Precast Concrete Box Beam Bridges: Connection Details*; The National Academies Press: Washington, DC, USA, 2009.
3. Naito, C.; Sause, R.; Hodgson, I.; Pessiki, S.; Macioce, T. Forensic Examination of a noncomposite adjacent precast prestressed concrete box beam bridge. *J. Bridge Eng.* **2010**, *15*, 408–418. [CrossRef]
4. Huckelbridge, A.A.; El-Esnawi, H.; Moses, F. Shear key performance in multibeam box girder bridges. *J. Perform. Constr. Facil.* **1995**, *9*, 271–285. [CrossRef]
5. Balakumaran, S.S.G.; Kassner, B.L.; Wyers, R.E. *Forensic Investigation of Two Voided Slab Bridges in the Virginia Department of Transportation's Richmond District*; Virginia Transportation Research Council: Charlottesville, VA, USA, 2017.
6. Frosch, R.J.; Williams, C.S.; Molley, R.T.; Whelchel, R.T. *Concrete Box Beam Risk Assessment and Mitigation: Volume 2—Evaluation and Structural Behavior*; Purdue University: West Lafayette, IN, USA, 2020.
7. Wang, Q.; Wu, Q.X.; Chen, B.C. Experimental study on failure mode of hinged joint in assembly voided slab bridge. *Eng. Mech.* **2014**, *31* (Suppl. 1), 115–120.
8. Yuan, J.; Graybeal, B. Full-scale testing of shear key details for precast concrete box-beam bridges. *J. Bridge Eng.* **2016**, *21*, 04016043. [CrossRef]
9. Yuan, J.; Graybeal, B.; Zmetra, K. *Adjacent Box Beam Connections: Performance and Optimization*; Federal Highway Administration Office of Infrastructure Research and Development: Washington, DC, USA, 2018.
10. Miller, R.A.; Hlavacs, G.M.; Long, T.; Greuel, A. Full-scale testing of shear keys for adjacent box girder bridges. *PCI J.* **1999**, *44*, 80–90. [CrossRef]
11. Leng, Y.; Zhang, J.; Jiang, R.; Cooper, S.; He, H.; Cheng, S. Experimental research on transverse load distribution of prefabricated hollow slab concrete bridges with hinge joint cracks. In Proceedings of the Transportation Research Board 94th Annual Meeting, Washington, DC, USA, 11–15 January 2015.
12. Grace, N.F.; Jensen, E.; Bebawy, M. Transverse post-tensioning arrangement for side-by-side box-beam bridges. *PCI J.* **2012**, *57*, 48–63. [CrossRef]
13. Hussein, H.H.; Walsh, K.K.; Sargand, S.M.; Al Rikabi, F.T.; Steinberg, E.P. Modeling the Shear Connection in adjacent box-beam bridges with ultrahigh-performance concrete joints. I: Model calibration and validation. *J. Bridge Eng.* **2017**, *22*, 04017043. [CrossRef]
14. Mutashar, R.; Sargand, S.; Al Rikabi, F.T.; Khoury, I. Response of a composite-adjacent box beam bridge with skewed beams under static and quasi-static loads. *J. Perform. Constr. Facil.* **2019**, *33*, 04019022. [CrossRef]
15. Huckelbridge, A.A.; El-Esnawi, H. *Evaluation of Improved Shear Key Designs for Multi-Beam Box Girder Bridges*; Department of Civil Engineering, Case Western Reserve University: Cleveland, OH, USA, 1997.
16. Ulku, E.; Attanayake, U.; Aktan, H.M. Rationally designed staged posttensioning to abate reflective cracking on side-by-side box-beam bridge decks. *Transp. Res. Rec.* **2010**, *2172*, 87–95. [CrossRef]
17. Lopez de Murphy, M.; Kim, J.; Sang, Z.; Xiao, C. *Determining More Effective Approaches for Grouting Shear Keys of Adjacent Box Beams*; Pennsylvania State University: State College, PA, USA, 2010.
18. Grace, N.F.; Jensen, E.; Matsagar, V.; Bebawy, M.; Soliman, E.; Hanson, J. *Use of Unbonded CFCC for Transverse Post-Tensioning of Side-by-Side Box-Beam Bridges*; Michigan Department of Transportation: Lansing, MI, USA, 2008.
19. Chen, L.F.; Graybeal, B.A. Modeling structural performance of second-generation ultrahigh-performance concrete pi-girders. *J. Bridge Eng.* **2012**, *17*, 634–643. [CrossRef]
20. Hussein, H.H.; Sargand, S.M.; Steinberg, E.P. Shape optimization of UHPC shear keys for precast, prestressed, adjacent box-girder bridges. *J. Bridge Eng.* **2018**, *23*, 04018009. [CrossRef]
21. Mutashar, R.N.; Sargand, S.; Khoury, I.; Al Rikabi, F.T. Influence of nonuniform box beam dimensions and bridge transverse slope on environmentally induced stresses in adjacent box beam bridges. *J. Perform. Constr. Facil.* **2018**, *32*, 04018081. [CrossRef]
22. Porter, S.D.; Julander, J.L.; Halling, M.W.; Barr, P.J.; Boyle, H.; Xing, S. Flexural testing of precast bridge deck panel connections. *J. Bridge Eng.* **2011**, *16*, 422–430. [CrossRef]
23. Habouh, M. Key Way Joint Strength of Precast Box-Beam Bridges. Ph.D. Thesis, University of Akron, Akron, OH, USA, 2015.
24. Gulyas, R.J.; Wirthlin, G.J.; Champa, J.T. Evaluation of keyway grout test methods for precast concrete bridges. *PCI J.* **1995**, *40*, 44–57. [CrossRef]
25. Gulyas, R.J.; Champa, J.T. Use of composite testing for evaluating of keyway grout for precast prestressed bridge beams. *ACI Mater. J.* **1997**, *94*, 244–250.
26. Issa, M.A.; Valle, C.L.R.D.; Abdalla, H.A.; Islam, S.; Issa, M.A. Performance of transverse joint grout materials in full-depth precast concrete bridge deck systems. *PCI J.* **2003**, *48*, 92–103. [CrossRef]
27. Porter, S.D.; Julander, J.L.; Halling, M.W.; Barr, P.J. Shear testing of precast bridge deck panel transverse connections. *J. Perform. Constr. Facil.* **2012**, *26*, 462–468. [CrossRef]
28. Ye, J.S.; Liu, J.S.; Yu, B.; Fu, Y.X. Experiment on shear property of hinge joints of concrete hollow slab. *J. Highw. Transp. Res. Dev.* **2013**, *30*, 33–39.

29. Hussein, H.H.; Sargand, S.M.; Al Rikabi, F.T.; Steinberg, E.P. Laboratory evaluation of ultrahigh-performance concrete shear key for prestressed adjacent precast concrete box girder bridges. *J. Bridge Eng.* **2017**, *22*, 04016113. [CrossRef]
30. Yi, S.T.; Yang, E.I.; Choi, J.C. Effect of specimen sizes, specimen shapes, and placement directions on compressive strength of concrete. *Nucl. Eng. Des.* **2006**, *236*, 115–127. [CrossRef]
31. Zhang, X.; Zhang, S.; Luo, Y.; Wang, L. Effects of Interface Orientations on Bond Strength between Old Conventional Concrete and New Self-Consolidating Concrete. *ACI Struct. J.* **2020**, *117*, 191–201.
32. Hambly, E.C. *Bridge Deck Behavior*, 2nd ed.; Chapman & Hall: London, UK, 1991; pp. 71–74.
33. Precast/Prestressed Concrete Institute (PCI). *PCI Bridge Design Manual*, 3rd ed.; PCI: Chicago, IL, USA, 2014.
34. American Association of State Highway and Transportation Officials (AASHTO). *AASHTO LRFD Bridge Design Specifications*, 8th ed.; American Association of State Highway and Transportation Officials, Inc.: Washington, DC, USA, 2017.
35. Canadian Standards Association (CSA). *Canadian Highway Bridge Design Code*; CSA Group: Toronto, ON, Canada, 2019.
36. Walsh, K.K.; Kelly, B.T.; Steinberg, E.P. Damage identification for prestressed adjacent box-beam bridges. *Adv. Civ. Eng.* **2014**, *2014*, 540363. [CrossRef]
37. Song, G.X. Research on Hinge Joint Damage Identification Method of Hollow Slab Bridge Based on Influence Line of Lateral Load Distribution. Master's Thesis, Dalian University of Technology, Dalian, China, 2020.
38. Zhan, J.; Zhang, F.; Siahkouhi, M.; Kong, X.; Xia, H. A damage identification method for connections of adjacent box-beam bridges using vehicle–bridge interaction analysis and model updating. *Eng. Struct.* **2021**, *228*, 111551. [CrossRef]
39. Zhejiang Provincial Department of Transport. *Manual for Prevention and Repairing of Typical Defects in Medium and Small Span Bridge*; China Communications Press: Beijing, China, 2011; pp. 1–31.
40. Ministry of Transport of the People's Republic of China. *JTG D60—2015*; General Specifications for Design of Highway Bridges and Culverts. China Communication Press: Beijing, China, 2015.
41. Ministry of Construction of the People's Republic of China. *GB/T 50081—2002*; Standard for Test Method of Mechanical Properties on Ordinary Concrete. China Architecture & Building Press: Beijing, China, 2003.
42. Ministry of Housing and Urban-Rural Development of the People's Republic of China. *GB/T 50107–2010*; Standard for Evaluation of Concrete Compressive Strength. China Architecture & Building Press: Beijing, China, 2010.
43. Ministry of Housing and Urban-Rural Development of the People's Republic of China. *GB 50107–2010*; Code for Design of Concrete Structures. China Architecture & Building Press: Beijing, China, 2015.
44. State General Administration of the People's Republic of China. *GB/T 1499.1—2017*; Steel for the Reinforcement of Concrete—Part 1: Hot Rolled Plain Bars. China Standard Press: Beijing, China, 2017.
45. State General Administration of the People's Republic of China. *GB/T 1499.2—2018*; Steel for the Reinforcement of Concrete—Part 2: Hot Rolled Ribbed Bars. China Standard Press: Beijing, China, 2018.
46. Chen, W.F.; Saleeb, A.F. *Constitutive Equations for Engineering Materials: Elasticity and Modeling*; Elsevier Science B.V.: Amsterdam, The Netherlands, 1994; pp. 257–259.
47. Zhou, Z.; Yuan, G.; Tian, Q. Evaluation method for hinge joint damage in multi-slab girder bridge based on stiffness of hinge joint. *China J. Highw. Transp.* **2013**, *26*, 26–121.
48. Rizkalla, S.H.; Serrette, R.L.; Heuval, J.S.; Attiogbe, E.K. Multiple shear key connections for precast shear wall panels. *PCI J.* **1989**, *34*, 104–120. [CrossRef]
49. Liu, J. Study on the mechanics performance of adherence of young on old concrete. Ph.D. Thesis, Dalian University of Technology, Dalian, China, 2000.
50. Harries, K.A.; Zeno, G.; Shahrooz, B. Toward an improved understanding of shear-friction behavior. *ACI Struct. J.* **2012**, *109*, 835–844.
51. Liu, J.; Fang, J.X.; Chen, J.J.; Xu, G. Evaluation of design provisions for interface shear transfer between concretes cast at different times. *J. Bridge Eng.* **2019**, *24*, 06019002. [CrossRef]
52. Dassault Systèmes. *Abaqus Analysis User's Guide*; Dassault Systèmes: Waltham, MA, USA, 2016.
53. Bažant, Z.P.; Becq-Giraudon, E. Statistical prediction of fracture parameters of concrete and implications for choice of testing standard. *Cem. Concr. Res.* **2002**, *32*, 529–556. [CrossRef]

materials

MDPI

Article

Analysis of Tensile Strength and Failure Mechanism Based on Parallel Homogenization Model for Recycled Concrete

Yijiang Peng, Semaoui Zakaria, Yucheng Sun *, Ying Chen * and Lijuan Zhang

Key Laboratory of Urban Security and Disaster Engineering, Ministry of Education, Beijing University of Technology, Beijing 100124, China; pengyijiang@bjut.edu.cn (Y.P.); Semaouizakaria@emails.bjut.edu.cn (S.Z.); zhanglijuan@bjut.edu.cn (L.Z.)
* Correspondence: sunyucheng@emails.bjut.edu.cn (Y.S.); chenying@emails.bjut.edu.cn (Y.C.);

Abstract: In this paper, a parallel homogenization model for recycled concrete was proposed. A new type of finite element method, the base force element method, based on the complementary energy principle and the parallel homogenization model, is used to conduct meso-level damage research on recycled concrete. The stress–strain softening curve and failure mechanism of the recycled concrete under uniaxial tensile load are analyzed using the nonlinear damage analysis program of the base force element method based on the parallel homogenization model. The tensile strength and destructive mechanisms of recycled concrete materials are studied using this parallel homogenization model. The calculation results are compared with the results of the experiments and meso-level random aggregate model analysis methods. The research results show that this parallel homogenization analysis method can be used to analyze the nonlinear damage analysis of recycled concrete materials. The tensile strength, stress–strain softening curve, and crack propagation process of recycled concrete materials can be obtained using the present method.

Keywords: mesoscopic damage; recycled concrete; parallel homogenization model; base force element method

Citation: Peng, Y.; Zakaria, S.; Sun, Y.; Chen, Y.; Zhang, L. Analysis of Tensile Strength and Failure Mechanism Based on Parallel Homogenization Model for Recycled Concrete. *Materials* **2022**, *15*, 145. https://doi.org/10.3390/ma15010145

Academic Editor: Dario De Domenico

Received: 28 October 2021
Accepted: 19 December 2021
Published: 25 December 2021

1. Introduction

Recycled aggregates are aggregates made from waste concrete through a series of processing methods. Recycled aggregate concrete is concrete made by replacing part, or all, of the natural aggregate with recycled aggregate. It has been widely valued as a green and environmentally friendly building material [1].

Many scholars have carried out a lot of experimental research on recycled concrete, and some research results have been obtained [1–5]. However, the test cycle is long, the cost is high, and it is difficult to measure the internal stress, strain, and failure mechanisms of the material. Therefore, it is very useful to carry out numerical simulation research on recycled concrete. In reference [6], a plastic-damage constitutive models are employed in numerical studies on recycled concrete under uniaxial compression and uniaxial tension loadings to predict the overall mechanical behavior, particularly the stress–strain relationship. In reference [7], a statistical analysis on its composition has been performed considering the randomness in properties of old adhered mortar around recycled aggregate. Peng et al. [8–11] proposed the base force element method and used this new type of finite element method to carry out a numerical simulation analysis on the recycled concrete, and conducted uniaxial tensile and compression tests under static and dynamic loads to study its mechanics and performance. In 2016, Rajendra [12] established a virtual crack model and a double-K fracture model, and determined the fracture parameters of recycled concrete with different coarse aggregate contents. In reference [13], a stochastic elastic FEM analyses model was established based on the Nano-indentation technique for recycled concrete at three different scales to obtain the ef-fective elastic moduli and Poisson's ratios, also the correlations of recycled concrete were studied. Anuruddha [14] investigated the influence of the old mortar content on the elastic module and the strength of recycled concrete, and found that the pressure strength of recycled concrete decreased with the increase in mortar content. Job et al. [15] used numerical simulation and regression analysis methods to study various mechanical properties of recycled concrete, and obtained the influence of different replacement rates of recycled aggregate on the strength of recycled concrete. In 2019, Tan [16] carried out a two-dimensional numerical simulation of recycled concrete based on the discrete element method, and mainly studied the influence of the weak link between the old and new interfaces on the damage and failure process of recycled concrete. Guo [17] has developed a creep coupling model for the heterogeneity of recycled concrete, to investigate the influence of recycled aggregate on the creeping of recycled concrete. Kazemian et al. [18] conducted experimental research on the compressive strength, flexural strength, and fracture energy of recycled concrete, and compared the mechanical properties of treated and untreated recycled concrete. There are also some scholars who have carried out research works in this field or other types of concrete [19–26].

In the paper, a homogenization analysis method will be used to establish a parallel homogenization model for recycled concrete materials. A new type of finite element method, the base force element method based on the complementary energy principle, is used to conduct meso-level damage research on recycled concrete. The stress–strain softening curve and failure mechanisms of recycled concrete under uniaxial tensile load are analyzed using the nonlinear damage analysis program of the base force element method.

2. Materials and Methods

2.1. Random Aggregate Model

The main difference between natural concrete and recycled concrete is that the outer layer of recycled concrete aggregate is attached with a layer of old mortar. The random aggregate model represents the recycled concrete in the form of each phase medium, so as to facilitate the subsequent mesh division and calculation at the meso level. The circular aggregate model was adopted to simplify the recycled aggregate into two concentric circles, as shown in Figure 1. The macro-mechanical properties of the whole structure are obtained through the analysis and calculation of the random aggregate model. It is the link between the macro-structure and the macro-mechanical properties of recycled concrete.

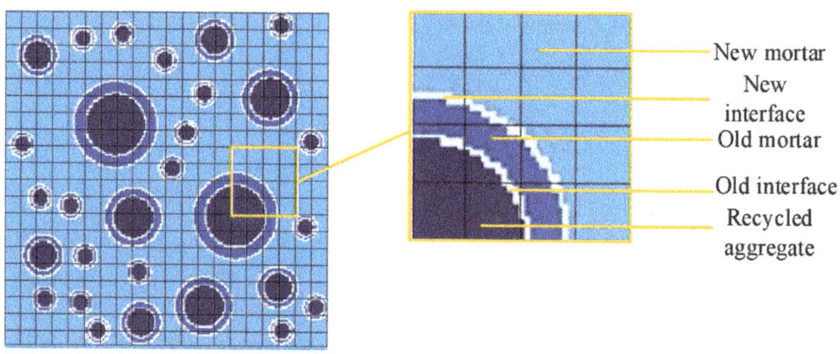

Figure 1. Each phase of random aggregate.

In Figure 1, the recycled concrete is treated as a 5-phase medium, including the aggregate, the old mortar, the new mortar, the old interface and the new interface. The aggregate center position is automatically generated by a computer program according to the Monte Carlo method. The aggregate size and particle number are calculated according to the grading of recycled concrete. The placement of aggregate should ensure that each aggregate cannot overlap.

The calculation and analysis are carried out on the basis of the random aggregate model of recycled concrete. First, the two-dimensional random aggregate is generated. After that, the mesh is divided and mapped to the model, the position of the element node is judged, and the attributes of each element are assigned. Aggregates are divided into coarse aggregates and fine aggregates. Fine aggregates refer to aggregate particles with a particle size of less than 5 mm. The influence of the fine aggregate is ignored, and the fine aggregate and mortar are regarded as a whole in the research of this article. The data of the particle size range of the coarse aggregate are obtained based on the experimental data.

2.2. Parallel Homogenization Model

2.2.1. Parallel Homogenization Model

A parallel method was used to homogenize the heterogeneous elements of recycled concrete, based on the basic idea of equivalent model of meso-damage element, and the equivalent model of meso-damage was established. The validation and numerical calculation of homogenized equivalent model are based on random aggregate model. A simple material strength equivalent method was adopted using the Kelvin–Voigt parallel model, ignoring transverse deformation, as shown in Figure 2. σ is the stress and E is the modulus of elasticity.

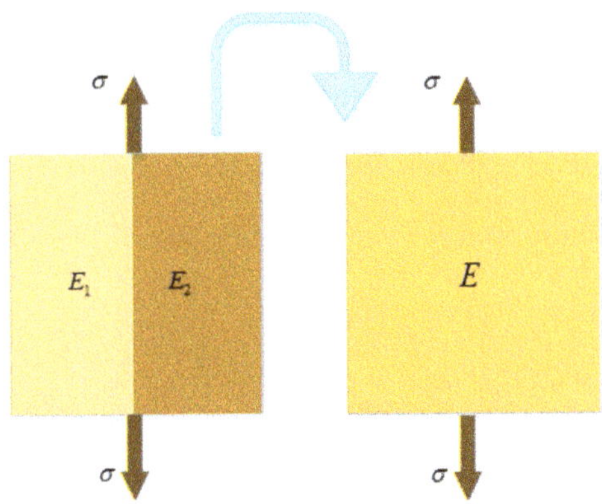

Figure 2. Voigt parallel model.

The equilibrium equation of element stress is as follows:

$$\sigma A = \sigma_1 A_1 + \sigma_2 A_2 \tag{1}$$

where A is the area of an equivalent element, A_1 and A_2 are the areas of two different media elements, respectively.

The strain compatibility equation is as follows:

$$\varepsilon = \varepsilon_1 = \varepsilon_2 \tag{2}$$

The mean stress–mean strain relationship is as follows:

$$\sigma = E\varepsilon \tag{3}$$

It can be obtained from the following formulas:

$$E\varepsilon A = E_1\varepsilon_1 A_1 + E_2\varepsilon_2 A_2 \tag{4}$$

$$EA = E_1 A_1 + E_2 A_2 \tag{5}$$

For the composite recycled concrete model, it can used the volume fraction to calculate $EV = E_1V_1 + E_2V_2$.

Let $c_1 = \frac{V_1}{V}$, $c_1 = \frac{V_1}{V}$ $(c_1 + c_2 = 1)$, then we can obtain the following formula:

$$E = E_1c_1 + E_2c_2 \quad (6)$$

where c_1 and c_2 are the percentage of the two different media elements to the total volume of the equivalent elements, respectively.

The equivalent Poisson's ratio is as follows:

$$v = v_1c_1 + v_2c_2 \quad (7)$$

2.2.2. Finite Element Meshing and Homogenization

The finite element method is the process of dividing a continuous object into simple and regular elements. It is an effective method for calculating and analyzing objects with heterogeneous materials. Scholars have proposed many meshing methods, as the finite element method is widely used in the scientific community. The most common elements are as follows: one-dimensional rod element, two-dimensional triangle element, quadrilateral element and three-dimensional tetrahedron element, pentahedron element and hexahedron element. The finer and denser the mesh, the more accurate the calculation result for the traditional finite element method. However, at the same time, the more computational elements, the lower the computational efficiency.

In this paper, the meshing method of the two-dimensional quadrilateral element is selected based on the base force element method of complementary energy principle. The mesh is divided according to the size of the specimen and the element size. The midpoint of each side of the quadrilateral element is used as the calculation point, as shown in Figure 3. Then the mesh is mapped to the random aggregate model.

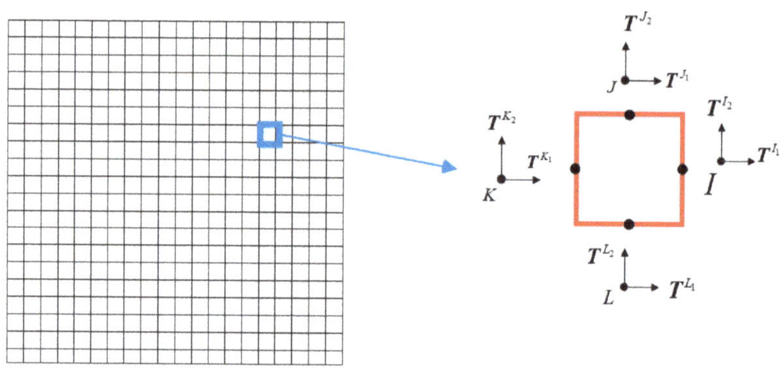

Figure 3. Mesh subsection diagram.

In Figure 4, the fine mesh of the random aggregate meso-model is divided first, and then the coarse mesh is formed by homogenization. The division method and judgment rule of coarse mesh are the same as those of fine mesh, and the judgment rule is determined according to the location of element nodes. The attribute of the element is judged as aggregate (or old mortar, or new mortar), when more than or equal to three of the four nodes of an element are projected on aggregate (or old mortar, or new mortar) medium. The element is defined as old interface element, when some element nodes fall in aggregate medium and some fall in old mortar. Similarly, the element is defined as the new interface when some nodes of an element fall on the old mortar medium and some nodes fall on the new mortar medium. After determining the attributes of fine elements, the number of small elements of each attribute was counted, the area was calculated according to the element size, and the proportion of each component in the coarse mesh element was calculated. The coarse mesh is equivalent to uniform single-attribute element by homogenization method.

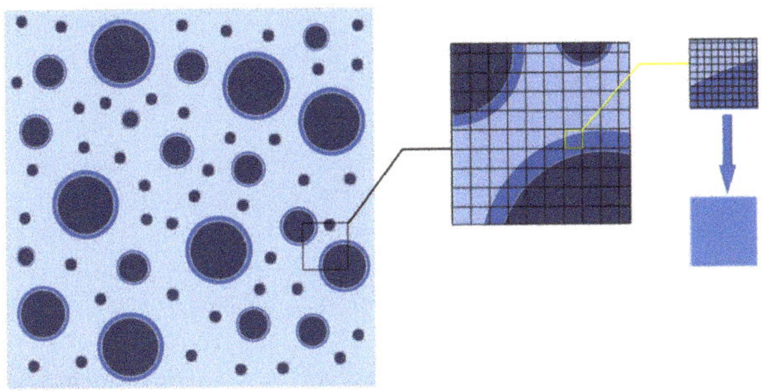

Figure 4. Mesoscopic model of regenerated concrete meshes.

The mesh division and attribute assignment of the elements were programmed to calculate, and the mesh node numbers and coordinates of all the specimens were obtained using Fortran language program. The attribute distribution of an element could be obtained from the element attribute file by the calculating program. The composition of the heterogeneous element containing heterogeneous medium could be obtained. The data are provided for the subsequent homogenization calculation.

The divided elements are mapped to the random aggregate model and then each large element is subdivided to obtain small elements. The proportions of the attributes of the small elements, the number of components, and equivalent parameter calculations are determined. The equivalent parameters are assigned to large elements.

The distribution of equivalent elastic modulus in the model is obtained. The equivalent elastic modulus distribution of the parallel equivalent method of the test specimen is obtained, as shown in Figure 5.

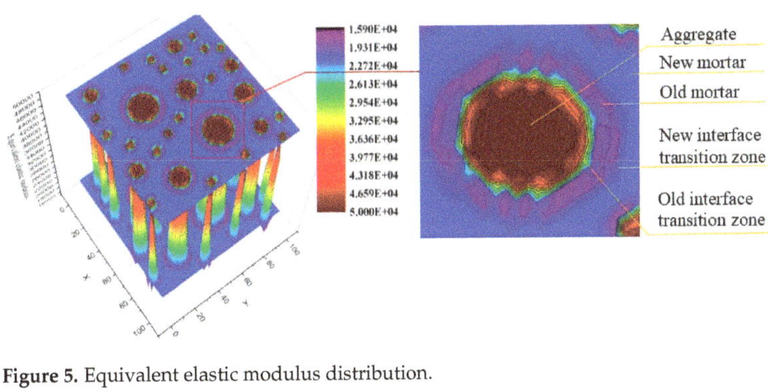

Figure 5. Equivalent elastic modulus distribution.

In the red area of Figure 5, all elements are aggregate. Therefore, the computer's automatic judgment belongs to the same medium. Computers do not use equivalent processing, that is, the elastic modulus is the elastic modulus of aggregate.

In the blue area in Figure 5, all elements are new mortar. Therefore, after the automatic judgment of the computer, the area belongs to the same medium. Computers do not perform the equivalent, that is, the elastic modulus is the elastic modulus of new mortar.

If all small elements inside the large element (equivalent elements) are old mortar, the area belongs to the same medium. There is no equivalent processing, that is, the elastic modulus is the elastic modulus of old mortar. This is shown in purple in Figure 5.

Equivalent treatment is required if small elements within large elements have aggregate elements and interface elements, or if large elements contain aggregate elements, mortar elements and interface elements. The old interface transition zone and new interface transition zone belong to equivalent elements, as shown in Figure 5. The elastic modulus varies in this region. There are different colors.

2.2.3. Damage Model

The stress–strain relationship of material is a very important and complex problem in the case of material damage. The constitutive nature of material is an old and still open question, starting with the papers of Hudson et al., Bažant and Chang [27–29]. Even more recently, Ferretti [30] has shown that the meso-scale curves (mean stress/mean strain and damage curves) are not constitutive, while it is possible to identify constitutive laws at the micro scale.

In this paper, the multi-line stress–strain relationship is adopted for calculation because the stress tends to be highly nonlinear when approaching the peak value under uniaxial tensile load, due to the non-uniformity of the recycled concrete. The material is damaged due to stretching. By introducing the scalar damage variable D, the relationship between the effective strain of the damaged material and Cauchy stress is as follows:

$$\sigma = E_0(1-D)\varepsilon \tag{8}$$

The elastic modulus after damage can be expressed by the initial elastic modulus, if the effect of damage on Poisson's ratio is neglected, as follows:

$$E = E_0(1-D) \tag{9}$$

where E represents the elastic modulus after damage, and E_0 represents the initial elastic modulus. Therefore, the damage elastic modulus of the five-phase medium in recycled concrete can be expressed as (aggregate (ag), old mortar (om), new mortar (m), old interface ($oitz$), new interface (itz)).

$$\begin{cases} E^{ag} = E_0{}^{ag}(1-D^{ag}) \\ E^{om} = E_0{}^{om}(1-D^{om}) \\ E^{m} = E_0{}^{m}(1-D^{m}) \\ E^{oitz} = E_0{}^{oitz}(1-D^{oitz}) \\ E^{itz} = E_0{}^{itz}(1-D^{itz}) \end{cases} \tag{10}$$

$$D_t = \begin{cases} 0 & \varepsilon_{\max} \le \varepsilon_{t0} \\ 1 - \frac{\varepsilon_{t0}}{\varepsilon_{\max}} + \frac{\varepsilon_{\max}-\varepsilon_{t0}}{\eta_t\varepsilon_{t0}-\varepsilon_{t0}}\frac{\varepsilon_{t0}}{\varepsilon_{\max}}(1-\mu) & \varepsilon_{t0} < \varepsilon_{\max} \le \eta_t\varepsilon_{t0} \\ 1 - \frac{\mu}{\xi_t-\eta_t}\frac{\varepsilon_{\max}-\eta_t\varepsilon_{t0}}{\varepsilon_{\max}} + \frac{\mu\varepsilon_{t0}}{\varepsilon_{\max}} & \eta_t\varepsilon_{t0} < \varepsilon_{\max} \le \xi_t\varepsilon_{t0} \\ 1 & \varepsilon_{\max} > \xi_t\varepsilon_{t0} \end{cases} \tag{11}$$

$$D_c = \begin{cases} 1 - \frac{\delta}{\omega} & \varepsilon_{\max} \le \lambda\varepsilon_{c0} \\ 1 - \frac{1-\delta}{1-\lambda}\frac{\varepsilon_{\max}-\lambda\varepsilon_{co}}{\varepsilon_{\max}} - \delta\frac{\varepsilon_{co}}{\varepsilon_{\max}} & \lambda\varepsilon_{c0} < \varepsilon_{\max} \le \varepsilon_{c0} \\ 1 - \frac{1-\omega}{1-\eta_c}\frac{\varepsilon_{\max}-\varepsilon_{co}}{\varepsilon_{\max}} - \frac{\varepsilon_{co}}{\varepsilon_{\max}} & \varepsilon_{c0} < \varepsilon_{\max} \le \eta_c\varepsilon_{c0} \\ 1 - \frac{\omega\varepsilon_{c0}}{\varepsilon_{\max}} & \eta_c\varepsilon_{c0} < \varepsilon_{\max} \le \xi_c\varepsilon_{c0} \\ 1 & \varepsilon_{\max} > \xi_c\varepsilon_{c0} \end{cases} \tag{12}$$

where ω is the residual compressive strength coefficient; ε_0 is the peak strain; η is the residual strain coefficient; λ is the elastic strain coefficient; δ is the elastic compressive strength coefficient; ξ is the limiting strain coefficient; μ is the residual tensile strength coefficient. Subscripts t represent the tension of the element.

The volume fraction of each phase medium can be simplified into area fraction for the two-dimensional random aggregate model. Assuming that the size of the large mesh is a and the size of the small mesh is $b(a > b)$, and there are n small mesh element attributes in the large mesh determined as aggregate, then the area fraction of aggregate is $c_1 = nb_2/a_2$.

Similarly, c_0, c_1, c_2, c_3 and c_4 are used to represent the area fraction of new mortar, aggregate, old interface, old mortar and new surface, respectively.

Based on the strain compatibility equation, we obtain the following:

$$\begin{aligned} \sigma = E_m\varepsilon &= c_0E_0\varepsilon_0 + c_1E_1\varepsilon_1 + c_2E_2\varepsilon_2 + c_3E_3\varepsilon_3 + c_4E_4\varepsilon_4 \\ E_m &= c_0E_0 + c_1E_1 + c_2E_2 + c_3E_3 + c_4E_4 \end{aligned} \tag{13}$$

The elastic modulus of the parallel equivalent element is as follows:

$$E_{eq} = c_0{E_0}^{m}(1 - D^m) + c_1{E_0}^{ag}(1 - D^{ag}) + c_2{E_0}^{oitz}(1 - D^{oitz}) + c_3{E_0}^{om}(1 - D^{om}) + c_4{E_0}^{itz}(1 - D^{itz}) \tag{14}$$

The thickness of the new and old interface is small for recycled concrete. The mesh element is larger when the homogenized equivalent model is used. A single element may contain multiphase media when the element size is larger than the thickness of the old mortar. Therefore, this article will use recycled concrete as an equivalent to the following three phases: mortar (m), aggregate (ag) and equivalent element (em). The meso-equivalent model of stress–strain relationship for three-phase medium is established, where $\varepsilon_0^{\mathrm{em}} \le \varepsilon_0^m \le \varepsilon_0^{ag} \le \varepsilon_r^{em} \le \varepsilon_r^m \le \varepsilon_r^{ag} \le \varepsilon_u^{em} \le \varepsilon_u^m \le \varepsilon_u^{ag}$. The formula of elastic damage model of parallel equivalent element of recycled concrete is as follows:

$$E^{\mathrm{eq}} = \begin{cases} c_0E_0^m + c_1E_0^{ag} + c_2E_0^{em} & \varepsilon \le \varepsilon_0^{\mathrm{em}} \\ c_0E_0^m + c_1E_0^{ag} + c_2E_r^{em} & \varepsilon_0^{em} < \varepsilon \le \varepsilon_0^m \\ c_0E_r^m + c_1E_0^{ag} + c_2E_r^{em} & \varepsilon_0^m < \varepsilon \le \varepsilon_0^{ag} \\ c_0E_r^m + c_1E_r^{ag} + c_2E_r^{em} & \varepsilon_0^{ag} < \varepsilon \le \varepsilon_r^{em} \\ c_0E_r^m + c_1E_r^{ag} + c_2E_u^{em} & \varepsilon_r^{em} < \varepsilon \le \varepsilon_r^m \\ c_0E_u^m + c_1E_r^{ag} + c_2E_u^{em} & \varepsilon_r^m < \varepsilon \le \varepsilon_r^{ag} \\ c_0E_u^m + c_1E_u^{ag} + c_2E_u^{em} & \varepsilon_r^{ag} < \varepsilon \le \varepsilon_u^{em} \\ c_0E_r^m + c_1E_r^{ag} & \varepsilon_u^{em} < \varepsilon \le \varepsilon_u^m \\ c_1E_r^{ag} & \varepsilon_u^m < \varepsilon \le \varepsilon_u^{ag} \\ 0 & \varepsilon_u^{ag} < \varepsilon \end{cases} \tag{15}$$

Formula (15) is the calculation formula of equivalent elastic modulus at each stage. This formula can be obtained by combining the stages in the constitutive model.

Below, we will deduce the calculation formula of equivalent tensile strength of recycled concrete according to energy equivalence. See Formula (16)–(18) for details. From this formula, the equivalent tensile strength of recycled concrete can be calculated. The equivalent element is a homogeneous element. The stored total strain energy W is $W = W_{ag} + W_{em} + W_{om}$ (the sum of the equivalent element, new mortar, and aggregate), when the equivalent element reaches the equivalent tensile strength ${f_t}^{eq}$.

Because the strain energy is $W = \int \frac{1}{2}{f_t}^{eq}\varepsilon dV$, where ε is the element strain, the following is obtained:

$$W = W_{ag} + W_{om} + W_{em} = \int \frac{1}{2}{f_t}^{ag}\varepsilon dV_{ag} + \int \frac{1}{2}{f_t}^{om}\varepsilon dV_{om} + \int \frac{1}{2}{f_t}^{em}\varepsilon dV_{em} \tag{16}$$

Substituting $\varepsilon = \frac{f_t}{E}$ into the above formula, we obtain the following:

$$(f_t^{eq})^2 E^{eq} V = (f_t^{ag})^2 E^{ag} V_{ag} + (f_t^{om})^2 E^{om} V_{om} + (f_t^{em})^2 E^{em} V_{em} \tag{17}$$

$$(f_t^{eq})^2 = E^{eq} \cdot ((f_t^{om})^2 E^{om} c_0 + (f_t^{ag})^2 E^{ag} c_1 + (f_t^{em})^2 E^{em} c_2) \tag{18}$$

The mean stress/mean strain and damage curves of equivalent element can be obtained after equivalence of different media, as shown in Figure 6.

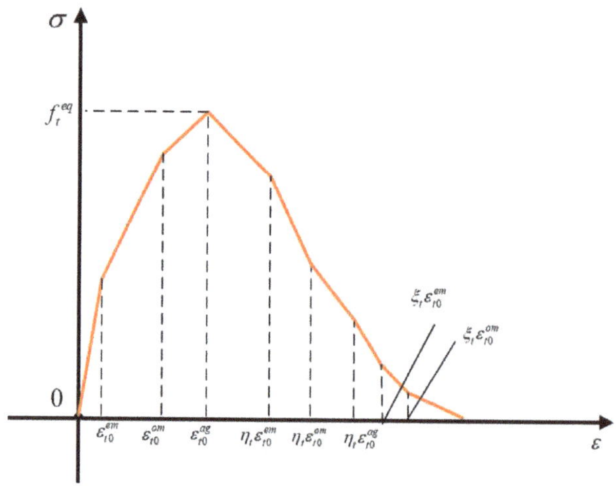

Figure 6. Mean stress/mean strain and damage curves under homogeneous equivalent tensile treatment.

In Figure 6, the stress–strain curve of homogenized equivalent material is a multi-broken line form. The y-coordinate is the equivalent strength and the x-coordinate is the equivalent strain. After the peak point, there is a period of strength decline, material is damaged and destroyed..

The slope of the curve can be calculated according to Formula (14) and (15).

Zhu et al. [31] assumes that the properties of each component material conform to the Weibull distribution, and considers the non-linear characteristics of material inhomogeneity, thus proposing a random mechanical model. The density function of the Weibull distribution is as follows:

$$f(u) = \frac{m}{u_0}\left(\frac{u}{u_0}\right)^{m-1} \exp\left[-\left(\frac{u}{u_0}\right)^{m}\right] \tag{19}$$

where m determines the shape of the Weibull distribution density function, and it represents the uniformity of the medium; u represents a random variable satisfying the Weibull distribution. This paper considers the random distribution of materials in each phase of recycled concrete, and its material parameters obey the Weibull distribution. The value of each parameter is shown in Table 1.

Table 1. Material parameter value.

Parameter	New Mortar	Recycled Aggregate	Old Bond Zone	Old Mortar	New Bond Zone
δ	0.25	0.65	0.23	0.25	0.23
ω	0.15	0.15	0.15	0.15	0.15
λ	0.13	0.25	0.15	0.13	0.15
μ	0.3	0.3	0.3	0.6	0.35
η_c/η_t	4	5	3	4	3
ξ_c/ξ_t	10	10	10	10	10

3. Calculation Results

3.1. Uniaxial Tensile Loading Model

The loading model of a cubic recycled concrete specimen is established by using the calculation program of the base force element method of the complementary energy principle in MATLAB programming. The specimen is selected as 100 mm × 100 mm × 100 mm to carry out the uniaxial tensile numerical simulation test. First, the cube model is simplified into a two-dimensional model with a cross-section size of 100 mm × 100 mm, and the loading model is shown in Figure 7. Vertical loading is adopted during loading. A static displacement-controlled loading condition is adopted step-by-step, with a displacement of 0.01 mm for each stage.

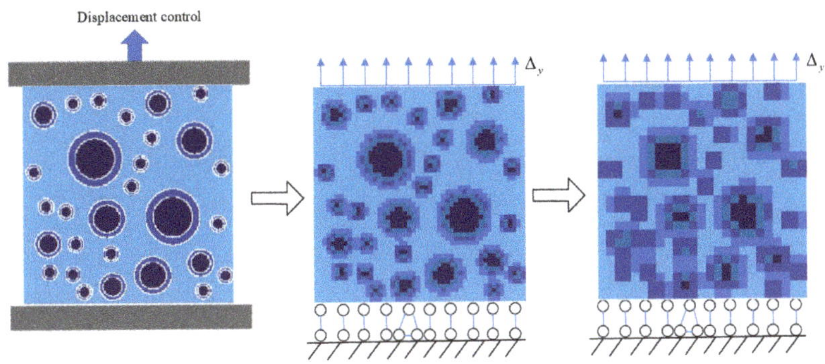

Figure 7. Load model.

3.2. Numerical Simulation Results

Different random numbers are selected to be put in the aggregate, in order to obtain specimens with different aggregate distribution positions in the random aggregate model. Three two-dimensional numerical models, with three groups of different aggregate distributions and the same aggregate particle number, are selected, as shown in Figure 8.

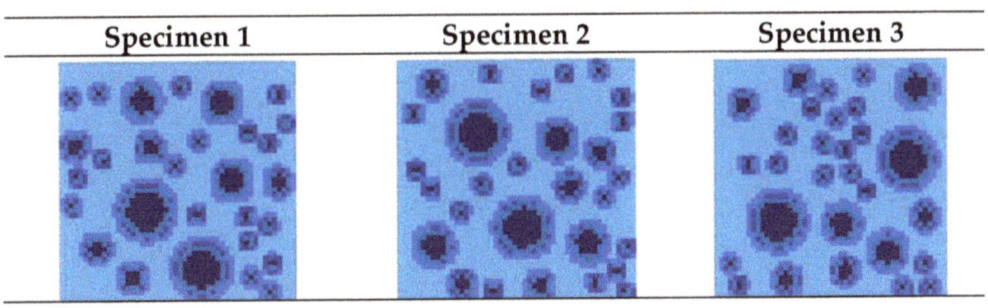

Figure 8. Two-dimensional diagram of homogenized model.

The basic force element method based on the principle of complementary energy is used to calculate and analyze the medium damage by the calculation program. The parallel equivalent homogenization model of the generated three random specimens is numerically simulated in a uniaxial tensile test. The calculation results of the three specimens, random aggregate results, and test data [32] are listed in Table 2. Meanwhile, the full stress–strain curve is drawn with strain as the abscissa and stress as the ordinate. The calculation result of the parallel equivalent is shown in Figure 9.

Table 2. Numerical simulation results data.

	Peak Strain (10^{-6})	Peak Stress (MPa)
Test data [32]	102	2.06
Parallel specimen 1	100	2.10
Parallel specimen 2	100	2.11
Parallel specimen 3	100	2.09

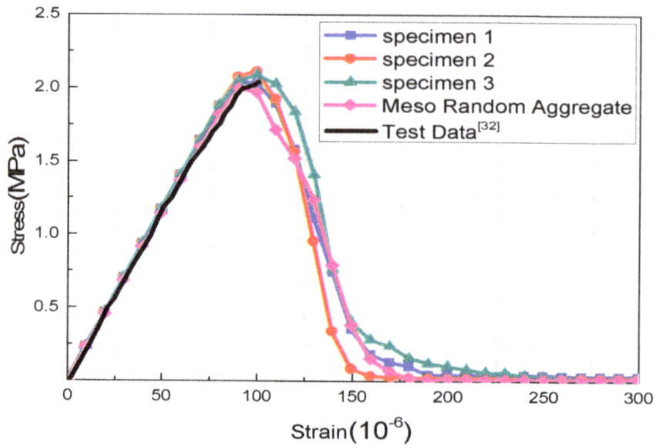

Figure 9. Uniaxial tensile stress–strain curve of parallel model.

The recycled concrete specimen is in the elastic stage at the initial stage, as shown in Figure 9. The stress begins to grow slowly, and reaches about 80% of the specimen's ultimate strength. When the strain of the local element is greater than the residual strain, it begins to enter the state of damage. As the strain increases, the stress decreases until it reaches zero.

The QuickWin module in Fortran is adopted to display the different stages of each element with different colors, in order to obtain the damage diagram of the numerical simulation calculation of recycled concrete clearly and simply. The equivalent element will be set as the same yellow–green color, the mortar as orange, the aggregate as blue, and the failure element is represented by the black block. The failure mode of the numerical simulation specimen is observed, as shown in Figure 10.

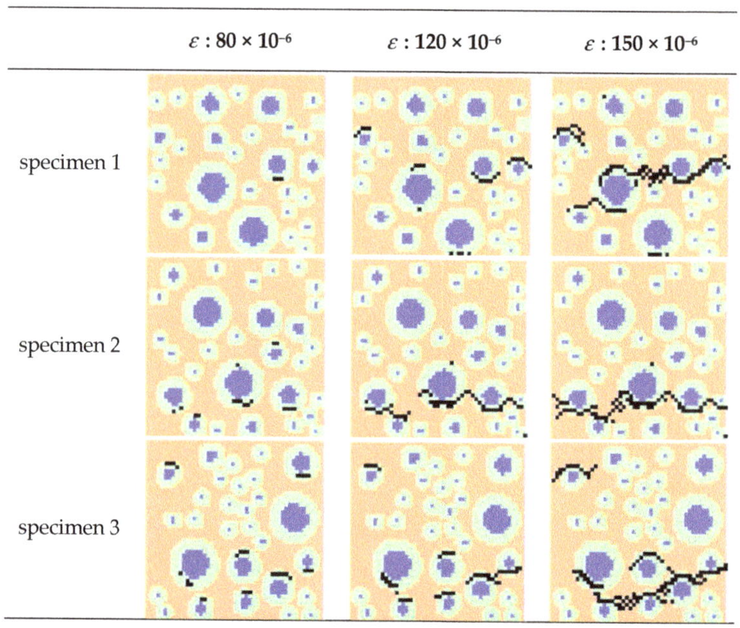

Figure 10. Damage diagram of uniaxial tension numerical simulation.

The law of tensile failure of recycled concrete can be observed from the damage state diagram of specimens in the process of uniaxial tensile loading. The local failure occurs first, when the element begins to be loaded under a certain strain. With the increase in loading strain, cracks gradually spread through the whole specimen, accompanied by the final failure of the specimen. It can be found that the location at which the cracks develop is generally the location where the aggregate is more concentrated in the failure diagram. The main reason for this is that the surrounding strength of the reclaimed aggregate is lower, and it is easier to reach the destruction stage first. The equivalent element contains a multiphase medium, and contains the old interface and the new interface with low tensile strength. Therefore, the element with the cracks that appear first is the equivalent element part. The crack development direction of the tensile failure of the homogenized specimen is perpendicular to the loading direction, and the failure state is basically a horizontal crack, which is consistent with the failure state of the random aggregate model, and is combined with the actual law.

4. Discussion

Using the homogenization model of recycled concrete proposed in this paper, the equivalent element is used to replace the tiny random aggregate model element, and the number of available elements is greatly reduced.

In this way, the computing speed is increased and the computer memory is reduced. Table 3 shows the comparison data.

Table 3. Model data comparison.

Model	Element Size (mm)	Element Number	Calculating Time of One Step (s)
Element of random aggregate model	0.5	40,000	3062
Element of homogenized equivalent model	2	2500	10.8

The homogenization model enlarges the size of the grid element of the calculation model, and the equivalent parameters of the homogenization element are obtained by using the parallel equivalent formula.

For the two-dimensional random aggregate recycled concrete model with a size of 100 mm × 100 mm, when the element size of the random aggregate model is 0.5 mm and the element mesh size of the homogenized model is 2 mm, the element mesh size increases by 4 times, the number of elements becomes 1/16 of the original, and the calculation time is reduced by about 300 times.

Obviously, the homogenized equivalent model can greatly save calculation time and improve calculation efficiency, which provides a new way for future numerical simulation analysis and calculation.

5. Conclusions

(1) The parallel equivalent stress–strain relationship of the homogenization model is derived. The multi-line damage model of recycled concrete materials is established by using the homogenization analysis method.

(2) The non-linear basic force element analysis software and the homogenization preprocessing software for the homogenization analysis of recycled concrete have been developed, based on the basic force element method of the complementary energy principle.

(3) A parallel-equivalent homogenization model was used to perform a numerical calculation and analysis on the uniaxial tensile test of recycled concrete. The stress–strain softening curve, and the damage and failure process were obtained.

(4) The feasibility and rationality of the model establishment are verified by comparing the results of this method with the experimental data.

(5) The calculation efficiency of the homogenization model has greatly improved. The calculation efficiency of this method is much higher than that of the mesoscopic damage

analysis method based on the random aggregate model, and it can guarantee a certain calculation accuracy.

(6) The research work in this paper shows that the base force element method based on the complementary energy principle and the parallel homogenization model can be used to analyze the meso-structure and mechanical properties of recycled concrete. It has the characteristics of high computational efficiency and can be used as an effective meso-analysis method for recycled concrete.

(7) In the future, we will study the base force element method based on the complementary energy principle for dynamic damage analysis and three-dimensional analysis of recycled concrete.

Author Contributions: Data curation, Y.C.; Formal analysis, Y.C.; Funding acquisition, Y.P.; Software, Y.C., S.Z. and Y.S.; Supervision, Y.P. and L.Z.; Writing—original draft, Y.C.; Writing—review & editing, Y.P., S.Z. and Y.S. All authors have read and agreed to the published version of the manuscript.

Funding: This work is supported by National Science Foundation of China (10972015, 11172015), the Beijing Natural Science Foundation (8162008), and the Pre-exploration Project of Key Laboratory of Urban Security and Disaster Engineering, Ministry of Education, Beijing University of Technology (USDE201404).

Institutional Review Board Statement: Not applicable.

Informed Consent Statement: Not applicable.

Data Availability Statement: The authors declare no conflict of interest.

Conflicts of Interest: The authors declare no conflict of interest.

References

1. Tam, V.W.Y.; Soomro, M.; Evangelista, A.C.J. A review of recycled aggregate in concrete applications (2000–2017). *Constr. Build. Mater.* **2018**, *172*, 272–292. [CrossRef]
2. Eguchi, K.; Teranishi, K.; Narikawa, M. Study on mechanism of drying shrinkage and water loss of recycled aggregate concrete. *J. Struct. Constr. Eng.* **2003**, *68*, 1–7. [CrossRef]
3. Park, S.B.; Seo, D.S.; Lee, J. Studies on the sound absorption characteristics of porous concrete based on the content of recycled aggregate and target void ratio. *Cem. Concr. Res.* **2005**, *35*, 1846–1854. [CrossRef]
4. Brito, J.D.; Ferreira, J.; Pacheco, J.; Soares, D.; Guerreiro, M. Structural, material, mechanical and durability properties and behaviour of recycled aggregates concrete. *J. Build. Eng.* **2016**, *6*, 1–16. [CrossRef]
5. Shahidan, S.; Azmi, M.A.M.; Kupusamy, K.; Zuki, S.S.M.; Ali, N. Utilizing construction and demolition waste as recycled aggregates in concrete. *Procedia Eng.* **2017**, *174*, 1028–1035. [CrossRef]
6. Xiao, J.Z; Li, W.G.; Corr, J.D.; Shah, P.S. Effects of interfacial transition zones on the stress–strain behavior of modeled recycled aggregate concrete. *Cem. Concr. Res.* **2013**, *52*, 82–99. [CrossRef]
7. Guo, M.H.; Grondin, F.; Loukili, A. Numerical analysis of the failure of recycled aggregate concrete by considering the random composition of old attached mortar. *J. Build. Eng.* **2020**, *28*, 101040. [CrossRef]
8. Ying, L.P.; Peng, Y.J; Kamel, M.M.A. Mesoscopic numerical analysis of dynamic tensile fracture of recycled concrete, *Eng. Comput.* **2020**, *37*, 1899–1922.
9. Wang, Y.; Peng, Y.J.; Kamel, M.M.A.; Gong, L.Q. Modeling interfacial transition zone of RAC based on a degenerate element of BFEM. *Constr. Build. Mater.* **2020**, *252*, 119063. [CrossRef]
10. Wang, Y.; Peng, Y.J.; Kamel, M.M.A.; Ying, L.P. Base force element method based on complementary energy principle for damage analysis of recycled aggregate concrete. *Int. J. Numer. Meth. Eng.* **2020**, *121*, 1484–1506. [CrossRef]
11. Ying, L.P.; Peng, Y.J.; Yang, H.M. Meso-analysis of dynamic compressive behavior of recycled aggregate concrete using BFEM. *Int. J. Comp. Meth.* **2020**, *17*, 1950013. [CrossRef]
12. Choubey, R.K.; Kumar, S.; Rao, M.C. Modeling of fracture parameters for crack propagation in recycled aggregate concrete. *Constr. Build. Mater.* **2016**, *106*, 168–178. [CrossRef]
13. Wang, C.; Wu, Y.H.; Xiao, J.Z. Three-scale stochastic homogenization of elastic recycled aggregate concrete based on nano-indentation digital images. *Front. Struct. Civ. Eng.* **2017**, *12*, 461–473. [CrossRef]
14. Anuruddha, J.; Adams, M.P.; Bandelt, M.J. Understanding variability in recycled aggregate concrete mechanical properties through numerical simulation and statistical evaluation. *Constr. Build. Mater.* **2018**, *178*, 301–312.
15. Thomas, J.; Thaickavil, N.N.; Wilson, P.M. Strength and durability of concrete containing recycled concrete aggregates. *J. Build. Eng.* **2018**, *19*, 349–365. [CrossRef]

16. Tan, X.; Li, W.G.; Zhao, M.H.; Tam, V.W.Y. Numerical Discrete-Element Method Investigation on Failure Process of Recycled Aggregate Concrete. *J. Mater. Civ. Eng.* **2019**, *31*, 04018353. [CrossRef]
17. Guo, M.H.; Grondin, F.; Loukili, A. Numerical method to model the creep of recycled aggregate concrete by considering the old attached mortar. *Cem. Concr. Res.* **2019**, *118*, 14–24. [CrossRef]
18. Kazemian, F.; Rooholamini, H.; Hassani, A. Mechanical and fracture properties of concrete containing treated and untreated recycled concrete aggregates. *Constr. Build. Mater.* **2019**, *209*, 690–700. [CrossRef]
19. Wang, C.H.; Xiao, J.Z.; Zhang, G.Z.; Li, L. Interfacial properties of modeled recycled aggregate concrete modified by carbonation. *Constr. Build. Mater.* **2016**, *105*, 307–320. [CrossRef]
20. Zhan, B.J.; Xuan, D.X.; Poon, C.S.; Scrivener, K.L. Characterization of interfacial transition zone in concrete prepared with carbonated modeled recycled concrete aggregates. *Cem. Concr. Res.* **2020**, *136*, 106175. [CrossRef]
21. Verian, K.P.; Ashraf, W.; Cao, Y. Properties of recycled concrete aggregate and their influence in new concrete production. *Resour Conserv Recycl.* **2018**, *133*, 30–49. [CrossRef]
22. Xue, C.; Li, W.; Li, J.; Wang, K. Numerical investigation on interface crack initiation and propagation behaviour of self-healing cementitious materials. *Cem. Concr. Res.* **2019**, *122*, 1–16. [CrossRef]
23. Nitka, M.; Tejchman, J. Meso-mechanical modelling of damage in concrete using discrete element method with porous ITZs of defined width around aggregates. *Eng. Fract. Mech.* **2020**, *231*, 107029. [CrossRef]
24. Nodehi, M. A comparative review on foam-based versus lightweight aggregate-based alkali-activated materials and geopolymer. *Innov. Infrastruct. Solut.* **2021**, *6*, 231. [CrossRef]
25. Nodehi, M.; Aguayo, F. Ultra high performance and high strength geopolymer concrete. *J. Build. Rehabil.* **2021**, *6*, 34. [CrossRef]
26. Nodehi, M.; Nodehi, S.E. Ultra high performance concrete (UHPC): Reactive powder concrete, slurry infiltrated fiber concrete and superabsorbent polymer concrete. *Innov. Infrastruct. Solut.* **2022**, *7*, 39. [CrossRef]
27. Hudson, J.A.; Brown, E.T.; Fairhurst, C. *Shape of the Complete Stress-Strain Curve for Rock. Proceedings of the 13th Symposium on Rock Mechanics*; University of Illinois: Urbana, IL, USA, 1971.
28. Bažant, Z.P.; Chang, T.P. Is Strain-Softening Mathematically Admissible? In *Unknown Host Publication Title*; ASCE: Reston, VA, USA, 1984.
29. Boresi, A. P.; Chong, K.P. *Elasticity in Engineering Mechanics*; John Wiley & Sons: New York, Ny, USA, 2000.
30. Ferretti, E. Shape-effect in the effective laws of plain and rubberized concrete. *CMC-Comput. Mater. Contin.* **2012**, *30*, 237–284.
31. Zhu, W.C.; Tang, C.A.; Wang, S.Y. Numerical study on the influence of mesomechanical properties on macroscopic fracture of concrete. *Struct. Eng. Mech.* **2005**, *19*, 519–533. [CrossRef]
32. Xiao, J.Z.; Lan, Y. Investigation on the tensile behavior of recycled aggregate concrete. *J. Build. Mater.* **2006**, *2*, 154–158. (in Chinese)

materials

Article

A Proposal to Improve the Effectiveness of the Deflection Control Method Provided by Eurocodes for Concrete, Timber, and Composite Slabs

Tommaso D'Antino * and **Marco Andrea Pisani**

Department of Architecture, Built Environment and Construction Engineering, Politecnico di Milano, 20133 Milan, Italy; marcoandrea.pisani@polimi.it
* Correspondence: tommaso.dantino@polimi.it

Abstract: Limited deflection of structural members represents an important requirement to guarantee proper functionality and appearance of building and infrastructures. According to Eurocodes, this requirement is ensured by limiting the maximum deflection of horizontal structural members to a fraction of their span. However, each Eurocode provides different maximum deflection limits, which are independent of the type of superstructures considered. Thus, the respect of these limits may not always guarantee the integrity of certain superstructures. In this paper, the reliability of the Eurocode deflection control methods, in guaranteeing the integrity of the superstructures, is assessed and discussed. First, different types of horizontal member, namely rib and clay (hollow) pot, composite steel–concrete, and timber beam slabs are designed to respect the deflection limit enforced by the Eurocodes. Then, the maximum curvature developed by these members is compared with the ultimate (limit) curvatures of various superstructures (e.g., ceramic and stone tile floorings). The results obtained show that the approach adopted by Eurocode 2 may provide non-conservative results, but also that the rules proposed by Eurocodes 4 and 5, albeit more reliable, do not always guarantee the integrity of the superstructure. Based on these results, an alternative method, based on the curvature control, is proposed and its advantages and limitations critically discussed. This method appears simpler and more reliable than the method currently adopted by the Eurocodes.

Keywords: deflection control; curvature; serviceability; reinforced concrete; timber; composite slab

Citation: D'Antino, T.; Pisani, M.A. A Proposal to Improve the Effectiveness of the Deflection Control Method Provided by Eurocodes for Concrete, Timber, and Composite Slabs. *Materials* **2021**, *14*, 7627. https://doi.org/10.3390/ma14247627

Academic Editors: Lizhi Sun, Dario De Domenico and Filippo Berto

Received: 7 September 2021
Accepted: 3 December 2021
Published: 11 December 2021

1. Introduction

Deflection control is crucial to guarantee proper functionality and good appearance under service loads of buildings and infrastructures [1–3]. However, the computation of the short- and long-term maximum deflection of a horizontal structural member can be a cumbersome task, and simplified analysis and verifications are often employed [4]. Although excessive deflection does not impair the structural safety, it seriously affects the serviceability of the structure. Cracking of the floor, due to the excessive deformability of the supporting slab, is a well-known issue, although often underestimated. Usually, cracking is attributed to the building settling, which may be responsible for the opening and widening of cracks over time, whereas the possibility that cracking is due to excessive slab deformability is rarely taken into account. Cracking caused by structure settling and slab deformability can be easily distinguished. In the former case, cracks occur and keep opening until the settling is complete. Afterwards, their width remains constant. In the latter case, the crack width varies with varying the service load (thus, the slab deflection). Therefore, while cracking produced by the structure settling can be repaired once the phenomenon is exhausted, cracking due to the excessive deformability, if repaired, will reoccur, since it is the result of a congenital deficiency of the structural element.

Controlling the maximum deflection is particularly hard in the case of RC structures, where the occurrence of concrete cracking and presence of concrete creep affects both the short- and long-time deflection of horizontal members. Since the limit values defined by Eurocode 2 [4] do not depend on the type of superstructures or finishes (including their application method [5]), they can provide non-reliable results, especially when stiff and brittle elements, such as ceramic or stone tiles (a quite common solution in Mediterranean countries), are used as floor finishing.

Although a proper computation of the member vertical deflection is quite difficult, the approaches currently provided by European standards are rather simplified. In general, deflection control is performed by enforcing a limit to the vertical displacement of horizontal members. This limit should depend on the intended use of the structure (residential, office, etc.), finishes, and superstructures (e.g., partitions). Eurocode 2 [4], Eurocode 4 [6], and Eurocode 5 [7], which are the European design and verification codes for reinforced concrete (RC), composite steel–concrete, and timber structures, respectively, limit the vertical displacement of horizontal members to a fraction of the associated span. However, each Eurocode provides a specific definition of the limiting deflection, and significant inhomogeneity can be found among these limiting values. This differentiation does not seem justified, considering that variable service loads, superstructures, and finishes could be the same, regardless of the type of structure.

This paper aims at critically reviewing the approaches provided by Eurocode 2 [4], Eurocode 4 [6], and Eurocode 5 [7], which are derived from ISO 4356 [8] (as stated by Eurocode 2), for the control of horizontal members deflection. The limits of these approaches and discrepancies among them are pointed out, and a new approach, that appears simpler than those provided by the Eurocodes, is proposed. To do so, some horizontal member types typical of the Mediterranean Basin and Alpine region, and of some countries in Latin America, are first designed according to the Eurocode indications. These horizontal members are:

- rib (clay pot or hollow block) slab;
- composite steel–concrete slab;
- traditional slab made of timber beams and planks.

Subsequently, the deformability of the following types of superstructures is studied, computing the limit curvature value associated with the absence of damage (cracking) in the following elements:

- partition walls;
- floorings (of various type).

These curvature values are then compared with those obtained by applying the maximum service load, i.e., the characteristic load combination [9], to the horizontal members designed according to the Eurocodes. The analysis focuses on horizontal members and superstructures frequently adopted in areas where Eurocodes apply. However, the same analysis could be extended to other structural systems, including innovative solutions, such as fiber-reinforced polymer (FRP) reinforced concrete members, for which the design process is controlled by the member deformability [10].

Comparison between curvature values, obtained by enforcing the limits imposed by the Eurocodes and corresponding superstructure limit curvatures, showed non-conservative results in some cases, which proves that Eurocode provisions do not always guarantee the integrity of the most rigid finishing elements. A new curvature control method to verify the horizontal member deflection under service loads is finally proposed and discussed. This method represents a performance approach, with some important advantages with respect to the approach currently adopted by Eurocodes, as discussed below.

2. Bending Limit Imposed by Superstructures and Finishes

A reliable deflection control method should provide deflection limit values capable of guaranteeing the absence of damage (cracking) in the superstructures. Since these limits vary depending on the type of superstructure, a separate analysis of the deflection limits associated to partition walls and various types of flooring was performed in this study.

Representative geometrical and mechanical properties of the superstructures considered were collected from available scientific literature and product technical sheets.

The behavior of ceiling finishes, among which the most brittle is the plaster finishing applied to ribbed concrete slabs [11], was not investigated because damage of ceiling plasters is prevented by limiting the slab crack opening through crack control.

2.1. Partition Walls

Several different alternatives, such as plasterboard walls or walls made of cellular concrete blocks, can be adopted to realize partition walls. Among all these possible solutions, hollow clay brick walls covered with plaster appear to be the most brittle solution (see, for instance, [12–14]). Although cracking of partition walls represents an important issue, when caused by excessive deformability of the supporting slab is usually preceded by cracking of the tiles, which is often associated with a slab curvature lower than that associated with cracking of the partitions. Indeed, partition walls have a significant strength when loaded in their mean plane and are able to compensate for limited differential settlements through the arch effect [15]. Furthermore, partition walls are often placed in the same positions in the different stories of a building, which limits the deflection of the slab in these positions, thus preventing the partition cracking.

It should be noted that facade walls do not usually have important deflection problems because they are supported by the perimeter beams or are themselves structural elements, as in the case of masonry buildings.

2.2. Floorings

Floorings can be made with many different materials. Among them, the most diffused are:

i carpet flooring (moquette);
ii synthetic materials (usually rubber, PVC, or linoleum);
iii resin;
iv timber;
v stone (e.g., marble, granite, or sandstone);
vi ceramic.

(i) Three main types of carpet can be identified, depending on the type of fabric (namely natural, polyamide, and polyester fiber fabric [16]). However, since the mechanical behavior of carpets is strictly related to the support to which the fabric is applied, and since this support is generally a synthetic material (usually a rubber), the same considerations regarding synthetic materials can be applied to carpets.

(ii) Rubber, PVC, or linoleum floorings have high deformability (see for instance [17]) and, therefore, are able to adapt to deformations and cracks of the slab.

(iii) Epoxy resins are generally employed for floorings [18]. These resins have good strength and high deformability, although their flexural strength decreases with increasing thickness. The values supplied by various manufacturers suggest that the thickness varies between 1.5 and 3 mm, the flexural strength between 25 and 60 MPa, the elastic modulus between 2800 and 4200 MPa, and the elongation at break between 7.5 and 8% [19,20]. These values allow the flooring to withstand significant flexural deformations of the slab.

(iv) Timber floors may be nailed or glued to the subfloor. In addition, floating timber flooring (i.e., flooring laid on an underlay that provides good noise insulation) can be found. The first two solutions are the most sensitive to bending of the slab because the timber floor is directly and firmly connected to it. Various types of timber board, which differ for geometry, color, and species, can be used in timber floorings. Among the various species available, one of the most employed is oak, which offers excellent mechanical properties and relatively low cost. Indeed, its elastic modulus is approximately 12,500 MPa, while the bending strength can reach 108 MPa (in the absence of defects), although these values may vary with ambient temperature and relative humidity [21]. The thickness of the boards generally varies between 10 and 22 mm, with the width between 70 and 250 mm and length between 250 and 2500 mm. Due to its elastic modulus, this type of flooring is able to adapt to deformations far greater than those generally considered acceptable for the slabs of a multi-story building [22].

(v) The bending strength of stone tiles depends on the thickness to width ratio, as well as on the type of stone [23]. The quality of these natural products, which is the result of physical (e.g., porosity) and mechanical properties of the stone, can vary significantly even for blocks extracted from the same quarry [24]. Furthermore, stone materials have a brittle behavior [25], which makes stone floorings particularly sensitive to bending of the slab. Floorings made by granite and marble tiles are among the most diffused solutions. In these cases, the tile thickness may vary between 10 and 30 mm, regardless of the size, which may vary from 50 × 50 mm^2 to 600 × 1200 mm^2. The minimum bending strength found in the literature for natural stones is 18 MPa for marble and 20 MPa for granite [26,27]. These values increase up to 64 MPa, both for marble and granite, when artificial stones are employed [28]. Artificial stones are made industrially starting from the same precursor materials of the corresponding natural stones and have a similar aspect and geometry (thickness and size) [23,29].

(vi) Ceramic tiles are subjected to the provisions of EN 14411 [30], which enforces specific characteristics and refers to EN ISO 10545-4 [31] for the determination of the tile flexural strength by means of a three-point bending test. Tiles can have very different sizes, with thickness up to 20 mm and size up to 1200 × 2400 mm^2. The behavior of ceramic tiles is always elastic-brittle and depends on their width [32,33]. The determination of their strength is complicated by the fact that many producers simply state that their ceramic tiles exceed the minimum value imposed by EN ISO 10545-4 [31]. However, bending strength values varying between a maximum of 55 MPa and minimum of 35 MPa, which can reduce to 15 MPa in the case of small-size tiles (thickness less than 15 mm and size not exceeding 200 × 250 mm^2) can be found on the market [28,34–37].

3. Limit Curvature Values

The limit curvature of a material subjected to bending is the curvature value associated with the attainment of the material tensile strength, i.e., with the occurrence of the first crack. Available standards and design guidelines provide criteria for guaranteeing absence of damage in brittle floorings [38]. In this paper, the limit curvatures of ceramic, marble, and granite tiles were considered, as these flooring types are brittle and largely used in residential and commercial buildings. For each of these materials, five different square sizes (i.e., 300 × 300 mm^2, 400 × 400 mm^2, 600 × 600 mm^2, 900 × 900 mm^2, and 1200 × 1200 mm^2), which reflect the sizes commonly used on the market, were considered. It should be noted that partition walls were not considered, since they rarely crack before floorings due to the presence of the arch effect (see Section 2.1). Therefore, flooring limit curvatures may be considered as lower bound values with respect to limit curvatures associated with partition walls cracking. However, when flooring cracking does not represent an issue (e.g., in the case of flexible floorings), the same analysis proposed for floorings in this section could be easily extended to various types of partition walls, where the limit curvature shall be determined by analyzing the partition wall behavior (see e.g., [39]). In this case, the limit curvature is the maximum curvature of the partition wall surface in contact with the slab.

According to [31], the strength of ceramic tiles can be expressed by the breaking strength (S) obtained by a three-point bending test:

$$S = \frac{FL}{b} \tag{1}$$

where F is the breaking force, L is the span between the supports, and b is the width of the specimen. Therefore, the associated bending moment at breaking (M_u) is:

$$M_u = \frac{FL}{4} \tag{2}$$

and the limit curvature of the tile (χ_{lim}) (a plate bent in one direction only) is:

$$\chi_{lim} = \frac{12(1-\nu^2)M_u}{bEs^3} = \frac{3(1-\nu^2)S}{Es^3} \quad (3)$$

where E is the material elastic modulus, ν the Poisson's ratio, and s the thickness of the specimen.

Considering ceramic tiles currently available on the market and setting E = 60 GPa and ν = 0.28 [40,41], Equation (3) was used to compute the limit curvatures for the tile sizes studied, which are reported in Table 1. In Table 1, the breaking strengths (S) considered are representative values currently available on the market for each specific tile size [28,34–37].

Table 1. Limit curvatures of ceramic tiles.

Size [mm]	s [mm]	S [N]	χ_{lim} [10^{-5} mm^{-1}]
300 × 300	9.5	750	4.03
400 × 400	9.5	900	4.84
600 × 600	10	900	4.15
900 × 900	10	1000	4.61
1200 × 1200	20	1200	6.91

The strength of marble and granite tiles is usually expressed by the bending strength f [31]:

$$f = \frac{3FL}{2bs^2} \quad (4)$$

Therefore, the limit curvature can be obtained as:

$$\chi_{lim} = \frac{12(1-\nu^2)M_u}{bEs^3} = \frac{2(1-\nu^2)f}{Es} \quad (5)$$

The limit curvatures computed for marble tiles (E = 123 GPa and ν = 0.25 [29]) are shown in Table 2, whereas Table 3 shows the limit curvatures computed for granite tiles (E = 90 GPa and ν = 0.25 [29]); f in Tables 2 and 3 are representative values of the bending strength for the specific tile size considered and available on the market.

Table 2. Limit curvatures of marble tiles.

Size [mm]	s [mm]	f [MPa]	χ_{lim} [10^{-5} mm^{-1}]
300 × 300	20	12	0.85
400 × 400	20	12	0.85
600 × 600	20	35	2.49
900 × 900	30	35	1.66
1200 × 1200	30	50	2.37

Table 3. Limit curvatures of granite tiles.

Size [mm]	s [mm]	f [MPa]	χ_{lim} [10^{-5} mm^{-1}]
300 × 300	20	12	1.18
400 × 400	20	12	1.18
600 × 600	20	25	2.47
900 × 900	30	25	1.64
1200 × 1200	30	35	2.30

The elastic moduli of these materials are extremely variable, in relation to the porosity of the material. Furthermore, stone tiles can be natural or artificial, i.e., obtained with a sintering process that maximizes their mechanical characteristics [28]. There are marble tiles with an elastic modulus between 57 GPa and 123 GPa, while the elastic modulus of ceramic tiles may vary between 40 GPa and 60 GPa. Since the elastic modulus is almost never declared by the manufacturers, in this paper the highest values of E found in the literature were conservatively considered.

4. Comparison between the Eurocode Limits and the Performance Requirements of the Flooring

In this section, the three member types considered, namely rib and clay pot, composite steel–concrete, and timber beam slabs, are designed to respect the deflection limits provided by the Eurocodes. The maximum curvature of these members is then compared with the limit curvature obtained for the floorings considered, in order to verify the reliability of this approach.

The deflection of a structural member depends on its geometry, mechanical properties, number of spans, type of constraints, and applied loads. Considering uniformly distributed applied loads, the case of a simply supported beam provides the highest maximum deflection. Therefore, this configuration is adopted here, although any different configuration could be analyzed depending on the specific case studied. According to [6], the uniformly distributed load was determined as the sum of the self-weight of the building slab, including flooring, of a load of 2 kN/m^2, which is the distributed (equivalent) load of partition walls made of hollow clay bricks, and of a variable load of 2 kN/m^2 or 5 kN/m^2, which are the variable loads for residential and commercial buildings, respectively. Four values of the span were considered, namely 4 m, 5 m, 6 m, and 7 m. These span values are representative of common values adopted for the slab types studied in this paper. While span values shorter than 4 m would lead to small maximum deflections, which are not likely to impair the integrity of the floor, spans larger than 7 m would lead to excessive slab depth, which would rarely be adopted in practice.

Different concrete classes were considered in the calculations, which provided consistent results regardless of the concrete strength. Indeed, it should be noted that the concrete strength does not play a fundamental role in the definition of the member deflection, where the elastic modulus (not the strength) is the crucial parameter. With increasing the concrete compressive strength, the elastic modulus only slightly increases (e.g., increasing the concrete class from C25/30 to C40/50 leads to an increase of the elastic modulus of only 17% [4]). Furthermore, provided a certain span and applied load, the concrete class affects the height of the designed cross-section, while the maximum deflection (enforced by the Eurocodes) remains the same. In general, an increase in the concrete strength leads to a decrease of the member cross-section height, which, in turn, may determine an increase of the slab deflection. However, high concrete strength is associated with low shrinkage and high concrete tensile strength, which entail for small cracked portions and a consequent decrease of the slab deflection. These contrasting effects (decrease of the cross-section height and high concrete properties) do not generally lead to significant differences in the deflection of slabs with concrete of different strength classes. Therefore, only the results obtained considering a concrete class C25/30 are provided in this paper.

4.1. Rib and Clay Pot Slab

A representative rib and clay pot slab cross-section with ceramic and stone floorings was considered in this study (Figure 1). The cross-section geometry, which was determined starting from the standardized geometry of the hollow blocks, was kept constant except for the height (H). Provided the applied load associated with the specific flooring and type of building (residential or commercial, see previous section), the minimum value of the height (H) was calculated in order to satisfy the deflection limit enforced by Eurocode 2 (independently of the height of the hollow blocks effectively available on the market). This H value was then considered to verify whether the maximum curvature along the member exceeds the limit values provided in Tables 1–3.

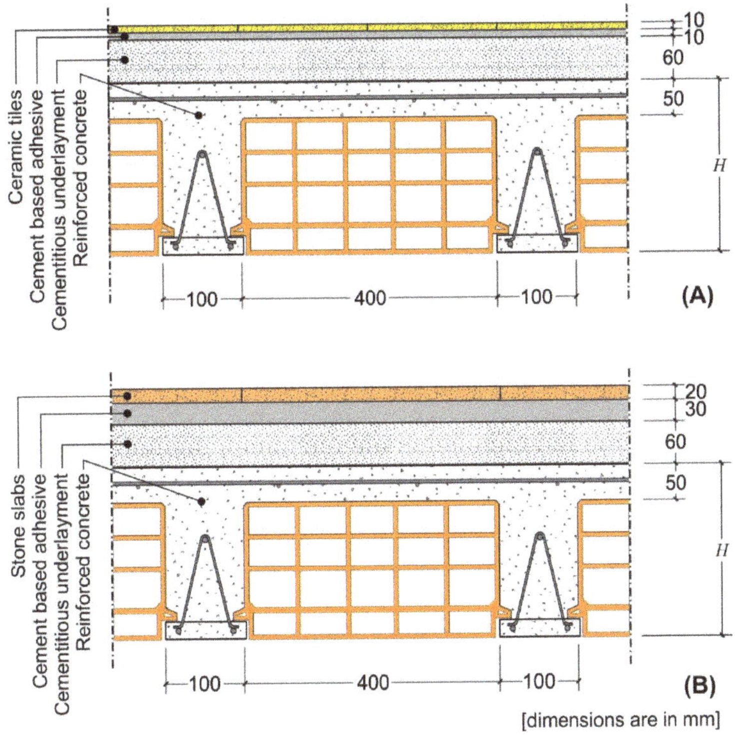

Figure 1. Cross-section of the rib slab with (**A**) ceramic and (**B**) stone floorings.

The rib and clay pot slabs were designed assuming no contribution of the cementitious underlayment and hollow blocks (Figure 1) to the slab structural response. The underlayment is usually made with a mixture of water, sand, and cement and has a low compressive strength, which can be neglected.

The hollow blocks are employed to decrease the weight of the slab and, unless specific cases where they have low percentage of voids and certain geometrical characteristics [42], do not contribute to the structural response of the slab.

Eurocode 2 allows two alternative verifications for the limit state of deflection, one based on a limit of the span/depth ratio and the other on a limit of the deflection. The deflection limits cannot be directly compared with the limit curvatures of the floorings. Moreover, two distinct limits for the deflection of horizontal members are provided, namely:

i "the appearance and general utility of the structure could be impaired when the calculated sag of a beam, slab or cantilever subjected to quasi-permanent loads exceeds span/250" [4];

ii "deflections that could damage adjacent parts of the structure should be limited. For the deflection after construction, span/500 is normally an appropriate limit for quasi-permanent loads" [4].

The presence of two different limits is confusing and deflection values lower than these limits do not always guarantee the absence of damage to the superstructures. In fact, these limits should be verified only with respect to the quasi-permanent load combination, without considering other load combinations associated with service loads (e.g., the characteristic combination) that might affect the construction appearance. However, limiting the deflection verification to the quasi-permanent load combination does not guarantee the integrity of the superstructures under the characteristic load combination, which will certainly occur during the service life of the structure.

The minimum cross-section height (H_{min}) that satisfies limits (i) or (ii) of Eurocode 2 was determined iteratively by matching the calculated sag with the sag limit value (w_{max}) (i.e., span/250 or span/500):

$$w(H_{min}) = w_{max} \quad (6)$$

After assigning the guess value of H, the phases of the iterative process were:

1. Determination of the self-weight of the member (slab).
2. Determination of the cross-section cracking moment, where the tensile strength of concrete was computed according to Eurocode 2 [4].
3. Identification of the structural element cracked segment (located at midspan) and of the two symmetrical uncracked segments (located at the supports).
4. Computation of the second moment of area (J) of the reinforced concrete cross-section of each segment:

$$J = \int_{A_c} y^2 dA_c + m \sum_i A_{si} y_{si}^2 \quad (7)$$

where A_c is the concrete un-cracked area (i.e., the entire cross-sectional area when J is associated with a bending moment lower than the cracking moment or concrete compressed area when J is associated with a bending moment higher than or equal to the cracking moment), y is the vertical distance measured from the cross-section neutral axis, A_{si} is the i-th longitudinal steel cross-sectional area, y_{si} the vertical distance between the i-th longitudinal steel cross-section centroid and the neutral axis, and m is the ratio between the steel elastic modulus E_s = 200 GPa [43] and concrete effective elastic modulus E_{ce}. To account for the long-term behavior of the RC cross-section under the quasi-permanent load combination, E_{ce} was computed as $E_{ce} = E_{cm}/[1 + \varphi(\infty, t_0)]$, where E_{cm} is the concrete elastic modulus at 28 days, and $\varphi(\infty, t_0)$ is the creep coefficient [4].

5. Determination of the sag $w(H)$ under the quasi-permanent load.
6. Comparison of the sag $w(H)$ with the limit w_{max} and determination of a new guess value of H, until the calculated sag matches the limit value.

The minimum cross-section height (H_{min}), obtained for the four spans considered in the case of rib and clay pot slab with ceramic and marble or granite tiles and the different variable loads selected, are shown in Figure 2.

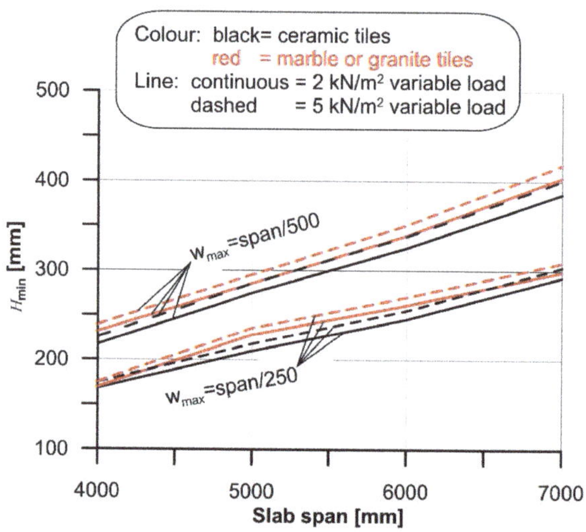

Figure 2. H_{min} computed for 4 m, 5 m, 6 m, and 7 m span of the rib and clay pot slab.

Once H_{min} was determined, the cross-section curvature at midspan due to the maximum service load, i.e., the maximum curvature χ_{max}, was computed and then compared with χ_{lim}. Since the flooring is applied when the structure is already bent due to the presence of the permanent load, χ_{max} was computed as the sum of curvature due to long-term permanent load $\chi_{max,pl}$ and the curvature due to variable load $\chi_{max,v}$, minus the curvature due to the permanent load acting when the flooring was applied $\chi_{max,p}$ (i.e., in the absence of concrete creep):

$$\chi_{max} = \chi_{max,pl} + \chi_{max,v} - \chi_{max,p} \tag{8}$$

$\chi_{max,v}$ and $\chi_{max,p}$ were computed considering the secant modulus of elasticity of concrete E_{cm}, whereas $\chi_{max,pl}$ was computed considering $E_{cm}/[1+\varphi(\infty,t_0)]$, where $\varphi(\infty,t_0)$ is the concrete creep coefficient:

$$\chi_{max,v} = \frac{M_{max,v}}{E_{cm}J} \tag{9}$$

$$\chi_{max,p} = \frac{M_{max,p}}{E_{cm}J} \tag{10}$$

$$\chi_{max,pl} = \frac{M_{max,pl}}{E_{cm}J}[1+\varphi(\infty,t_0)] \tag{11}$$

where $M_{max,v}$, $M_{max,p}$, and $M_{max,pl}$ are the maximum bending moments associated with the variable load, permanent load acting when the flooring was applied, and long-term permanent load, respectively, and J is the second moment of area of the cross-section considered. It should be noted that, although the use of this approach to account for the long-term behavior of concrete is only an approximation, it provides conservative results [4,44,45] and a refined and complex analysis (see for instance [46]) would not markedly affect the calculated curvature.

The effect of slab shrinkage was neglected in this paper because it was assumed that, when the flooring was placed, the slab already underwent most of the drying shrinkage and no significant further shrinkage would occur. However, even in those cases where the slab is still undergoing shrinkage when the flooring is applied, shrinkage will not play a significant role in the slab curvature. Shrinkage may induce curvature to RC members due to the eccentricity of the steel longitudinal reinforcement with respect to the cross-section centroid [47]. However, RC members usually have both tension and compression longitudinal steel reinforcement, which limits the effect of shrinkage on the cross-section curvature. Furthermore, the effect of shrinkage on the cross-section curvature is significantly lower than that of creep. As an example, considering the 5 m span slab made by concrete with f_{ck} = 35 MPa and where the top (or bottom) face is fully constrained, when the deflection limit is set to $L/500$, the corresponding minimum slab height is 275 mm and the curvature induced by shrinkage after 10,000 days (member loaded after 3 days from casting, notional size conservatively assumed equal to 275 mm, RH = 75%) is $\chi_{sh} = 1.345 \cdot 10^{-6}$, according to the approach provided by Model Code 2010 [48]. The comparison between this curvature and that obtained for the same cross-section with Equation (8), considering a span L = 5 m and a variable load q = 2 kN/m^2, $\chi_{max} = 1.580 \cdot 10^{-5}$, shows that neglecting shrinkage would lead to an underestimation of the cross-section curvature of only 8.5%. This underestimation decreases to 3.5% if the deflection limit is set to $L/250$ and a variable load q = 5 kN/m^2 is considered. These examples—which overestimated the shrinkage contribution to the cross-section curvature, since one side of the cross-section was assumed fully constrained—support the decision of neglecting shrinkage in the computation of the cross-section maximum curvature.

The results of the comparison between the curvature χ_{max} and the flooring limit curvature χ_{lim} are summarized in Figure 3, where red markers indicate cases where the flooring cracks, although the specific Eurocode 2 limit was respected. When the span/250 limit was adopted, flooring cracking occurred in the majority (76.7%) of the cases. However, even when the span/500 limit was adopted, 34.1% of cases led to flooring cracking. The comparison between the maximum curvature (χ_{max}) and limit curvature (χ_{lim}) showed that damage of the floorings occurred more frequently for small spans than for large spans. This result is due to the use of the deflection limits provided by the Eurocode to design the slab thickness. In the case of a simply supported beam with a parabolic bending moment along its axis, the relationship between the deflection limit provided by the Eurocode was $w_{max} = L/k$, where k is a dimensionless parameter related to the specific structure type, and the maximum curvature χ_{max} can be expressed as:

$$w_{max} = \frac{L}{k} = \frac{5}{48}\chi_{max}L^2 \Rightarrow \chi_{max} = \frac{48}{5kL} \quad (12)$$

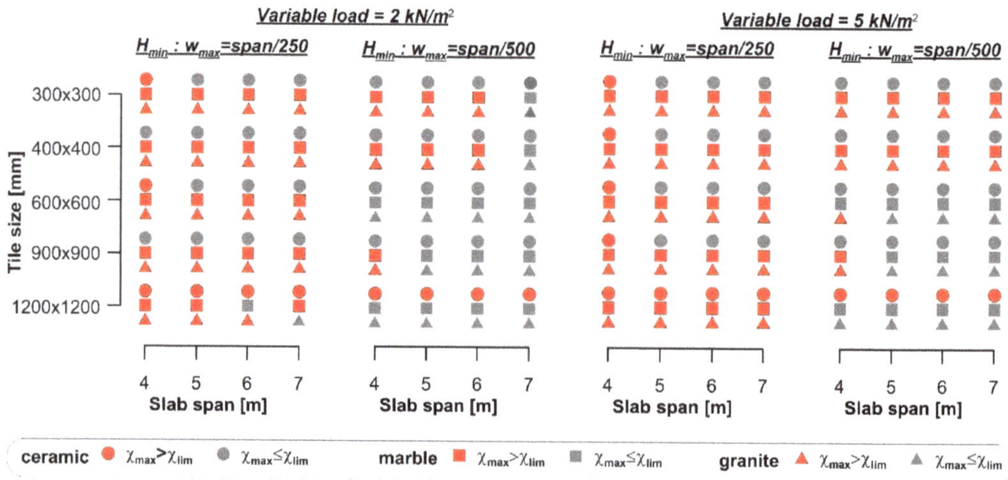

Figure 3. Comparison between the maximum and limit curvatures of floorings for rib and clay pot slabs.

Equation (12) clearly shows that, provided the parameter k, the maximum curvature that complies with the Eurocodes decreases for increasing spans. Since the limit curvature is constant (provided a certain type of tyle), a low value of L implies higher probability of damage of flooring.

The limit curvature of tiles in Figure 3 depends on their bending stiffness, which was obtained from experimental tests performed according to specific standards and reported in the datasheet. It should be noted that varying the dimension of the tile does not entail for a linear variation of its thickness, whereas the elastic modulus remains constant. Therefore, the limit curvature does not vary linearly with the variation of the tile dimension, which explains why flooring cracking seems independent from the tile dimensions in Figure 3.

4.2. *Composite Steel–Concrete Slab*

The design and requirements for composite steel–concrete slabs are provided by Eurocode 4 [6]. According to it, the maximum deflection of these types of structural member should comply with the limits enforced by Eurocode 3 [49], which, in turn, refers to the national annex. Therefore, the deflection limit span/250 under the characteristic load combination [9] adopted by the Italian code [50] will be considered in this section.

A representative composite steel–concrete slab cross-section with ceramic and stone floorings was considered in this study (Figure 4).

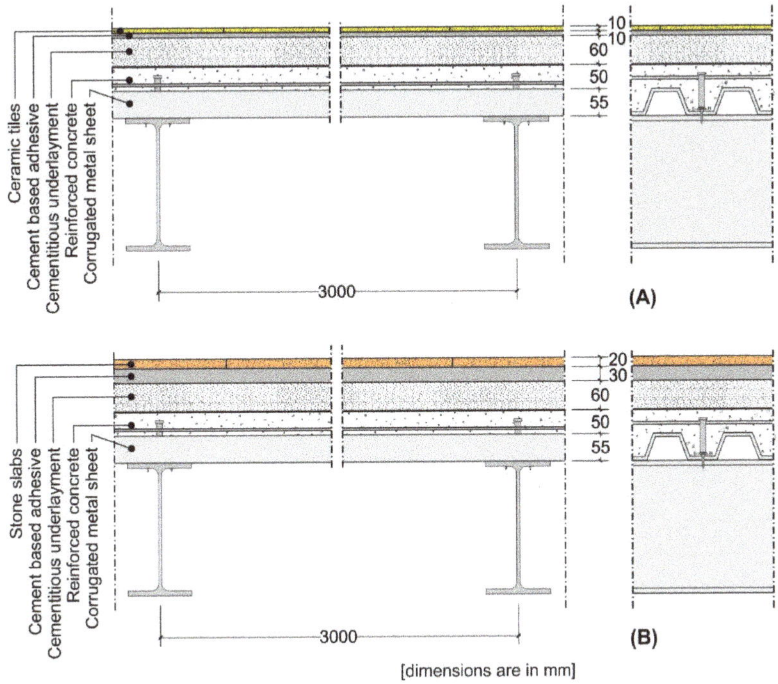

Figure 4. Cross-section of the composite steel–concrete slab with (**A**) ceramic and (**B**) stone floorings.

The cross-section geometry, which was determined starting from standardized I-shaped steel beam geometry, was kept constant, except for the second moment of area of the steel beam. Provided the applied load associated with the specific flooring and type of building (residential or commercial), the minimum value of the steel beam second moment of area J_{min} was (iteratively), computed to match the deflection limit span/250. The presence of creep of concrete and effective width (b_{eff}) were taken into account in the calculations, following the procedure proposed in [6]. J_{min} obtained for the four spans considered with ceramic and marble or granite tiles and the different variable loads selected are shown in Figure 5.

The results of the comparison between the curvature (χ_{max}) and the flooring limit curvature (χ_{lim}) are reported in Figure 6, where red markers indicate cases where the flooring cracks. Figure 6 shows that, although the deflection limit span/250 was fulfilled, large ceramic tiles cracked, despite the fact that the adopted steel beams were commercially available products and, therefore, their second moment of area was higher than the minimum required (J_{min}).

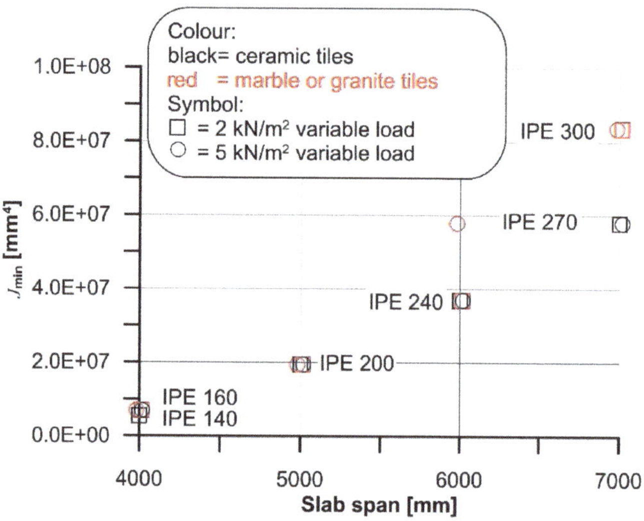

Figure 5. J_{min} computed for 4 m, 5 m, 6 m, and 7 m span of the composite steel–concrete slab.

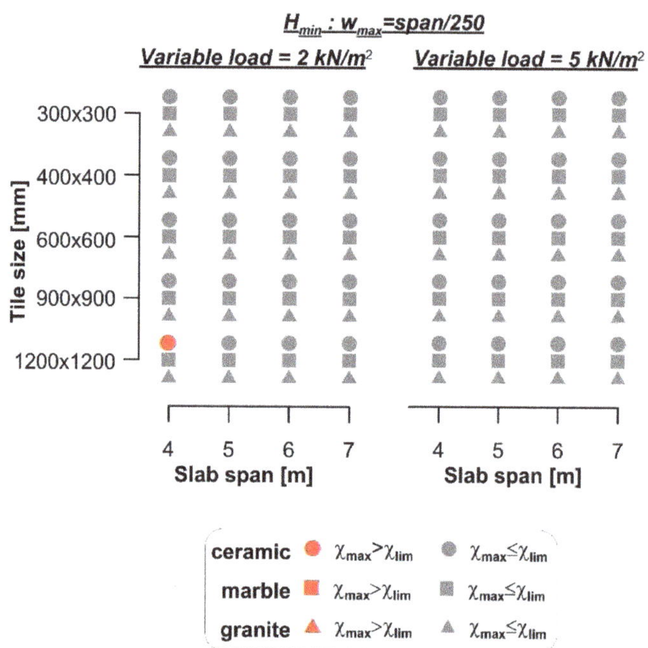

Figure 6. Comparison between the maximum and limit curvatures of floorings for composite steel–concrete slabs.

4.3. Traditional Slab Made of Timber Beams and Planks

The approach of Eurocode 5 [7] to the deflection analysis of timber structures is complex, since limits on the instant sag (w_{inst}), final net sag ($w_{net,fin}$), and final sag (w_{fin}) are enforced. For each sag considered, a range of limit values within which the specific sag limit should be determined based on the determined acceptable level of member deformation, is provided. For the sake of simplicity, only the $w_{net,fin}$ limit, which ranges between span/250 and span/350 for simply supported beams [7], is considered in this section. However, the same approach described for $w_{net,fin}$ can be applied in the case of w_{inst} and w_{fin}.

A representative cross-section of a traditional slab made by timber beams and planks with ceramic and stone floorings was considered in this study (Figure 7). Similar to the case of rib slabs (Section 4.1), the cross-section was kept constant, except for the height H of the timber beams (joists), which was iteratively computed to match the minimum and maximum limit deflections of the range provided by Eurocode 5 [7], i.e., span/250 and span/350, respectively.

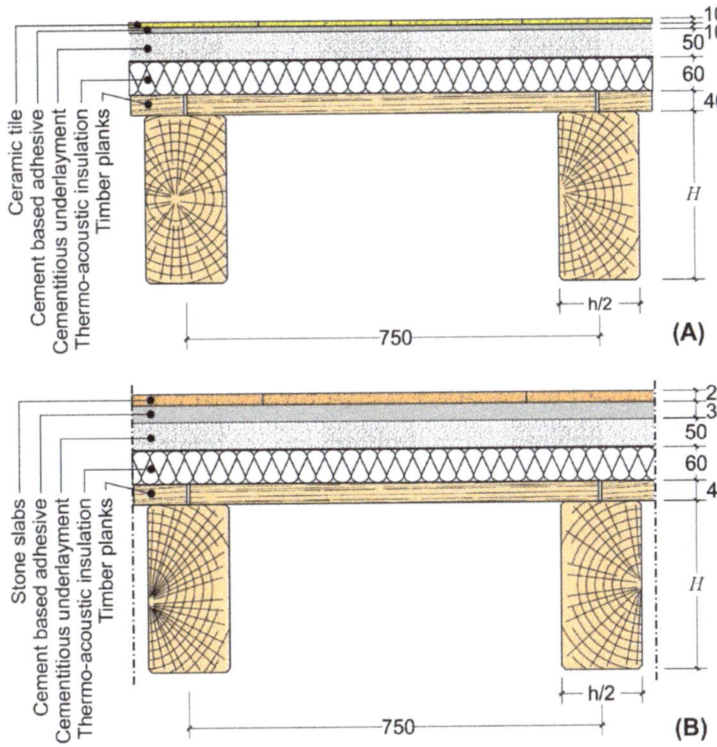

Figure 7. Cross-section of the timber beam and plank slab with (**A**) ceramic and (**B**) stone floorings.

Timber beams develop time-dependent (creep) deformation when subjected to long-term applied loads. Equation (13) was used in this study to compute the final net sag, accounting for the beam creep and shear deformations [7]:

$$w(H_{min}) = g\Delta(1 + k_{def}) + q\Delta(1 + \psi_{2.1}k_{def}) = w_{net,fin} \tag{13}$$

$$\Delta = \frac{5}{16} \times \frac{L^4}{E_m H_{min}^4} + \alpha\frac{L^2}{4G_m H_{min}^2} \tag{14}$$

where g is the permanent load, q the variable load, k_{def} = 0.6 a deformation factor that accounts for creep deformations, $\psi_{2.1} = 0.7$ the factor for quasi-permanent value of a variable action [9], α = 1.2 is the form factor [51], and E_m = 11.6 GPa and G_m = 720 MPa are the timber elastic and shear modulus, respectively [52].

The minimum cross-section height (H_{min}) obtained for the four spans considered with ceramic and marble or granite tiles and the different variable loads selected, are shown in Figure 8. The results of the comparison between the curvature (χ_{max}) and flooring limit curvature (χ_{lim}) are reported in Figure 9, where red markers indicate cases where the flooring cracks. Figure 9 shows that ceramic tiles with large size (i.e., 1200 × 1200 mm^2) cracked in almost all cases (93.8%) considered, whereas 5% of marble tiles and no granite tiles cracked.

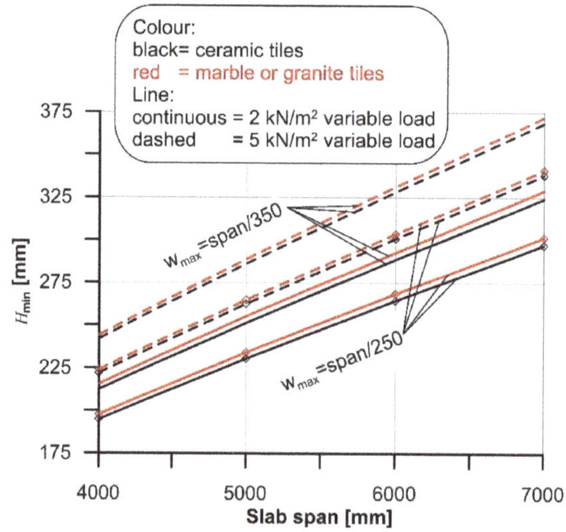

Figure 8. H_{min} computed for 4 m, 5 m, 6 m, and 7 m span of the timber slab.

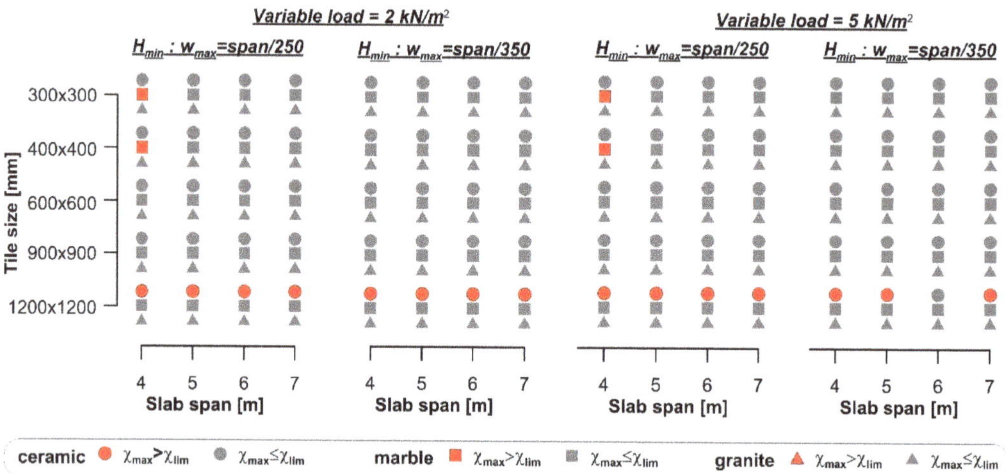

Figure 9. Comparison between the maximum and limit curvatures of floorings for traditional timber beam and plank slabs.

4.4. Effect of the Underlayment

The analyses described in Sections 4.1–4.3 were carried out assuming a rigid cementitious underlayment (Figures 1, 4 and 7). To verify the reliability of this assumption, a non-linear finite element (FE) model of a strip 1 m wide and 4 m long of the rib and clay pot slab, designed in Section 4.1, was used to obtain the flooring curvature considering a deformable underlayment. The same analysis was carried out for the composite steel–

concrete and timber beam slabs of Sections 4.2 and 4.3, but the results were not described here for brevity. All models were developed in the FE software Abaqus [53].

Three types of underlayment [54–57] and three types of flooring, with mechanical properties assumed according to the values provided by [58], were considered in the FE model. The elastic modulus and Poisson's ratio of the underlayment and floorings modeled are reported in Table 4.

Table 4. Mechanical properties of underlayment and floorings modeled.

Material	Elastic Modulus [GPa]	Poisson's Ratio
Concrete	32	0.20
Underlayment 1 (cementitious)	11	0.20
Underlayment 2 (cementitious)	18	0.20
Underlayment 3 (cementitious)	25	0.20
Flooring: ceramic tiles	60	0.28
Flooring: marble tiles	132	0.25
Flooring: granite tiles	90	0.25

The C25/30 concrete was modeled using the concrete damaged plasticity model (CDP) available in Abaqus, whereas the cementitious underlayment, which was assumed to behave as a granular-like soil, was modeled using a linear Drucker–Prager model [53]. The parameters needed for the CDP model were defined following the procedure suggested in [59], which provided a dilation angle $\psi = 36°$ [53], whereas the concrete secant elastic modulus $E_{cm} = 31$ GPa provided by the Eurocode 2 for a C25/30 concrete was considered. The underlayment was modeled considering a linear behavior (friction angle $\varphi = 30°$) up to the minimum tensile strength required by EN 13813 [58], i.e., 5 MPa, which was followed by a softening behavior to account for possible material failure [60]. The softening branch was defined following the softening curve generally adopted for cohesive materials proposed by [61], which conservatively did not account for compaction of the granular-like underlayment. A 950 N force, which simulates the force induced by the foot of a bookcase, was applied as a uniformly distributed load on a 40 mm diameter circular surface. Different mesh sizes, of either 8-node solid elements or 4-node tetrahedral elements, were used to investigate their effect on the model convergence and time required to obtain the solution. This study resulted in a FE model discretized using 8-node solid elements with approximate dimensions of $20 \times 20 \times 20$ mm^3 (see Figure 10).

Figure 10. Discretization of the FE model.

The cross-section deformations, obtained by the FE model, were used to obtain the curvatures of the flooring and of the supporting slab. The results showed that the maximum difference between the curvature of the flooring and of the supporting slab was always lower than 2%, in all the cases considered (Figure 11). This confirms that assuming a rigid underlayment did not significantly affect the result of the analytical procedure adopted in this paper. The same conclusion was obtained for composite steel–concrete and timber beam slabs.

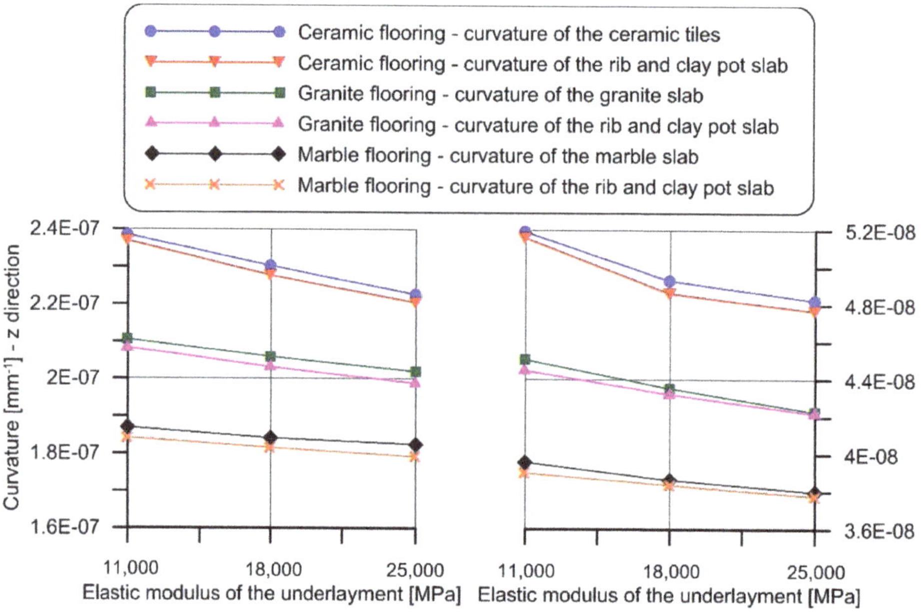

Figure 11. Comparison between the FE model curvatures of the rib and clay pot slab and those of the flooring.

5. Discussion

The analysis carried out on the different slab and flooring types showed that the deflection limits provided by the Eurocodes do not always guarantee the integrity of the superstructures. The approach adopted by the Eurocodes could be improved by reducing the maximum displacement allowed. Nevertheless, the study carried out highlights some limitations of the deflection control method, at least in the case of reinforced concrete or rib and clay pot slabs, where the calculation of the deflection is so complex that it is either oversimplified or not performed at all. Indeed, under service conditions, the simultaneous presence of cracked and non-cracked areas, which are also affected by time-dependent (viscous) phenomena and oligo-cyclic variable loads, makes correctly computing the member deflection extremely difficult [62]. The temporal evolution of the behavior of a reinforced concrete cracked section is already a quite complex problem (see for instance [43]). If then the effect of tension stiffening has to be accounted for, together with its temporal evolution due to creep, and the integration of all these phenomena over the whole length of the beam has to be carried out, the problem becomes unreasonable for a professional engineer, who must necessarily maintain a correct cost-benefit ratio in the design work. Design standards should, as far as possible, provide rules that respect this ratio. To overcome the issues associated with the deflection control method, a curvature control method, as a verification method of the horizontal member deflection under service loads, is proposed and discussed in the next sections.

5.1. Curvature Control Method

In this section, the possibility of using the criterion adopted to verify the reliability of the Eurocode deflection limits, i.e., the comparison between the maximum member curvature (χ_{max}) and limit curvature of the specific finishing elements (χ_{lim}) (or conservatively of the stiffest finishing elements), is discussed.

This method, referred to as curvature control method, is more reliable than the deflection control method currently provided by the Eurocodes, since it allows a direct comparison with the limit curvature of the finishing elements. Furthermore, computing the maximum member curvature is simpler than computing the corresponding maximum deflection. Indeed, the member deflection depends on several factors, including the load type and distribution, member static scheme, and flexural stiffness of the various elements that compose it, which determines the need of a double integration of the curvature along the entire member extent to compute its maximum deflection. This calculation becomes particularly complex when dealing with cracked concrete, since, in this case, the flexural stiffness varies along the member.

The curvature control method, unlike the deflection control method, requires a local analysis at the cross-section with the highest bending moment, independent of the member static scheme. The curvature is independent of the member span, which is required only to define the maximum service bending moment. Therefore, provided the maximum service bending moment and cross-section bending stiffness, the cross-section curvature can be simply computed without any integration along the member axis. For those cases where a deflection control is still required (e.g., to guarantee adequate drainage), the member maximum deflection could be computed by double integration of the limit curvature, which could be conservatively assumed constant along the longitudinal axis of the member considered. This computation would not require the knowledge of the curvature along the member and provide conservative results.

As an example of the advantages of adopting a curvature control rather than a deflection control method, the case of the Generali Tower, a 44-story building by Zaha Hadid Architects, built in Milan (Italy), between 2014 and 2017, is examined. The building structure consists of columns and a central core supporting structural slabs, all made of reinforced concrete. When performing the deflection control in point A of Figure 12, according to Eurocode 2 [4], many different spans (indicated with arrows in Figure 12) can be considered, each providing a specific deflection limit. Although considering the shortest span would lead to the smallest maximum deflection possible, this value could be either applied to the entire slab or only to the span selected, while other values could be computed for the other spans. By performing the deflection control based on the curvature limit, this issue does not exist, since a unique value of the maximum curvature under service loads is associated to point A. A simple finite element model of the building (needed due to the complexity of the structure) would be sufficient to determine the cross-sections with the highest applied stress and corresponding bending moments in two orthogonal directions, which can be used to compute the cross-section maximum curvature.

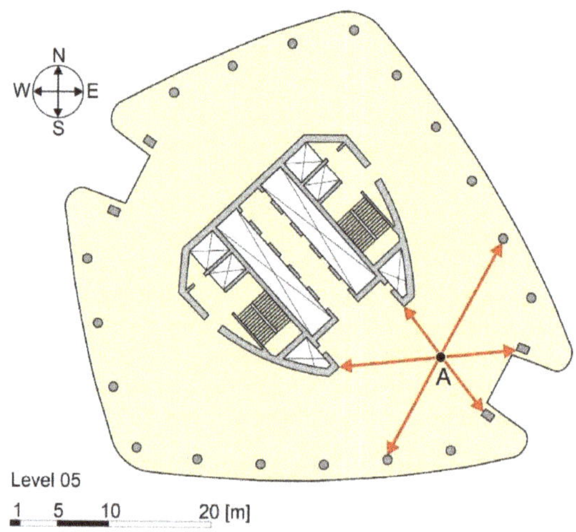

Figure 12. Plan of the fifth floor of the Generali Tower (by Zaha Hadid Architects) in Milan.

5.2. Deflection Control Based on the Limit Curvature

Assuming that the flexural stiffness along the member longitudinal axis is constant and equal to the flexural stiffness of the cross-section associated with the maximum bending moment, the limit curvature could be used to conservatively estimate the member maximum deflection, such that no damage occurs to the superstructures. Considering the case of a generic simply supported ribbed slab where the shear deformability is neglected under service loads, the maximum curvature can be computed as the sum of the curvature due to permanent loads (indicated with the subscript g) and curvature due to variable loads (indicated with the subscript q):

$$\chi_{\lim} \leq \frac{M_{\max,g}}{(EJ)_g} + \frac{M_{\max,q}}{(EJ)_q} \tag{15}$$

where $(EJ)_g$ and $(EJ)_q$ are the cross-section flexural stiffness under the permanent and variable loads, respectively, which may differ due to the cross-section applied stress and short- or long-term material properties considered. Equation (15) provides a simple solution to compute the cross-section curvature. Flexural stiffnesses under permanent and variable loads were defined separately to properly account for concrete creep, which is associated with the permanent load. Following the approach of Eurocode 2 [4], the flexural stiffnesses under permanent loads can be defined as:

$$(EJ)_g = E_{cm}J/[1 + \varphi(t, t_0)] \tag{16}$$

where $\varphi(\infty, t_0) \approx 2$ and J is the cross section second moment of area computed with Equation (7). Although more refined approaches can be adopted (see e.g., [46]), they would require complex numerical solutions that do not appear suitable for current practice applications. Note that creep plays a major role in the definition of the limit curvature and should not be neglected. As an example, in the case of rib and clay pot slabs with span varying from 4 m to 7 m (see Figure 2), neglecting creep (i.e., setting $\varphi(\infty, t_0) = 0$) would result in an average slab maximum curvature ($\chi_{\max}$) 61% and 40% lower than that obtained when considering creep when the variable load is $q = 2$ kN/m^2 and $q = 5$ kN/m^2, respectively.

Assuming uniformly distributed applied loads, Equation (15) can be rewritten as:

$$\chi_{\lim} \leq \frac{gL^2}{8(EJ)_g} + \frac{qL^2}{8(EJ)_q} = \frac{L^2}{8}\left[\frac{g}{(EJ)_g} + \frac{q}{(EJ)_q}\right] \Rightarrow \left[\frac{g}{(EJ)_g} + \frac{q}{(EJ)_q}\right] \leq \frac{8\chi_{\lim}}{L^2} \tag{17}$$

Similarly, the maximum vertical displacement ($w_{\max}$) of the same generic simply supported ribbed slab can be obtained as the sum of the displacement induced by the permanent loads and displacement induced by the variable loads:

$$w_{\max} = \frac{5}{384}\left[\frac{gL^4}{(EJ)_g} + \frac{qL^4}{(EJ)_q}\right] = \frac{5L^4}{384}\left[\frac{g}{(EJ)_g} + \frac{q}{(EJ)_q}\right] \tag{18}$$

$w_{\max}$ can be expressed as a function of the limit curvature by substituting Equation (17) into Equation (18):

$$w_{\max} \leq \frac{5\chi_{\lim}L^2}{48} \tag{19}$$

According to the Eurocodes, the maximum deflection should respect the inequality in Equation (20) (see Section 4.1):

$$w_{\max} \leq \frac{L}{k} \tag{20}$$

where k is a dimensionless parameter related to the specific structure type. Therefore, rearranging Equation (20) and substituting $w_{\max}$ provided by Equation (19) into it, k can be expressed as a function of the limit curvature:

$$k \leq \frac{48}{5\chi_{\lim}L} \tag{21}$$

Equation (21) shows that k is independent of the flexural stiffness of the element and applied loads, which makes it suitable for applications to any type of structure. Furthermore, the ratio 48/5 in Equation (21) accounts for the specific applied load distribution (this ratio is equal to 12 in the case of a simply supported beam with a concentrated load at midspan), whereas the approach adopted by the Eurocodes [i.e., Equation (20)] is independent from it. If a check based on the curvature control method is adopted, the structural scheme of the slab would not affect the results.

These considerations indicate that the approach adopted by the Eurocodes, based on the vertical displacement limit, may lead to uncertainties in the evaluation of the maximum displacement allowed and provide non-conservative results for certain slab configurations, whereas the approach based on the curvature control method appears more rapid and reliable. Moreover, adopting the curvature control method would allow us to define a general limit curvature for all types of flooring that would work as a minimum product performance target for the manufacturers and, at the same time, guarantee the absence of cracking in the floorings. Finally, the curvature control method could be also conveniently applied in specific problems associated with the use of innovative technologies that are emerging in the world of construction, such as the case of bridge slabs reinforced with glass fiber-reinforced polymer (GFRP) bars. In this case, the slab design is controlled by the deformability (rather than by the strength), which should be limited to ensure the integrity of the asphalt pavement under the characteristic load combination [10]. Therefore, a simple comparison between the maximum curvature allowed for the asphalt and corresponding slab curvature would be sufficient to verify the slab deformability.

6. Conclusions

This paper analyzed the approach provided by the Eurocodes to limit the deflection of horizontal structural members and, in turn, guarantee the integrity of the superstructures. Different types of horizontal member, namely rib and clay pot (or hollow block), composite steel–concrete, and timber beam slabs were designed to respect the deflection limit enforced by the Eurocodes. The maximum curvature of these members was compared with the limit curvatures of various types of flooring to verify the occurrence of damage. The results obtained allowed for drawing the following conclusions:

- The deflection limit method adopted by the Eurocodes is complex and does not always guarantee the absence of damage to the floorings. Although the rules enforced by the Eurocodes for a reinforced concrete or a rib and clay pot slab were respected (except for tension stiffening that was neglected), up to 76.7% of the ceramic, marble, and granite floorings cracked. Furthermore, when dealing with a reinforced concrete or a rib and clay pot slab, the Eurocode 2 approach requires taking into account cracking, concrete creep, and tension stiffening, which make it extremely complex and hardly applicable for a professional engineer.
- The curvature control method is much simpler than the deflection control method adopted by the Eurocodes, since a direct verification on the curvature limits is performed. The curvature control method only requires a cross-section analysis, whereas the deflection control method requires the integration of curvature along the entire member axis.
- The curvature control method considers the constraint of the slab by computing the maximum curvature from the maximum service bending moment, calculated considering geometry, constraints, and intended use of the slab. Similarly, the limit imposed to the displacement in the deflection limit method seems independent from the constraints acting on the slab (it depends just on the span), which are accounted for in the computation of the maximum deflection.
- The curvature control method would allow for defining a general limit curvature value for floorings that could be adopted as minimum performance level in standards and would be able to guarantee the absence of flooring cracking. Furthermore, it appears promising for applications to specific problems arising with the use of innovative technologies, as in the case of bridge slabs reinforced with GFRP bars in which the design is controlled by the slab deformability rather than by its strength.

In conclusion, this research tries, through an alternative proposal, to open the discussion on the theme of deflection control in horizontal members, which is traditionally considered well-established, although it presents unsolved issues.

Author Contributions: Conceptualization, T.D. and M.A.P.; data curation, M.A.P.; investigation, T.D. and M.A.P.; writing—original draft, M.A.P.; writing—draft review, T.D. All authors have read and agreed to the published version of the manuscript.

Funding: This research received no external funding.

Data Availability Statement: The data presented in this study are available on request from the corresponding author.

Conflicts of Interest: The authors declare no conflict of interest.

References

1. Gilbert, R.I. The Serviceability Limit States in Reinforced Concrete Design. *Procedia Eng.* **2011**, *14*, 385–395.
2. De Domenico, D.; Messina, D.; Recupero, A. A Combined Experimental-Numerical Framework for Assessing the Load-Bearing Capacity of Existing PC Bridge Decks Accounting for Corrosion of Prestressing Strands. *Materials* **2021**, *14*, 4914. [CrossRef]
3. The Concrete Society. *TR 58. Deflections in Concrete Slabs and Beams*; The Concrete Society: Camberley, UK, 2019.
4. European Committee for Standardization. *Eurocode 2: Design of Concrete Structures—Part 1-1: General Rules and Rules for Buildings—EN 1992-1-1*; European Committeee for Standardization: Bruxelles, Belgium, 2010.
5. Topolář, L.; Kocáb, D.; Šlanhof, J.; Schmid, P.; Daněk, P.; Nováček, J. Testing the Influence of the Material Bonding System on the Bond Strength of Large-Format Tiles Installed on Concrete Substrate under Mechanical Loading. *Materials* **2020**, *13*, 3200. [CrossRef]
6. European Committee for Standardization. *Eurocode 4: Design of Composite Steel and Concrete Structures—Part 1-1: General Rules and Rules for Buildings—EN 1994-1-1*; European Committeee for Standardization: Bruxelles, Belgium, 2009.
7. European Committee for Standardization. *Eurocode 5: Design of Timber Structures—Part 1-1: General-Common Rules and Rules for Buildings—EN 1995-1-1:2004+A1*; European Committeee for Standardization: Bruxelles, Belgium, 2008.
8. International Organization for Standardization. *ISO 4356:1977 Bases for the Design of Structures—Deformations of Buildings at the Serviceability Limit States*; International Organization for Standardization: Geneva, Switzerland, 2014.
9. European Committee for Standardization. *Eurocode—Basis of Structural Design—EN 1990:2002*; European Committee for Standardization: Brussels, Belgium, 2005.

10. D'Antino, T.; Pisani, M.A. Influence of sustained stress on the durability of glass FRP. *Constr. Build. Mater.* **2018**, *187*, 474–486. [CrossRef]
11. Meille, S.; Saâdaoui, M.; Reynaud, P.; Fantozzi, G. Mechanisms of crack propagation in dry plaster. *J. Eur. Ceram. Soc.* **2003**, *23*, 3105–3112. [CrossRef]
12. Berardi, U.; Tronchin, L.; Manfren, M.; Nastasi, B. On the Effects of Variation of Thermal Conductivity in Buildings in the Italian Construction Sector. *Energies* **2018**, *11*, 872. [CrossRef]
13. Kumar, S.; Arun Prakash, S.; Pandiyarajan, V.; Geetha, N.; Antony Aroul Raj, V.; Velraj, R. Effect of phase change material integration in clay hollow brick composite in building envelope for thermal management of energy efficient buildings. *J. Build. Phys.* **2020**, *43*, 351–364. [CrossRef]
14. Masia, M.J.; Kleeman, P.W.; Melchers, R.E. Modeling Soil/Structure Interaction for Masonry Structures. *J. Struct. Eng.* **2004**, *130*, 641–649. [CrossRef]
15. Brodsky, A.; Rabinovitch, O.; Yankelevsky, D.Z. Determination of the interaction between a masonry wall and a confining frame. *Eng. Struct.* **2018**, *167*, 214–226. [CrossRef]
16. Besana Moquette. Available online: https://www.besanamoquette.com/ (accessed on 29 January 2020).
17. Biagiotti, J.; Iannoni, A.; Lopez-Manchado, M.A.; Kenny, J.M. Cure characteristics, mechanical properties, and morphological studies of linoleum flour-filled NBR compounds. *Polym. Eng. Sci.* **2004**, *44*, 909–916. [CrossRef]
18. Tian, N.; Ning, R.; Kong, J. Self-toughening of epoxy resin through controlling topology of cross-linked networks. *Polymer* **2016**, *99*, 376–385. [CrossRef]
19. Kerakoll Design House. Available online: https://kerakolldesignhouse.com/ (accessed on 29 January 2020).
20. Mapei Mapei—Adhesives, Sealants, Chemical Products for Building. Available online: https://www.mapei.com/en (accessed on 29 January 2020).
21. Ayrilmis, N.; Buyuksari, U.; As, N. Bending strength and modulus of elasticity of wood-based panels at cold and moderate temperatures. *Cold Reg. Sci. Technol.* **2010**, *63*, 40–43. [CrossRef]
22. Ruman, D.; Záborský, V.; Svoboda, T.; Kašičková, V.; Rodrová, V. Identifying the Characteristics of Laminated Wood Based on the Values of Deflection Measured during its Bending. *BioResources* **2017**, *12*, 2592–2608. [CrossRef]
23. Kaklis, K.; Maurigiannakis, S.; Agioutantis, Z.; Istantso, C. Influence of Specimen Shape on the Indirect Tensile Strength of Transversely Isotropic Dionysos Marble Using the Three-Point Bending Test. *Strain* **2009**, *45*, 393–399. [CrossRef]
24. Garzonio, C.A.; Giovannini, P.; Cavallucci, F. Analyses of the physical parameters correlated to bending phenomena in marble slabs. In Proceedings of the 9th International Congress on Deterioration and Conservation of Stone, Venice, Italy, 19–24 June 2000; Fassina, V., Ed.; Elsevier Science B.V.: Amsterdam, The Netherlands, 2000; pp. 89–97, ISBN 978-0-444-50517-0.
25. Liu, Z.; Shao, J. Strength Behavior, Creep Failure and Permeability Change of a Tight Marble Under Triaxial Compression. *Rock Mech. Rock Eng.* **2017**, *50*, 529–541. [CrossRef]
26. Granitmarmi. Available online: https://www.granitmarmi.it/ (accessed on 29 January 2020).
27. Sa.Ge.Van. Marmi. Available online: https://www.sagevanmarmi.com/ (accessed on 29 January 2020).
28. FMG FMG Book. Available online: https://www.irisfmg.it/doc/cataloghi_catalogues/2general_fmg_standard_size.pdf (accessed on 29 January 2020).
29. Secchiari, L. *Stone Materials: The Techniques of Use in Advanced Façade Systems*; Alinea Editrice s.r.l.: Florence, Italy, 2012; ISBN 978-88-6055-722-3. (In Italian)
30. CEN. *EN-14411:2016—Ceramic Tiles—Definition, Classification, Characteristics, Assessment and Verification of Constancy of Performance and Marking*; European Committee for Standardization: Brussels, Belgium, 2016.
31. CEN. *EN ISO 10545 4:2019—Ceramic Tile. Part 4: Determination of Modulus of Rupture and Breaking Strength*; European Committee for Standardization: Brussels, Belgium, 2019.
32. Fragassa, C.; Ubertini, F.; Pavlovic, A.; Conti, P. *Limits of Applicability of Regulations on Ceramics in the Case of Pirodeformed Elements*; AIAS—Associazione Italiana per l'Analisi delle Sollecitazioni: Bologna, Italy, 2014; pp. 433–443. (In Italian)
33. Hegedűsová, L.; Ceniga, L.; Dusza, J. Bending and Contact Strength of Monolithic Ceramic Materials. *Int. J. Damage Mech.* **2012**, *21*, 293–305. [CrossRef]
34. FAP. FAP Tile Company: Quality Ceramic Floor and Wall Tiles Suppliers. Available online: https://www.fapceramiche.com/en/ (accessed on 29 January 2020).
35. GranitiFiandre. Italian Floor Tiles, Porcelain Tile Flooring and Walls—Fiandre. Available online: http://www.granitifiandre.com/ (accessed on 29 January 2020).
36. Ceramica Sant'Agostino. Porcelain Stoneware Floor and Wall Tiles. Available online: https://www.ceramicasantagostino.it/en/ (accessed on 29 January 2020).
37. Ceramiche Refin. Porcelain Tiles | Floor Tile | Italian Ceramic Tile. Available online: https://www.refin-ceramic-tiles.com/ (accessed on 29 January 2020).
38. Tile Council of North America (TCNA). *Handbook for Ceramic, Glass, and Stone Tile Installation*; Tile Council of North America Inc.: Anderson, SC, USA, 2017.
39. de Sousa, R.; Sousa, H. Analysis of reinforcement techniques for partition walls subjected to vertical deformations of concrete slabs. In Proceedings of the XII International Conference on Structural Repair and Rehabilitation, Porto, Portugal, 26 October 2016.

40. Abreu, M.; Leitao, V.; Lucas, J.C. *Modelling the Behaviour of Ceramic Tile Coverings*; Colegio Oficial de Ingenieros Superiores Industriales: Castellón, Spain, 2004; Volume 3, pp. 3–17.
41. Diolaiti, E. Physical-Mechanical Characterizations of Large Format Ceramic Tiles. Master's Thesis, Alma Mater Studiorum (University of Bologna), Bologna, Italy, 2016. (In Italian).
42. CEN. *EN 771-1:2011+A1:2015—Specification for Masonry Units. Part 1: Clay Masonry Units*; European Committee for Standardization: Brussels, Belgium, 2011.
43. Toniolo, G.; Di Prisco, M. *Reinforced Concrete Design: To Eurocode 2*; Springer: Berlin/Heidelberg, Germany, 2017.
44. European Committee for Concrete. *CEB FIP Manual of Buckling and Instability. Bulletin 123*; CEB FIP: Paris, France, 1977.
45. Gilbert, R.I.; Ranzi, G. *Time-Dependent Behaviour of Concrete Structures*; Spon Press: London, UK, 2010.
46. Pisani, M.A. Strength of FRP RC sections after long-term loading. *Struct. Eng. Mech.* **2003**, *15*, 345–365. [CrossRef]
47. Bradford, M.A. Generic modelling of composite steel–concrete slabs subjected to shrinkage, creep and thermal strains including partial interaction. *Eng. Struct.* **2010**, *32*, 1459–1465. [CrossRef]
48. Federation Internationale du Beton. *Fib Model Code for Concrete Structures 2010*; Ernst & Sohn GmbH & Co.: Lausanne, Switzerland, 2013; ISBN 978-3-433-03061-5.
49. European Committeee for Standardization. *Eurocode 3: Design of Steel Structures—Part 1-1: General Rules and Rules for Buildings—EN 1993-1-1*; European Committeee for Standardization: Bruxelles, Belgium, 2006.
50. Ministero delle Infrastrutture e dei Trasporti. *Norme Tecniche per le Costruzioni*; MIT: Rome, Italy, 2018.
51. Gere, J.M.; Timoshenko, S.P. *Mechanics of Materials*, 2nd ed.; Brooks/Cole Engineering Div: Monterey, CA, USA, 1984; ISBN 978-0-534-03099-5.
52. Forest Products Laboratory. Mechanical properties of wood. In *Wood Handbook—Wood as an Engineering Material*; Department of Agricolture, Forest Service: Madison, WI, USA, 1999; p. 463.
53. Simulia. Abaqus 6.14 Extended Functionality Online Documentation. Available online: http://130.149.89.49:2080/v6.14/ (accessed on 15 January 2016).
54. Fermacel-Sottofondi. Available online: https://www.fermacell.it/Sottofondi (accessed on 29 January 2020).
55. Knauf. Available online: http://www.knauf.it/Default.aspx (accessed on 29 January 2020).
56. Laterlite. Available online: https://www.laterlite.com/ (accessed on 29 January 2020).
57. Saint-Gobain Weber. Available online: https://www.it.weber/ (accessed on 29 January 2020).
58. CEN. *prEN 13813:2017 E—Screed Material and Floor Screeds—Screed Material—Properties and Requirements*; European Committee for Standardization: Brussels, Belgium, 2017.
59. Jankowiak, T.; Lodygowsky, T. Identification of parameters of concrete damage plasticity constitutive model. *Found. Civ. Environ. Eng.* **2005**, *6*, 53–69.
60. de Borst, R.; Groen, A.E. Computational strategies for standard soil plasticity models. In *Modeling in Geomechanics*; John Wiley & Sons, Ltd.: Hoboken, NJ, USA, 2000; pp. 23–50.
61. Reinhardt, H.W.; Cornelissen, H.; Hordijk, D.A. Tensile Tests and Failure Analysis of Concrete. *J. Struct. Eng.* **1986**, *112*, 2462–2477. [CrossRef]
62. Marí, A.; Bairán, J.-M.; Oller, E.; Duarte, N. Modeling serviceability performance and ultimate capacity of corroded reinforced and prestressed concrete structures. *Struct. Concr.* **2021**, 1–10. [CrossRef]

materials MDPI

Article

The Development of a New Phosphogypsum-Based Construction Material: A Study of the Physicochemical, Mechanical and Thermal Characteristics

Hela Garbaya [1,2], Abderraouf Jraba [1,2], Mohamed Amine Khadimallah [3,*] and Elimame Elaloui [1]

1. Laboratory of Materials Applications in Environment, Water and Energy, Faculty of Sciences, University of Gafsa, Gafsa 6029, Tunisia; helagarbaya@hotmail.fr (H.G.); jrabaraoof@gmail.com (A.J.); limamealoui@gmail.com (E.E.)
2. Faculty of Sciences of Gabes, University of Gabes, Gabes 6029, Tunisia
3. Civil Engineering Department, College of Engineering, Prince Sattam Bin Abdulaziz University, Al-Kharj 16273, Saudi Arabia

* Correspondence: mohamedamine.khadimallah@fsgf.rnu.tn

Citation: Garbaya, H.; Jraba, A.; Khadimallah, M.A.; Elaloui, E. The Development of a New Phosphogypsum-Based Construction Material: A Study of the Physicochemical, Mechanical and Thermal Characteristics. *Materials* **2021**, *14*, 7369. https://doi.org/10.3390/ma14237369

Academic Editors: Dario De Domenico and Luís Filipe Almeida Bernardo

Received: 8 October 2021
Accepted: 29 November 2021
Published: 1 December 2021

Publisher's Note: MDPI stays neutral with regard to jurisdictional claims in published maps and institutional affiliations.

Abstract: Phosphogypsum (PG) is a waste (or by-product) of the production of phosphoric acid, a basic constituent in the manufacturing of modern fertilizers. The annual production of phosphogypsum in Tunisia is currently estimated to be 10 million tons. Its storage in slag in close proximity to production plants generates pollution problems; however, valorization may be a solution. The present paper proposes a simple process for the valorization of this by-product into a construction material. Several physicochemical characterizations are used to prove the characteristics of samples. The chemical composition shows that PG is a gypsum compound with several impurities. The morphological analyses show that the powder materials are mesoporous with a lower specific area. The structural characterizations show that these solids play the role of a water pump as the degree of hydration changes from 2 to 0 and vice versa, depending on the temperature. Mechanical and thermal analyses show that the prepared formulation is brittle and insulating, which presents opportunities for it to be used as a decoration material.

Keywords: phosphogypsum; valorization; construction material; mechanical and thermal properties

1. Introduction

Phosphogypsum (PG) represents the majority of the solid waste produced by the phosphate industry. Phosphoric acid and calcium sulfate dehydrate phosphogypsum are produced by the decomposition of extracted raw phosphate rocks with concentrated sulfuric acid at a temperature range of 75–80 °C. The chemical reaction of PG production is [1]:

$$(Ca_3(PO_4)_2)_3CaF_2 + 10H_2SO_4 + 20H_2O \rightarrow 6H_3PO_4 + 10CaSO_4.2H_2O + 2HF$$

Phosphoric acid is mainly used in the production of phosphorus fertilizers such as DAP (diammonium phosphate) and MAP (monoammonium phosphate). For every ton of P_2O_5 produced as phosphoric acid, five tons of dry mass phosphogypsum are produced. This equates to an annual quantity of waste of 10 million tons. Phosphogypsum is usually deposited in large stockpiles without any treatment [2–4].

For decades, the only way of recycling PG was by using it as an additive in agriculture. Mesic et al. demonstrated the positive effects of PG for soil, water and plants [5]. Waste PG is used mainly in agriculture with several methods of recycling for the fertilization and amelioration of soil rentability. It is used as a fertilizer in agriculture because of its high volumes of calcium, phosphorus and sulfur [6].

Many studies have suggested that phosphogypsum can be used as a substitute for natural gypsum to control the hydration reaction rate in Portland cement production [7]. PG has also been used in the manufacturing of bricks: the incorporation of 30% of PG into annealed clay bricks provides a product that successfully satisfies the standard requirements [8].

In 2021, Ajam [9] proved that the use of Tunisian PG in non-load-bearing brick fabrication requires a low amount of energy and consumes a large amount of waste, which largely reduces environmental pollution, in addition to the high socioeconomic benefits. In addition, this study shows that the radioactive emission of the components of this brick is below the limit values recommended by the standards, and therefore its use is safe. Moreover, Hamdi et al. 2020 [10] prove that Tunisian PG have potential uptake in the material construction industry as paving blocks.

Despite the recycling routes presented above, the large quantities of phosphogypsum generated pose a problem of space especially in urban areas. Over 85% of PG is stored in close proximity to phosphoric acid production units in stockpiles that can reach tens of meters in height. The remaining 15% is either reused or thrown into the sea [11,12]. This management of PG waste presents an extreme threat to human and marine life. The study by Rouis et al. [13] shows that phosphogypsum is a by-product that harms the environment if not stored properly; the storage and recycling of phosphogypsum present the main challenges of the phosphate industry in many countries [8].

In light of the above, the objective of the present paper is presented in four parts. The first part is the study of the radioactivity of generated phosphogypsum, aiming to ensure that PG is safe to be used as construction material. The second part examines the physicochemical characteristics of raw, washed and treated PG as well as prepared samples (PGM) using FTIR, XRD, SEM, EDX, adsorption/desorption of N_2 at 77 K and XRF analyses. The last section discusses the mechanical and thermal behaviors of the prepared formulation.

2. PG Radioactivity

Usually, PG contains radioactive elements. The radioactivity of PG (in particular (α)) is due to the radium content resulting from the decomposition of uranium (present in the phosphate ore). To ensure that the use of PG as a construction material does not pose any danger to users, we have chosen the study of the radioactivity of two types of PG (PG from Croatia and PG from Tunisia).

The activities and concentrations of the different radionuclides in Croatia PG (Lonjsko Polje Nature Park) [14] are presented in Table 1. The major sources of radioactivity in PG are ^{238}U and ^{232}Th [15]. Uranium is the main environmental radiotoxic element associated with phosphoric acid production; it is transferred from a non-mobile fraction in the phosphate rock to a bioavailable fraction in phosphogypsum [16].

Table 1. Radioactive nuclides in PG.

Nuclide	^{238}U	^{226}Ra	^{228}Ra	^{228}Th	^{330}Th	^{232}Th
Value (Bq/g)	0.05–0.21	0.004–1.48	0.07–0.5	0.001–0.63	0.1–2.9	0.004–0.5

The radiation dose resulting from phosphogypsum piles or received by workers is negligible compared with the average annual effective dose from natural sources [17]. The resulting radiation dose caused by phosphogypsum used as a construction or plaster material can be considered to be negligible [18,19].

Sfar et al. [20] measured the activity of natural radioelements in three Tunisian PGs with different storage times using gamma spectrometry. They noted a decreasing trend of the concentrations of ^{238}U and ^{232}Th from the most recent to the oldest phosphogypsum, respectively (Table 2). This reduction was most likely due to leaching by natural processes, mainly rainwater. The concentration of ^{226}Ra in phosphogypsum remained constant during storage. The measurements of thorium confirmed that ^{232}Th preferentially passed into phosphoric acid during the manufacturing process [20].

Table 2. Average activities of the principal natural radioelements of PG.

	Fresh	10 Years	50 Years
^{238}U (Bq/kg)	65.9 ± 1.7	41.2 ± 2.2	35.2 ± 1.9
^{232}Th(Bq/kg)	19.7 ± 1.7	16.0 ± 1.4	8.2 ± 1.2
^{226}Ra(Bq/kg)	209.4 ± 6.0	209.8 ± 6.9	219.6 ± 6.3

On the basis of the previously cited studies, it can be concluded that phosphogypsum does not exhibit any nuclear activity that is harmful to humans or to the environment. As a result, its recovery as a construction or insulation material presents a solution for environmental decontamination and not a new danger.

3. Experimental Procedure

3.1. Sample Preparation

Wet natural phosphogypsum (PGF) was directly obtained from the slag heap of the Tunisian Chemical Group M'dhilla Plant (Gafsa, Tunisia). The washed phosphogypsum (PGW) was obtained by washing the PGF; this was placed in a large sieve and washed several times until the wash water obtained a neutral pH (between 6.7 and 7). The PGW was left to dry in the open air for 7 days. The dried PGW was ground by a mechanical grinder equipped with an 0.5 mm sieve to obtain a fine and uniform powder. The PGT was obtained by the thermal treatment of the PGW powder at 200 °C for 12 h [14]. The PGM 1/1 and PGM 1/2 samples were prepared by mixing PGT with water at a PGT/water ratio of 1:1 and 1:2, respectively. The mixture was then poured into molds as needed. The geometric dimensions of the prepared materials were 4 × 4 × 1 cm for the thermal test and 4 × 4 × 16 cm for the mechanical test (Figure 1).

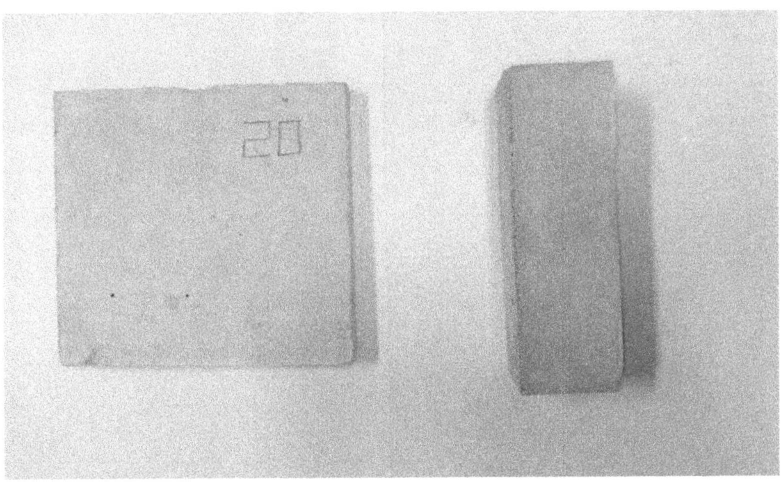

Figure 1. Geometric forms of the PGM 1/1 and PGM 1/2 samples.

3.2. Characterization

The characterization of the phosphogypsum samples was measured by different techniques. The morphological analysis was determined by environmental scanning electron microscopy (SEM) (Quanta 200-FEI) with an accelerating voltage of 15 kV coupled to an EDAX probe. The textural analysis was obtained using a micromeritics instrument (model ASAP 2020 V4.03). The porosity and specific surface area were measured at 77 °K after degassing for 4 h at 105 °C under a vacuum (10 μm/Hg). The FTIR spectra were recorded in KBr pellets using a Shimadzu S400 instrument. The spectra of the solids were obtained using KBr pellets. Prior to the measurements, PG and KBr were mixed at a quality ratio of 1:100. The vibrational transition frequencies were reported in transmittance versus the wave numbers (cm^{-1}). The structural properties of the samples were determined using a PANanalytical X'Pert Pro wide-angle X-ray powder diffractometer equipped with a copper anticathode that produced 15,418 Å Cu Kα radiation. An X-ray fluorescence (XRF) Philips sequential wavelength dispersion unit (model PW-1404) was used to determine the elemental composition. The mechanical properties were obtained using a universal ZWICK/ROELL machine. The test bench was equipped with self-tightening jaws and a force cell with a capacity of 5 kN. It was controlled by TEST EXPERT software, which logged the test parameters, acquired, and processed the data. The thermal conductivity and diffusivity coefficient measurements were obtained using a Hot-Disc TP 2500 apparatus. A probe (reference 5465) with a radius of 3189 mm was used. A heating power of 80 mW was applied for 20 s. Further details concerning this method are available in [21–25].

4. Results and Discussion

4.1. X-ray Fluorescence

As seen in Table 3, the phosphogypsum in these various states was formed mainly by gypsum ($CaSO_4$) with the presence of other elements with a low percentage such as Si, Ti, Na, Mg and Fe. The presence of these elements could be attributed to the ore of the phosphate used in the industrial process. This composition was similar to that cited by Mechi et al. [26]. The treatment of PGF did not significantly affect the chemical composition but it increased the CaO/SO_3 ratio from 0.83 to 0.95.

Table 3. Chemical composition of the different phosphogypsum samples.

	PGF	PGW	PGT
Fe_2O_3 (%)	0.298	0.319	0.366
MgO (%)	1.460	1.489	1.276
Na_2O (%)	0.586	0.203	0.303
SO_3 (%)	49.170	50.203	46.489
SiO_2 (%)	7.638	6.015	6.905
CaO (%)	40.826	41.750	44.302
TiO_2 (%)	0.0198	0.0193	0.209
CaO/SO_3	0.830	0.831	0.953

4.2. Powder X-ray Diffraction

The registered X-ray diffractograms of the different phosphogypsum samples studied showed only the presence of the characteristic peaks of $CaSO_4$. These results confirmed the gypsum aspect of PG observed by XRF. The characteristic peaks of the different hydration degrees of the gypsum ($CaSO_4$, $CaSO_4.1/2H_2O$ and $CaSO_4.2H_2O$) appeared. Figure 2 indicates that only the characteristic peaks of anhydrous $CaSO_4$ appeared in the spectrum of PGT; the lack of coordinate water in this sample could be explained by the thermal treatment at 200 °C. The diffractograms of the other two samples (PGF and PGW) showed the presence of two degrees of hydration ($2H_2O$ and $1/2H_2O$). The presence of these degrees of hydration was because they were mixed with water both from the industrial process for PGF and from the washing water for PGW. During the process of preparing PG for use as a building and/or decoration material, phosphogypsum changed from a hydration level of 1/2 and 2 (PGF and PGW, respectively) to an anhydrous state of 0 (PGT). In its finished state, the prepared material sample (PGM) regained the hydration levels of 1/2 and 2 (Figure 2); this was due to being mixed with water and the low temperature of the treatment (T = 40 °C).

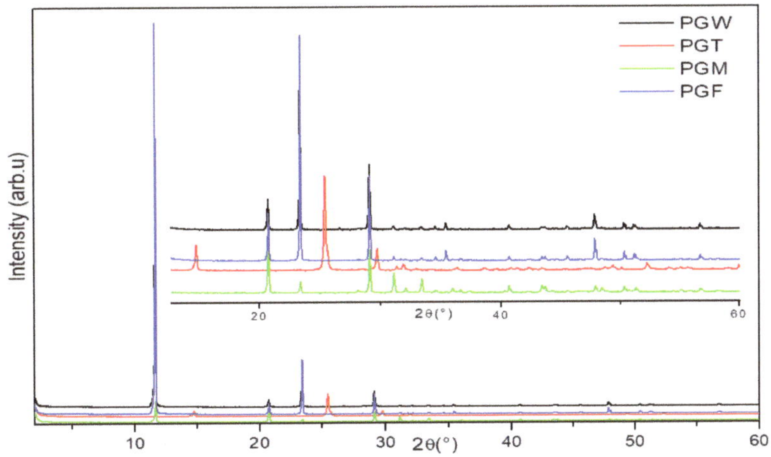

Figure 2. X-ray diffractograms of the PGF, PGW, PGT and PGM samples.

It should be noted that the PG with these different degrees of hydration presented a hydration–dehydration phenomenon. The PG in its natural or washed form (PGF, PGW) was gypsum hydrated at 1/2 and/or 2 H_2O. After a thermal treatment at 200 °C, it became anhydrous ($CaSO_4$) and it obtained the degrees of 1/2 and/or 2 H_2O by being mixed with water (PGM). This reversible hydration–dehydration phenomenon that the PG presented in its different states allowed us to qualify it as water pump, as shown in Figure 3.

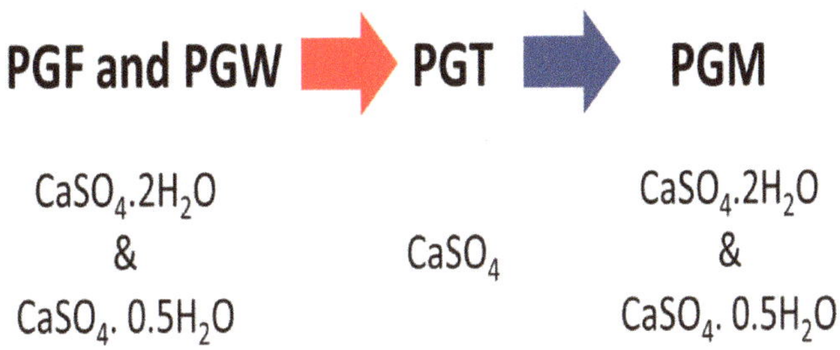

Figure 3. Hydration–dehydration of PG.

4.3. FTIR Analysis

The FTIR spectra of the PG in the different states (Figure 4) confirmed two results (one qualitative and the other quantitative) shown by the X-ray diffractograms, namely, that the PG was mainly composed of gypsum. This was represented by the peaks of $SO_4{}^{2-}$ lying at 594 cm^{-1}, 1099 cm^{-1} and 2132 cm^{-1} [26–28], as shown in Figures 5 and 6. The second result consisted of the quantitative variation in the peaks of the water at 1620 cm^{-1}, which was attributed to the vibrations of the OH groups of the water. At 3590 cm^{-1}, this corresponded with the elongation of the internal OH groups [29,30]. The spectrum of PGT presented the least important peaks compared with those of the states mixed with water.

The resulting isotherms of the surface analysis of the treated and modeled phosphogypsum (Figure 5) were type IV, which corresponded with mesoporous solids [31]. This isotherm type corresponded with multimolecular adsorption or a gradual increase in the adsorbed layer thickness. The presence of a type B hysteresis curve was characteristic of slot-shaped porosities [31].

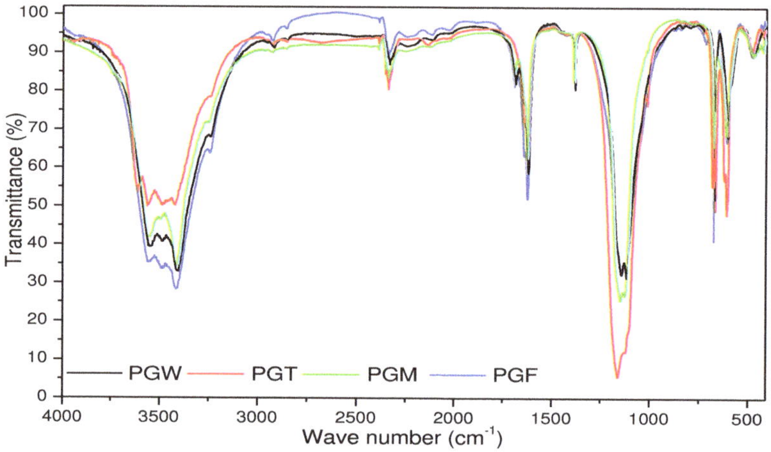

Figure 4. FTIR spectra of the PGF, PGW, PGT and PGM samples.4.4. BET Analysis.

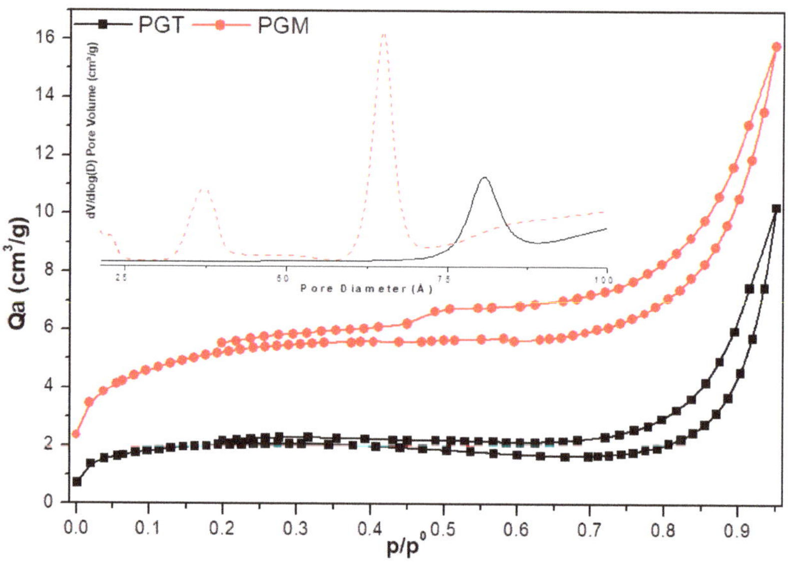

Figure 5. Nitrogen adsorption/desorption isotherms and the pore distribution of the PGT and PGM samples.

The studied samples could be classified as a mesoporous solid with a low specific surface area of 6.7 m^2/g for PGT and 17.5 m^2/g for PGM (Table 4). The pore distribution of these two samples, as presented in Table 4, confirmed the mesoporous properties of the materials with an average pore size of 10.9 nm for PGM and 18.1 nm for PGT. The decrease in the pore size from 18.1 nm for PGT to 10.9 nm for PGM was responsible for the increase in the specific surface area from 6.7 m^2/g for PGT to 17.2 m^2/g for PGM.

Table 4. Textural properties of the PGM and PGT samples.

	BET Surface Area m^2/g	Pore Size nm	Pore Volume cm^3/g	Nanoparticle Size nm
PGM	17.5	10.9	0.0186	342.6
PGT	6.7	18.1	0.0134	893.8

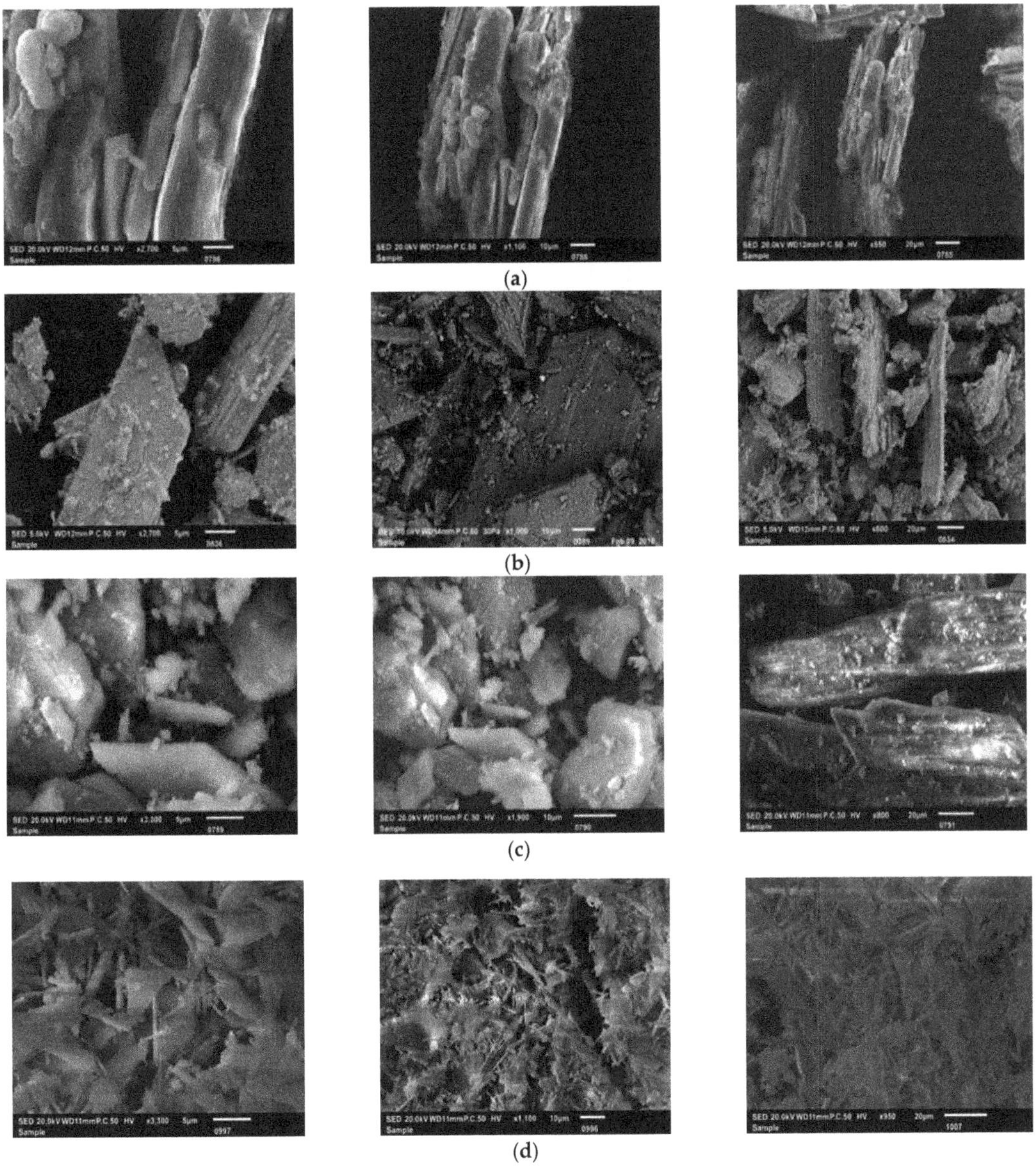

Figure 6. SEM micrographs of: PGF (**a**), PGW (**b**), PGT (**c**) and PGM (**d**).

4.4. SEM Analysis

The SEM micrographs (Figure 6a–d) presented shapes of crystallites in sticks and hexagonal structures for PGT and PGW. For PGF, the SEM micrographs exhibited a fibrous aspect. These results may be due to the fact that after washing we removed the soluble impurities and we approached the crystallization form of $CaSO_4$. After the modeling of the material (the PGM samples), the SEM micrographs demonstrated that the particles had a regular shape and form. This could have been the result of the recrystallization of $CaSO_4$ via the hemi and bihydrate process [32].

Two samples were analyzed by the EDS technique: PGT and PGW (Figures 7 and 8). The results of the two samples were quite close and confirmed that the majority of the composition was $CaSO_4$ with a few impurities such as carbon (C), sodium (Na), fluorine (F), phosphorus (P), silicon (Si) and aluminum (Al).

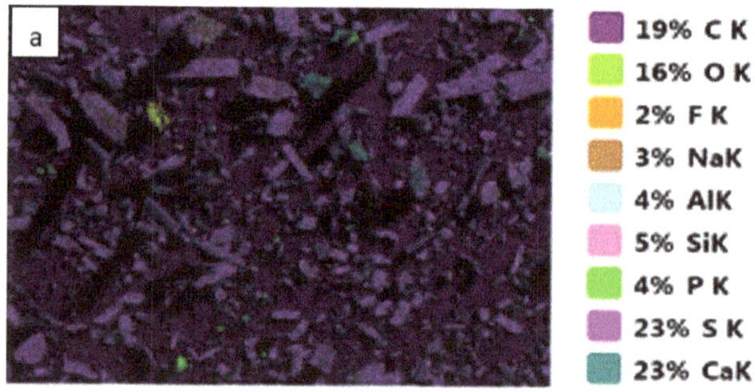

(**a**)

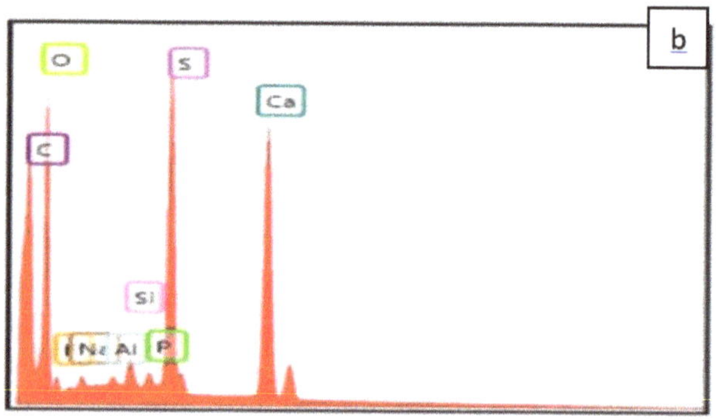

(**b**)

Figure 7. EDS results of PGW: Element map (**a**) and the quantitative ratio of the elements (**b**).

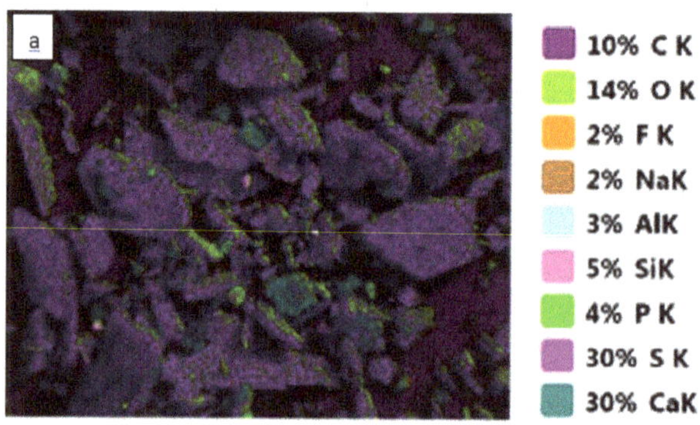

(**a**)

Figure 8. *Cont.*

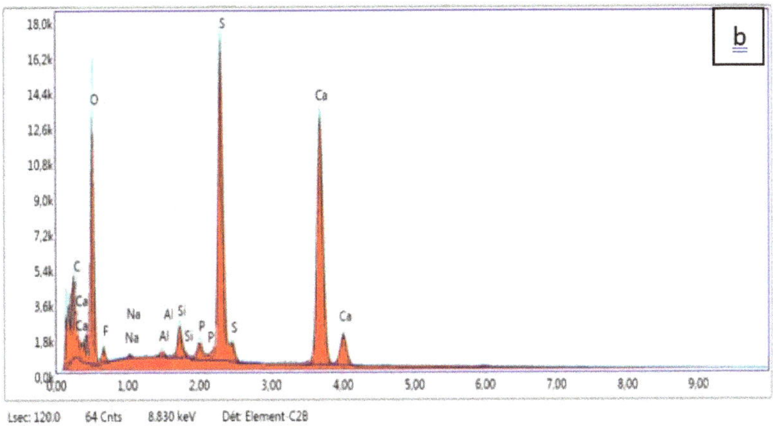

(**b**)

Figure 8. EDS results of PGT: Element map (**a**) and the quantitative ratio of the elements (**b**).

These maps confirmed the quantitative results of the XRF, which clearly demonstrated that the phosphogypsum samples were formed mainly from gypsum ($CaSO_4$). The other elements present such as F, Na, Si and Al that were found with very low quantities could be classified as impurities in the two samples (Table 5).

Table 5. Quantitative ratio of the elements in PGW and PGT.

	PGT		PGW	
Element	**Weight Ratio %**	**Atomic Ratio %**	**Weight Ratio %**	**Atomic Ratio %**
C	9.63	16.82	20.39	32.11
O	40.13	52.62	37.90	44.80
F	2.47	2.72	2.42	2.41
Na	0.34	0.31	0.55	0.45
Al	0.35	0.27	0.37	0.26
Si	1.30	0.97	0.92	0.62
P	0.99	0.67	0.78	0.48
S	16.64	10.88	13.29	7.84
Ca	28.17	14.74	23.38	11.03

The carbon contained in the phosphogypsum originated from the organic fraction of the mineral phosphate, which is generally in the form of humic acid [33]. This quantity decreased after the treatment of the sample at 200 °C for 24 h. This decrease could be attributed to the incomplete oxidation of the organic matter.

4.5. Thermal Conductivity

The thermal conductivity of the materials was the function of the material density. For the prepared samples, the thermal conductivity was 0.160 W/mK and 0.282 W/mK for the PGM 1/2 (d = 486.4 kg/m^3) and PGM 1/1 (848.4 kg/m^3) formulation, respectively (Table 6). These values were similar to those of gypsum mentioned by Ayse et al. [34]. These results confirmed that our prepared samples could be classified as thermal insulation materials.

Gypsum is a hygroscopic salt that has hydration–dehydration equilibrium from low temperatures (between 40 and 70 °C) [35,36]. Its dehydration reaction is endothermic, as shown in Equation (1); therefore, its hydration is exothermic. This equilibrium can participate in the conditioning of the ambient air by a process of water exchange between

$CaSO_4.2H_2O$ and $CaSO_4.0.5\ H_2O$, which is accompanied by a heat exchange; this is the gypsum water pump phenomenon that was verified by DRX and FTIR.

$$CaSO_4.2H_2O + heat \rightleftarrows CaSO_4.\frac{1}{2}H_2O + \frac{2}{3}H_2O_{vapor} \quad (1)$$

$$\Delta H^{\circ}{}_{(25\ ^{\circ}C)} = 83.26\ kJ.mol^{-1}$$

Table 6. Thermal conductivity and density of PGM 1/2 and 1/1.

Samples	Thermal Conductivity (W/mK)	Density (kg/m^3)
PGM 1/2	0.160	486.4
PGM 1/1	0.282	848.4

4.6. Mechanical Properties

Figure 9 shows that the mechanical properties depended on the density. The higher quantity of water used for the preparation considerably decreased the density of the material and caused the fall in the mechanical properties, thus making the material brittle. These results (Table 7) demonstrated that this material derived from phosphogypsum could be used as a building material for non-loadbearing structures or as a plaster or separation plate.

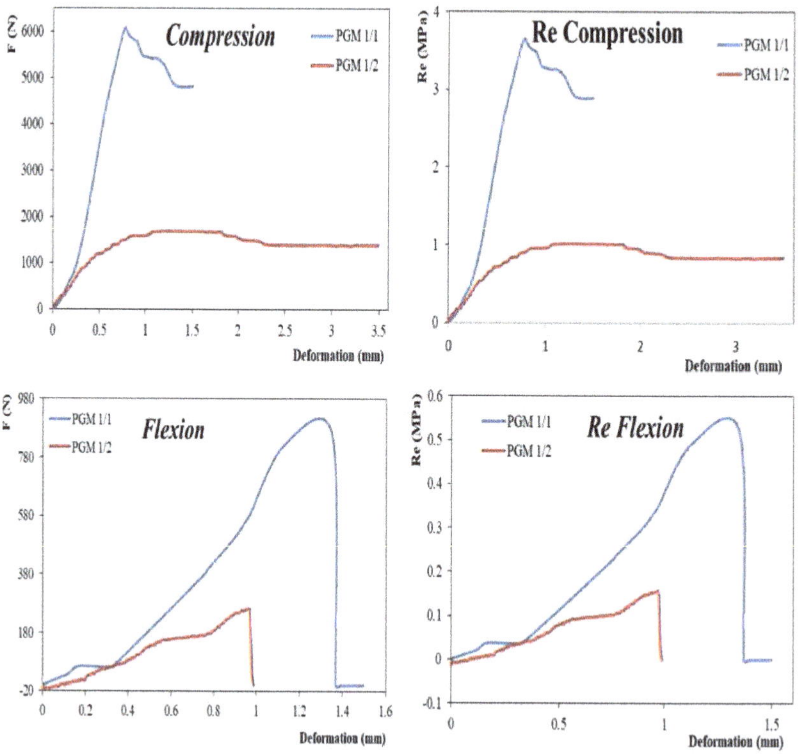

Figure 9. Compressive force, compressive strength, flexion force and flexion strength curves of the prepared materials (PGM 1/1 and 1/2).

Table 7. Mechanical characteristics of the prepared materials.

Density		Compression			Flexion	
		F_{max} N	R_e (MPa)	E (MPa)	F_{max} N	$R_e$max (MPa)
PGM 1/1	848.4	6083	3.08	253.88	862.5	0.54
PGM 1/2	486.4	1696	1.06	3.14	261.81	0.16

5. Conclusions

Based on our results, phosphogypsum can be considered to be a source of gypsum-based material because, in its natural state, it is mainly composed of $CaSO_4$ at two degrees of hydration with a few mineral and organic impurities from the phosphate rock. The physicochemical characterization of phosphogypsum demonstrated that the different degrees of hydration that this material possesses allows it to exchange water with the external environment by creating a water pump that helps to condition the ambient air. It should be noted that the mechanical properties of the prepared material PGM 1/1 and PGM 1/2 closely depended on the density. Due to the lower Young's modulus of these materials, they could be used as construction materials for non-loadbearing structures or as decoration materials. The thermal properties demonstrated that the prepared materials were suitable for insulation in building construction with thermal conductivities lower than 0.3W/mK. It was also concluded that the materials prepared were not suitable for supporting structures.

Author Contributions: Conceptualization, H.G. and A.J.; methodology, H.G., A.J. and E.E.; software, H.G.; validation, H.G., A.J., E.E. and M.A.K.; formal analysis, H.G.; investigation, H.G.; resources, H.G. and E.E.; data curation, H.G. and A.J.; writing—original draft preparation, H.G.; writing—review and editing, H.G., A.J. and M.A.K.; supervision, E.E.; project administration, E.E.; funding acquisition, M.A.K. All authors have read and agreed to the published version of the manuscript.

Funding: This research was funded by the Deanship of Scientific Research at Prince Sattam Bin Abdulaziz University, research project No 16794/01/2020 and the APC was funded by the Deanship of Scientific Research at Prince Sattam Bin Abdulaziz University.

Institutional Review Board Statement: Not applicable.

Informed Consent Statement: Not applicable.

Acknowledgments: This project was supported by the Deanship of Scientific Research at Prince Sattam Bin Abdulaziz University under the research project No 16794/01/2020.

Conflicts of Interest: The authors declare no conflict of interest.

References

1. Sfar Felfoul, H.; Clastres, P.; Ben Ouezdou, M.; Carles-Gibergues, A. Proprietes et perspectives de valorisation du phosphogypse l'exemple de la Tunisie Properties and perspectives of valorization of phosphogypsum the example of Tunisia. In Proceedings of the International Symposium on Environmental Pollution Control and Waste Management, (EPCOWM'2002), Tunis, Tunisia, 7–10 January 2002; pp. 510–520.
2. Lvignes, L. *Centre de Ressources Pédagogique de Chimie (Economie –Industrie-EXC-1ENS de Cachan—94235 Cachan Cedex), CEM-CNRS-94400 Vitry-sur-Seine, ESSADDAM H IPEST-La Marsa-Tunisie et DALIGAND D.*; Syndicat National des Industries du plâtre: Paris, France, 1997.
3. Kurandt, H.F. Emploi de gypse de l'acide phosphorique dans l'industrie du bâtiment. In Proceedings of the Congrès technique de l'IFA, Rio de Janeiro, Brazil, 11–13 November 1980; pp. 10–13.
4. Choura, M. *Evaluation Environnementale de l'activité de Transformation de Phosphate Côtière Sud du Grand Sfax, Projet SMAP III-TUNISIE (2006–2008), Stratégie de Gestion Intégrée de la Zone Côtière sud du Grand Sfax*; Ministère de l'Environnement et du Développement Durable Direction Générale de l'Environnement et de la Qualité de Vie: Tunis, Tunisie, 2008.
5. Mesic, L.; Brezinšcak, M.; Zgorelec, Z.; Percin, A.; Šestak, I.; Bilandžija, D.; Trdenic, M.; Lisac, H. The application of phosphogypsum in agriculture. *Agric. Conspect. Sci.* **2016**, *81*, 7–13.

6. Gennari, R.F.; Garcia, I.; Medina, N.H.; Silveira, M.G. Phosphogypsum analysis: Total content and extractable element concentrations. In Proceedings of the International Nuclear Atlantic Conference, Belo Horizonte, Brazil, 24–28 October 2011; pp. 24–28.
7. Islam, G.M.S.; Chowdhury, F.H.; Raihan, M.T.; Amit, S.K.S.; Islam, M.R. Effect of Phosphogypsum on the Properties of Portland Cement. *Proc. Eng.* **2017**, *171*, 744–751. [CrossRef]
8. Ajam, L.; Ben Ouezdou, M.; Felfoul, H.S.; El Mensi, R. Characterization of the Tunisian phosphogypsum and its valorization in clay bricks. *Constr. Build. Mater.* **2009**, *23*, 3240–3247. [CrossRef]
9. Kammoun, Z. Pressed non-fired bricks from phosphogypsum waste for non-load bearing wall. *Civil Eng. J.* **2021**, *30*, 716–728. [CrossRef]
10. Hamdi, A.; Ben Jamaa, N.; Kammoun, I.K. Potential use of phosphogypsum in paving blocks. *Green Mater.* **2020**, *9*, 97–107. [CrossRef]
11. Xiaohong, R.; Baoguo, M.; Jian, H.; Yun, H. Phosphogypsum transition to α-calcium sulfate hemihydrate in the presence of omongwaite in NaCl solutions under atmospheric pressure. *J. Am. Ceram. Soc.* **2012**, *95*, 3478–3482.
12. Li, J.; Yu, S.; Ma, L. Feasibility analysis for decomposition of phosphogypsum in cement precalciner. *Env. Prog. Sust. Energy* **2011**, *30*, 44–49.
13. Rouis, M.J.; Bensalah, A. phosphogypsum in Tunisia: Environmental and required solution. In Proceeding of the Third International Symposium on Phodphogypsum, Orlando, FL, USA, 6 December 1990; pp. 87–105.
14. Bituh, T.; Petrinec, B.; Skoko, B.; Vučić, Z.; Marović, G. Measuring and modeling the radiological impact of a phosphogypsum deposition site on the surrounding environment. *Arch. Ind. Hyg. Toxicol.* **2015**, *66*, 31–40.
15. Lysandrou, M.; Pashalidis, I. Uranium chemistry in stack solutions and leachates of phosphogypsum disposed at a coastal area in Cyprus. *J. Environ. Radioact.* **2008**, *99*, 359–366. [CrossRef]
16. Pérez-López, R.; Álvarez-Valero, A.M.; Nieto, J.M. Changes in mobility of toxic elements during the production of phosphoric acid in the fertilizer industry of Huelva (SW Spain) and environmental impact of phosphogypsum wastes. *J. Hazard. Mater.* **2007**, *148*, 745–750. [CrossRef]
17. Sahu, S.; Ajmal, P.; Bhangare, R.; Tiwari, M.; Pandit, G. Natural radioactivity assessment of a phosphate fertilizer plant area. *J. Radiat. Res. Appl. Sci.* **2014**, *7*, 123–128. [CrossRef]
18. Ali, K.K.; Awad, Y.D. Radiological assessment of Iraqi phosphate rock and phosphate fertilizers. *Arab. J. Geosci.* **2015**, *8*, 9481–9488. [CrossRef]
19. Folek, S.; Walawska, B.; Wilczek, B.; Miśkiewicz, J. Use of phosphogypsum in road construction. *Pol. J. Chem. Technol.* **2011**, *13*, 18–22. [CrossRef]
20. Sfar, H.F.; Reguigui, N.; Ben Ouezdou, M.; Clastres, P. Radioactivité Naturelle Du Phosphogypse Tunisien. In Proceedings of the First Symposium on Nuclear Sciences and Technology JSTN-2005, Hammamet, Tunisia, 8 December 2005.
21. Hamdaoui, O.; Ibos, L.; Mazioud, A.; Safi, M.; Limam, O. Thermophysical characterization of Posidonia Oceanica marine fibers intended to be used as an insulation material in Mediterranean buildings. *Constr. Build. Mater.* **2018**, *180*, 68–76. [CrossRef]
22. Gustafsson, S.E. Transient plane source techniques for thermal conductivity and thermal diffusivity measurements of solid materials. *Rev. Sci. Instruments* **1991**, *62*, 797–804. [CrossRef]
23. Gustavsson, M.; Karawacki, E.; Gustafsson, S.E. Thermal conductivity, thermal diffusivity, and specific heat of thin samples from transient measurements with hot disk sensors. *Rev. Sci. Instrum.* **1994**, *65*, 3856–3859. [CrossRef]
24. Log, T.; Gustafsson, S.E. Transient plane source (TPS) technique for measuring thermal transport properties of building materials. *Fire Mater.* **1995**, *19*, 43–49. [CrossRef]
25. Hamdaoui, O.; Limam, O.; Ibos, L.; Mazioud, A. *Construction and Building Materials*; Elsevier: Amsterdam, The Netherlands, 2021; Volume 269, p. 121339.
26. Mechi, N.; Ammar, M.; Loungou, M.; Elaloui, E. Thermal Study of Tunisian Phosphogypsum for Use in Reinforced Plaster. *Br. J. Appl. Sci. Technol.* **2016**, *16*, 1–10. [CrossRef]
27. Hassen, S.; Anna, Z.; Elaloui, E.; Belgacem, M.N.; Mauret, E. Study of the valorization of phosphogypsum in the region of Gafsa as filler in paper. In Proceedings of the IOP Conference Series: Materials Science and Engineering, Mahdia, Tunisia, 4–7 November 2012; pp. 12–18.
28. Lamzougui, G.; Nafai, H.; Bouhaouss, A.; Bchitou, R. Détermination de la teneur maximale des métaux lourds dans le Phosphogypse. *J. Mater. Environ. Sci.* **2016**, *7*, 2161–2169.
29. Qliha, A.; Dhimni, S.; Melrhaka, F.; Hajjaji, N.; Srhiri, A. Caractérisation physico-chimique d'une argile Marocaine Physico-chemical characterization of a morrocan clay. *J. Mater. Environ. Sci.* **2016**, *7*, 1741–1750.
30. Aarfane, A.; Salhi, A.; El Krati, M.; Tahiri, S.; Monkade, M.; Lhadi, E.K.; Bensitel, M. Etude cinétique et thermodynamique de l'adsorption des colorants Red195 et Bleu de méthylène en milieu aqueux sur les cendres volantes et les mâchefers; Kinetic and thermodynamic study of the adsorption of Red195 and Methylene blue dyes on fly ash and bottom ash in aqueous medium. *J. Mater. Environ. Sci.* **2014**, *6*, 1927–1939.
31. Jraba, A.; Anna, Z.; Elaloui, E. Physicochemical properties of La^{3+}-doped TiO_2 monolith prepared by sol–gel approach: Application to adsorption and solar photodegradation of ibuprofen. *J. Mater. Sci. Mater. Electron.* **2019**, *31*, 1072–1083. [CrossRef]
32. Cabañas, M.V.; Rodríguez-Lorenzo, L.M.; Vallet-Regí, M. Setting Behavior and in Vitro Bioactivity of Hydroxyapatite/Calcium Sulfate Cements. *Chem. Mater.* **2002**, *14*, 3550–3555. [CrossRef]

33. Khoualdia, B.; Loungou, M.; Elaloui, E. Adsorption of organic matter from industrial phosphoric acid (H 3 PO 4) onto activated bentonite. *Arab. J. Chem.* **2017**, *10*, S1073–S1080. [CrossRef]
34. Bicer, A.; Kar, F. Thermal and mechanical properties of gypsum plaster mixed with expanded polystyrene and tragacanth. *Therm. Sci. Eng. Prog.* **2017**, *1*, 59–65. [CrossRef]
35. Triollier, M. Hydratation du Sulfate de Calcium-Hémi Hydratation. Ph.D. Thesis, Institut National Polytechnique de Grenoble, Grenoble, France, 1979.
36. Singh, N.; Middendorf, B. Calcium sulphate hemihydrate hydration leading to gypsum crystallization. *Prog. Cryst. Growth Charact. Mater.* **2007**, *53*, 57–77. [CrossRef]

materials MDPI

Article

Nonlinear ABAQUS Simulations for Notched Concrete Beams

Ahmed Bahgat Tawfik [1,*], Sameh Youssef Mahfouz [1] and Salah El-Din Fahmy Taher [2]

1 Construction and Building Engineering Department, College of Engineering and Technology, Arab Academy for Science, Technology and Maritime Transport (AASTMT), B 2401 Smart Village, Giza 12577, Egypt; symahfouz@aast.edu
2 Professor of Concrete Structures, Structural Engineering Department, Tanta University, Tanta 31527, Egypt; salah.taher@f-eng.tanta.edu.eg
* Correspondence: ahmedbahagt@aast.edu

Abstract: The numerical simulation of concrete fracture is difficult because of the brittle, inelastic-nonlinear nature of concrete. In this study, notched plain and reinforced concrete beams were investigated numerically to study their flexural response using different crack simulation techniques in ABAQUS. The flexural response was expressed by hardening and softening regime, flexural capacity, failure ductility, damage initiation and propagation, fracture energy, crack path, and crack mouth opening displacement. The employed techniques were the contour integral technique (CIT), the extended finite element method (XFEM), and the virtual crack closure technique (VCCT). A parametric study regarding the initial notch-to-depth ratio (a_o/D), the shear span-to-depth ratio (S.S/D), and external post-tensioning (EPT) were investigated. It was found that both XFEM and VCCT produced better results, but XFEM had better flexural simulation. Contrarily, the CIT models failed to express the softening behavior and to capture the crack path. Furthermore, the flexural capacity was increased after reducing the (a_o/D) and after decreasing the S.S/D. Additionally, using EPT increased the flexural capacity, showed the ductile flexural response, and reduced the flexural softening. Moreover, using reinforcement led to more ductile behavior, controlled damage propagation, and a dramatic increase in the flexural capacity. Furthermore, CIT showed reliable results for reinforced concrete beams, unlike plain concrete beams.

Keywords: ABAQUS; finite element analysis (FEA); concrete damage plasticity (CDP); extended finite element method (XFEM); external post-tensioning (EPT)

Citation: Tawfik, A.B.; Mahfouz, S.Y.; Taher, S.E.-D.F. Nonlinear ABAQUS Simulations for Notched Concrete Beams. *Materials* **2021**, *14*, 7349. https://doi.org/10.3390/ma14237349

Academic Editors: Dario De Domenico and Luís Filipe Almeida Bernardo

Received: 29 October 2021
Accepted: 28 November 2021
Published: 30 November 2021

1. Introduction

Employing numerical finite element (FE) simulations for different concrete elements reduces the need for further physical testing and helps researchers undertake complex parametric studies precisely. The complex nonlinear behavior of concrete in both tension and compression makes it challenging to simulate. The complexity stems from the brittle response of concrete, leading to failure due to cracking or crushing when subjected to tension or compression, respectively. Also, it is difficult to simulate crack initiation and propagation due to the tensile damage.

Various constitutive numerical models were presented to simulate the concrete behavior in tension and compression [1–5]. Y. Nikaido [6] improved a constitutive numerical model to simulate concrete behavior by considering compression stiffness recovery. Furthermore, different studies were conducted to simulate the nonlinear behavior of different plain and reinforced concrete elements [7–11]. Zhang et al. [12] compared different numerical crack simulation techniques to simulate a notched concrete beam using the unified FE software package ABAQUS [13] without experimental result validation. These techniques included the virtual crack closure technique (VCCT) and the extended finite element method (XFEM). It was deduced that both VCCT and XFEM can capture the softening regime for concrete fracture well. An experimental work achieved by Yin et al. [14] was conducted on notched plain concrete beams having different spans. The experimental scope was set to determine the effect of changing the span on different concrete fracture parameters.

Many studies integrated the use of different composite materials with notched concrete beams through various experiments [15–18]. This integration aimed to enhance the flexural performance and to investigate the fracture and bond behavior of concrete beams. Zhong et al. [15] experimentally investigated the effect of installing a channel steel plate ahead of a concrete beam notch on the strengthening performance. It was found that placing a channel steel plate at the notch tip of the concrete beam can significantly increase the load carrying capability against mode 1 fractures. Additionally, an increase in fracture energy and ductile softening were captured after placement. Furthermore, the fracture failure of the tested beams was significantly influenced by the debonding and slippage behavior between the steel plate channel and the concrete beam. It was noted that steel plate size has no significant effect on flexural capacity. Sun et al. [16] studied the effect of using different volumetric dosages of basalt fibers on the fracture resistance of precast notched concrete beams. It was deduced that the peak load, the initiation toughness, and the fracture energy significantly increased after increasing the basalt fiber dosage. Moreover, the fracture energy and the unstable toughness had no size effect, unlike the initiation toughness which increased with increased specimen height.

De Domenico et al. [17] investigated the interfacial characteristics of the carbon fiber-reinforced polymers (CFRP) system and the fiber-reinforced cementitious matrix with poly-benzoxole and cement-based mortar (PBO-FRCM) system. Both systems were adhesively bonded at the bottom of the notched concrete beams to study the effect of environmental conditions on the bond and the ultimate capacity for each system. It was found that the PBO-FRCM system was not affected by environmental conditions. Conversely, the CFRP system was affected by the curing conditions, as more than a 30% reduction in the average peak load was noticed compared to non-conditioned beams.

Chen et al. [18] conducted three-point bending tests on notched steel fiber-reinforced concrete (SFRC) beams containing single or novel multiple hooked-end steel fibers to study the effects of fiber aspect ratio, fiber length, and fiber shape on flexural tensile strength. It was concluded that the limit of proportionality, which is related to the initial concrete cracking, is mainly dependent on the concrete strength, and not on the fiber properties. On the other hand, it was found that increasing the fiber dosage, length, aspect ratio, and number of hooked-ends enhanced the residual flexural tensile strength in the post-cracking stage.

The previously mentioned studies focused on the incorporation of different composite materials with notched concrete beams to enhance flexural tensile capacity and to improve cracking damage resistance. However, studies scrutinizing the effects of geometrical characteristics on the flexural performance of notched concrete beams are lacking. Along this research line, this paper provides numerical investigations regarding the effects of different geometrical aspects, such as loading scheme and notch height, on flexural performance. Additionally, this paper provides further numerical investigations on methods of enhancing the flexural tensile capacity and damage control performance of notched concrete beams. Unlike the studies mentioned in the literature that addressed this problem using different composite materials, the current study implemented steel reinforcement and un-bonded post-tensioning. Moreover, the simulations conducted in the current work were achieved by employing different numerical techniques in a comparative study to determine the most appropriate technique for concrete fracture simulation.

The current study utilizes the experimental work conducted by Yin et al. [14]. The current work aims to study the flexural behavior and fracture initiation and propagation of notched plain and reinforced concrete beams. Additionally, different parametric studies were conducted to understand the influence of changing different parameters on the flexural response. The studied parameters were the initial notch-to-depth ratio (a_o/D), the shear span-to-depth ratio (S.S/D), and external post-tensioning (EPT). The flexural behavior, damage initiation and propagation of six notched plain concrete beams having various span-to-depth ratios (S/D) were numerically investigated. All investigated beams were subjected to a three-point bending setup. Their response was captured as a relationship between loading level (P) and crack mouth opening displacement (CMOD). Three techniques for crack modeling in ABAQUS, such as the contour integral technique (CIT), XFEM and VCCT were considered, to determine the most suitable crack modeling technique. These techniques employed different material damage criteria: concrete damage plasticity (CDP), maximum principal stress (MAXPS) and critical energy release rate (G_c). The numerical results were compared with those of the experimental investigation to find out the most suitable technique for expressing the behavior and the damage of all tested beams.

Section 2 of this paper illustrates the details of the experimental works and reviews their outcomes. Section 3 discusses the details of developing the numerical FE models. Section 4 presents the numerical simulation outcomes. Section 5 illustrates the parametric studies conducted, their results, and discussion. Section 6 provides a summary of the obtained conclusions.

2. Synopsis of Experimental Data

According to the experimental work done by Yin et al. [14], a three-point-bending test was carried out, as indicated in Figure 1. This test was conducted on six series of notched plain concrete beams having different S/D ratios. Each beam series included four typical beams. The depth (D) and the breadth (B) of each beam were 150 mm and 100 mm, respectively. The length (L) of beams ranged from 375 mm to 1100 mm, while the corresponding span (S) ranged from 300 mm to 900 mm. These span ranges were selected to achieve S/D ratios ranging from 2 to 6. An initial notch (a_o) of 60 mm at the middle of all beams was considered. Table 1 illustrates the details of beams. The behavior of each beam was captured in the form of a P-CMOD relationship. These relationships, along with the crack propagation path after cracking, are also provided.

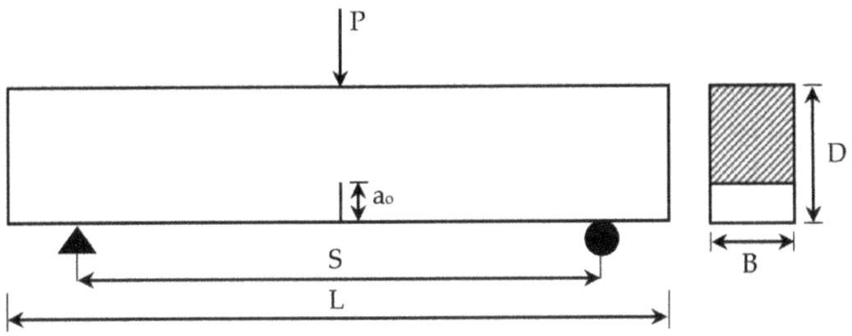

Figure 1. Experimental test loading scheme used.

Table 1. Concrete beams studied.

Beam Series	Breadth (B) mm	Depth (D) mm	Length (L) mm	Span (S) mm	Notch Length (a_o) mm	S/D
T2	100	150	375	300	60	2
T2.5			475	375		2.5
T3			550	450		3
T4			750	600		4
T5			950	750		5
T6			1100	900		6

3. Evolution of the Numerical FE Models

This section explains the details of developing ABAQUS FE models employing CIT, VCCT and XFEM.

3.1. Contour Integral Technique

The contour integral technique (CIT) was used to study the commencement of damage in quasi-static problems [13]. To use CIT, the crack front region, crack line, and extension direction were specified. Moreover, a region with separate faces was modeled to be free to move apart as the crack mouth began to open. These faces were created using the seam command in ABAQUS. To simulate the nonlinearity of concrete, the CDP model was employed. The CDP model was utilized for its capability of expressing the inelastic responses of concrete and quasi-brittle materials, as well as damage characteristics in both compression and tension.

3.1.1. Concrete Tensile Behavior

The uniaxial behavior of concrete in tension was defined using a relationship between the tensile stresses and their corresponding cracking strains. Defining this relationship makes it necessary to convert the tensile strain (ε_t) to the cracking strain ($\varepsilon_t^{\sim ck}$) using Equation (1).

$$\varepsilon_t^{\sim ck} = \varepsilon_t - \varepsilon_{ot}^{el} \tag{1}$$

where ε_{ot}^{el} refers to the tensile elastic strain of the undamaged material condition and equals σ_{t0}/E_o; σ_{t0} is the maximum elastic tensile stress and E_o is the concrete modulus of elasticity.

A modified tension stiffening material model suggested by Wahalathantri et al. [19] was used in the present work to obtain the uniaxial tensile stress–strain relationship. The modifications were made to overcome convergence problems and to avoid ABAQUS solution errors. This model is based on the tension stiffening model of Nayal and Rasheed [20].

3.1.2. Concrete Compressive Behavior

The uniaxial behavior of concrete in compression was defined using a relationship between the compressive stresses and their corresponding inelastic strains. Based on the ABAQUS documentation, the compressive strain (ε_c) was converted to the inelastic strain ($\varepsilon_c^{\sim in}$) to define this relationship by employing Equation (2).

$$\varepsilon_c^{\sim in} = \varepsilon_c - \varepsilon_{oc}^{el} \tag{2}$$

where ε_{oc}^{el} refers to the compressive elastic strain condition of the undamaged material and equals σ_c/E_o, and σ_c is the maximum elastic compressive stress.

A numerical material model for concrete developed by Hsu et al. [21] was employed to obtain the stress–strain relationship in compression. This material model can be used for concrete material with a concrete cube compressive strength (σ_{cu}) up to 62 MPa [19,21]. It considers that the stress–strain relationship behaves linearly up to 50% of σ_{cu} in the hardening portion of the relationship. Afterward, numerical material model equations were used to describe the relationship until the softening occurred and reached the value of 30% of the σ_{cu} in the softening portion. The numerical material model equations can be expressed as follows:

$$\sigma_{comp.} = \left(\frac{\beta(\varepsilon_c/\varepsilon_o)}{\beta - 1 + (\varepsilon_c/\varepsilon_o)^{\beta}} \right) \sigma_{cu} \tag{3}$$

$$\beta = \frac{1}{1 - \left[\frac{\sigma_{cu}}{\varepsilon_o E_o} \right]} \tag{4}$$

$$\varepsilon_o = 8.9 \times 10^{-5}\, \sigma_{cu} + 2.114 \times 10^{-3} \tag{5}$$

$$E_o = 1.2431 \times 10^{2}\, \sigma_{cu} + 3.28312 \times 10^{3} \tag{6}$$

where $\sigma_{comp.}$ is the compressive stress. The parameter β is a parameter that depends on the stress–strain diagram shape, and ε_o is the compressive strain at peak stress.

3.1.3. Concrete Damage Parameters

The concrete damage plasticity (CDP) model expresses concrete damage utilizing a damage parameter ranging between zero (indicating the intact material state) and one (indicating a complete loss of material strength). Thus, the operative modulus of elasticity after tensile and compressive damage can be evaluated using Equations (7) and (8), respectively.

$$E_t = E_o \times (1 - d_t) \tag{7}$$

$$E_c = E_o \times (1 - d_c) \tag{8}$$

where d_t and d_c refer to the tensile and compressive damage parameters, respectively, and can be evaluated using Equations (9) and (10), respectively.

$$d_t = 1 - \left(\frac{\sigma_t}{\sigma_t'}\right) \tag{9}$$

$$d_c = 1 - \left(\frac{\sigma_c}{\sigma_c'}\right) \tag{10}$$

where σ_c' and σ_t' are the effective compressive and tensile strength, respectively.

Along with stress–strain relationship and damage parameter data, five plasticity parameters are needed to define the CDP model in ABAQUS. These parameters can be described as follows:

1. The dilation angle (Ψ) measured in the p–q plane at high confining pressure and is necessary to evaluate the potential plastic flow, which uses the Drucker–Prager hyperbolic function. The dilation angle ranges between 0° to 56° [5]. In the current research, a value of 31° was used, according to Hafezolghorani et al. [22].
2. Flow potential eccentricity (ϵ)designates the rate at which the hyperbolic Drucker–Prager function reaches the linear Drucker–Prager function. A value of 0.1 was used. This value guarantees that the material has nearly the same dilation angle among a wide band of confining pressure values [13].
3. The ratio of initial equibiaxial compressive strength (f_{b0}) to the initial uniaxial compressive strength (f_{c0}) is responsible for the evolution of yield surfaces. This ratio contributes to the evaluation of the yield function proposed by Lubliner et al. [23] and modified by Lee and Fenves [24]. This parameter ranges between 1.10 and 1.16 [13]. The default value of 1.16 is used by many researchers [25] and was adopted in the current study.
4. The ratio (K_c) of the second stress invariant on the tensile meridian (q_{TM}) to that on the compressive meridian (q_{CM}) contributes to evaluating the yield function. It ranges between 0.5 and 1 [26]. The default value of 2/3 was employed in the present simulation as per many researchers [25].
5. The viscosity parameter (μ) helps to make the tangent stiffness of the degrading material have a positive value for small time increments. This value is achieved by allowing stresses to be outside the developed yield surfaces. Using small values of μ compared to the characteristic analysis time increment tends to enhance the rate of solution convergence in the softening regime. The default value of zero [13] caused premature termination of the analysis due to the damage that occurred in the element. According to Tao et al. [5], μ has no significant effect on the analysis precision. Thus, a value of 0.0007 was utilized.

In the current research, a reduced integration quadratic 20-node brick element (C3D20R) was used. This element is a second-order element type, which gives higher accuracy and is effective in bending dominated problems [13]. As result of a sensitivity study for mesh size, a fine mesh size of 10 mm was used for the different models made for the present study.

3.2. Virtual Crack Closure Technique

The virtual crack closure technique (VCCT) employs the principles of linear elastic fracture mechanics (LEFM) [13,27] and Irwin's criterion [28]. Rybicki and Kanninen [29] presented this technique, and it was improved by Raju [30] as higher-order interpolation elements were added. This technique assumes that once the crack is opened to a specific extent, it releases the same amount of strain energy needed to close the crack by the same extent [13]. As shown in Figure 2, nodes 1, 2, 3, and 4 were debonded nodes, while the others were bonded nodes. The energy release rate of mode I fractures (G_I) for 4-noded elements can be estimated using Equation (11).

$$G_I = \lim_{\Delta a \to 0} \frac{1}{2b\Delta a} F_{5,6}(v_3 - v_4) \tag{11}$$

where Δa and b are the length and the width of the elements at the crack front, respectively. The symbol $F_{5,6}$ denotes the force between nodes 5 and 6. The displacements of nodes 3 and 4 are v_3 and v_4, respectively.

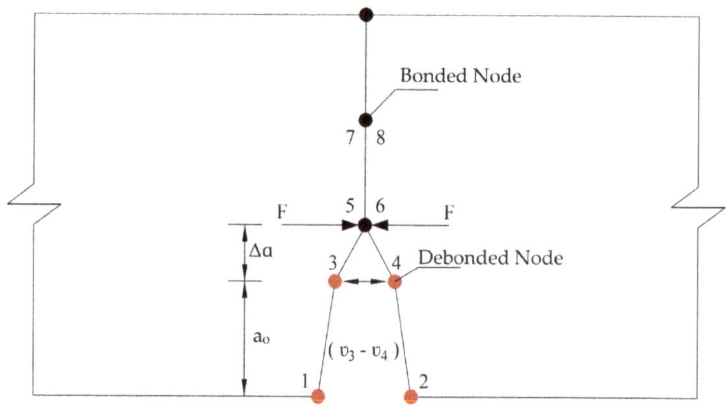

Figure 2. Virtual crack closure technique (VCCT) debonding at crack location.

The VCCT technique requires a predefined crack path along which to propagate [31]. To define the crack path, the beam was modeled as two separate identical parts. A contact interaction was assigned to the bonded mutual nodes of the two parts. This interaction included a fracture criterion factor (f) that depends on the overall rate of energy release (G_T) and is fulfilled when the critical energy value (G_T^C) is achieved. The fracture criterion factor (f) is evaluated using Equation (12).

$$f = \frac{G_T}{G_T^C} \tag{12}$$

A value of 0.06 N/mm was adopted for G_T^C [14]. Based on the mesh sensitivity study, a fine mesh of reduced integration continuum 8-node biquadratic plane stress elements (CPS8R) was used for the VCCT models.

3.3. Extended Finite Element Method Technique

The extended finite element method (XFEM) utilizes the principles of LEFM until the crack is initiated. This technique was first proposed by Belytschko and Black [32], and was modified by Moës et al. [33]. The XFEM technique facilitates the study of crack propagation along an arbitrary path that depends on the solution. In this technique, the separation is performed by providing supplementary freedom degrees to elements around the crack path and the crack tip. It uses the partition of the unity property FE method of Melenk and Babuška [34], in which the summation of all shape functions equals one.

The displacement function expressed in Equation (13) allows the propagation of the crack through the meshed elements without the need for remeshing [35].

$$u = \sum_{I=1}^{S_I} N_I(x)u_I + \sum_{c=1}^{S_c} N_c(x)H(x)a_c + \sum_{t=1}^{S_t} N_t(x) \sum_{\alpha=1}^{4} F_\alpha(r,\theta)b_t^\alpha \tag{13}$$

where S_I expresses the node count in the elements that contain the fracture, S_c represents the node count within the elements that include the fracture line, and S_t, is the node count in the elements enclosing the fracture tip. The node shape functions for elements enclosing the crack tip, elements including crack line and elements containing a fracture are denoted by the symbols N_t, N_c and N_I, respectively. The standard displacement of node I is denoted by u_I. Both a_c and b_t^α are the coefficients that express the degrees of nodal enhanced freedom for the nodes related to the elements that enclose the crack line and the tip, respectively. $H(x)$ is the Heaviside function across the crack surfaces and $F_\alpha(r,\theta)$ is the crack tip asymptotic enrichment function.

To initiate a discontinuity among the damaged elements to represent a crack, the Heaviside function $H(x)$ is employed as expressed in Equation (14).

$$H(x) = \begin{cases} 1 & if\ (x - x^*) \cdot n \geq 0 \\ -1 & otherwise \end{cases} \tag{14}$$

where x is a sample (Gauss) point, x^* is the point on the crack closest to x and n is the unit outward normal to the crack at x^*.

The asymptotic enrichment function $F_\alpha(r,\theta)$ was adopted to allow the fracture to grow and propagate. This function adds supplementary freedom degrees to the nodes related to the element containing the fracture tip using Equation (15). The symbol α represents the node number within the element that encloses the crack tip while r and θ express the distance and the angle of the fracture within the element enclosing the fracture tip, respectively [36].

$$F_\alpha(r,\theta) = \left\{ \sqrt{r}\cos\left(\frac{\theta}{2}\right), \sqrt{r}\sin\left(\frac{\theta}{2}\right), \sqrt{r}\sin\left(\frac{\theta}{2}\right)\sin(\theta), \sqrt{r}\cos\left(\frac{\theta}{2}\right)\sin(\theta) \right\} \tag{15}$$

It is essential to specify the crack domain and the initial crack location within the selected domain [13]. Moreover, it is critical to define a material damage initiation criterion, such as the maximum principal strain (MAXPE) or the maximum principal stress (MAXPS) [13]. That is why the MAXPS damage criterion factor (z) was employed and can be expressed using Equation (16).

$$z = \left\{ \frac{\sigma_{max}}{\sigma_{max}^o} \right\} \tag{16}$$

where σ_{max}^o is the maximum allowable principal stress. The $\langle\rangle$ symbol denotes the Macaulay bracket to indicate that pure compressive stress cannot cause or commence damage. Instead, the damage is set to start if the maximum principal stress ratio reaches 1. The maximum allowable principal stress σ_{max}^o was 2.2 MPa to express the concrete tensile strength.

4. FE Models Verification and Discussion

All models were developed using CIT, VCCT, and XFEM. Based on the results deduced from the T6 simulations, the most suitable technique was determined by using five different statistical indicators that are used by researchers in different fields [37–39]. These statistical indicators are: 1. root mean square error (RMSE); 2. Nash–Sutcliffe efficiency (NSE) [40]; 3. modified index of agreement (md) [41]; 4. coefficient of determination (R^2); and 5. Kling–Gupta efficiency (KGE) [42]. The RMSE measures the differences between the experimental and the numerical data, RMSE has an optimal value of zero. The NSE determines the relative magnitude of the residual variance in the numerical data compared to the variance in the experimental data. The md estimates the additive and proportional differences in the means and variances of the experimental and numerical data. The R^2 assesses the degree of collinearity between the numerical and experimental data. The KGE assesses the correlation, bias, and variability between the experimental and numerical data, thus providing a complete similarity estimation. The last four statistical indicators have an optimal value of one. The RMSE, NSE, md, R^2, and KGE were computed as shown below:

$$\text{RMSE} = \sqrt{\frac{\sum_{i=1}^{N}(\hat{x}_i - x_i)^2}{N}} \tag{17}$$

$$\text{NSE} = 1 - \left[\frac{\sum_{i=1}^{N}(\hat{x}_i - x_i)^2}{\sum_{i=1}^{N}(\hat{x}_i - x^{mean})^2}\right] \tag{18}$$

$$\text{md} = 1 - \frac{\sum_{i=1}^{N}|x_i - \hat{x}_i|}{\sum_{i=1}^{N}(|\hat{x}_i - x^{mean}| + |x_i - x^{mean}|)} \tag{19}$$

$$\text{R}^2 = \left(\frac{\sum_{i=1}^{N}[(x_i - x^{mean})(\hat{x}_i - \hat{x}^{mean})]}{\sqrt{\sum_{i=1}^{N}[\hat{x}_i - \hat{x}^{mean}]^2}\sqrt{\sum_{i=1}^{N}[x_i - x^{mean}]^2}}\right)^2 \tag{20}$$

$$\text{KGE} = 1 - \sqrt{(P_c - 1)^2 + \left(\frac{\hat{x}^{mean}}{x^{mean}} - 1\right)^2 + \left(\frac{S.\hat{D}/\hat{x}^{mean}}{S.D/x^{mean}} - 1\right)^2} \tag{21}$$

where N is the number of data points, x_i is the actual experimental data value, $\hat{x}_i$ is the numerical data value, x^{mean} is the experimental data mean value, $\hat{x}^{mean}$ is the numerical data mean value, P_c is the Pearson's correlation coefficient, $S.D$ is the standard deviation of the experimental data, and $S.\hat{D}$ is the standard deviation of the numerical data. For the beam T6, the results were monitored as shown in Figure 3. From this figure, it can be deduced that, for CIT model, the flexural capacity conforms with the experimental findings. On the other hand, the post-failure stage does not reflect the experimental results. Both XFEM and VCCT were found to capture the flexural response and simulate the softening part more precisely. Table 2 shows the results of RMSE, NSE, md, R^2, and KGE for CIT, VCCT, and XFEM and shows that both XFEM and VCCT have better correlation to the experimental results. The XFEM managed to achieve the closest optimal value for each statistical indicator, indicating that XFEM offered the best flexural simulation.

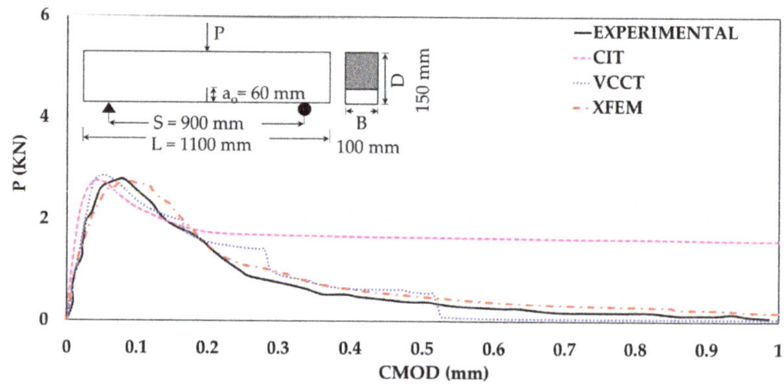

Figure 3. Loading level (P) versus crack mouth opening displacement (CMOD).

Table 2. Results of statistical indicators.

Statistical Indicator	CIT	VCCT	XFEM	Optimal Value
RMSE	1.000	0.210	0.170	0
NSE	−3.740	0.950	0.960	1
md	0.370	0.890	0.900	1
R^2	0.570	0.950	0.970	1
KGE	0.000	0.910	0.910	1

The fracture energy $\left(G_f\right)$ was evaluated by dividing the work (W_o) by the ligament area (A_L) of the notched beam, as given in Equation (22). From Table 3, it can be concluded that both VCCT and XFEM provide better fracture energy estimates than CIT. Moreover, XFEM gives better a estimation of the experimental result.

$$G_f = \frac{W_o}{A_L} \tag{22}$$

Table 3. Fracture energy $\left(G_f\right)$ for different analyses versus experimental results for beam T6.

	CIT	VCCT	XFEM	Experimental
W_o (N.m)	3.380	1.175	1.260	-
A_L (m^2)	0.009	0.009	0.009	0.009
G_f (N/m)	375.556	130.556	140.000	175.960

According to the experimental results [14], the crack propagation path indicated the dominance of mode I fractures among all tested beam series. It can be noticed from Figure 4. That, for CIT analysis, the crack propagation path could not be captured. Instead, only a tensile damage region was captured. This is due to the stationary nature of CIT cracks. Conversely, XFEM and VCCT models showed a crack initiation and propagation path that followed the experimental behavior precisely. The initiation and propagation of concrete damage is illustrated in Figure 4.

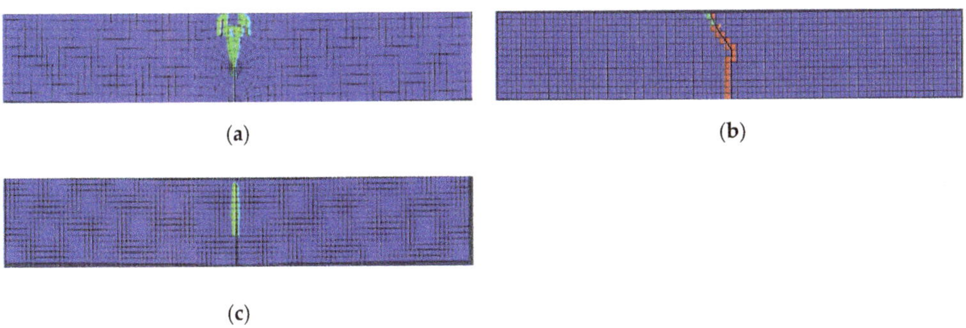

Figure 4. Concrete damage initiation and propagation paths; (**a**) tensile damage for contour integral technique (CIT) model; (**b**) extended finite element method (XFEM) model; (**c**) VCCT model.

It can be concluded that both XFEM and CIT are the most suitable and least suitable techniques for plain concrete beam fracture simulation, respectively. To support this conclusion and to identify the discrepancy between the two techniques, both XFEM and CIT were considered to verify the experimental results of the other beams stated in Table 1.

Figure 5 shows the verification of all beams using both XFEM and CIT as a relationship between P and CMOD. It can be noted that both XFEM and CIT can express the flexural capacity of all beams in a good correspondence with the experimental outcomes. As for

XFEM, it can simulate the experimental flexural behavior and the post-failure softening precisely for all beams unlike CIT which shows an obvious discrepancy. These findings support the conclusion that XFEM is the most suitable approach for plain concrete beam fracture simulation.

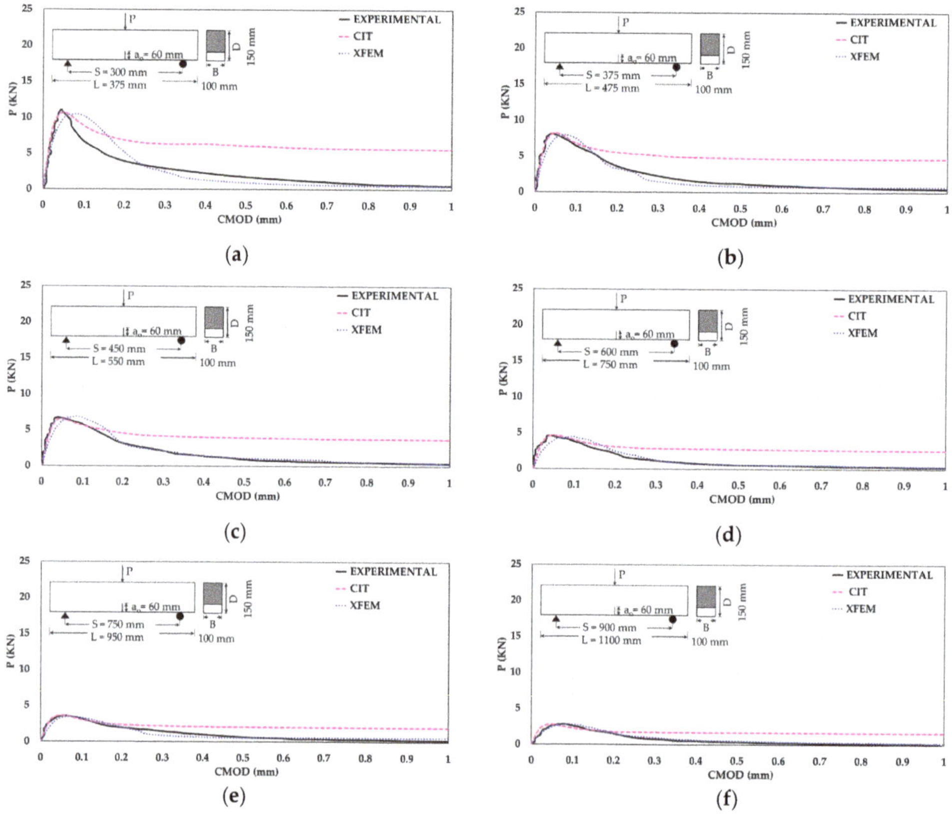

Figure 5. The P-CMOD relationships for beams (**a**) T2; (**b**) T2.5; (**c**) T3; (**d**) T4; (**e**) T5; (**f**) T6.

5. Parametric Study and Results

In the current study, three parameters were investigated for plain and reinforced concrete beams to study their effect on the flexural behavior. These parameters were: 1. the initial notch-to-depth ratio; 2. the shear span-to-depth ratio; 3. the use of external post-tensioning. This investigation was carried out on beam T6. For reinforced concrete beams, the reinforcement consists of two top corner bars as stirrup hangers, two bottom corner bars as the main reinforcements, and shear stirrups. All bars had a diameter of 6 mm. The spacing between the stirrups was 100 mm. The reinforcement steel grade used was the standard ASTM (A 615M/A 615) grade 300. This steel grade has a minimum yield strength of 300 MPa. A constant concrete cover of 10 mm was maintained in the current work.

5.1. *Influence of Changing Initial Notch-to-Depth Ratio*

Initial notch-to-depth ratio (a_o/D) values of 0.3, 0.4, and 0.5 were studied. These values imply that the corresponding values of initial notch height (a_o) were 45 mm, 60 mm, and 75 mm, respectively. The FE models for plain and reinforced beams were executed. Figures 6 and 7 show P-CMOD relationships when considering different values of a_o/D for plain and reinforced concrete beam models, respectively. For plain concrete beams, it can be noted that the flexural capacity increases with reduction of the a_o/D. This is because reducing the a_o/D increases the ligament area of the notched beam, causing increased resistance to flexural damage leading to higher flexural capacity. Furthermore, a complete flexural softening can be noticed for all XFEM results unlike CIT. For the reinforced concrete beam, it can be noted that the behavior became more ductile as there was no sudden failure. Instead, a failure plateau took place. Also, the flexural capacity was drastically increased due to the presence of reinforcement. Moreover, the flexural capacity of the reinforced concrete beam seems to be nearly constant regardless the a_o/D value. This means that even with a light amount of reinforcement, damage initiation and propagation can be controlled, causing stability. Additionally, it is noted that both the CIT and XFEM models gave very similar results, as no softening occurred in the reinforced beam. Thus, it can be concluded that CIT can give reliable results for reinforced concrete beams, unlike plain concrete beams.

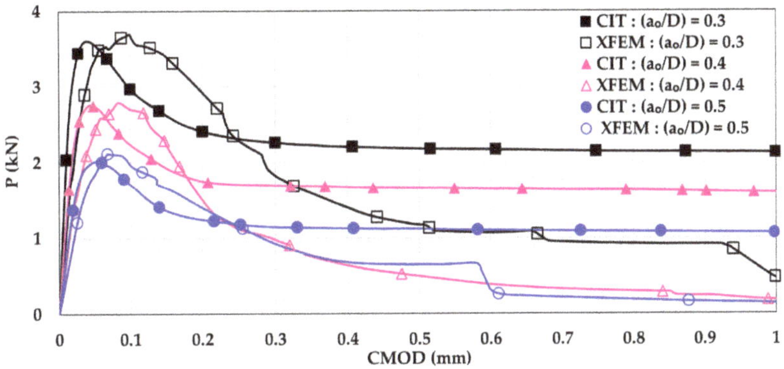

Figure 6. Plain concrete beam P-CMOD relationships for different initial notch-to-depth ratios (a_o/D).

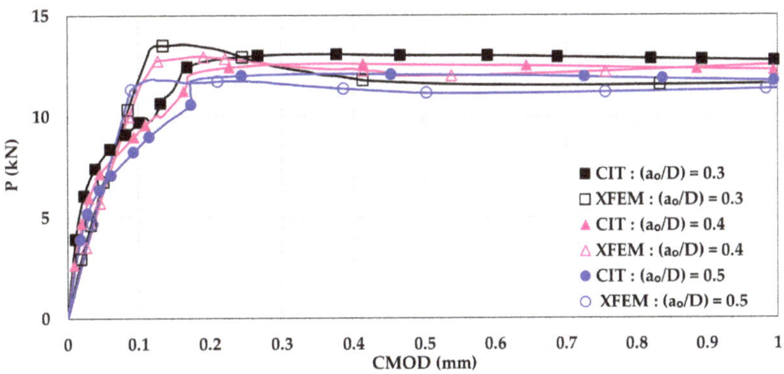

Figure 7. Reinforced concrete beam P-CMOD relationships for different initial notch-to-depth ratios (a_o/D).

5.2. Influence of Changing the Shear Span-to-Depth Ratio

In order to study the effect of changing the shear span-to-depth ratio (S.S/D), the model was modified to follow the four-point loading setup. As seen in Figure 8, the beam was subjected to two equal loads (0.5 P). Both loads were located at the same distance (X) from the support. In the current study, the three different distances (X) considered were X = 150 mm, X = 300 mm, and X = 450 mm. These distances imply to S.S/D values of 1, 2, and 3, respectively. The P-CMOD relationships of the plain and reinforced concrete beams are shown in Figures 9 and 10, respectively. These figures indicate that increasing the S.S/D decreases the flexural capacity. This is because a more direct influence is delivered to the mid-notch, causing faster crack propagation and lower flexural capacity. For plain concrete beams, XFEM showed lower flexural capacity than CIT for lower S.S/D. This means that XFEM has higher sensitivity to damage initiation and propagation than CIT. It is noted that implementing light reinforcement led to higher flexural capacity and increased ductility. It can be also noted that, after using reinforcement, the results obtained when utilizing XFEM were almost identical to the corresponding results when using CIT. This supports the conclusion that CIT is reliable for simulating reinforced concrete beams, unlike plain concrete beams.

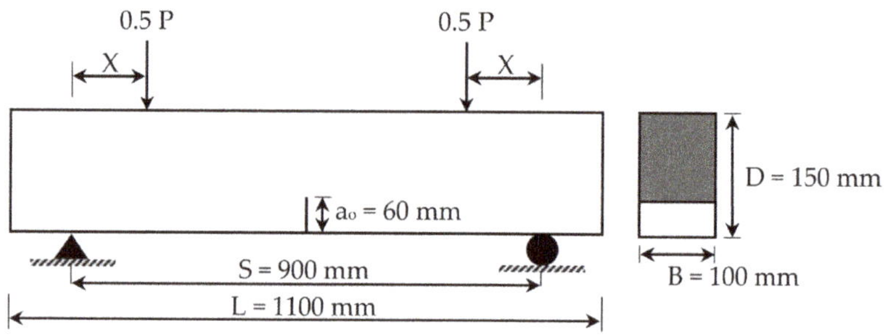

Figure 8. Four-point loading scheme for T6.

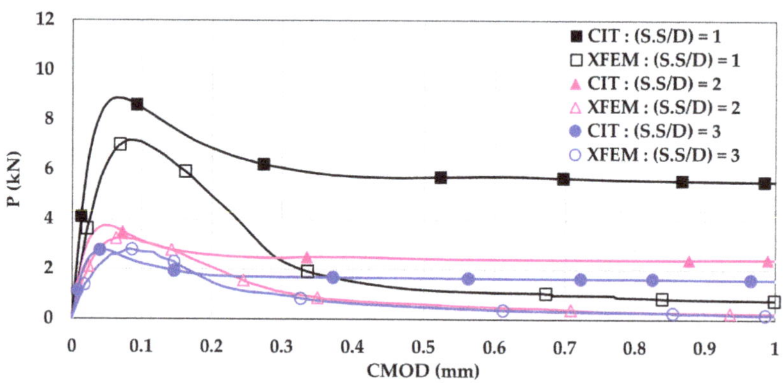

Figure 9. Plain concrete beam P-CMOD relationships for different shear span-to-depth ratios (S.S/D).

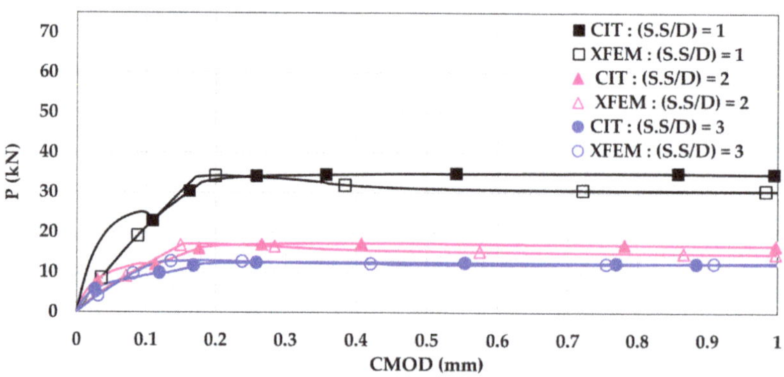

Figure 10. Reinforced concrete beam P-CMOD relationships for different shear span-to-depth ratios (S.S/D).

5.3. Influence of Using External Post-Tensioning

External post-tensioning (EPT) rods were used in a three-point bending loading scheme at three locations and four levels of post-tensioning stress. This was study their effects on the flexural behavior of the plain and reinforced T6 concrete beams. These rods had a yield strength of 900 MPa [43]. The EPT configurations were different in the manner of rod location relative to the crack height, as shown in Figure 11. These locations were: 1. at the crack mouth (X= 10 mm); 2. at the middle of the crack height (X = 30 mm); and 3. at the crack tip (X = 60 mm).

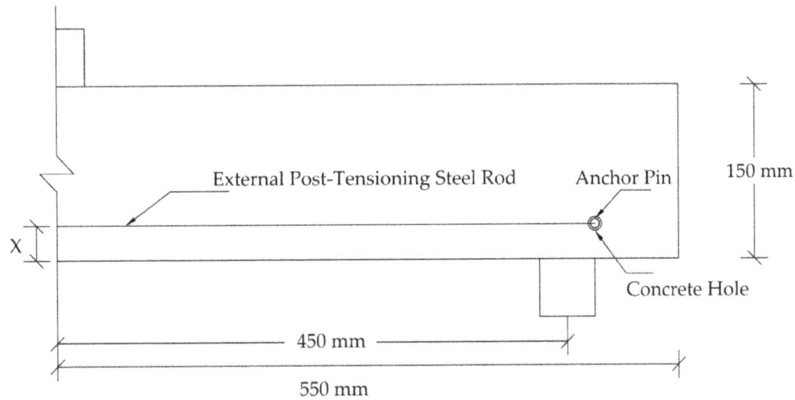

Figure 11. Schematic external post-tensioning (EPT) setup for beam T6.

Each configuration was investigated at four levels of post-tensioning stress, as follows: 1. rods are present but without post-tensioning; 2. rods are post-tensioned with 25% of their yield strength; 3. rods are post-tensioned with 50% of their yield strength; 4. rods are post-tensioned with 75% of their yield strength.

Two cylindrical steel anchor pins were placed at both ends of the concrete beam to connect the post-tensioning rods at both beam sides. Surface-to-surface contact interaction was used for the mutual surfaces between the steel anchor and the concrete beam. The motion of the steel anchor pins was constrained to the motion of the prestressing rod ends to apply the prestressing effect on the beam.

For all EPT beam models, the following findings can be drawn:

1. Placing the EPT rods towards the crack mouth and increasing the EPT stress gives a better effect. This is because increasing EPT towards the crack mouth enhances the role of post-tensioning in resisting the crack opening, resulting in improved crack control.
2. For plain concrete beam models, Figure 12 shows that EPT rods increase the flexural capacity and show more ductile flexural response. The increased capacity and ductility are attributed to the contribution of post-tensioning in handling the applied stresses.
3. For plain concrete beam models, EPT rods reduce flexural softening when they are placed closer to the crack mouth. The reduction of the flexural softening occurred because the stresses were handed over to the EPT rods earlier when they were closer to the crack mouth. As result, more softening and degradation was captured for the beams having EPT rods closer to the crack tip.
4. For plain concrete beam models, increasing the EPT stress to reach 75% of the rod's yield strength reduced flexural softening even with a rod location at the crack tip. This indicates that this amount of post-tensioning stress is capable of enhancing damage control even if the EPT rods are not located at crack mouth.
5. For plain concrete beam models, a clear discrepancy can be noticed between the CIT and XFEM results. The discrepancy is attributed to the stationary nature of a CIT

crack that does not grow, preventing a complete failure regime. Conversely, XFEM is capable of representing complete damage softening.

6. For the reinforced concrete beam models, Figure 13 shows higher flexural capacity due to reinforcement presence.
7. For the reinforced concrete beam models, the flexural results are close as the reinforcement managed to control the damage initiation and propagation.
8. For the reinforced concrete beam models, both the CIT and XFEM results are in good agreement. This supports the conclusion that CIT can give trustworthy results for reinforced concrete beams, unlike plain concrete beams. Additionally, this agreement reveals that using reinforcement has more performance in controlling damage than using EPT only. This is due to the complete bond with the concrete, the bond between the reinforcement bar and the concrete is achieved along the entire bar. Conversely, EPT rods are placed outside the concrete section and are bonded to the concrete at the end anchorages only.

Based on all previous simulations, beneficial practical implications can be interpreted. For both plain and reinforced concrete fracture simulations, the XFEM model is favored over the CIT model. However, the CIT model still can be used effectively to simulate reinforced concrete fractures. Additionally, using a minimal bonded steel reinforcement is capable of controlling the fracture damage propagation and enhancing the flexural performance significantly. Moreover, decreasing shear spans to eliminate shear stress at notched sections can contribute to the escalation of the flexural capacity and fracture control performance. For rehabilitation and renovation purposes, using unbonded external post-tensioning at the crack mouth at higher stresses can improve flexural tensile capacity and tensile damage resistance.

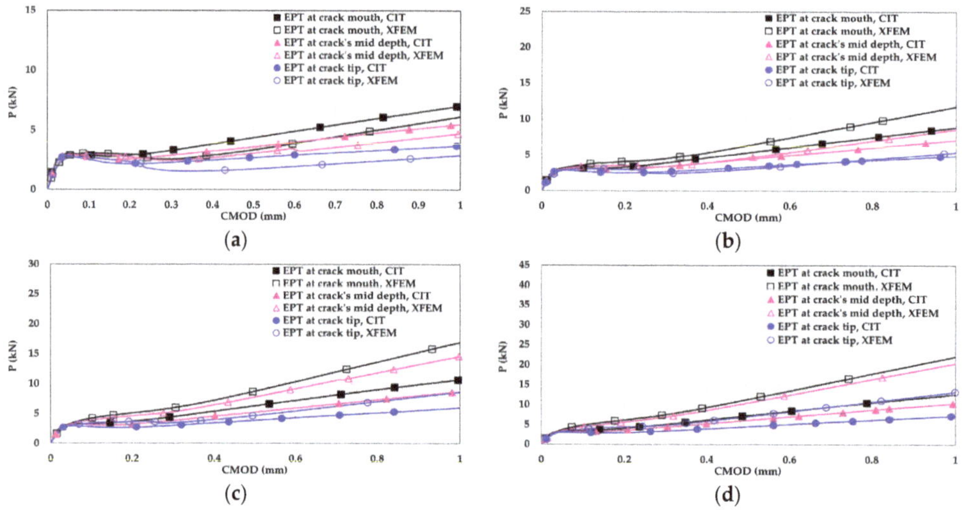

Figure 12. Plain concrete beam P-CMOD relationships for different EPTs at different locations: (**a**) 0%; (**b**) 25%; (**c**) 50%; (**d**) 75%.

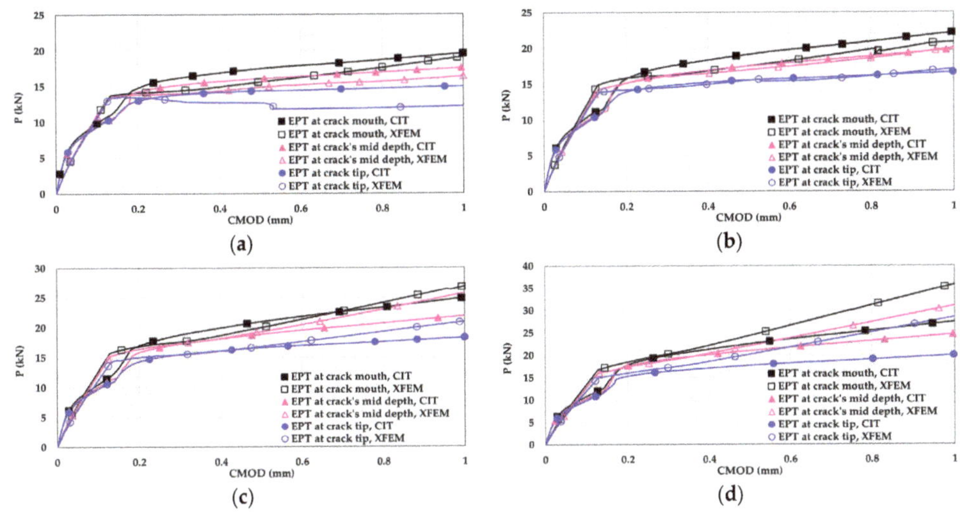

Figure 13. Reinforced concrete beam P-CMOD relationships for different EPTs at different locations: (**a**) 0%; (**b**) 25%; (**c**) 50%; (**d**) 75%.

6. Conclusions

This study aimed to develop numerical models using different notch modeling techniques built in ABAQUS. Notched plain and reinforced concrete beams subjected to three-point-bending and four point-bending loading setups were simulated. These notch modeling techniques were CIT, VCCT, and XFEM. The outcomes were compared with those of a pervious experimental study. Additionally, the influence of changing some parameters, such as a_o/D, S.S/D, and EPT, was investigated. The XFEM and VCCT models simulated flexural response in good agreement with the experimental outcomes. The CIT model showed a discrepancy for plain concrete simulations but yielded reliable results for reinforced concrete beams. Furthermore, the XFEM model was considered the most suitable, as it had better fracture energy estimation, solution-dependent crack path, and lower RMSE. It was found that increasing the a_o/D and S.S/D decreased the flexural capacity. Reinforcement implementation controlled concrete damage and increased the ductility and the flexural capacity for all studied parameters. Additionally, using EPT rods and increasing EPT stress increased the flexural capacity and ductility. Also, placing EPT rods near the crack mouth reduced flexural softening. However, for higher EPT stresses, flexural softening was reduced even if EPT rods were at the crack tip. For future research, it is recommended to extend this study by investigating the effect of using different strengthening materials to replace the bonded steel reinforcement and the unbonded steel external post-tensioning. The strengthening materials to be studied may include carbon fibers, glass fibers, and shape memory alloys.

Author Contributions: Conceptualization, S.E.-D.F.T., S.Y.M., A.B.T.; methodology, S.E.-D.F.T., S.Y.M., A.B.T.; validation, A.B.T.; software, A.B.T.; visualization, A.B.T.; writing—original draft preparation, S.E.-D.F.T., S.Y.M., A.B.T.; writing—review and editing, S.Y.M., A.B.T.; supervision, S.E.-D.F.T., S.Y.M.; project administration, S.E.-D.F.T., S.Y.M. All authors have read and agreed to the published version of the manuscript.

Funding: This research received no external funding.

Institutional Review Board Statement: Not applicable.

Informed Consent Statement: Not applicable.

Data Availability Statement: The FE data used to support the findings of this study are available upon request.

Conflicts of Interest: The authors declare no conflict of interest.

References

1. Jankowiak, T.; Lodygowski, T. Identification of parameters of concrete damage plasticity constitutive model. *Found. Civ. Environ.* **2005**, *6*, 53–69.
2. Sümer, Y.; Aktaş, M. Defining parameters for concrete damage plasticity model. *Chall. J. Struct. Mech.* **2015**, *1*, 149–155. [CrossRef]
3. Demin, W.; Fukang, H. Investigation for plastic damage constitutive models of the concrete material. *Procedia Eng.* **2017**, *210*, 71–78. [CrossRef]
4. Kmiecik, P.; Kamiński, M. Modelling of reinforced concrete structures and composite structures with concrete strength degradation taken into consideration. *Arch. Civ. Mech. Eng.* **2011**, *11*, 623–636. [CrossRef]
5. Tao, Z.; Wang, Z.-B.; Yu, Q. Finite element modelling of concrete-filled steel stub columns under axial compression. *J. Constr. Steel Res.* **2013**, *89*, 121–131. [CrossRef]
6. Nikaido, Y.; Mihara, Y.; Sawada, S.; Takahashi, Y. Improvement and enhancement of concrete damage plasticity model. In Proceedings of the Simulia Community Conference Proceedings, Berlin, Germany, 18–21 May 2015; pp. 1–10.
7. Nazem, M.; Rahmani, I.; Rezaee-Pajand, M. Nonlinear fe analysis of reinforced concrete structures using a tresca-type yield surface. *Sci. Iran.* **2009**, *16*, 512–519.
8. Yu, T.; Teng, J.G.; Wong, Y.L.; Dong, S.L. Finite element modeling of confined concrete-I: Drucker–Prager type plasticity model. *Eng. Struct.* **2010**, *32*, 665–679. [CrossRef]
9. Chaudhari, S.V.; Chakrabarti, M.A. Modeling of Concrete for Nonlinear Analysis using Finite Element Code ABAQUS. *Int. J. Comput. Appl.* **2012**, *44*, 14–18. [CrossRef]
10. Earij, A.; Alfano, G.; Cashell, K.; Zhou, X. Nonlinear three–dimensional finite–element modelling of reinforced–concrete beams: Computational challenges and experimental validation. *Eng. Fail. Anal.* **2017**, *82*, 92–115. [CrossRef]
11. Dere, Y. Nonlinear FE Modeling of Reinforced Concrete. *Int. J. Struct. Civ. Eng. Res.* **2017**, *6*, 71–74. [CrossRef]
12. Zhang, C.; Cao, P.; Cao, Y.; Li, J. Using finite element software to simulation fracture behavior of three-point bending beam with initial crack. *J. Softw.* **2013**, *8*, 1145–1150. [CrossRef]
13. Systèmes, D. ABAQUS 6.14 Analysis User's Manual. 2014. Available online: http://130.149.89.49:2080/v6.14/books/usb/default.htm (accessed on 28 December 2020).
14. Yin, Y.; Qiao, Y.; Hu, S. Determining concrete fracture parameters using three-point bending beams with various specimen spans. *Theor. Appl. Fract. Mech.* **2020**, *107*, 102465. [CrossRef]
15. Zhong, J.; Zhou, Y.; Bao, Q.; Wang, E.; Li, Q. Strengthening mechanism of channel steel plate for notched concrete beams against fracture: Test and numerical study. *Eng. Fract. Mech.* **2017**, *180*, 132–147. [CrossRef]
16. Sun, X.; Gao, Z.; Cao, P.; Zhou, C.; Ling, Y.; Wang, X.; Zhao, Y.; Diao, M. Fracture performance and numerical simulation of basalt fiber concrete using three-point bending test on notched beam. *Constr. Build. Mater.* **2019**, *225*, 788–800. [CrossRef]
17. De Domenico, D.; Urso, S.; Borsellino, C.; Spinella, N.; Recupero, A. Bond behavior and ultimate capacity of notched concrete beams with externally-bonded FRP and PBO-FRCM systems under different environmental conditions. *Constr. Build. Mater.* **2020**, *265*, 121208. [CrossRef]
18. Chen, G.; Gao, D.; Zhu, H.; Song Yuan, J.; Xiao, X.; Wang, W. Effects of novel multiple hooked-end steel fibres on flexural tensile behaviour of notched concrete beams with various strength grades. *Structures* **2021**, *33*, 3644–3654. [CrossRef]
19. Wahalathantri, B.L.; Thambiratnam, D.P.; Chan, T.H.T.; Fawzia, S. A material model for flexural crack simulation in reinforced concrete elements using ABAQUS. In Proceedings of the First International Conference on Engineering, Designing and Developing the Built Environment for Sustainable Wellbeing, Brisbane, Australia, 27–29 April 2011; pp. 260–264.
20. Nayal, R.; Rasheed, A. Tension Stiffening Model for Concrete Beams Reinforced with Steel and FRP Bars. *J. Mater. Civ. Eng.* **2006**, *18*, 831–841. [CrossRef]
21. Hsu, L.S.; Hsu, C.-T. Complete stress—Strain behaviour of high-strength concrete under compression. *Mag. Concr. Res.* **1994**, *46*, 301–312. [CrossRef]
22. Hafezolghorani, M.; Hejazi, F.; Vaghei, R.; Jaafar, M.S.B.; Karimzade, K. Simplified Damage Plasticity Model for Concrete. *Struct. Eng. Int.* **2017**, *27*, 68–78. [CrossRef]
23. Lubliner, J.; Oliver, J.; Oller, S.; Oñate, E. A plastic-damage model for concrete. *Int. J. Solids Struct.* **1989**, *25*, 299–326. [CrossRef]
24. Lee, J.; Fenves, G.L. Plastic-damage model for cyclic loading of concrete structures. *J. Eng. Mech.* **1998**, *124*, 892–900. [CrossRef]
25. Lee, S.H.; Abolmaali, A.; Shin, K.J.; Lee, H. Du ABAQUS modeling for post-tensioned reinforced concrete beams. *J. Build. Eng.* **2020**, *30*. [CrossRef]
26. Seow, P.E.C.; Swaddiwudhipong, S. Failure surface for concrete under multiaxial load—A unified approach. *J. Mater. Civ. Eng.* **2005**, *17*, 219–228. [CrossRef]
27. Janssen, M.; Zuidema, J.; Wanhill, R. *Fracture Mechanics*; Spon Press: Abingdon, UK, 2004.
28. Irwin, G.R. Analysis of stresses and strains near the end of a crack transversing a plate. *Trans. ASME Ser. E J. Appl. Mech.* **1957**, *24*, 361–364. [CrossRef]

29. Rybicki, E.F.; Kanninen, M.F. A finite element calculation of stress intensity factors by a modified crack closure integral. *Eng. Fract. Mech.* **1977**, *9*, 931–938. [CrossRef]
30. Raju, I.S. Calculation of strain-energy release rates with higher order and singular finite elements. *Eng. Fract. Mech.* **1987**, *28*, 251–274. [CrossRef]
31. Krueger, R. Virtual crack closure technique: History, approach, and applications. *Appl. Mech. Rev.* **2004**, *57*, 109–143. [CrossRef]
32. Belytschko, T.; Black, T. Elastic crack growth in finite elements with minimal remeshing. *Int. J. Numer. Methods Eng.* **1999**, *45*, 601–620. [CrossRef]
33. Moës, N.; Dolbow, J.; Belytschko, T. A finite element method for crack growth without remeshing. *Int. J. Numer. Methods Eng.* **1999**, *46*, 131–150. [CrossRef]
34. Melenk, J.M.; Babuška, I. The partition of unity finite element method: Basic theory and applications. *Comput. Methods Appl. Machanics Eng.* **1996**, *139*, 289–314. [CrossRef]
35. Khoei, A.R. *Extended Finite Element Method: Theory and Applications*; John Wiley & Sons: Hoboken, NJ, USA, 2014; ISBN 1118869680.
36. Karmakov, S.; Cepero-Mejías, F.; Curiel-Sosa, J.L. Numerical analysis of the delamination in CFRP laminates: VCCT and XFEM assessment. *Compos. Part C Open Access* **2020**, *2*, 100014. [CrossRef]
37. Hamed, M.M.; Nashwan, M.S.; Shahid, S. A Novel Selection Method of CIMP6 GCMs for Robust Climate Projection. *Int. J. Climatol.* **2021**. [CrossRef]
38. Hamed, M.M.; Nashwan, M.S.; Shahid, S. Performance evaluation of reanalysis precipitation products in Egypt using fuzzy entropy time series similarity analysis. *Int. J. Climatol.* **2021**, *41*, 5431–5446. [CrossRef]
39. Hamed, M.M.; Nashwan, M.S.; Shahid, S.; bin Ismail, T.; Wang, X.J.; Dewan, A.; Asaduzzaman, M. Inconsistency in historical simulations and future projections of temperature and rainfall: A comparison of CMIP5 and CMIP6 models over Southeast Asia. *Atmos. Res.* **2022**, *265*, 105927. [CrossRef]
40. Nash, J.E.; Sutcliffe, J. V River flow forecasting through conceptual models part I—A discussion of principles. *J. Hydrol.* **1970**, *10*, 282–290. [CrossRef]
41. Willmott, C.J. On the validation of models. *Phys. Geogr.* **1981**, *2*, 184–194. [CrossRef]
42. Gupta, H.V.; Kling, H.; Yilmaz, K.K.; Martinez, G.F. Decomposition of the mean squared error and NSE performance criteria: Implications for improving hydrological modelling. *J. Hydrol.* **2009**, *377*, 80–91. [CrossRef]
43. Sigrist, V.; Bäurich, A. The Development of Form in Prestressed Concrete Bridge Design. In Proceedings of the International Conference on Bridges, Dubrovnik, Croatia, 21–24 May 2006.

materials

Article

Mechanical Properties of Recycled Aggregate Concretes Containing Silica Fume and Steel Fibres

Soheil Jahandari [1,*], Masoud Mohammadi [1], Aida Rahmani [1,*], Masoumeh Abolhasani [2], Hania Miraki [3], Leili Mohammadifar [4], Mostafa Kazemi [5], Mohammad Saberian [6] and Maria Rashidi [1]

1 Centre for Infrastructure Engineering, Western Sydney University, Penrith, NSW 2751, Australia; m.mohammadi@westernsydney.edu.au (M.M.); m.rashidi@westernsydney.edu.au (M.R.)
2 Department of Civil and Environmental Engineering, Alaodoleh Semnani Institute of Higher Education, Garmsar 5815, Iran; masoumeh.abolhasani88@gmail.com
3 Department of Civil Engineering, Iran University of Science and Technology, Tehran 6846, Iran; hania_miraki@alumni.iust.ac.ir
4 Department of Architectural Engineering, Kerman Branch, Islamic Azad University, Kerman 1167, Iran; leilimohammadifar@gmail.com
5 GeMMe Building Materials, Urban and Environmental Engineering, University of Liège, 4000 Liège, Belgium; mostafa.kazemi@uliege.be
6 School of Engineering, RMIT University, Melbourne, VIC 3000, Australia; mohammad.boroujeni@rmit.edu.au
* Correspondence: S.jahandari@westernsydney.edu.au (S.J.); aida.rahmani63@gmail.com (A.R.)

Citation: Jahandari, S.; Mohammadi, M.; Rahmani, A.; Abolhasani, M.; Miraki, H.; Mohammadifar, L.; Kazemi, M.; Saberian, M.; Rashidi, M. Mechanical Properties of Recycled Aggregate Concretes Containing Silica Fume and Steel Fibres. *Materials* **2021**, *14*, 7065. https://doi.org/10.3390/ma14227065

Academic Editors: Dario De Domenico and Luís Filipe Almeida Bernardo

Received: 26 October 2021
Accepted: 18 November 2021
Published: 21 November 2021

Abstract: In this study, the impact of steel fibres and Silica Fume (SF) on the mechanical properties of recycled aggregate concretes made of two different types of Recycled Coarse Aggregates (RCA) sourced from both low- and high-strength concretes were evaluated through conducting 60 compressive strength tests. The RCAs were used as replacement levels of 50% and 100% of Natural Coarse Aggregates (NCA). Hook-end steel fibres and SF were also used in the mixtures at the optimised replacement levels of 1% and 8%, respectively. The results showed that the addition of both types of RCA adversely affected the compressive strength of concrete. However, the incorporation of SF led to compressive strength development in both types of concretes. The most significant improvement in terms of comparable concrete strength and peak strain with ordinary concrete at 28 days was observed in the case of using a combination of steel fibres and SF in both recycled aggregate concretes, especially with RCA sourced from high strength concrete. Although using SF slightly increased the elastic modulus of both recycled aggregate concretes, a substantial improvement in strength was observed due to the reinforcement with steel fibre and the coexistence of steel fibre and SF. Moreover, existing models to predict the elastic modulus of both non-fibrous and fibrous concretes are found to underestimate the elastic modulus values. The incorporation of SF changed the compressive stress-strain curves for both types of RCA. The addition of steel fibre and SF remarkably improved the post-peak ductility of recycled aggregates concretes of both types, with the most significant improvement observed in the case of RCA sourced from a low-strength parent concrete. The existing model to estimate the compressive stress-strain curve for steel fibre-reinforced concrete with natural aggregates was found to reasonably predict the compressive stress-strain behaviour for steel fibres-reinforced concrete with recycled aggregate.

Keywords: compressive behaviour; elastic modulus; recycled aggregates; steel fibres; silica fume

1. Introduction

Concrete is considered the most widely used construction material in the world [1–13]. Rapid urbanisation due to population growth results in the redevelopment of housing sectors and infrastructures in many cities around the world [14–18]. These redevelopments generate huge amounts of demolition waste due to the destruction of existing infrastructure, such as buildings and bridges [19,20]. New construction activities also generate concrete and building waste. Therefore, a significant amount of construction and demolition (C&D) waste is generated, and only a small amount is recycled in road bases while the rest goes to landfills [21–24]. Conversely, in many cities, land areas for C&D wastes disposal are scarce, and the landfill levy to dump these C&D wastes is also increasing every year. Hence, the additional cost is paid off by contractors and asset owners. The use of waste materials, such as C&D wastes, as aggregates in concrete is a sustainable and economical practice in the construction industry [23,25,26].

The properties of concrete made by recycled aggregates have been analysed in many research studies [27,28]. Therefore, there is a good understanding of the mechanical and durability properties of recycled aggregate concrete. As a result, the partial replacement of natural aggregates with recycled aggregates in concrete has been adopted in many projects around the world [29,30]. However, previous studies on the use of the recycled coarse aggregates (RCA) in concrete indicate the lower mechanical and durability characteristics of recycled aggregate concrete in comparison with the natural aggregate concrete, which is due to the weaker properties of RCA in comparison to the natural coarse aggregate (NCA). The weaker performance of RCA is due to the presence of attached mortar and inferior interfacial transition zone (ITZ).

Like concrete-containing natural aggregates, recycled aggregate concrete also exhibits brittle behaviour in tension and flexure [31]. Therefore, various fibres are used to reinforce recycled aggregate concrete to improve its mechanical properties [32–35]. Among many fibres, steel fibre is one of the most effective materials to enhance the tensile strength of recycled aggregate concrete [36–38]. Most studies have investigated the improvement in the mechanical properties of recycled aggregate concrete containing steel fibres by measuring the compressive, tensile and flexural strengths.

To mitigate the weaker performance of RCA and make it more comparable to conventional concrete, multiple approaches have been utilized in previous research, including the addition of supplementary cementitious materials (SCMs) such as fly ash, electric arc furnace slag, ground granulated blast furnace slag, and SF [39]. Such SCMs contribute to strength enhancement through eliminating the inferiority of RCA and make it comparable to natural aggregate concrete. For instance, the latent hydraulic property of ground granulated blast furnace slag as well as its pozzolanic characteristics contributes to the mitigation of the adverse mechanical impacts of RCA [40].

In addition to the desirable impact of the mentioned materials, previous research studies indicate the superb performance of SF in the enhancement of mechanical and durability properties of recycled concrete [41]. The addition of SF improves the mechanical and durability properties of recycled aggregate concrete in two ways. First, SF fills out RCA pores, which later improves the microstructure of the interfacial transition zone; second, hydration products fill the micro-cracks initially present in the RCA during crushing [42]. The incorporation of SF also improves the behaviour of fibre-reinforced concrete [43]. However, there is still a need to better understand the stress-strain behaviour of steel fibre-reinforced recycled aggregate concrete in the construction of structures once SF is used as a supplementary cementitious substance [44,45]. While determining the compressive strength is necessary for calculating the strength of structural components, the stress-strain curve is required for evaluating the toughness resistance to determine the ductility of structures made with sustainable materials [46–48].

Carneiro et al. [49] measured the compressive stress-strain behaviour of steel fibre-reinforced concrete containing recycled aggregates replaced by 25% of natural aggregates. The results showed that steel fibres affect the stress-strain behaviour of recycled aggregate concrete and increase its toughness. The behaviour of steel fibre-reinforced recycled aggregate concrete under compression was similar to that of fibre-reinforced natural aggregate concrete. However, to maximise the use of RCA in concrete and increase its sustainability, high amounts of RCA as a replacement for natural coarse aggregate are required.

Meesala [50] studied the effects of various types of fibres, such as woollen fibres, glass fibres, and steel fibres, on the mechanical and durability properties of recycled aggregate concretes. The experimental results showed that the incorporation of fibres could significantly improve the mechanical properties of recycled aggregate concrete. However, steel fibres showed the best performance in enhancing the mechanical properties of concrete. In another study [16], the axial stress-strain behaviour of macro-polypropylene fibres reinforced recycled aggregate concrete was investigated. Test results indicated that the peak stress, peak strain, and ultimate strain of concrete specimens increased with an increase in the fibres dosage, and the addition of fibres had a positive effect on the ductility of recycled aggregate concrete. Additionally, it has been reported that lower aspect ratio of fiber could lead to strength reduction [51]. This is due to the weak bond properties between the cement matrix and the fibres at lower aspect ratios. Furthermore, it has been reported that when the aspect ratio is higher than a specific value, with the addition of steel fibres, the ductility increases rather than the strength of concrete [52].

A better understanding of the compressive stress-strain behaviour and elastic modulus of recycled aggregate concrete containing steel fibre, SF, and their combination needs to be accepted by many designers, contractors, and policymakers as a sustainable alternative to conventional concrete. Therefore, the current study aims to evaluate the impacts of steel fibre and SF and their combination on the compressive stress-strain behaviour and elastic modulus of different recycled aggregate concretes. In published research, it is proven that the replacement levels of up to 30% of NCA by RCA does not significantly jeopardise the mechanical properties of concrete. A recent study reported a 5.0–9.3% reduction in compressive strength when different amounts of RCA were utilized [53]. However, due to the poor mechanical properties of RCA, increasing the replacement levels of NCA by RCA to over 30% can adversely affect the strength properties of concrete once no other additives, such as SF, are added into the mixes [16,54,55]. Therefore, the RCA replacement levels in this study were considered at 50% and 100%, and two different types of RCAs sourced from both low- and high-strength concretes were prepared and tested to investigate the improvement in mechanical properties.

2. Experimental Program

2.1. Raw Materials

The cementitious materials used in this study were ordinary Portland cement (OPC), equivalent to ASTM Type I, and SF. Their chemical compositions and physical properties are summarised in Table 1. The water quality used to make concrete specimens has a significant impact on concrete strength properties [56–59]. Therefore, distilled water was utilised for the characterisation tests and tap water for moulding the specimens [60–69]. Furthermore, the workability of the concrete mixtures was adjusted by using a Sika HRF-2 superplasticiser. Hooked-end steel fibres with a 50 mm length, 0.85 mm diameter, aspect ratio of 60, and tensile strength of 1309 MPa were used. The RCAs, with an angular shape, were obtained by crushing two laboratory concretes with low and high strength levels labelled as "Type A" and "Type B", with water/cement ratios of 0.60 and 0.40, respectively. The compressive strength of the Type A and B concretes cured for 28 days were 27 MPa and 41 MPa, respectively. The sieve analysis and physical properties of the used aggregates are presented in Figure 1 and Table 2, respectively. The attached mortar was obtained according to the thermal method, as recommended by other researchers [70,71]. In this method, before removing all the impurities, such as asphalt, plastics, and bricks, the prepared sample of recycled aggregate (mi) was immersed in water for 2 h to fully saturate the attached mortar. Next, the recycled aggregate sample was placed in a muffle at 500 °C to dry before being immersed in the cold water. This sudden cooling procedure caused cracks and stress generation, leading to easy removal of the mortar from the recycled aggregates. To remove the remaining attached mortar, a rubber hammer was used. Finally, to screen the recycled aggregate sample, a 4 mm sieve was used. Equation (1) was used for calculating the attached mortar:

$$\%\,attached\ mortar = (mi - mf)/mi \times 100 \tag{1}$$

where *mi* and *mf* are the initial and final masses of the sample, respectively.

Table 1. Characteristics of cementitious materials [71].

Properties	Cement	Silica Fume
SiO_2	21.66	90.01
Al_2O_3	4.21	1.29
Fe_2O_3	3.10	1.09
CaO	63.41	-
MgO	2.82	1.80
SO_3	2.61	-
Loss of ignition	0.81	-
Relative density, g/cm^3	3.11	2.20
Specific surface, cm^2/g	2950	20.700

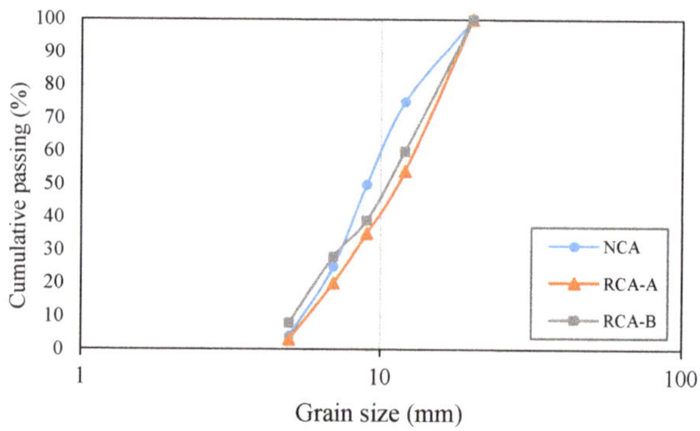

Figure 1. Sieve analysis of natural aggregates and RCAs.

Table 2. Physical characteristics of the aggregates.

Type of Aggregates	Crushing Value (%)	Density (kg/m^3)	Attached Mortar (%)	Water Absorption (%)
RCA (A)	27.3	2440	25.34	4.45
RCA (B)	27.1	2470	33.51	4.07
SAND	-	2510	-	0.91
NCA	26.5	2630	-	0.47

2.2. Mixture Proportions

The mix design of the concrete samples is shown in Table 3. In total, twenty mixtures were prepared, which were divided into four main groups. According to previous research studies [15,71], the optimum percentages of steel fibres and SF that provide sufficient mechanical strength for concrete mixes are 1% (by volume) and 8% (by cement weight), respectively. Therefore, the first group consisted of five control concrete mixtures containing NCA and RCA (including types A and B); the second group included 8% SF used as a partial replacement for OPC. In the third group, 1% steel fibres by volume were added to the mixtures. "S" and "F" were denoted at the beginning of the names of groups two and three, respectively. In the fourth group, the concretes contained 8% SF as a partial replacement for OPC and 1% steel fibres by volume, and "FS" was denoted at the beginning of the mixtures.

Table 3. Mixture design of concrete samples.

Group	Mix Code	Cement (Kg/m^3)	Water/Binder	Steel Fibres (Kg/m^3)	SF (Kg/m^3)	Sand (Kg/m^3)	RCA (Kg/m^3)	NCA (Kg/m^3)	SP (Kg/m^3)
Control	NC	380	0.40	-	-	910	-	910	2.30
	RC50-A	380	0.40	-	-	910	455	455	2.30
	RC100-A	380	0.40	-	-	910	910	-	2.30
	RC50-B	380	0.40	-	-	910	455	455	2.30
	RC100-B	380	0.40	-	-	910	910	0	2.30
Silica fume	SNC	350	0.40	-	30	910	-	910	2.30
	SRC50-A	350	0.4.0	-	30	910	455	455	2.30
	SRC100-A	350	0.40	-	30	910	910	-	2.30
	SRC50-B	350	0.40	-	30	910	455	455	2.30
	SRC100-B	350	0.40	-	30	910	910	0	2.30
Steel fibre	FNC	380	0.40	78	-	900	-	900	4.40
	FRC50-A	380	0.40	78	-	900	450	450	4.40
	FRC100-A	380	0.40	78	-	900	900	-	4.40
	FRC50-B	380	0.40	78	-	900	450	450	4.40
	FRC100-B	380	0.40	78	-	900	900	0	4.40
Steel fibre and silica fume	FSNC	350	0.40	78	30	900	-	900	4.40
	FSRC50-A	350	0.40	78	30	900	450	450	4.40
	FSRC100-A	350	0.40	78	30	900	900	-	4.40
	FSRC50-B	350	0.40	78	30	900	450	450	4.40
	FSRC100-B	350	0.40	78	30	900	900	-	4.40

F: concrete containing steel fibre, S: concrete containing silica fume, NC: normal concrete, FS: concrete containing steel fibres and silica fume, RC: recycled aggregate concrete, FSRC100-B: steel fibre-reinforced concrete containing 100% RCA and silica fume.

A constant water-to-binder ratio equal to 0.4 was used in all the mixtures. Two different RCA contents included the partial replacement of NCA (50% by mass) and full replacement (100% by mass). It should be noted that the replacements were made by mass because the RCA featured a different density compared with the NCA.

The trial-and-error method was used to find the suitable mixing procedure. First, the fine aggregates and binders were mixed using a Hobart mixer for one minute until a homogenous mixture was obtained. Next, half of the mixing water and super-plasticiser were added to the mix of binder and aggregates and were mixed for two minutes. The coarse aggregates and the other half of the water were then added, and the mixing process was resumed for five minutes. Finally, the fibres were added, and the mixing was resumed for five minutes.

2.3. Sample Preparation and Test Methods

The freshly mixed concrete was poured into cylindrical moulds with a 100 mm diameter and 200 mm in height to undergo the compressive strength tests and determine the stress-strain curves [45,72]. Next, 24 h after casting, the specimens were demoulded and cured in a basin with 100% relative humidity at 23 °C for 28 days [72,73]. In total, 60 cylindrical samples were prepared, and the compressive strength tests were carried out on them. Three replicate samples were prepared for each test to increase the accuracy of the test results. The uniaxial compressive strength and the stress-strain curves were automatically measured via a data logger connected to a compressive strength test machine with a maximum capacity of 2000 kN, and the loading rate was set to 24 MPa/min.

The slump values of all the mix designs adopted in this study were set to be between 50 mm and 75 mm, which is a reasonable value for practical applications. Moreover, the workability of the samples was slightly reduced while replacing the natural aggregates with both types of recycled concrete aggregates, or by using higher amounts of silica fume. Detailed information regarding the sample preparation, compression tests, and the used standards can be found in recent research studies conducted by the authors [45,74]. The elastic modulus for each specimen was measured by calculating the slope of the linear portion of the compressive stress-strain curve [46]. In other words, the concrete elastic modulus (Es) was calculated from the stress-strain curves according to Equation (2) [75]:

$$E_s = \frac{\sigma_2 - \sigma_1}{\varepsilon_2 - 0.005\%} \tag{2}$$

where σ_2 is equivalent to the 40% of the peak load, σ_1 corresponds to the strain at 0.005%, and ε_2 is the strain when the stress is equal to σ_2.

3. Results and Discussion

3.1. Compressive Stress-Strain Behaviour

The compressive stress-strain behaviours of the non-fibrous concretes containing RCA types A and B are shown in Figure 2a,b respectively. The stress-strain curves in both types only contain the ascending branch and the peak stress at which the specimen suddenly fractured and failed. This was due to the fact that in the absence of fibre, the samples exhibited brittle behaviour and failed after reaching their peak strength. When comparing the ascending branch of concrete containing NCA with those of the concretes containing RCA types A and B, the slope in the latter cases is less stiff than that of the former. In addition, between concretes containing two types of RCA, the concrete containing RCA type A was less stiff than that of concrete containing RCA type B. The results indicated that the fracture strain of the concretes containing two types of RCA was higher than that of the control concrete containing NCA. This may have been due to the fact that the total ITZ of RCA is higher than that of concrete containing NCA. The increased interfacial zone may give rise to the progressive development of micro-cracks at these interfaces and lead to reduced strength. Naturally, the ITZs of RCA type A, which produced from the low strength concrete, were more extensive than those of the RCA type B, and this may be the reason for the low strength and higher fracture strain of the concrete containing RCA type A. The lower strength of the samples containing RCA type A can also be attributed to the lower strength of the parent concrete. The effect of the SF addition on the ascending branch of the compressive stress-strain behaviour of all the concretes is also shown in Figure 3. Irrespective of coarse aggregate type, the compressive strength and the stiffness of the slopes of all the concretes increased by adding SF. The reason for the improved behaviour in the control concrete containing NCA is pore refinement owing to the particle packing and the formation of additional calcium-silica-hydrate due to the pozzolanic reaction of the SF. The SF also decreased the pores and densified the matrix in the ITZ between the RCAs and the matrix. Previous studies have also reported improvement in the case of ordinary concrete containing NCA [31].

Figure 3 presents the impact of steel fibre inclusion on the compressive stress-strain behaviour of all the concretes. The descending branch of the stress-strain curves was due to the contribution of steel fibres, which increased both the toughness and the ductility of the specimens. The compressive strength and the stiffness of the ascending branch of the stress-strain curve of all the concretes also increased due to the addition of steel fibres. In the case of the concretes containing RCA, the improvement was more prominent. It was observed that the coexistence of steel fibre and SF compensated for the negative effect of RCA in the concretes with the highest strength values of 58.38 and 58.23 MPa for FSRC 100-A and FSRC 50-B, respectively. The results also indicate that the impact of SF was more significant in the fibrous concrete compared with the non-fibrous concretes. This could be tied to the better bonding of steel fibre with the matrix, as observed recently in another study [76]. Additionally, the failure pattern of the fibrous specimens changed from brittle to ductile. The peak strain of the fibrous concretes increased approximately 10 times compared to that of the non-fibrous concretes. The addition of SF also improved the stiffness of the ascending branch slope of the fibrous recycled aggregates concretes, with significant improvement in the case of the concrete containing RCA type B.

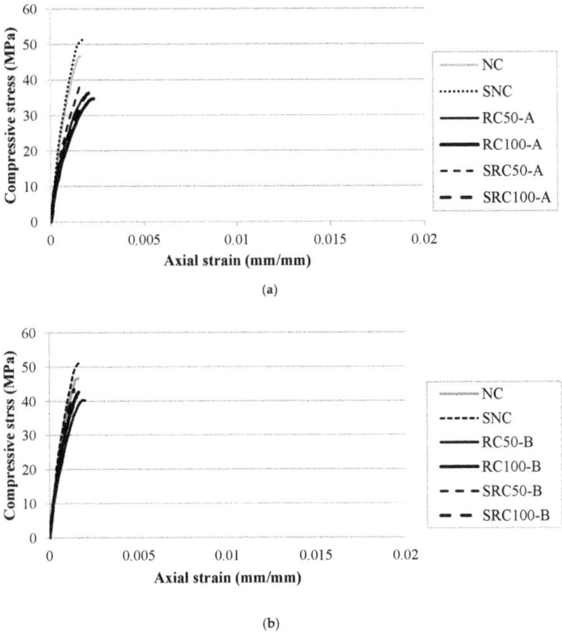

Figure 2. Compressive stress-strain behaviour of non-fibrous concretes containing RCA: (**a**) type A and (**b**) type B.

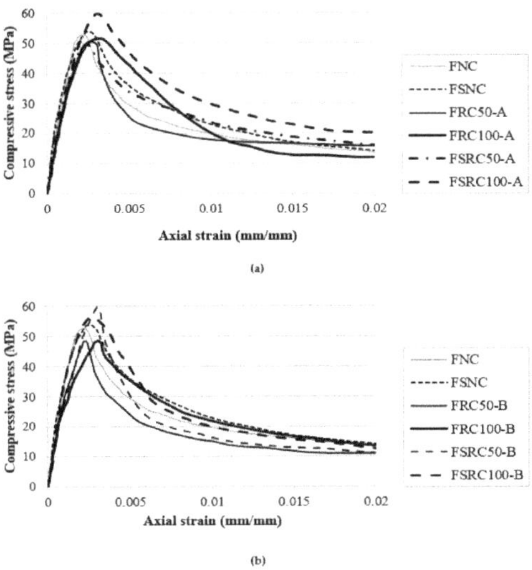

Figure 3. Compressive stress-strain performance of steel fibre-reinforced concretes containing RCA: (**a**) type A, (**b**) type B.

The typical failure patterns of all the concrete samples are shown in Figure 4. All the non-fibrous samples exhibited brittle failure, including those containing RCAs. The addition of steel fibres changed the failure pattern of the cylinders from brittle to ductile, as evidenced from the shear-type failure plane in the specimens, which was also similar to the samples prepared with the combination of SF and steel fibres. By comparing the failure patterns of the concretes containing steel fibre with those containing both SF and steel fibres, more minor damage was seen in the latter than in the former.

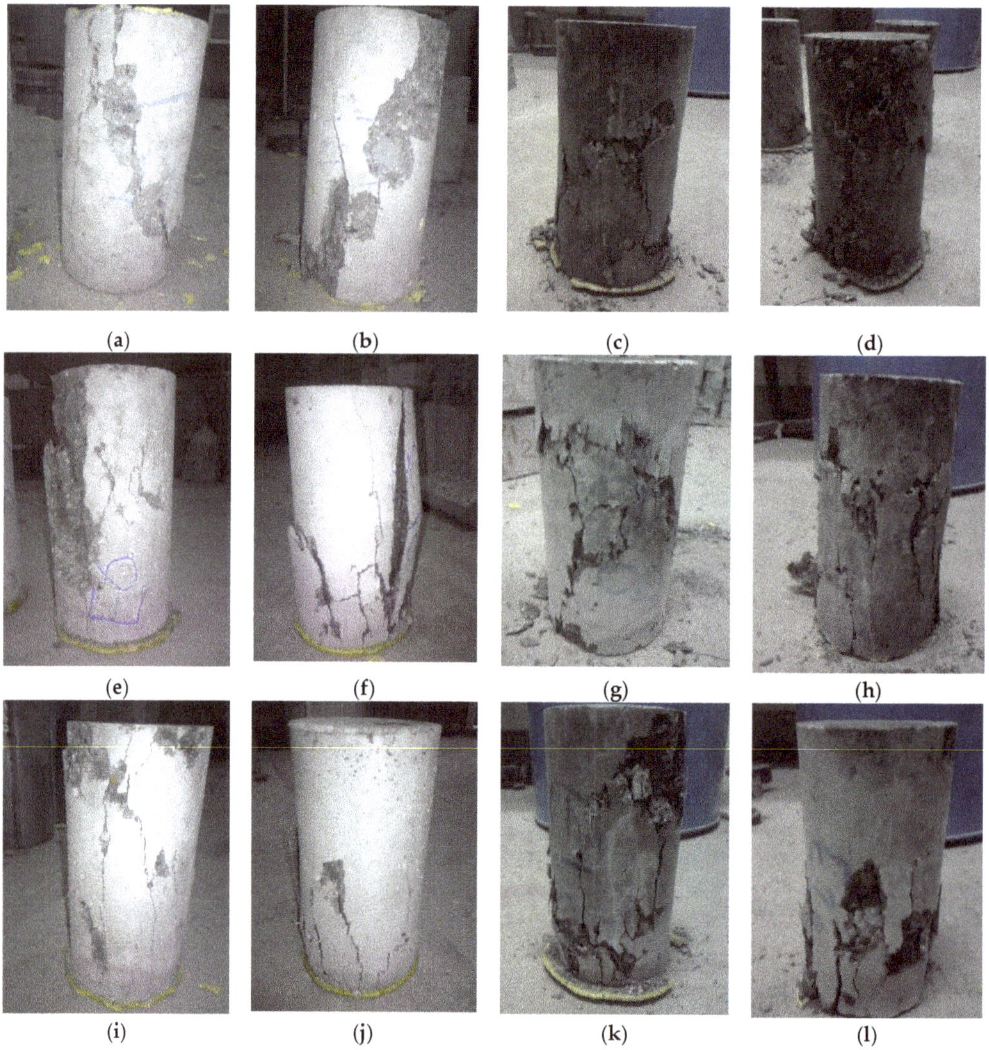

Figure 4. Failure patterns of control concrete and recycled aggregate concretes containing SF, steel fibre, and combination of SF and steel fibre: (**a**) NC; (**b**) SNC; (**c**) FNC; (**d**) FSNC; (**e**) RC100-A; (**f**) SRC100-A; (**g**) FRC100-A; (**h**) FSRC100-A; (**i**) RC100-B; (**j**) SRC100-B; (**k**) FRC100-B; (**l**) FSRC100-B.

3.2. Modulus of Elasticity

The calculated elastic modulus values of all the concretes are presented in Figure 5. It can be observed that the modulus of elasticity of the recycled aggregate concretes decreased with an increase in the RCA content. This change could have been due to the lower elastic modulus of the RCA than that of the NCAs and the weaker ITZ of the RCA. Similar results were achieved by Xiao et al. [75] and Salem and Burdette [77]. The replacement with RCA at 50% of both types reduced the modulus of elasticity by about 25%. With 100% replacement of the NCA with RCA types A and B, the elastic modulus decreased by about 40% and 10%, respectively, compared to that of normal concrete. The addition of SF increased the modulus of elasticity of the mixtures in comparison to conventional concrete. This increase could have been due to the pozzolanic activity of SF, which improved the ITZ of the concrete and thus enhanced the modulus of elasticity. Similar results were also reported by Corinaldesi and Moriconi [42]. The addition of steel fibre reduced the elasticity modulus of both the recycled aggregate concretes. For instance, through the introduction of steel fibres, the elasticity modulus of the RC50-A sample was reduced by approximately 19% (from 36.37 to 29.46 GPa in the FRC50-A sample). These results are in line with the findings of Altun et al. [78], who concluded that the modulus of elasticity decreases by increasing the percentage of steel fibre volume. However, the combination of steel fibre and SF had no significant impact on the mixtures containing recycled aggregates type A, but reduced the modulus of elasticity by about 18% in the mixtures containing recycled aggregates type B.

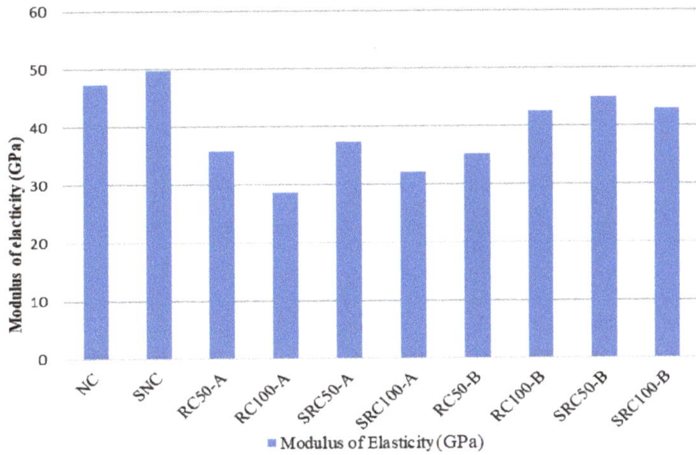

(a)

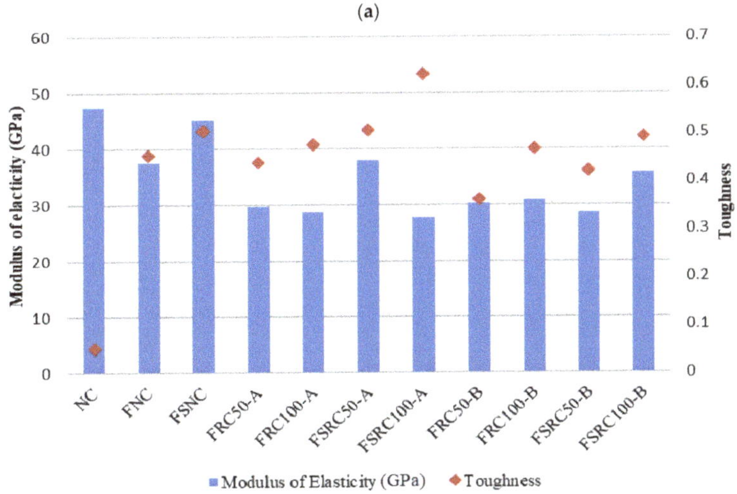

(b)

Figure 5. Modulus of elasticity for (**a**) non-fibrous concretes and (**b**) steel fibre-reinforced recycled aggregate concretes.

A correlation between the compressive strength and the modulus of elasticity of the non-fibrous recycled aggregates concretes was established, as shown in Figure 6. A reliable correlation was obtained with R^2 equal to 0.87. The measured elastic modulus values were compared with those predicted by existing models for both the non-fibrous and steel fibre-reinforced concretes to examine the feasibility of using existing models. In the case of the non-fibrous concrete, the models proposed by Warner et al. [79] and Thomas and Ramaswamy [80] for steel fibre-reinforced concrete were considered and compared with the measured values.

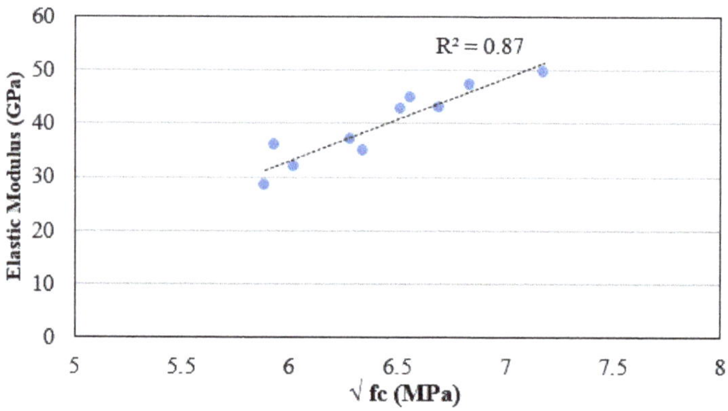

Figure 6. Relationship between elastic modulus and compressive strength of non-fibrous recycled aggregate concretes.

Figure 7a,b show the correlations between the experimentally measured and the model-predicted elastic modulus of non-fibrous and steel fibre-reinforced recycled aggregates concretes, respectively. A good correlation can be seen in both cases, with the slight deviation of a few experimentally measured elastic modulus values.

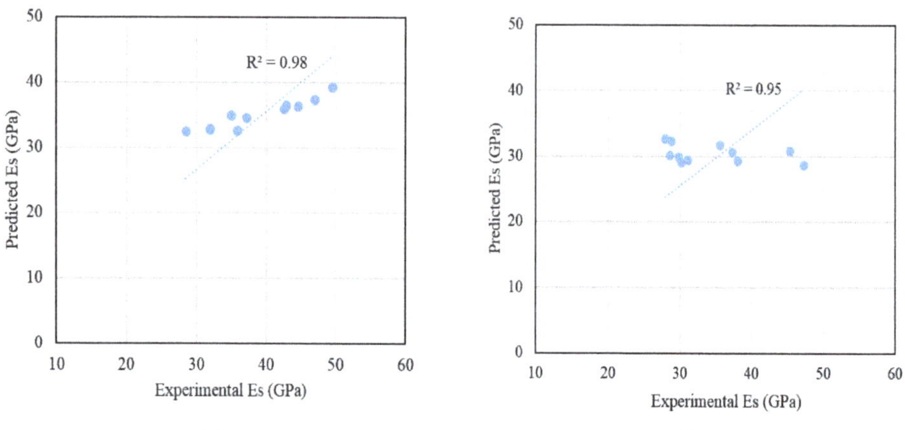

Figure 7. Comparison between experimentally measured elastic modulus and predicted modulus from: (**a**) AS 3600 for non-fibrous recycled aggregate concretes; and (**b**) for steel fibre-reinforced recycled aggregate concretes [80].

The effect of SF addition on the toughness of both types of steel fibre-reinforced recycled aggregates concretes was calculated from the area under the compressive stress-strain curve in each concrete. The results are summarised in Figure 5b. The toughness values of all the non-fibrous samples were less than 0.1; hence, they are not indicated in the Figure. The toughnesses of the steel fibre-reinforced recycled aggregate concretes containing RCA's types A and B were comparable with those of the steel fibre-reinforced concrete containing NCA. The addition of SF improved the toughness of both steel fibre-reinforced recycled aggregate concretes such that the toughness increased about 31% (from 0.47 in FRC100-A to 0.62 in FSRC100-A). Similar results were observed in the case of concrete containing NCA. This could have been due to the densification of the ITZ of the steel fibre in the cement matrix, which improved the steel fibre bond in the matrix and hence better post-peak ductility in the concrete.

The consequences of adding SF on the peak compressive strain and compressive strength of both non-fibrous and fibrous recycled aggregates concretes are shown in Figure 8.

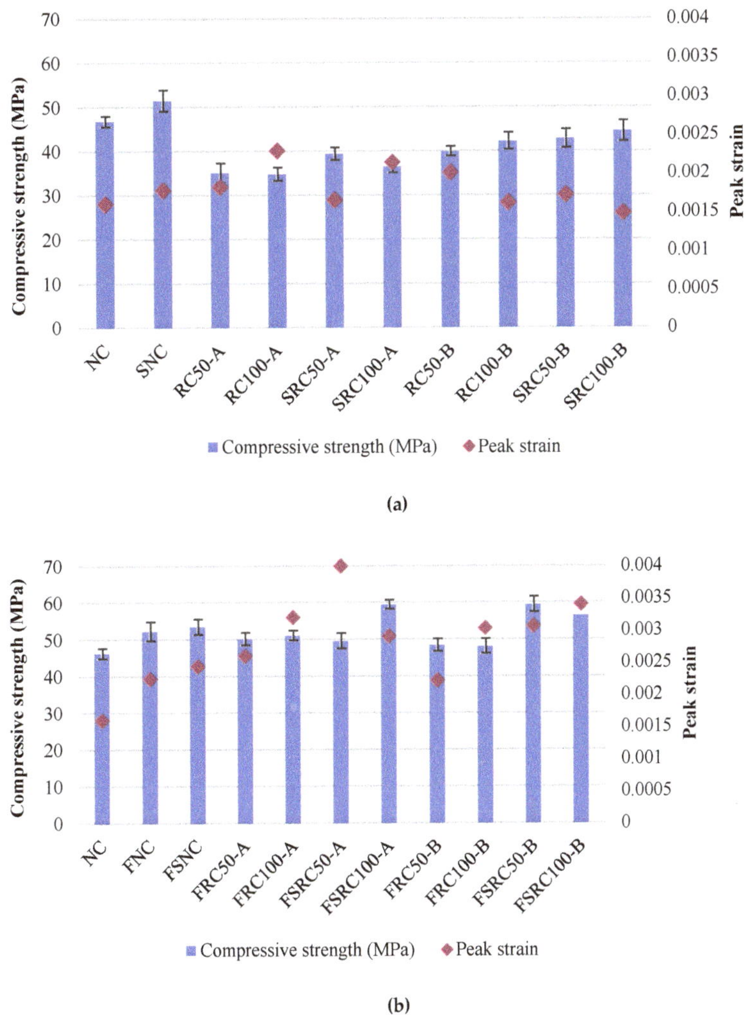

Figure 8. Impact of SF and steel fibre and coexistence of steel fibre and SF on the peak strain and compressive strength of recycled aggregate concretes. (**a**) non-fibrous concretes and (**b**) steel fibre-reinforced recycled aggregate concretes.

As shown in Figure 8, the peak strain of both types of recycled aggregates concretes was slightly decreased due to the addition of SF. However, in the case of the steel fibres reinforced recycled aggregates concrete, a significant improvement in the peak strain was observed. This improvement can be attributed to the bridging of micro-cracks by the steel fibres. The addition of SF led to a slight improvement in the peak strain of the steel fibre-reinforced recycled aggregate concretes; however, this amount was not very insignificant.

3.3. Modelling of Stress-Strain Behaviour of Recycled Aggregates Concretes Containing Steel Fibre and Combination of Steel Fibre and SF

The prediction of compressive stress-strain behaviour of concrete helps to model the structural behaviour of concrete structures. Various models that predict the compressive stress-strain behaviour of concrete containing natural aggregates and fibre-reinforced concretes can be found in previous research, Ezeldin and Balaguru [81] proposed the following model Equation (3) to predict the compressive stress-strain behaviour of ordinary concrete containing steel fibres:

$$\frac{f_c}{f_{cf}} = \frac{\beta \frac{\varepsilon_c}{\varepsilon_{co}}}{\beta - 1 + \left(\frac{\varepsilon_c}{\varepsilon_{co}}\right)^{\beta}} \tag{3}$$

$$\beta = 1.093 + 0.7132\,(RI)^{-0.926} \tag{4}$$

$$RI = V_f \frac{l}{ø} \tag{5}$$

where f_{cf} is the compressive strength of fibre concrete; ε_{co} is the strain corresponding to the compressive strength (f_c), and ε_c is the strain value in the compressive stress-strain curve. The value β is the material parameter and RI is a reinforcing index combining the effect of the steel fibre volume fractions, where V_f is the volume fraction of fibers, and l and ø are the length and diameter of fibers, respectively [82].

The comparison between the experimental compressive stress-strain curve of RCA and that predicted by the aforementioned model proposed by Ezeldin and Balaguru [81] for ordinary concrete containing steel fibres is shown in Figure 9. The model for steel fibre-reinforced concrete containing natural aggregates agrees well with the ascending branch of the stress-strain curve for all recycled aggregates concretes. However, a slight variation in the post-peak behaviour between the model predicted and the experimentally observed curve for recycled aggregates concretes can be seen. Nevertheless, the existing model proposed for steel fibre-reinforced ordinary concrete can be used to predict the compressive stress-strain behaviour of steel fibre-reinforced recycled aggregates concretes even with RCA from different grades of concrete.

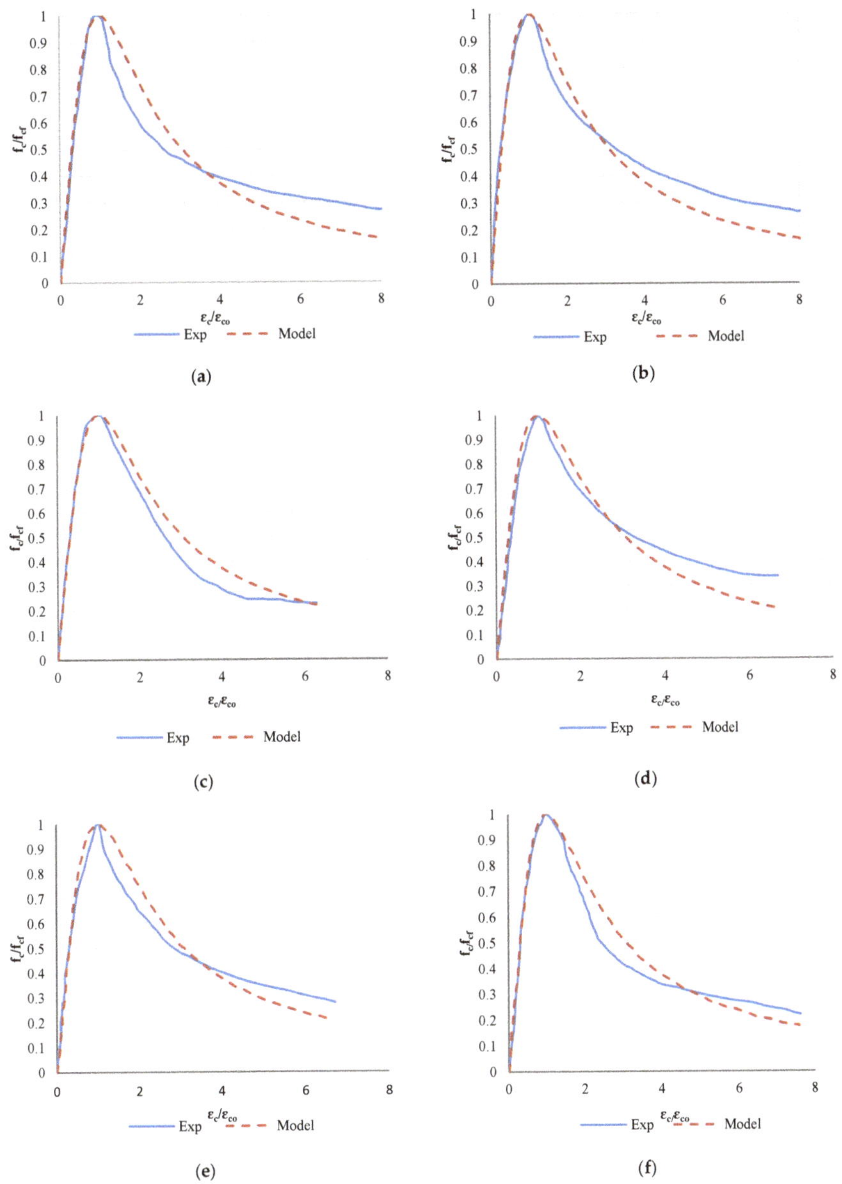

Figure 9. Comparison between the experimental compressive stress-strain curve of steel fibre-reinforced RAC and the predicted model for steel fibre-reinforced NAC proposed by Ezeldin and Balaguru (1992): (**a**) FNC, (**b**) FSNC, (**c**) FRC100-A, (**d**) FSRC100-A, (**e**) FRC100-B, (**f**) FSRC100-B.

4. Conclusions

The effects of steel fibres, silica fume (SF), and the combined use of steel fibres and SF on the mechanical properties of recycled aggregate concretes containing 50% and 100% recycled coarse types aggregates (RCA), sourced from both low- and high-strength concretes, were investigated. The following main conclusions were drawn based on the experimental and prediction studies:

1. The discrete addition of SF and steel fibre slightly increased the compressive strength of concretes containing both types of RCA. The combined use of SF and steel fibre significantly improved the compressive strength of recycled aggregates concretes, especially with RCA sourced from high-strength concrete. Similar behaviour was also observed in both recycled aggregate concretes in the case of peak strain.
2. The addition of SF slightly increased the elastic modulus of both recycled aggregate concretes; however, a significant improvement was observed due to the addition of steel fibre and a combination of steel fibre and SF. Existing models underestimate the elastic modulus of both non-fibrous and fibrous concretes at higher magnitudes.
3. The addition of SF improved the ascending branch of the compressive stress-strain curve of the concretes containing both types of RCA. No significant changes in the ascending branch of the compressive stress-strain curve were observed due to the addition of SF in the recycled aggregate concretes containing steel fibre. The addition of steel fibres and the combined addition of SF and steel fibre significantly improved the post-peak ductility of the recycled aggregate concretes of both types, with the most significant improvement, in the case of RCA, sourced from the low-strength parent concrete.
4. The existing model reasonably predicts the compressive stress-strain behaviour of steel fibre-reinforced concrete containing both natural aggregates and recycled aggregates. This indicates the applicability of the existing model for steel fibre-reinforced recycled aggregates concretes with and without SF.

For future research studies, it is recommended to explore the effects of different water/cement ratios on the same mix designs. The investigation of the impact of using other types of fibres on the engineering properties of the mix designs adopted in this research is also suggested.

Author Contributions: Conceptualization, data curation, investigation, methodology, writing—review and editing, resources, supervision, S.J.; formal analysis, writing original draft, validation, project administration, M.M.; writing original draft, investigation, writing—review and editing, supervision, A.R.; writing original draft, writing—review and editing, investigation, M.A.; data curation, writing—review and editing, investigation, formal analysis, H.M.; writing—review & editing, investigation, visualization, resources, L.M.; conceptualization, data curation, methodology, M.K.; investigation, writing—review and editing, M.S.; project administration, supervision, writing—review and editing, M.R. All authors have read and agreed to the published version of the manuscript.

Funding: This study was funded by Chem Concrete Pty Ltd. and Abadgaran Negin Jonoobshargh Company [2009A12].

Institutional Review Board Statement: Not applicable.

Informed Consent Statement: Not applicable.

Data Availability Statement: Data are available upon reasonable request from the corresponding authors.

Acknowledgments: This study was supported by Chem Concrete Pty Ltd., Abadgaran Negin Jonoobshargh Company (ANJ Co.), and SAHM research team at Western Sydney University. The authors would also like to thank Mojgan Akbari for her valuable advice on the test program.

Conflicts of Interest: The authors declare no conflict of interest.

References

1. Kazemi, M.; Courard, L.; Hubert, J. Coarse recycled materials for the drainage and substrate layers of green roof system in dry condition: Parametric study and thermal heat transfer. *J. Build. Eng.* **2021**, *45*, 103487. [CrossRef]
2. Kazemi, M.; Courard, L. Modelling hygrothermal conditions of unsaturated substrate and drainage layers for the thermal resistance assessment of green roof: Effect of coarse recycled materials. *Energy Build.* **2021**, *250*, 111315. [CrossRef]
3. Kazemi, M.; Courard, L.; Hubert, J. Heat Transfer Measurement within Green Roof with Incinerated Municipal Solid Waste Aggregates. *Sustainability* **2021**, *13*, 7115. [CrossRef]
4. Rasekh, H.; Joshaghani, A.; Jahandari, S.; Aslani, F.; Ghodrat, M. Rheology and workability of SCC. In *Self-Compacting Concrete: Materials, Properties and Applications*; Woodhead Publishing: Cambridge, UK, 2020; pp. 31–63.

5. Khalilpasha, M.H.; Sadeghi-Nik, A.; Lotfi-Omran, O.; Kimiaeifard, K.; Amirpour-Molla, M. Sustainable development using recyclable rubber in Self-Compacting Concrete. In Proceedings of the 3rd International Conference on Construction in Developing Countries (Advancing Civil, Architectural and Construction Engineering & Management), Bangkok, Thailand, 4–6 July 2012; pp. 580–585.
6. Lotfi-Omran, O.; Nikbin, I.; Manssouri, A.R.; Sadeghi-Nik, A.; Rabbanifar, S.; Fallahnejad, H. Propagation of ultrasonic waves in self-compacting -concrete and investigation of distribution of dynamic modulus of elasticity in concrete. *J. Basic. Appl. Sci. Res.* **2011**, *1*, 2319–2323.
7. Nik, A.S.; Bahari, A.; Nik, A.S. Investigation of nano structural properties of cement-based materials. *Am. J. Sci. Res* **2011**, *25*, 104–111.
8. Bahari, A.; Sadeghi-Nik, A.; Roodbari, M.; Sadeghi-Nik, A.; Mirshafiei, E. Experimental and theoretical studies of ordinary Portland cement composites contains nano LSCO perovskite with Fokker-Planck and chemical reaction equations. *Constr. Build. Mater.* **2018**, *163*, 247–255. [CrossRef]
9. Farhangi, V.; Karakouzian, M. Effect of fiber reinforced polymer tubes filled with recycled materials and concrete on structural capacity of pile foundations. *Appl. Sci.* **2020**, *10*, 1554. [CrossRef]
10. Mehrabi, P.; Honarbari, S.; Rafiei, S.; Jahandari, S.; Bidgoli, M.A. Seismic response prediction of FRC rectangular columns using intelligent fuzzy-based hybrid metaheuristic techniques. *J. Ambient. Intell. Humaniz. Comput.* **2021**, *12*, 10105–10123. [CrossRef]
11. Feng, Y.; Mohammadi, M.; Wang, L.; Rashidi, M.; Mehrabi, P. Application of Artificial Intelligence to Evaluate the Fresh Properties of Self-Consolidating Concrete. *Materials* **2021**, *14*, 4885. [CrossRef] [PubMed]
12. Shariatmadari, N.; Hasanzadehshooiili, H.; Ghadir, P.; Saeidi, F.; Moharami, F. Compressive Strength of Sandy Soils Stabilized with Alkali-Activated Volcanic Ash and Slag. *J. Mater. Civ. Eng.* **2021**, *33*, 04021295. [CrossRef]
13. Ghadir, P.; Zamanian, M.; Mahbubi-Motlagh, N.; Saberian, M.; Li, J.; Ranjbar, N. Shear strength and life cycle assessment of volcanic ash-based geopolymer and cement stabilized soil: A comparative study. *Transp. Geotech.* **2021**, *31*, 100639. [CrossRef]
14. Kazemi, M.; Madandoust, R.; de Brito, J. Compressive strength assessment of recycled aggregate concrete using Schmidt rebound hammer and core testing. *Constr. Build. Mater.* **2019**, *224*, 630–638. [CrossRef]
15. Madandoust, R.; Kazemi, M.; Talebi, P.K.; de Brito, J. Effect of the curing type on the mechanical properties of lightweight concrete with polypropylene and steel fibres. *Constr. Build. Mater.* **2019**, *223*, 1038–1052. [CrossRef]
16. Kazmi, S.M.S.; Munir, M.J.; Wu, Y.-F.; Patnaikuni, I.; Zhou, Y.; Xing, F. Axial stress-strain behavior of macro-synthetic fiber reinforced recycled aggregate concrete. *Cem. Concr. Compos.* **2019**, *97*, 341–356. [CrossRef]
17. Fatehi, H.; Bahmani, M.; Noorzad, A. Strengthening of Dune Sand with Sodium Alginate Biopolymer. In Proceedings of the of the 8th International Conference on Case Histories in Geotechnical Engineering (Geo-Congress 2019), Philadelphia, PA, USA, 24–27 March 2019; pp. 157–166.
18. Bahmani, M.; Fatehi, H.; Noorzad, A.; Hamedi, J. Biological soil improvement using new environmental bacteria isolated from northern Iran. *Environ. Geotech.* **2019**, *40*, 1–13. [CrossRef]
19. Miraki, H.; Shariatmadari, N.; Ghadir, P.; Jahandari, S.; Tao, Z.; and Siddique, R. Clayey soil stabilization using alkali-activated volcanic ash and slag. *J. Rock Mech. Geotech. Eng.* **2021**, in press. [CrossRef]
20. Ghadir, P.; Ranjbar, N. Clayey soil stabilization using geopolymer and Portland cement. *Constr. Build. Mater.* **2018**, *188*, 361–371. [CrossRef]
21. Saberian, M.; Li, J.; Kilmartin-Lynch, S.; Boroujeni, M. Repurposing of COVID-19 single-use face masks for pavements base/subbase. *Sci. Total. Environ.* **2021**, *769*, 145527. [CrossRef] [PubMed]
22. Saberian, M.; Li, J.; Perera, S.T.A.M.; Zhou, A.; Roychand, R.; Ren, G. Large-scale direct shear testing of waste crushed rock reinforced with waste rubber as pavement base/subbase materials. *Transp. Geotech.* **2021**, *28*, 100546. [CrossRef]
23. Saberian, M.; Li, J.; Boroujeni, M.; Law, D.; Li, C.-Q. Application of demolition wastes mixed with crushed glass and crumb rubber in pavement base/subbase. *Resour. Conserv. Recycl.* **2020**, *156*, 104722. [CrossRef]
24. Kazemi, M.; Hajforoush, M.; Talebi, P.K.; Daneshfar, M.; Shokrgozar, A.; Jahandari, S.; Saberian, M.; Li, J. In-situ strength estimation of polypropylene fibre reinforced recycled aggregate concrete using Schmidt rebound hammer and point load test. *J. Sustain. Cem.-Based Mater.* **2020**, *9*, 289–306. [CrossRef]
25. Li, J.; Saberian, M.; Nguyen, B.T. Effect of crumb rubber on the mechanical properties of crushed recycled pavement materials. *J. Environ. Manag.* **2018**, *218*, 291–299. [CrossRef]
26. Afshar, A.; Jahandari, S.; Rasekh, H.; Shariati, M.; Afshar, A.; Shokrgozar, A. Corrosion resistance evaluation of rebars with various primers and coatings in concrete modified with different additives. *Constr. Build. Mater.* **2020**, *262*, 120034. [CrossRef]
27. de Brito, J.; Pereira, A.; Correia, J. Mechanical behaviour of non-structural concrete made with recycled ceramic aggregates. *Cem. Concr. Compos.* **2005**, *27*, 429–433. [CrossRef]
28. De Brito, J.; Ferreira, J.; Pacheco, J.; Soares, D.; Guerreiro, M. Structural, material, mechanical and durability properties and behaviour of recycled aggregates concrete. *J. Build. Eng.* **2016**, *6*, 1–16. [CrossRef]
29. de Brito, J.; Agrela, F.; Silva, R.V. Legal regulations of recycled aggregate concrete in buildings and roads. In *New Trends in Eco-efficient and Recycled Concrete*; Woodhead Publishing: Cambridge, UK, 2018; pp. 509–526.
30. Liu, J.; Mohammadi, M.; Zhan, Y.; Zheng, P.; Rashidi, M.; Mehrabi, P. Utilizing Artificial Intelligence to Predict the Superplasticizer Demand of Self-Consolidating Concrete Incorporating Pumice, Slag, and Fly Ash Powders. *Materials* **2021**, *14*, 6792. [CrossRef]

31. Ardalan, R.B.; Emamzadeh, Z.N.; Rasekh, H.; Joshaghani, A.; Samali, B. Physical and mechanical properties of polymer modified self-compacting concrete (SCC) using natural and recycled aggregates. *J. Sustain. Cem. Based Mater.* **2020**, *9*, 1–16. [CrossRef]
32. Vytlačilová, V.; Vodička, J. Properties of fiber reinforced concrete using recycled aggregates. In Proceedings of the European Conference of Chemical Engineering, and European Conference of Civil Engineering, and European Conference of Mechanical Engineering, and European Conference on Control, Tenerife, Spain, 30 November–2 December 2010; pp. 71–75.
33. Kumar, D.S.; Vikranth, J. Experimental Study on Strength and Durability Characteristics of Fibre Reinforced Recyled Aggregate Concrete. *J. Eng. Res. Appl.* **2013**, *3*, 1883–1892.
34. Nam, J.; Kim, G.; Yoo, J.; Choe, G.; Kim, H.; Choi, H.; Kim, Y. Effectiveness of fiber reinforcement on the mechanical properties and shrinkage cracking of recycled fine aggregate concrete. *Materials* **2016**, *9*, 131. [CrossRef]
35. Akinkurole, O.O. Experimental Investigation on the Influence of Steel Fiber on the Compressive and Tensile Strength of Recycled Aggregates Concrete. *JEAS* **2010**, *5*, 264–268. [CrossRef]
36. Kosior-Kazberuk, M.; Grzywa, M. Recycled aggregate concrete as material for reinforced concrete structures. *J. Sustain. Archit. Civ. Eng.* **2014**, *7*, 60–66. [CrossRef]
37. Ramana, N.V. Performance of crimped steel fibre-reinforced recycled aggregate concrete. *Int. J. Adv. Technol. Eng. Sci.* **2016**, *4*, 152–159.
38. Singh, S.P.; Bansal, R.S. Strength evaluation of steel fibre-reinforced concrete with recycled aggregates. *Int. J. Tech. Res. Appl.* **2016**, *4*, 171–175.
39. De Domenico, D.; Faleschini, F.; Pellegrino, C.; Ricciardi, G. Structural behavior of RC beams containing EAF slag as recycled aggregate: Numerical versus experimental results. *Constr. Build. Mater.* **2018**, *171*, 321–337. [CrossRef]
40. Majhi, R.K.; Nayak, A.N. Production of sustainable concrete utilising high-volume blast furnace slag and recycled aggregate with lime activator. *J. Clean. Prod.* **2020**, *255*, 120188. [CrossRef]
41. Roychand, R.; Li, J.; De Silva, S.; Saberian, M.; Law, D.; Pramanik, B.K. Development of zero cement composite for the protection of concrete sewage pipes from corrosion and fatbergs. *Resour. Conserv. Recycl.* **2021**, *164*, 105166. [CrossRef]
42. Corinaldesi, V.; Moriconi, G. Influence of mineral additions on the performance of 100% recycled aggregate concrete. *Constr. Build. Mater.* **2009**, *23*, 2869–2876. [CrossRef]
43. Zhang, P.; Li, Q.; Zhang, H. Combined effect of polypropylene fiber and silica fume on mechanical properties of concrete composite containing fly ash. *J. Reinf. Plast. Compos.* **2011**, *30*, 1349–1358. [CrossRef]
44. Mehrabi, P.; Shariati, M.; Kabirifar, K.; Jarrah, M.; Rasekh, H.; Trung, N.T.; Shariati, A.; Jahandari, S. Effect of pumice powder and nano-clay on the strength and permeability of fiber-reinforced pervious concrete incorporating recycled concrete aggregate. *Constr. Build. Mater.* **2021**, *287*, 122652. [CrossRef]
45. Toghroli, A.; Mehrabi, P.; Shariati, M.; Trung, N.T.; Jahandari, S.; Rasekh, H. Evaluating the use of recycled concrete aggregate and pozzolanic additives in fiber-reinforced pervious concrete with industrial and recycled fibers. *Constr. Build. Mater.* **2020**, *252*, 118997. [CrossRef]
46. Jahandari, S.; Li, J.; Saberian, M.; Shahsavarigoughari, M. Experimental study of the effects of geogrids on elasticity modulus, brittleness, strength, and stress-strain behavior of lime stabilized kaolinitic clay. *GeoResJ* **2017**, *13*, 49–58. [CrossRef]
47. Saberian, M.; Jahandari, S.; Li, J.; Zivari, F. Effect of curing, capillary action, and groundwater level increment on geotechnical properties of lime concrete: Experimental and prediction studies. *J. Rock Mech. Geotech. Eng.* **2017**, *9*, 638–647. [CrossRef]
48. Jahandari, S.; Mojtahedi, S.F.; Zivari, F.; Jafari, M.; Mahmoudi, M.R.; Shokrgozar, A.; Kharazmi, S.; Vosough Hosseini, B.; Rezvani, S.; Jalalifar, H. The impact of long-term curing period on the mechanical features of lime-geogrid treated soils. *Geomech. Geoengin.* **2020**, 1–13. [CrossRef]
49. Carneiro, J.A.; Lima, P.R.L.; Leite, M.B.; Toledo Filho, R.D. Compressive stress–strain behavior of steel fiber reinforced-recycled aggregate concrete. *Cem. Concr. Compos.* **2014**, *46*, 65–72. [CrossRef]
50. Meesala, C.R. Influence of different types of fiber on the properties of recycled aggregate concrete. *Struct. Concr.* **2019**, *20*, 1656–1669. [CrossRef]
51. Gao, D.; Zhang, L. Flexural performance and evaluation method of steel fiber reinforced recycled coarse aggregate concrete. *Constr. Build. Mater.* **2018**, *159*, 126–136. [CrossRef]
52. Abbass, W.; Khan, M.I.; Mourad, S. Evaluation of mechanical properties of steel fiber reinforced concrete with different strengths of concrete. *Constr. Build. Mater.* **2018**, *168*, 556–569. [CrossRef]
53. Tran, D.; Allawi, A.; Albayati, A.; Cao, T.; Elzohairy, A.; Nguyen, T.H.Y. Recycled Concrete Aggregate for Medium-Quality Structural Concrete. *Materials* **2021**, *14*, 4612. [CrossRef] [PubMed]
54. Ferreira, L.; De Brito, J.; Barra, M. Influence of the pre-saturation of recycled coarse concrete aggregates on concrete properties. *Mag. Concr. Res.* **2011**, *63*, 617–627. [CrossRef]
55. Kazemi, M.; Li, J.; Harehdasht, S.L.; Yousefieh, N.; Jahandari, S.; Saberian, M. Non-linear behaviour of concrete beams reinforced with GFRP and CFRP bars grouted in sleeves. *Structures* **2020**, *23*, 87–102. [CrossRef]
56. AzariJafari, H.; Kazemian, A.; Rahimi, M.; Yahia, A. Effects of pre-soaked super absorbent polymers on fresh and hardened properties of self-consolidating lightweight concrete. *Constr. Build. Mater.* **2016**, *113*, 215–220. [CrossRef]
57. AzariJafari, H.; Tajadini, A.; Rahimi, M.; Berenjian, J. Reducing variations in the test results of self-consolidating lightweight concrete by incorporating pozzolanic materials. *Constr. Build. Mater.* **2018**, *166*, 889–897. [CrossRef]

58. Hajforoush, M.; Madandoust, R.; Kazemi, M. Effects of simultaneous utilization of natural zeolite and magnetic water on engineering properties of self-compacting concrete. *Asian J. Civ. Eng.* **2019**, *20*, 289–300. [CrossRef]
59. Madandoust, R.; Bazkiyaei, Z.F.Z.; Kazemi, M. Factor influencing point load tests on concrete. *Asian J. Civ. Eng.* **2018**, *19*, 937–947. [CrossRef]
60. Mohammadi, M.; Kafi, M.A.; Kheyroddin, A.; Ronagh, H.R. Experimental and numerical investigation of an innovative buckling-restrained fuse under cyclic loading. *Structures* **2019**, *22*, 186–199. [CrossRef]
61. Xu, C.; Zhang, X.; Haido, J.H.; Mehrabi, P.; Shariati, A.; Mohamad, E.T.; Hoang, N.; Wakil, K. Using genetic algorithms method for the paramount design of reinforced concrete structures. *Struct. Eng. Mech.* **2019**, *71*, 503–513.
62. Shariati, M.; Rafie, S.; Zandi, Y.; Fooladvand, R.; Gharehaghaj, B.; Mehrabi, P.; Shariat, A.; Trung, N.T.; Salih, M.N.; Poi-Ngian, S. Experimental investigation on the effect of cementitious materials on fresh and mechanical properties of self-consolidating concrete. *Adv. Concr. Constr.* **2019**, *8*, 225–237.
63. Jahandari, S.; Tao, Z.; Alim, M.A. Effects of different integral hydrophobic admixtures on the properties of concrete. In Proceedings of the 30th Biennial National Conference of the Concrete Institute of Australia, Perth, Australia, 5–8 September 2021.
64. Jahandari, S.; Tao, Z.; Saberian, M.; Shariati, M.; Li, J.; Abolhasani, M.; Kazemi, M.; Rahmani, A.; Rashidi, M. Geotechnical properties of lime-geogrid improved clayey subgrade under various moisture conditions. *Road Mater. Pavement Des.* **2021**, 1–19. [CrossRef]
65. Sadeghian, F.; Haddad, A.; Jahandari, S.; Rasekh, H.; Ozbakkaloglu, T. Effects of electrokinetic phenomena on the load-bearing capacity of different steel and concrete piles: A small-scale experimental study. *Can. Geotech. J.* **2021**, *58*, 741–746. [CrossRef]
66. Parsajoo, M.; Armaghani, D.J.; Mohammed, A.S.; Khari, M.; Jahandari, S. Tensile strength prediction of rock material using non-destructive tests: A comparative intelligent study. *Transp. Geotech.* **2021**, *31*, 100652. [CrossRef]
67. Mohammadi, M.; Kafi, M.A.; Kheyroddin, A.; Ronagh, H. Performance of innovative composite buckling-restrained fuse for concentrically braced frames under cyclic loading. *Steel Compos. Struct. Int. J.* **2020**, *36*, 163–177.
68. Sadeghian, F.; Jahandari, S.; Haddad, A.; Rasekh, H.; Li, J. Effects of variations of voltage and pH value on the shear strength of soil and durability of different electrodes and piles during electrokinetic phenomenon. *J. Rock Mech. Geotech. Eng.* **2021**, in press. [CrossRef]
69. Mehdizadeh, B.; Jahandari, S.; Vessalas, K.; Miraki, H.; Rasekh, H.; Samali, B. Fresh, mechanical and durability properties of self-compacting mortar incorporating alumina nanoparticles and rice husk ash. *Materials* **2021**, *14*, 6778. [CrossRef]
70. De Juan, M.S.; Gutiérrez, P.A. Study on the influence of attached mortar content on the properties of recycled concrete aggregate. *Constr. Build. Mater.* **2009**, *23*, 872–877. [CrossRef]
71. Nazarimofrad, E.; Shaikh, F.U.A.; Nili, M. Effects of steel fibre and silica fume on impact behaviour of recycled aggregate concrete. *J. Sustain. Cem.-Based Mater.* **2017**, *6*, 54–68. [CrossRef]
72. Jahandari, S.; Toufigh, M.M.; Li, J.; Saberian, M. Laboratory study of the effect of degrees of saturation on lime concrete resistance due to the groundwater level increment. *Geotech. Geol. Eng.* **2018**, *36*, 413–424. [CrossRef]
73. Jahandari, S.; Saberian, M.; Zivari, F.; Li, J.; Ghasemi, M.; Vali, R. Experimental study of the effects of curing time on geotechnical properties of stabilized clay with lime and geogrid. *Int. J. Geotech. Eng.* **2019**, *13*, 172–183. [CrossRef]
74. Jahandari, S.; Saberian, M.; Tao, Z.; Mojtahedi, S.F.; Li, J.; Ghasemi, M.; Rezvani, S.S.; Li, W. Effects of saturation degrees, freezing-thawing, and curing on geotechnical properties of lime and lime-cement concretes. *Cold Reg. Sci. Technol.* **2019**, *160*, 242–251. [CrossRef]
75. Xiao, J.; Li, J.; Zhang, C. Mechanical properties of recycled aggregate concrete under uniaxial loading. *Cem. Concr. Res.* **2005**, *35*, 1187–1194. [CrossRef]
76. Shaikh, F.; Shafaei, Y.; Sarker, P. Effect of nano and micro-silica on bond behaviour of steel and polypropylene fibres in high volume fly ash mortar. *Constr. Build. Mater.* **2016**, *115*, 690–698. [CrossRef]
77. Salem, R.M.; Burdette, E.G. Role of chemical and mineral admixtures on the physical properties and frost-resistance of recycled aggre-gate concrete. *Mater. J.* **1998**, *95*, 558–563.
78. Altun, F.; Haktanir, T.; Ari, K. Effects of steel fiber addition on mechanical properties of concrete and RC beams. *Constr. Build. Mater.* **2007**, *21*, 654–661. [CrossRef]
79. Warner, R.; Rangan, B.; Hall, A.; Faulkes, K. *Concrete Structures*; Addison Wesley Longman: Melbourne, Australia, 1998.
80. Thomas, J.; Ramaswamy, A. Mechanical properties of steel fiber-reinforced concrete. *J. Mater. Civ. Eng.* **2007**, *19*, 385–392. [CrossRef]
81. Ezeldin, A.S.; Balaguru, P.N. Normal-and high-strength fiber-reinforced concrete under compression. *J. Mater. Civ. Eng.* **1992**, *4*, 415–429. [CrossRef]
82. Ou, Y.-C.; Tsai, M.-S.; Liu, K.-Y.; Chang, K.-C. Compressive behavior of steel-fiber-reinforced concrete with a high reinforcing index. *J. Mater. Civ. Eng.* **2012**, *24*, 207–215. [CrossRef]

materials

Article

Influence of CNT Incorporation on the Carbonation of Conductive Cement Mortar

Gun-Cheol Lee [1,†], Youngmin Kim [1,†], Soo-Yeon Seo [1], Hyun-Do Yun [2] and Seongwon Hong [3,*]

1 Department of Architectural Engineering, Korea National University of Transportation, Chungbuk, Chungju 27469, Korea; gclee@ut.ac.kr (G.-C.L.); imkym97@ut.ac.kr (Y.K.); syseo@ut.ac.kr (S.-Y.S.)
2 Department of Architectural Engineering, Chungnam National University, Daejeon 34134, Korea; wiseroad@cnu.ac.kr
3 Department of Safety Engineering, Korea National University of Transportation, Chungbuk, Chungju 27469, Korea
* Correspondence: shong@ut.ac.kr; Tel.: +82-43-841-5339
† Both authors contributed equally to this manuscript and are dual first authors.

Abstract: This study analyzed the influence of carbon nanotubes (CNTs) on the carbonation conductive cementitious composites. Two powder types of CNT, multi-walled and single-wall CNTs, were employed to give the cement mortar the conductivity, and four tests including the accelerated carbonation, compressive and flexural strength, electrical resistance, and porosity tests were carried out. To intentionally accelerate the carbonation, the prismatic specimens of conductive cement composites were fabricated and stored in the controlled environmental chamber at a constant temperature of 20 ± 2 °C, constant relative humidity of 60 ± 5%, and carbon dioxide (CO_2) concentration of 5% for 12 weeks. It was observed that carbonation resulted in only chemical damage so that there was no change in the electrical resistance value of conductive cementitious mortar that had undergone a carbonation attack.

Keywords: multi-walled and single-walled carbon nanotubes; carbonation; electrical resistance; porosity; strength

Citation: Lee, G.-C.; Kim, Y.; Seo, S.-Y.; Yun, H.-D.; Hong, S. Influence of CNT Incorporation on the Carbonation of Conductive Cement Mortar. *Materials* **2021**, *14*, 6721. https://doi.org/10.3390/ma14216721

Academic Editor: Dario De Domenico

Received: 30 September 2021
Accepted: 2 November 2021
Published: 8 November 2021

1. Introduction

The quality degradation of concrete structures is generally caused by the continuous occurrence of various types of environmental factors [1]. In particular, the carbonation problems of concrete structures that are common phenomena in metropolitan areas, seaside cities, or underground spaces have been gradually raised since the 1980s, and the durability due to the carbonation has significantly decreased as the result of global climate changes [2–9]. In general, carbonation is the result of a chemical reaction in which calcium hydroxide, a hydrate in the concrete structures, is changed into calcium carbonate as carbon dioxide (CO_2) diffuses, and carbonation lowers the pH in concrete or cementitious composites, leading to neutralization. As the carbonized concrete or cementitious composite is neutralized, the reinforcing bars inserted inside reinforced concrete structures are prone to corrosion, and the corroded reinforcement generates internal cracks due to its volume expansion. Furthermore, as this process progresses for long periods of time, such cracks that occur on the surface of the structure pose a great threat to structural safety. Particularly, underground structures, such as a subway structure, are exposed to a CO_2 concentration of more than 660 ppm and high humidity at 60–70%, which promotes carbonation [10,11].

Recent research in the fields of concrete structure has been focused on the development of structural health monitoring systems (SHM) using self-monitoring or self-sensing composites [12]. Choi E.K. et al. [13] used steel fibers both to embed the self-sensing system into the concrete structures and to strengthen the tensile strength. In addition to the incorporation of steel fiber in concrete mixtures, using carbon nanotubes (CNTs) has been investigated by many researchers and scholars because CNTs can give the concrete composites the conductivity [14–16]. Generally, the dosage of CNTs was up to 2.0% of the weight of the binder because of its low density of 1.3–1.4 g/cm^3 [17,18]. Moreover, poor dispersion of CNTs in the composites induced by van der Waals forces between the CNT particles has become a major issue because it caused low mechanical performance of the concrete structures [19]. To address this problem, many studies have been conducted. Collins F. et al. [20] used aqueous solutions containing CNTs with several types of admixtures such as air-entraining agents based on alkylbenzene sulfonic acid, styrene butadiene rubber copolymer latex, and aliphatic propylene glycol ether including ethoxylated alkylphenol, polycarboxylate, calcium naphthalene sulfonate, naphthalene sulphonic acid derivative, and lignosulfonate. Sobolkina A. et al. [21] investigated the effects of sonication on CNT dispersion with anionic and nonionic surfactants. Various surfactants including cetrimonium bromide, sodium dodecyl benzene sulfonate, and triton X-100 were used as admixtures to uniformly disperse the CNTs in the mixtures [22]. To develop the appropriate dispersion of CNT in cementitious composites, Gao et al. [23] added graphene oxide and employed an ultrasonication technique. As the CNT mixtures have been uniformly dispersed, it has not only proper strength but also excellent self-sensing performance. Despite numerous published articles, investigations, and studies on the carbonation of concrete structures and SHM using CNT above, to the authors' knowledge, at present there is no strong consensus in the literature regarding the analysis of carbonated conductive cementitious composites. To bridge this gap, physical, mechanical, and electrical characteristics of CNT incorporated cementitious mixtures that have undergone carbonation attack are first obtained through various experiments, and the data and results of experiments can be used to update how carbonation influences the performance of conductive cementitious composites.

The remainder of this research is organized as follows; the ordinary Portland cement (OPC), standard sand, and two powder types of CNT used in the experiment were first analyzed in terms of physical and chemical properties, the mixture proportions of conductive cementitious composites were explained in detail, four experiments including compressive and flexural strength, electrical resistance, and porosity tests were conducted to investigate the CNT effects on the carbonation and finally, through the data from electrical resistance and porosity experiments, meaningful findings were reached.

2. Experimental Program

2.1. Materials

In this experiment, ordinary Portland cement (OPC, Type I KSL 5201 [24]) and standard sand (KS L ISO 679 [25]) were employed. Table 1 and Figure 1 show the chemical and physical properties of OPC and particle size distribution curve of standard sand (KSL ISO 679 [25]), respectively. To make conductive composites, two powder types of CNTs, multi-wall CNT (MW) and single-wall CNT (SW) (Tuball, OCSiAI, Leudelange, Luxembourg), were added to the mixtures and presented in Figure 2, and their physical porosities are summarized in Table 2.

Table 1. Properties of ordinary Portland cement.

Chemical Properties (%)					Physical Properties	
SiO_2	Al_2O_3	Fe_2O_3	CaO	MgO	Density (g/cm^3)	Specific Surface Area (cm^2/g)
22.23	5.21	3.38	64.58	2.3	3.15	3300

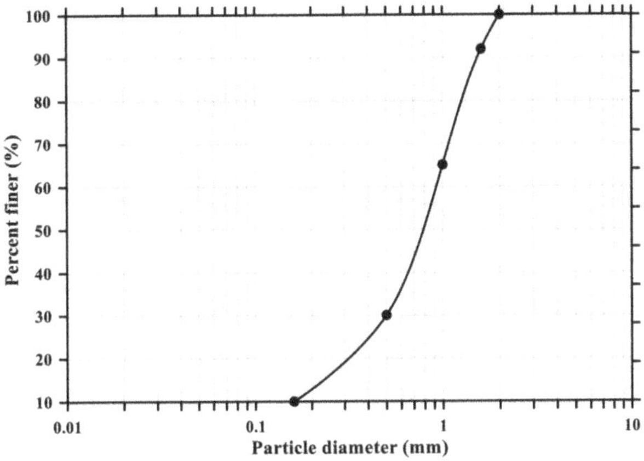

Figure 1. Particle size distribution curve of standard sand (KS L ISO 679).

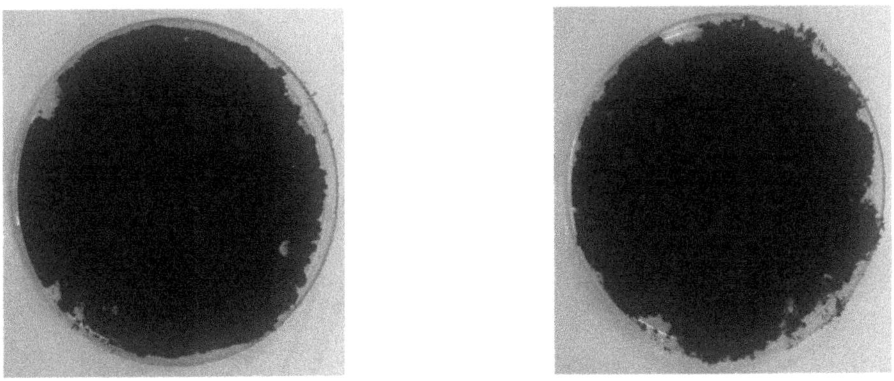

(**a**) (**b**)

Figure 2. Picture of powder-type carbon nanotubes: (**a**) multi-walled carbon nanotubes; (**b**) single-walled carbon nanotubes.

Table 2. Physical properties of multi-walled and single-walled carbon nanotubes.

	MW	SW
Electrical resistance ($\Omega \cdot m^2$)	5.1×10^{-6}	10×10^{-4}
Diameter (nm)	5–100	1.2–3.0
Length (μm)	10	10
Specific surface area (m^2/g)	130~160	700~900
Tension (GPa)	50	45
Thermal conductivity (W/m·K)	3000	6000

2.2. Mixture Proportions

The mixture proportions are shown in Table 3. The CNTs used in this study are divided into MW and SW and prepared at three levels (0, 1.0, and 2.0% mass fraction), which are determined based on the literature review [17,18]. The specimens were fabricated in accordance with KS L ISO 679 [25]. Admixture (poly carboxylate-based high-performance water reducing agent, KS F 2560 [26]) was added to the mixture to improve workability and homogenous dispersion, and its amounts at MW 1.0, SW 1.0, MW 2.0, and SW 2.0 were 2.0, 6.0, 4.0, and 14.0%, respectively.

Table 3. Mixture proportion of conductive cement mortar.

Sample	W/C (%)	Weight (g)			
		Cement	Water	Sand	CNT
Plain	50	450	255	1 350	0
MW 1.0 or SW 1.0					4.50
MW 2.0 or SW 2.0					9.00

2.3. Experimental Method

To analyze the conductive properties of the cement mortar incorporating CNTs undergone carbonation, the accelerated carbonation, compressive and flexural strength, electrical resistance, and porosity tests were carried out. External factors affecting the carbonation of cementitious composites are generally temperature, humidity, and carbon dioxide concentrations. In this study, to accelerate the carbonation of the conductive cement mortar, the prismatic specimens with a cross-section of 100 × 100 mm^2 and a length of 400 mm were fabricated and stored in a carbonation acceleration chamber (SSENES Lab & Scientific Instrument, NEX1200) at a constant temperature of 20 ± 2 °C, constant relative humidity of 60 ± 5%, and CO_2 concentration of 5% until the target age (up to 12 weeks), as demonstrated in Figure 3. The control specimens with the same dimension were cured in a water tank at 20 ± 2 °C for the same number of days. The specimens were cut into two halves and sprayed with a 1% phenolphthalein solution on the measurement surface, and the carbonation depth was then determined by measuring the distance from the edge of the specimen to the color boundary in accordance with the carbonation depth measurement method of cementitious composites (KS F 2596 [27]). The carbonation velocity coefficient (A) was calculated based on Equation (1).

$$x_c = A\sqrt{t} \tag{1}$$

where x_c is carbonation depth (mm), and t is accelerated carbonation period (week).

(a) (b)

Figure 3. *Cont.*

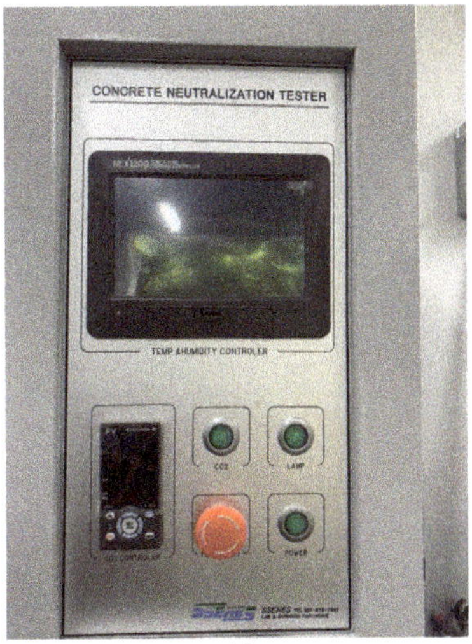

(c)

Figure 3. Carbonation acceleration chamber: (**a**) outside chamber; (**b**) inside chamber; (**c**) control system.

In accordance with KS L ISO 679 [25], three prismatic specimens with a cross-section of 40 × 40 mm^2 and a length of 160 mm were fabricated, demolded after 24 h, and cured in a water tank maintained at 20 ± 1 °C until the target day. The compressive and flexural strengths were measured at 3, 7, and 28 days. Figure 4 displays the prismatic specimen dimensions of conductive cement mortar for the electrical resistance test, which were the same dimension as the strength tests. To properly measure the resistance, copper plates were installed at both ends of the specimen and a two-probe method with the DAQ970A data acquisition system with the BenchVue program was used and alternating current (AC) was used. Due to the moisture content effect of cementitious composites [28], the specimens were dried in an environmental chamber for 24 h at a temperature of 80 ± 1 °C. After drying, 5 V AC power was supplied to the copper plates at both ends of the cementitious composites. To stabilize the resistance value, it was measured approximately 20 min after the supply of current. Since pore size and its distribution can influence mechanical properties of conductive cementitious composites which have undergone a carbonation attack, pore distribution curves are necessarily obtained [29–31]. For this, approximately 2 g of sample was collected from the top surface of the specimen, and mercury porosimetry analysis was performed by using the Mercury Porosimeter (ATS Scientific Inc., Autopore V 9600, Burlington, ON, Canada). The pressure range of Autopore V 9600 is between 50 and 60,000 psi and it can measure pore sizes of 0.003–900 μm.

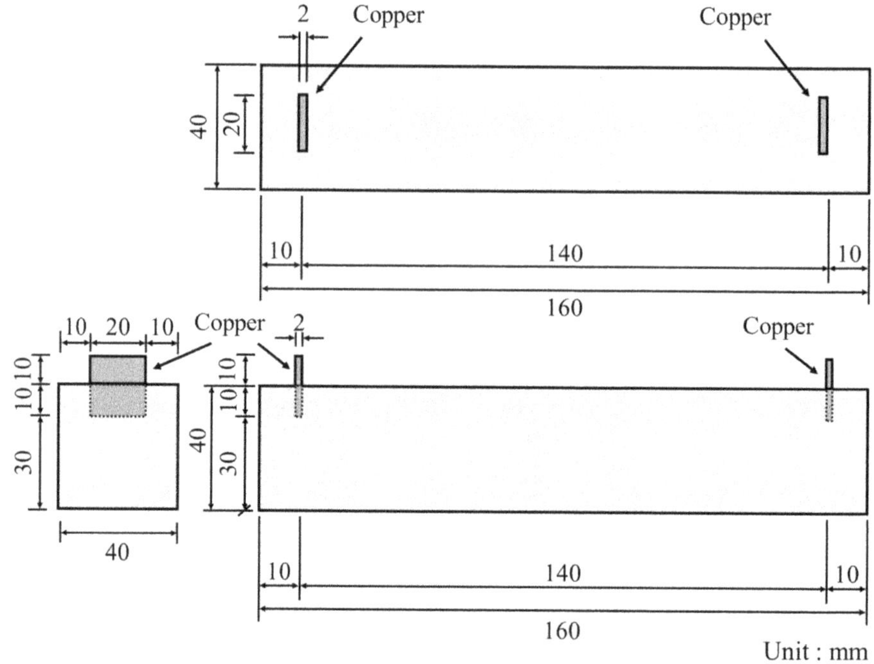

Figure 4. Specimen dimension of conductive cement mortar for electrical resistance measurement.

3. Test Results and Discussion

3.1. Carbonation Depth of Conductive Cement Mortar

Carbonation occurs in the cement mortar as the calcium hydroxide ($Ca(OH)_2$) in cement reacts with CO_2 from the atmosphere and water in the pore. First, carbon acid (H_2CO_3) is the result of a reaction between carbon dioxide and water. The generated carbon acid reacts with the calcium hydroxide to form calcium carbonate ($CaCO_3$) and water, as shown in Equation (2). In general, the volume of calcium carbonation increases by about 11.7% compared to that of calcium hydroxide. Consequently, as carbonation proceeds in plain cement mortar, the carbonated region forms a dense pore structure resulting in the prevention of the CO_2 penetration and restraint of the CO_2 effect [32,33]. However, the conductive cement mortar has a larger pore than that of plain composites so that more diffusion of CO_2 in the CNT cementitious composites through these pores takes place. Figure 5 shows the results of carbonation of conductive cement mortar assessed by using alkalinity indicator, Phenolphthalein, after 12 weeks of the accelerated carbonation test. It was confirmed that CNT cementitious materials accelerated the carbonation.

$$Ca(OH)_2 + H_2CO_3 \rightarrow CaCO_3 + 2H_2O \quad (2)$$

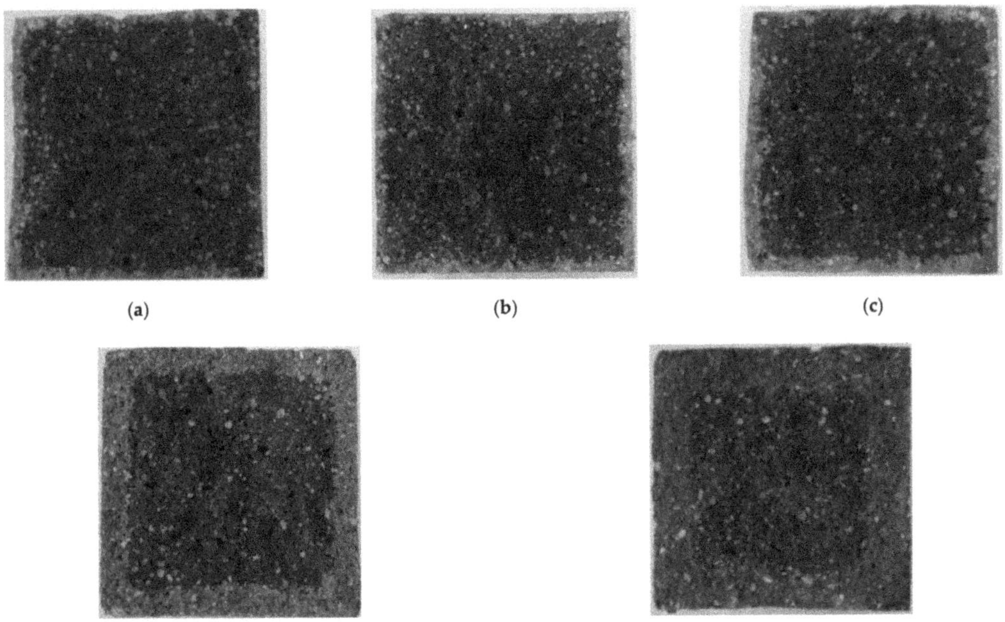

Figure 5. The results of carbonation of conductive cement mortar (12 weeks): (**a**) plain; (**b**) MW 1.0; (**c**) MW 2.0; (**d**) SW 1.0; (**e**) SW 2.0.

Figure 6 provides the results of carbonation depths of conductive cement mortar after the carbonation attack. In the case of the plain specimen, the carbonization depth increased to 2 mm with no coefficient of variation (COV) for 12 weeks. The depth of the MW 1.0 and MW 2.0 specimens was 5.3 mm with COV of 11% and 8 mm with COV of 25%, respectively, for 12 weeks whereas the carbonation depth of SW 1.0 and SW 2.0 was observed to be 12.6 mm with COV of 9% and 19.6 mm with COV of 8%, respectively, for 12 weeks. It signified that CNT clearly influenced accelerating the carbonation of conductive cement mortar.

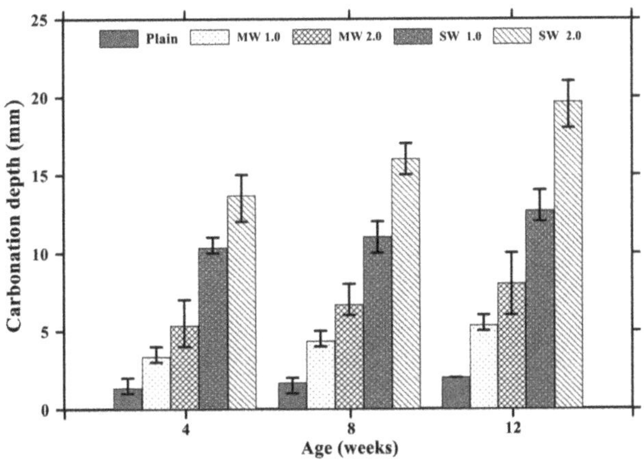

Figure 6. Carbonation depths of conductive cementitious composites after carbonation.

Figure 7 exhibits the relationship between carbonation depths and time. The carbonation velocity coefficient was 0.60 with an R-squared (R^2) of 98.9% for the plain specimen without CNTs. The carbonation velocity coefficient of the MW 1.0 and MW 2.0 specimens was measured to be 1.56 with an R^2 of 99.7% and 2.38 with an R^2 of 98.9%, respectively, while that of SW 1.0 and SW 2.0 was 3.99 with an R^2 of 93.0% and 5.86 with an R^2 of 98.0%, respectively. It was observed that the carbonation velocity coefficient of SW was 250% greater than that of MW and the carbonation rate of conductive cement mortar increased by 150% as the amount of CNT was doubled in the mixture. Therefore, it was confirmed that the incorporation of CNTs accelerated the carbonation of cement mortar.

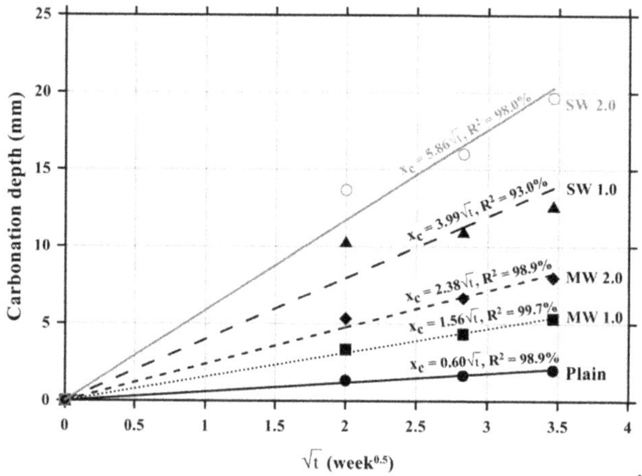

Figure 7. Relationship between carbonation depths and time of conductive cement mortar.

3.2. Compressive and Flexural Strength of CNT Cementitious Materials

The flexural and compressive strength results of conductive cement mortar are shown in Figures 8 and 9, respectively. The specimens incorporating CNTs showed a further decrease in compressive and flexural strengths compared with the plain specimen. Particularly, when the amount of incorporation of CNTs increased, the compressive and flexural strengths decreased because CNT is a hydrophobic composite, so it is impossible to be dispersed properly in the mortar mixtures, and it exists as pores in the composites, which results in decreases in the mechanical performance of CNT-embedded cementitious composites. This is also the reason why conductive cement mortar accelerated carbonation. Since SW 2.0 was not completely cured by the age of 7 days, its strength was zero. Because excessive chemical admixture (14%) was used, it was considered that it delayed the curing. It was clearly observed that compressive and flexural strength decreased as the amount of CNT incorporation increased, and the decrease of SW was greater than that of MW.

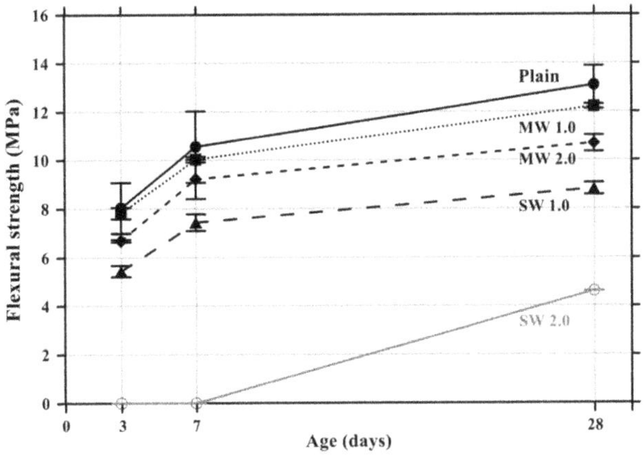

Figure 8. Flexural strength of conductive cementitious composites.

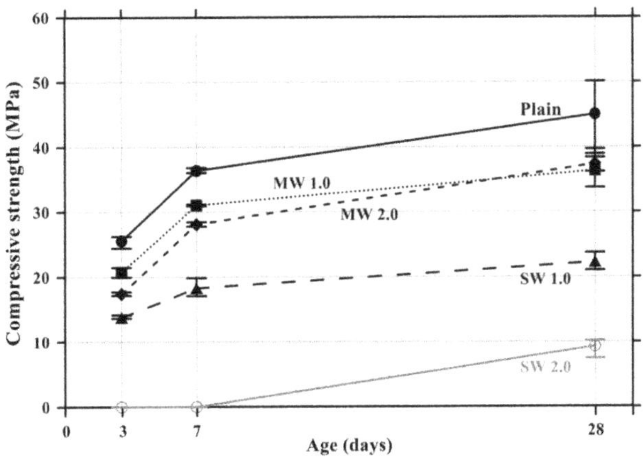

Figure 9. Compressive strength of conductive cement mortar.

3.3. Electrical Resistance Properties of Conductive Cementitious Composites

Figure 10 displays the electrical resistance properties of conductive cementitious before and after the accelerated carbonation. The electrical properties of the specimen before deterioration damage showed that as the CNTs were incorporated, the resistance value suddenly decreased to about 80–90%. By comparing MW with SW, it was clearly measured that the electrical resistance of SW was lower than that of MW. Since CNTs are conductive nanomaterials, conductivity can be given to the cement mortar and the small resistance is measured in spite of the bundle phenomenon induced by the van der Waals force in the composites. In other words, cement mortar mixing with CNTs has the capability to exhibit electrical performance. In addition, it was found that there was no significant change in the electrical resistance value as carbonation proceeded. The carbonation of cementitious composites did not result in internal damage, but it was a chemical change in which calcium hydroxide reacts with CO_2 to produce calcium carbonate, which made only dense pore structures in the mixtures. It signified that the carbonation of conductive cementitious materials caused only mechanical damage so that there was no change in electrical performance.

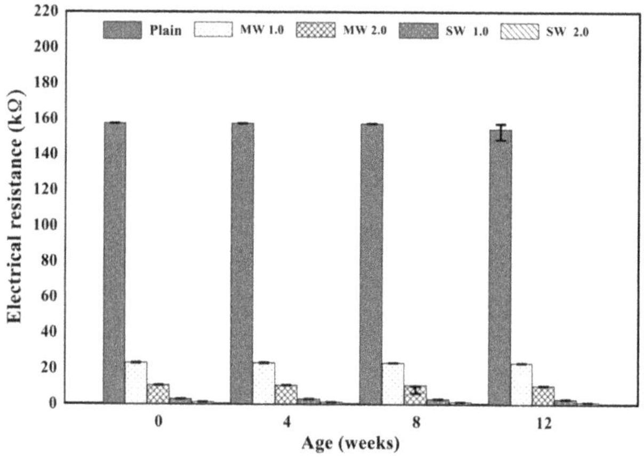

Figure 10. Electrical resistance of conductive cement mortar before and after carbonation.

3.4. Pore Distribution Characteristics of Conductive Mortar

Figures 11 and 12 present the pore size distribution and its cumulative pore volume of the conductive cementitious specimen before and after the carbonation attack, respectively. Before the carbonation in Figure 11a, it was observed that relatively large pores with sizes ranging between 370 μm and 80 μm increased with increasing the dosage of CNTs. In particular, in the range from 370 μm to 80 μm, the largest cumulative pore volume of SW 2.0 specimens was measured (see Figures 11a and 12b) and the pore size of conductive cement mortar incorporating SW was relatively larger than that of the composites with MW, which are clearly different pore size characteristics from the plain specimen. The filling effect due to the diameter of 5–100 ηm and 1.2–3.0 ηm for MW and SW, respectively, resulted in no pore distribution between 0.1 μm and 0.05 μm for MW 2.0 and SW 2.0. However, it was obviously observed that the specimen of MW 1.0 and SW 1.0 had pores from 0.1 μm and 0.05 μm. These are because that CNTs were hydrophobic and difficult to be uniformly distributed in the composites. In the case of plain mortar, the pore size was between 370 μm and 35 μm and there were no micro-pore characteristics. These pores from 370 μm to 35 μm were considered to be the ones left as the remaining water evaporated after the water reacted with cement. Figure 11b provides the pore distribution curve of the cementitious composites after carbonation. The maximum pore distribution of the conductive mortar was the same as ones before the carbonation attack. Carbonation of concrete structure implied loss of alkali and it became neutralization. Carbonation did not result in the internal and/or external damage of the concrete structures caused by freeze–thaw or sulfuric acid erosion. Instead of physical change, the generated calcium carbonate caused the filling effect as shown in Figures 11 and 12. Therefore, it was considered that no physical damage occurred in the conductive cementitious composites that had undergone carbonation so that the connection between CNTs was maintained and conductivity was not physically damaged. Moreover, as the CNT incorporation dosage increased, the distribution of large pores was measured. This is the evidence supporting the above results of compressive and flexural strength as well.

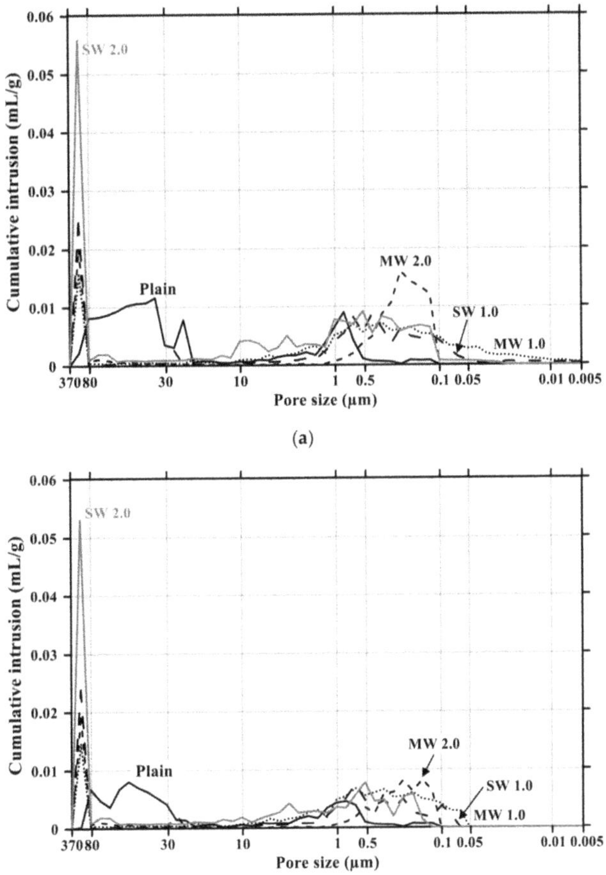

(a)

(b)

Figure 11. Pore size distribution of conductive cementitious composites before and after carbonation: (**a**) before carbonation and (**b**) after carbonation.

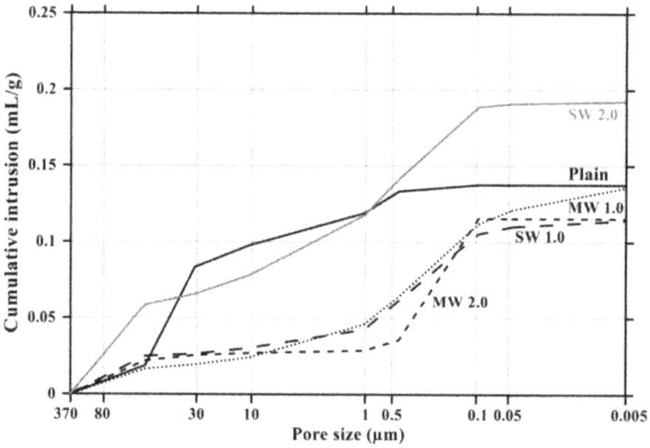

(**a**)

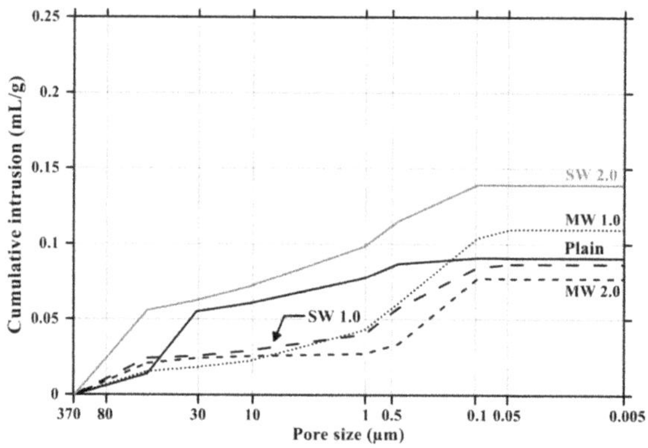

(**b**)

Figure 12. Cumulative pore volume of conductive cement mortar before and after carbonation: (**a**) before carbonation and (**b**) after carbonation.

4. Conclusions

To evaluate the change in the conductivity of the cementitious mortar incorporating MW and SW due to the carbonation, four laboratory tests such as the accelerated carbonation, compressive and flexural strength, electrical resistance, and porosity tests were performed and the following findings were drawn;

1. The acceleration rate of carbonation of conductive cementitious composite increased with an increasing amount of incorporation of CNTs because the large pores generated from the incorporation of CNTs facilitated the penetration of CO_2 in the mortar. It was found that the carbonation velocity coefficient of SW was 2.5 times greater than that of MW and the carbonation rate of conductive cement mortar increased by 1.5 times as the dosage of CNT was doubled in the mixture.
2. When CNTs were mixed with the cement mortar, the compressive and flexural strengths decreased compared to those of the plain mortar due to an increase in the internal pore volume. In particular, it was measured that relatively large pores with sizes ranging from 370 μm to 80 μm occurred due to the van der Waals force resulting from the incorporation of CNTs. These pores resulted in the degradation of mechanical properties.
3. The electrical resistance value of the conductive cement mortar was about 10–20% of the plain specimen, signifying that it had conductivity performance. In addition, the

decrease in resistance value was greater in SW than in MW, indicating that SW had better electrical properties than NW. Furthermore, there was no significant change in the electrical properties due to the carbonation. It denoted that carbonation only led to chemical change without causing any physical damage to the inside of the cement mortar, and the connection of CNTs was thus unimpaired.

4. Through the test results of the pore distribution curve, it was worth noting that large pores with sizes ranging between 370 μm and 80 μm increased with the increase in the amount of CNTs. The filling effect due to the diameter of 1.2–100 ηm for CNTs caused no micro-pore distribution in a range between 0.1 μm and 0.05 μm for MW 2.0 and SW 2.0, while it was obviously detected that the MW 1.0 and SW 1.0 composites had micro-pores. It implies that hydrophobic CNTs were difficult to be uniformly dispersed in the mixtures and CNT incorporation clearly resulted in a decrease in the mechanical performance of cement mortar. After carbonation the pore distribution curves were clearly changed because the pore created by CNTs would be the penetration route of CO_2 into the inside of cementitious composites, causing acceleration of carbonation. The generated calcium carbonate resulted in the filling effect and chemical change in the composites such that the connection between CNTs were not damaged and the conductive cementitious composite had a self-sensing performance.

Author Contributions: Conceptualization, G.-C.L., S.-Y.S. and H.-D.Y.; methodology, G.-C.L. and S.-Y.S.; validation, G.-C.L., S.-Y.S., H.-D.Y., Y.K. and S.H.; formal analysis, G.-C.L., H.-D.Y. and S.H.; investigation, G.-C.L., S.-Y.S. and S.H.; resources, G.-C.L. and Y.K.; data curation, G.-C.L. and Y.K.; writing—original draft preparation, G.-C.L., Y.K., S.-Y.S. and S.H.; writing—review and editing, G.-C.L., S.-Y.S., H.-D.Y. and S.H.; visualization, S.H.; supervision, G.-C.L., S.-Y.S. and S.H.; project administration, G.-C.L. and S.-Y.S.; funding acquisition, G.-C.L., S.-Y.S., H.-D.Y. and S.H. All authors have read and agreed to the published version of the manuscript.

Funding: This research was supported by Basic Science Research Program through the National Research Foundation of Korea (NRF) funded by the Ministry of Education (No. 2021R1A4A2001964) and by NRF grant funded by the Korea government (MSIT) (No. 2020R1F1A104824112).

Institutional Review Board Statement: Not applicable.

Informed Consent Statement: Not applicable.

Data Availability Statement: Data contained within the article.

Acknowledgments: Not Applicable.

Conflicts of Interest: The authors declare no conflict of interest.

References

1. Piqueras, M.A.; Company, R.; Jódar, L. Numerical analysis and computing of free boundary problems for concrete carbonation chemical corrosion. *J. Comput. Appl. Math.* **2018**, *336*, 297–316. [CrossRef]
2. Yoo, H.-S.; Sung, W.-Y.; Yoon, S.-J.; Kim, Y.-H.; Joo, S.-K. Novel Triode-Type Field Emission Arrays and Appropriate Driving Method for Flat Lamp Using Carbon Nanofibers Grown by Plasma Enhanced Chemical Vapor Deposition. *Jpn. J. Appl. Phys.* **2007**, *46*, 4381–4385. [CrossRef]
3. Talukdar, S.; Banthia, N.; Grace, J.R. Carbonation in concrete infrastructure in the context of global climate change—Part 1: Experimental results and model development. *Cem. Concr. Compos.* **2012**, *34*, 924–930. [CrossRef]
4. Talukdar, S.; Banthia, N.; Grace, J.R.; Cohen, S. Carbonation in concrete infrastructure in the context of global climate change: Part 2—Canadian urban simulations. *Cem. Concr. Compos.* **2012**, *34*, 931–935. [CrossRef]
5. Talukdar, S.; Banthia, N. Carbonation in concrete infrastructure in the context of global climate change: Development of a service lifespan model. *Constr. Build. Mater.* **2012**, *40*, 775–782. [CrossRef]
6. Talukdar, S.; Banthia, N. Carbonation in Concrete Infrastructure in the Context of Global Climate Change: Model Refinement and Representative Concentration Pathway Scenario Evaluation. *J. Mater. Civ. Eng.* **2016**, *28*, 04015178. [CrossRef]
7. Kim, K.T.; Eom, Y.S.; Son, I. Fabrication Process and Thermoelectric Properties of CNT/Bi2(Se,Te)3Composites. *J. Nanomater.* **2015**, *2015*, 1–6. [CrossRef]
8. Mizzi, B.; Wang, Y.; Borg, R.P. Effects of climate change on structures; analysis of carbonation-induced corrosion in Reinforced Concrete Structures in Malta. *IOP Conf. Ser. Mater. Sci. Eng.* **2018**, *442*, 012023. [CrossRef]

9. Chen, X.; Bao, R.; Yi, J.; Fang, D.; Tao, J.; Li, F. Enhancing mechanical properties of pure copper-based materials with CrxOy nanoparticles and CNT hybrid reinforcement. *J. Mater. Sci.* **2021**, *56*, 3062–3077. [CrossRef]
10. Kim, T.H.; Kwon, S.J. Probabilistic service life analysis of GGBFS concrete exposed to carbonation cold joint and loading con-ditions. *Korea Inst. Struct. Maint. Insp.* **2020**, *24*, 39–46. [CrossRef]
11. Hwang, S.-H.; Yoon, Y.-S.; Kwon, S.-J. Carbonation Behavior of GGBFS Concrete Considering Loading Conditions and Cold Joint. *J. Korea Concr. Inst.* **2019**, *31*, 365–373. [CrossRef]
12. Bao, Y.; Chen, Z.; Wei, S.; Xu, Y.; Tang, Z.; Li, H. The State of the Art of Data Science and Engineering in Structural Health Monitoring. *Engineering* **2019**, *5*, 234–242. [CrossRef]
13. Choi, E.K.; Yuan, T.F.; Lee, J.; Yoon, Y. Self-sensing Properties of Concrete with Electric Arc Furnace Slag and Steel Fiber. *J. Korean Soc. Hazard Mitig.* **2019**, *19*, 265–274. [CrossRef]
14. Han, B.; Yu, X.; Ou, J. *Self-Sensing Concrete in Smart Structures*; Butterworth-Heinemann, Elsevier: Kidlington, UK, 2014. [CrossRef]
15. Yoon, H.; Jang, D.J.; Lee, H.K.; Nam, I.W. Influence of carbon fiber additions on the electromagnetic wave shielding characteristics of CNT-cement composites. *Constr. Build. Mater.* **2021**, *269*, 121238. [CrossRef]
16. Choi, I.J.; Kim, J.H.; Chung, C.W. Mechanical properties of cement paste with nano mateirals. *Korean. Inst. Build. Constr.* **2020**, *20*, 193–194.
17. Youn, D.-A.; Kim, J.-H.; Lee, G.-C.; Seo, S.-Y.; Yun, H.-D. Tensile and Strain-sensing Properties of Hybrid Fibers Reinforced Strain-hardening Cement Composite (Hy-SHCC) with Different Carbon Nanotube (CNT) Dosages. *J. Korea Concr. Inst.* **2020**, *32*, 285–293. [CrossRef]
18. Lee, G.-C.; Kim, Y.M.; Hong, S.W. Influence of Powder and Liquid Multi-Wall Carbon Nanotubes on Hydration and Dispersion of the Cementitious Composites. *Appl. Sci.* **2020**, *10*, 7948. [CrossRef]
19. Camacho, M.D.C.; Galao, O.; Baeza, F.J.; Zornoza, E.; Garcés, P. Mechanical Properties and Durability of CNT Cement Composites. *Materials* **2014**, *7*, 1640–1651. [CrossRef]
20. Collins, F.; Lambert, J.; Duan, W.H. The influences of admixtures on the dispersion, workability, and strength of carbon nanotube–OPC paste mixtures. *Cem. Concr. Compos.* **2012**, *34*, 201–207. [CrossRef]
21. Sobolkina, A.; Mechtcherine, V.; Khavrus, V.; Maier, D.; Mende, M.; Ritschel, M.; Leonhardt, A. Dispersion of carbon nanotubes and its influence on the mechanical properties of the cement matrix. *Cem. Concr. Compos.* **2012**, *34*, 1104–1113. [CrossRef]
22. Ha, S.-J.; Kang, S.-T.; Lee, J.-H. Strength of CNT Cement Composites with Different Types of Surfactants and Doses. *J. Korea Inst. Struct. Maint. Insp.* **2015**, *19*, 99–107. [CrossRef]
23. Gao, Y.; Jing, H.; Zhou, Z.; Chen, W.; Du, M.; Du, Y. Reinforced impermeability of cementitious composites using graphene oxide-carbon nanotube hybrid under different water-to-cement ratios. *Constr. Build. Mater.* **2019**, *222*, 610–621. [CrossRef]
24. KS L 5201. *Portland Cement*; Korean Agency for Technology and Standards: Maengdong-myeon, Korea, 2016; pp. 3–15.
25. KS L ISO 679. *Methods of Testing Cements—Determination of Strength*; Korean Agency for Technology and Standards: Maeng-dong-myeon, Korea, 2016; pp. 12–16.
26. KS F 2560. *Chemical Admixtures for Concrete*; Korean Agency for Technology and Standards: Maengdong-myeon, Korea, 2007; pp. 1–3.
27. KS F 2596. *Method for measuring carbonation depth of concrete*; Korean Agency for Technology and Standards: Maengdong-myeon, Korea, 2019.
28. Kwon, S.-J.; Maria, Q.F.; Na, U.-J. An Experimental Study on Characteristics of Averaged Electromagnetic Properties considering Moisture Changes in Cement Mortar. *J. Korea Concr. Inst.* **2009**, *21*, 199–207. [CrossRef]
29. Ortega, J.M.; Esteban, M.D.; Sánchez, I.; Climent, M. Ángel Performance of Sustainable Fly Ash and Slag Cement Mortars Exposed to Simulated and Real In Situ Mediterranean Conditions along 90 Warm Season Days. *Materials* **2017**, *10*, 1254. [CrossRef] [PubMed]
30. Ortega, J.M.; Esteban, M.D.; Rodríguez, R.R.; Pastor, J.L.; Ibanco, F.J.; Sánchez, I.; Climent, M. Ángel Long-Term Behaviour of Fly Ash and Slag Cement Grouts for Micropiles Exposed to a Sulphate Aggressive Medium. *Materials* **2017**, *10*, 598. [CrossRef] [PubMed]
31. Ghahari, S.A.; Ramezanianpour, A.M.; Ramezanianpour, A.A.; Esmaeili, M. An Accelerated Test Method of Simultaneous Carbonation and Chloride Ion Ingress: Durability of Silica Fume Concrete in Severe Environments. *Adv. Mater. Sci. Eng.* **2016**, *2016*, 1–12. [CrossRef]
32. Choi, S.; Lee, K.-M.; Jung, S.-H.; Kim, J.-H. A Study on the Carbonation Characteristics of Fly Ash Concrete by Accelerated Carbonation Test. *J. Korea Concr. Inst.* **2009**, *21*, 449–455. [CrossRef]
33. Ishida, M.; Fujita, J.; Ochiai, Y.; Yamamoto, H.; Touno, S. Fourier analysis of line-edge roughness in calixarene fine patterns. In Proceedings of the 2001 International Microprocesses and Nanotechnology Conference—Digest of Papers, Shimane, Japan, 31 October–2 November 2001; IEEE: Manhattan, NY, USA, 2002; Volume 01EX468, pp. 276–277. [CrossRef]

materials

MDPI

Article

A Combined Experimental-Numerical Framework for Assessing the Load-Bearing Capacity of Existing PC Bridge Decks Accounting for Corrosion of Prestressing Strands

Dario De Domenico *, **Davide Messina** and **Antonino Recupero**

Department of Engineering, University of Messina, Villaggio S. Agata, 98166 Messina, Italy; davide.messina@unime.it (D.M.); antonino.recupero@unime.it (A.R.)
* Correspondence: dario.dedomenico@unime.it; Tel.: +39-0906765921

Citation: De Domenico, D.; Messina, D.; Recupero, A. A Combined Experimental-Numerical Framework for Assessing the Load-Bearing Capacity of Existing PC Bridge Decks Accounting for Corrosion of Prestressing Strands. *Materials* **2021**, *14*, 4914. https://doi.org/10.3390/ma14174914

Academic Editor: Jorge de Brito

Received: 6 August 2021
Accepted: 27 August 2021
Published: 29 August 2021

Abstract: Bridges constitute important elements of the transportation network. A vast part of the Italian existing infrastructural system dates to around 60 years ago, which implies that the related bridge structures were constructed according to past design guidelines and underwent a probable state of material deterioration (e.g., steel corrosion, concrete degradation), especially in those cases in which proper maintenance plans have not been periodically performed over the structural lifetime. Consequently, elaborating rapid yet effective safety assessment strategies for existing bridge structures represents a topical research line. This contribution presents a systematic experimental-numerical approach for assessing the load-bearing capacity of existing prestressed concrete (PC) bridge decks. This methodology is applied to the Longano PC viaduct (southern Italy) as a case study. Initially, natural frequencies and mode shapes of the bridge deck are experimentally identified from vibration data collected in situ through Operational Modal Analysis (OMA), based on which a numerical finite element (FE) model is developed and calibrated. In situ static load tests are then carried out to investigate the static deflections under maximum allowed serviceability loads, which are compared to values provided by the FE model for further validation. Since prestressing strands appear corroded in some portions of the main girders, numerical static nonlinear analysis with a concentrated plasticity approach is finally conducted to quantify the effects of various corrosion scenarios on the resulting load-bearing capacity of the bridge at ultimate limit states. The proposed methodology, encompassing both serviceability and ultimate conditions, can be used to identify critical parts of a large infrastructure network prior to performing widespread and expensive material test campaigns, to gain preliminary insight on the structural health of existing bridges and to plan a priority list of possible repairing actions in a reasonable, safe, and costly effective manner.

Keywords: bridges; prestressed concrete bridge decks; operational modal analysis; dynamic identification; corrosion; corroded strands; structural vulnerability assessment; structural health monitoring; nonlinear static analysis

1. Introduction

Most of the existing bridges and viaducts in Italy were constructed around 60 years ago, thus complying with past design guidelines. Many of these bridges were realized with prestressed concrete (PC) deck (with either pretensioned strands or post-tensioned tendons) [1,2] and reinforced concrete (RC) piers. In addition to the differences between design regulations in force at the time of bridge construction and the current design standards, the vulnerability of these structures is further increased by material deterioration phenomena occurred over the structural lifetime, primarily corrosion of steel element (both reinforcement bars of RC piers and prestressing strands of PC main girders). Durability issues of these structures, which are related not only to construction details, but also (and most importantly) to the frequency and extensiveness of inspections and related maintenance interventions, have been often neglected and not considered as key performance objectives in the past few decades. However, recent collapses of bridges occurred in Italy in recent years have revealed the importance of such aspects for the risk mitigation of existing bridges and for the safety of the overall infrastructure network [3]. A representative example is given by the Polcevera viaduct failure in Genoa [4] occurred on 14 August 2018, causing social and economic losses estimated in 43 deaths and around 2 years of downtime. Consequently, the managing bodies of the road network (be them national agencies or private companies) are currently performing a comprehensive survey of the bridge structures through experimental measurements accompanied by numerical investigations aimed at safety assessment of existing bridges. The aim is to rapidly identify potential critical portions of the road network without performing extensive material tests, based on which specific plans of retrofitting actions and, above all, intervention priorities can be adopted in a timely and precise manner [5]. Retrofitting interventions to extend bridge service life can be either localized to specific, weak zones, such as applying composite strips to bridge deck slabs [6], or designed to control the overall bridge structural behavior, such as via the introduction of passive energy dissipation systems (e.g., viscous dampers) when an insufficient seismic capacity is detected [7].

Until one year ago, inspection activities on Italian bridges were regulated by guidelines established in a Technical Circular issued by the Italian Ministry of Public Works and dating back to 1967 [8]. The increasing importance of safety assessment of existing bridges has led to the development of a new document entitled "Guidelines for classification and risk management, safety assessment, and structural health monitoring of existing bridges" [9], approved in April 2020 and compliant with the Italian Technical Rules for Constructions [10] (in sequel simply referred to as NTC2018) in terms of load combinations, safety checks, load models and partial safety factors for the structural analysis of bridge structures. This document has allowed an extensive classification of the entire existing bridge stock, based on a simplified risk analysis combined with a multi-hazard methodology (including structural, seismic, geotechnical, and hydraulic hazards) to define an overall attention class of the bridge.

One of the most critical structural systems identified in the above-mentioned guidelines [9] and in recent inspections of existing bridges concerns PC girders that may be vulnerable to a series of structural issues, such as construction defects in the anchorage zones, prestressing steel relaxation losses [11], durability problems and deterioration phenomena, primarily corrosion on steel tendons [12]. Indeed, RC and PC structures placed in aggressive environments (e.g., featured by high concentration of sulphates and/or chlorides from marine environments, deicing salts, etc.) suffer from corrosion problems related to carbonation of concrete and electrochemical oxidization of steel tendons. Apart from the reduction of the steel area, the iron oxides and rust, occupying a volume six times higher than the original metal, cause transversal tensile stresses in steel [13], thus leading to diffuse microcracking in the surrounding concrete and resulting in spalling of concrete cover, deterioration of the bond links between concrete and steel [14,15], reduction of the sectional capacity [16] and consequent modification of the structural behavior [17]. These effects are more impactful on the load-bearing capacity of PC structures than ordinary RC structures, because steel strands operate at higher levels of stress (up to five times) than those of mild steel re-bars. Consequently, evaluating to what extent corrosion of steel strands (in pretensioned PC girders) or steel tendons (in post-tensioned PC girders) affects the load-bearing capacity of the bridge structure is of utmost importance.

The load-carrying capacity assessment of existing bridge structures represents a topical research line, as confirmed by the wide number of experimental [18], numerical [19], and combined experimental-numerical works [20] in the literature, including multi-level approaches [21] and probabilistic, time-dependent methods [22,23] accounting for degradation of material parameters, especially corrosion of steel. Generally, numerical analysis is assisted by field test results such as diagnostic load tests [24,25], or other non-destructive testing techniques like impact echo [26].

This contribution falls into this research line and proposes a rapid yet effective approach for the assessment of the safety conditions of existing bridge structures under both serviceability and ultimate loading conditions. In particular, the focus of the work is on PC bridge decks representing a common structural scheme for medium span bridges. The proposed methodology consists in a combined experimental-numerical framework that is applied to the case study of the PC bridge deck of the Longano viaduct, Barcellona P.G., Italy, and whose main steps are summarized as follows. First, dynamic identification of natural frequencies and mode shapes of the bridge deck is performed through operational modal analysis (OMA) based on vibration data collected in situ. These measurements are subsequently used for calibrating a numerical FE model of the bridge deck using 1D beam elements for girders and transverse diaphragms. Experimental static load tests on the bridge deck are then carried out in situ under increasing loads levels (in three subsequent phases) up to the maximum allowed serviceability loads, and measuring the corresponding deflections associated with each load step. The static tests performed in situ are also simulated through the previously calibrated FE model, by applying appropriate tributary loads on the beam elements and comparing the obtained numerical deflections with those measured experimentally for further validation. Once the FE model is calibrated and validated, additional numerical analyses are performed to investigate the bridge behavior at ultimate limit states. Macroscopic observations on the bridge deck reveal a state of advanced corrosion in some prestressing steel strands, which may significantly reduce the cross-sectional resisting capacity and may seriously affect the ultimate structural behavior of the bridge deck. To this aim, static nonlinear analysis with a concentrated plasticity approach is finally performed to assess the influence of various corrosion scenarios of the prestressing strands on the resulting load-bearing capacity of the bridge deck. The proposed experimental-numerical framework makes it possible to preliminarily assess the structural behavior of the bridge under simultaneous service and ultimate load conditions prior to performing extensive material test campaigns, unlike other methods from the literature. This preliminary assessment is a crucial goal to plan appropriate retrofitting interventions in a large infrastructure network. Moreover, the presented methodology enables one to investigate the influence of various corrosion scenarios of prestressing strands under extreme loading conditions. The latter information can be extremely useful to evaluate, in an effective and rapid manner, the structural vulnerability of bridge structures in those circumstances in which experimental measures of the actual strand corrosion rate are available, or when imminent inspections are designed to investigate material degradation phenomena in situ.

2. Description of the Longano Viaduct

The Longano viaduct was built in 1970. It is located in the municipality of Barcellona Pozzo di Gotto, Sicily (southern Italy), approximately 1 km in front of the Tyrrhenian Sea. This viaduct belongs to the A20 infrastructure network connecting the provinces of Messina and Palermo; location and photograph of the viaduct are provided in Figure 1.

With an overall length of 88 m, the viaduct is articulated in two twin parts, each representing an independent roadway, with a straight longitudinal configuration and modest altimetric variations. Each roadway (one following the traffic direction from Messina to Palermo and the other in the opposite direction) is composed of: 2 circular piers of heights 5.50 m and 6.00 m, having diameter 2.20 and equipped with spiral reinforcement and Dywidag prestressing steel bars with diameter 32 mm; 2 abutments with height 6 m and 7 m in RC; 3 spans with a statically determinate scheme in PC, having length equal to 29 m (two external spans) and 30 m (central span); RC slab having thickness 20 cm and width 11 m, including curbs and parapets. The structural scheme is composed of an assembly of 4 longitudinal I-shaped PC girders, spaced 2.75 m, with 42 pretensioned steel 0.6″ strands, of which 26 in the bottom flange, 2 in the web, 4 in the top flange (all of them arranged in straight configuration) and the remaining 10 inclined from the upper portion of the section (near the girder supports) to the bottom flange via intermediate deviators located

8 m from the supports (see Figure 2a,b) that show just one half of the girder reinforcement configuration up to midspan, the other half being perfectly symmetric). The longitudinal I-shaped PC girders, of height 1.65 m and width 0.70 m, are mutually connected through transverse PC (post-tensioned) diaphragms (5 per each span), having rectangular section (20 cm × 80 cm) and spaced 7.00 m (see Figure 2c,d) showing the horizontal and transverse sections of the bridge deck). The PC girders are simply supported on neoprene bearings having dimensions 0.60 m × 0.45 m × 0.064 m. In the transverse direction, the bridge deck has a couple of RC seismic restraints realized in the pier cap and in the abutments. As to the materials adopted, the original design drawings report concrete grade C35/45 (cylindrical characteristic strength ≥ 35 MPa), prestressing strands with ultimate tensile strength greater than 1667 MPa and reinforcement bars (of RC slab) with yielding stress greater than 430 MPa and ultimate strength greater than 540 MPa.

Figure 1. Location and photograph of the Longano viaduct, Barcellona P.G., Italy analyzed in this paper as a case study. (Base map © 2021 Basarsoft US Department of State Geographer Data, U.S. Navy, GEBCO Image Landsat/Copernicus).

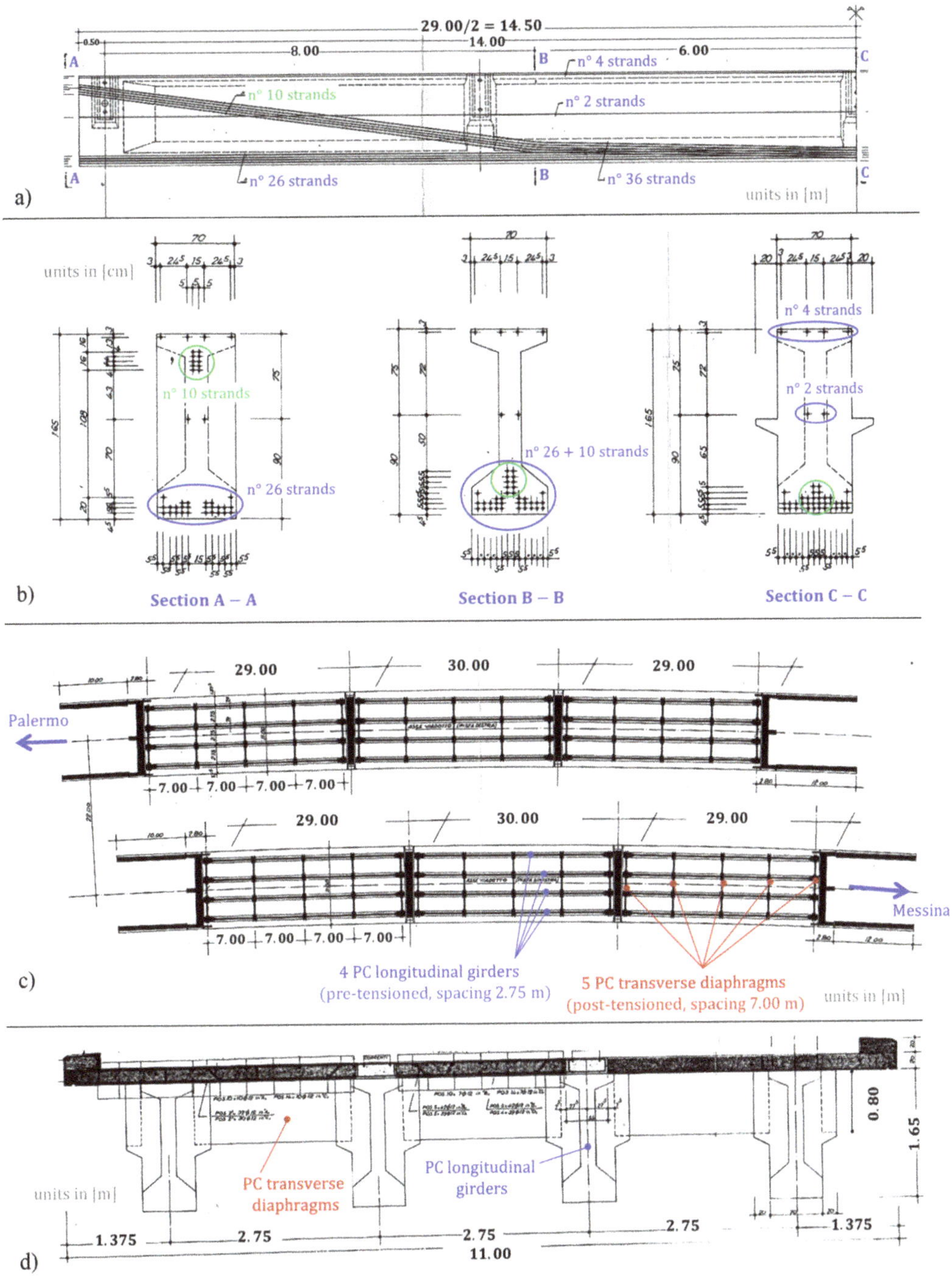

Figure 2. Original design drawings of the Longano viaduct: (**a**) configuration of the prestressing strands in longitudinal girders; (**b**) cross-sections of PC girders; (**c**) horizontal section of the bridge deck (two independent roadways); (**d**) transverse section of the bridge deck.

3. Visual Inspection and Motivations of the Experimental Campaign

Considering the location of the Longano viaduct (just 1 km from the sea) and the fact that the bridge was built more than 50 years ago, some deterioration phenomena might have occurred in the materials, mainly ascribed to aggressive agents like chlorides from the marine environment. These hypothetical considerations are supported by the bridge deterioration state actually detected by visual inspection, whose outcomes are briefly illustrated here and form the basis of the experimental campaign planned afterwards.

According to the multi-level approach recommended by the Italian Guidelines for bridges [9], after the level-0 process of data collection of the bridge, including location, design drawings to identify structural system, check for past reports of retrofit interventions, etc. [27], the next level 1 concerns a visual inspection on the bridge with the aim to detect existing damage signs, including (among others) cracking spots, spalling of concrete cover and, most importantly, steel corrosion. A set of photographs taken on different parts of the Longano viaduct is reported in Figure 3. It is clearly noticed that there is a widespread deterioration state along the bridge deck with concrete cover completely removed in extensive portions of the girder soffit and marked corrosion phenomena in prestressing strands (more pronounced in the strands located in the bottom part of the I-shaped section) as well as in the mild steel bars of the RC slab. On the contrary, the inspection does not reveal any damage state in the transverse diaphragms (neither cracking, nor concrete spalling, nor corrosion of steel tendons).

Based on the preliminary data acquired during the visual inspection, it is reasonable to think that the observed corrosion phenomena of prestressing strands might have reduced the sectional capacity of the PC longitudinal girders. Depending on the initial prestressing force considered in the design stage and the extensiveness of the corrosion process (which are two unknown parameters), theoretically speaking, the reduced steel area of the strands may either affect the ultimate limit states only, or also influence the sectional capacity under serviceability loads. Indeed, it is not easy to foresee whether the corrosion rate is such to produce the cracking state under service loading conditions. Should this critical scenario of cracking under service loads occur, the evolution of damage state detected from the visual inspection might evolve more and more rapidly under service traffic loads. It is, therefore, of extreme importance to check the structural behavior of the bridge under serviceability loads (up to the maximum loads allowed by the Italian regulations NTC2018 [10]). To this aim, static load tests on the bridge deck are planned to reproduce such extreme service loading conditions corresponding to the characteristic combination of loads reported in the Eurocode 0 [28] §6.5.3, with unitary combination coefficients for variable actions (normally used for irreversible limit states). However, before performing such static load tests, it is decided to preliminarily identify natural frequencies, mode shapes and damping ratios of the bridge through OMA (collecting free vibration data). These dynamic characteristics serve as useful indicators of the bridge structural health and are, therefore, checked and compared prior to and after the static load tests to assess whether some irreversible damage is caused by the applied loads. The description of the in situ experimental campaign is reported in the next section.

Figure 3. Widespread corrosion state of the Longano viaduct detected from visual inspection.

4. In Situ Experimental Dynamic and Static Tests under Service Loads

4.1. Operational Modal Analysis and Identification of Dynamic Parameters

The aim of dynamic tests performed in this experimental campaign is to identify dynamic parameters (natural frequencies and mode shapes, as well as damping ratios) of the bridge deck through free vibration analysis. As already said, the knowledge of such dynamic parameters is strictly related to the geometric and mechanical properties of the Longano viaduct and is, therefore, useful for a rapid control of the bridge structural health in operating conditions. Modal identification is typically performed through either operational modal analysis (OMA) [29], which is an output-only technique in which the vibrations of the structure are measured without knowing the excitation characteristics, or experimental modal analysis (EMA) [30], which is an input-output technique measuring the structural response to a given (known) dynamic excitation (e.g., artificial exciter or vibrodine). In structural health monitoring, OMA is more widely used because of the ease of realization, low cost of execution, and possibility of exerting a relatively weak excitation, which avoids exceeding the linear elastic regime of the structure. This technique is used in this work to identify natural frequencies and mode shapes of the Longano viaduct.

Six accelerometers with vertical axis are installed on the bridge deck (following an identical scheme in the two independent roadways). The accelerometers are placed symmetrically with respect to the roadway width and are rigidly anchored to the RC slab at 1/4, 1/2 and 3/4 of the span, cf. Figure 4. The signal acquisition is performed with sampling rate $\Delta t = 0.001$ s (sampling frequency $f_s = 1$ kHz). The range of accelerometer sensitivity is (912–1006) mV/g for the six devices employed, with g being the acceleration of gravity.

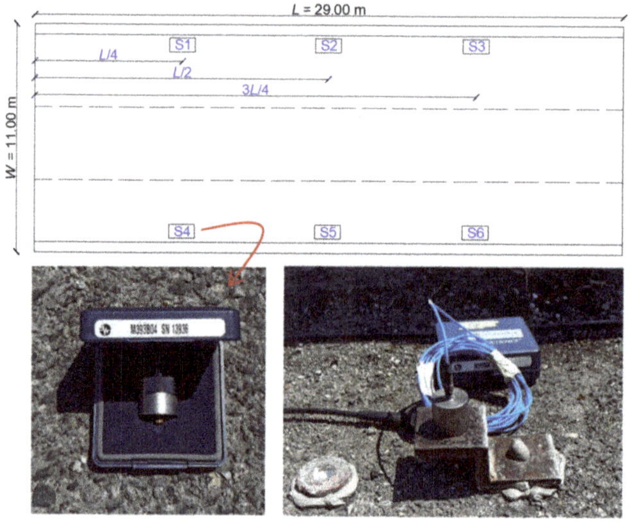

Figure 4. Position of the six accelerometers (S1–S6) on the bridge deck (**top**) and representative accelerometer photograph (**bottom**).

Two different excitations are considered to trigger free vibrations of the bridge deck and to identify the largest possible number of modal parameters: (1) pulse load exerted by a 4 kg hammer, as illustrated in Figure 5, which is applied to three different positions, namely at a quarter of the span (from either side) and at midspan; (2) step load exerted by the transit of a three-axle heavy truck (with gross weight of 347 kN) passing on a 12 cm step, as depicted in Figure 6, which is applied to three different positions, namely at a quarter of the span (from either side) and at midspan.

In the postprocessing phase of the recorded signals, the acceleration time histories are subjected to a Butterworth filter of order 6 to remove the frequencies falling outside the range of interest, here selected as (3–40) Hz. Subsequently, each recording signal is processed in the frequency domain through the fast Fourier transformation (FFT), a representative example of which is shown in Figure 7. It is worth noting that the two types of excitations (hammer pulse and truck transit) affect the modal response to a rather different extent: the hammer blow pulse generally amplifies a wider range of frequencies, including higher order modes, whereas the transit of the heavy truck mainly excites the first (low frequency) modes of vibration, cf. again Figure 7.

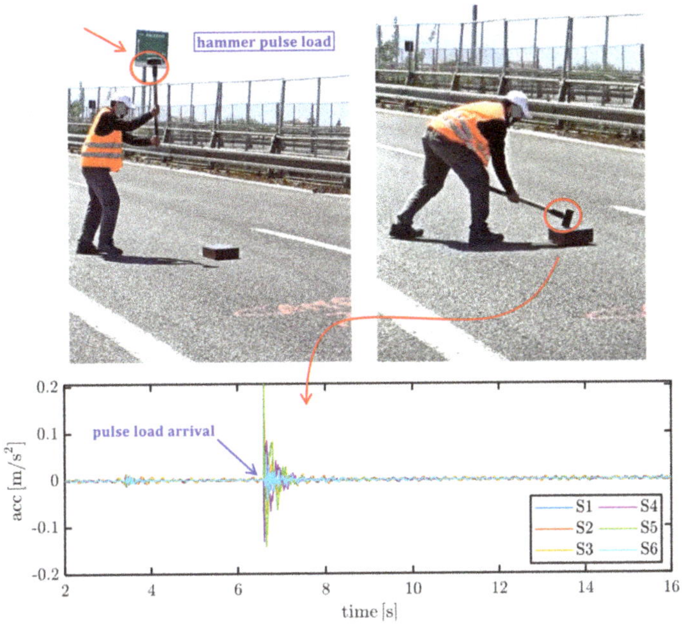

Figure 5. Dynamic excitation of the Longano viaduct with a pulse load produced by a 4 kg hammer blow and related filtered acceleration signals.

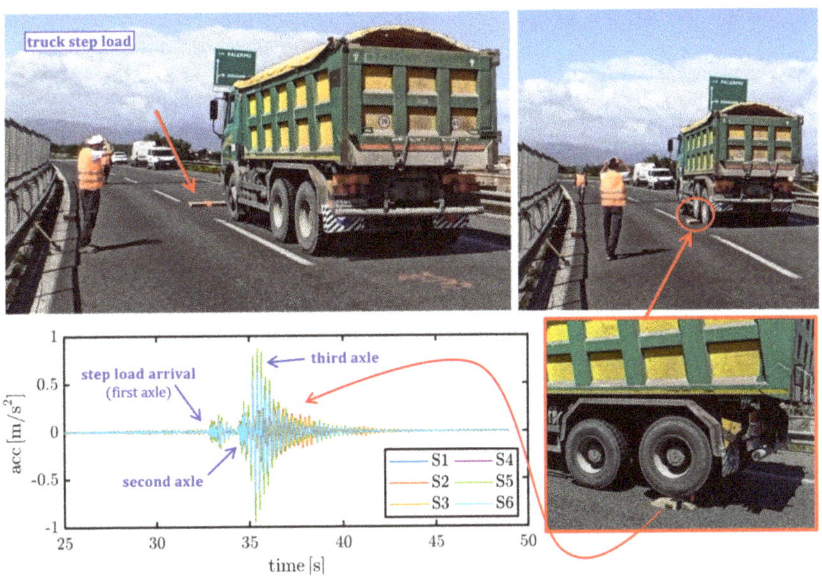

Figure 6. Dynamic excitation of the Longano viaduct realized with the transit of a three-axle 347 kN gross-weight truck on a 12 cm step and related filtered acceleration signals.

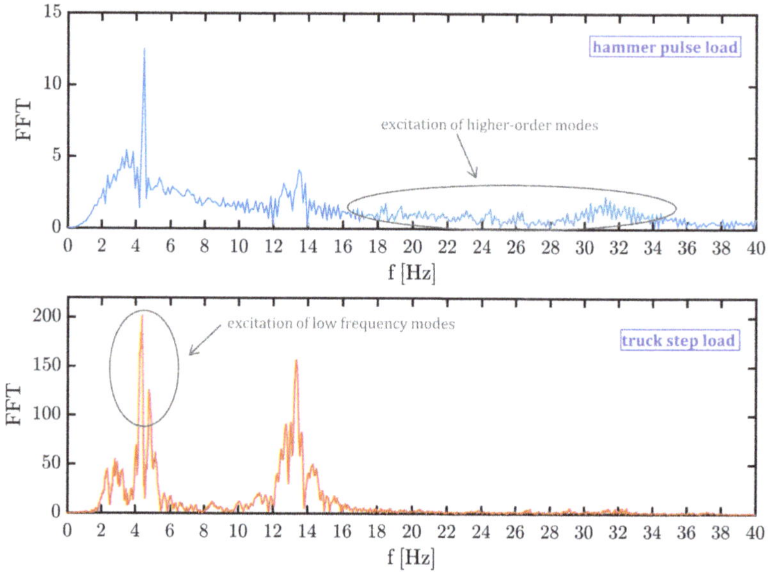

Figure 7. FFT of the S1 signal under hammer pulse load (**top**) and truck step load (**bottom**).

Due to the vicinity of the first frequencies to each other, the natural frequencies and mode shapes are identified through the frequency domain decomposition (FDD) technique [31]. The FDD is based on the calculation of the power spectral density (PSD) matrix of the output (i.e., of the structural response of the bridge deck $\boldsymbol{y}(\mathrm{t})$) $\boldsymbol{G}_{yy}(\omega)$ that, under the assumption of a white noise input (broad-banded excitation) and lightly damped system, is proportional to the frequency response function matrix and, consequently, is useful to identify dynamic parameters of the structure. According to the FDD, the spectral matrix is decomposed into a series of auto spectral density functions, each associated with a single degree of freedom (SDOF) system. The PSD matrix $\boldsymbol{G}_{yy}$ provides information on how the signal power is distributed along the frequency spectrum; this matrix has three dimensions $\left[m \times m \times N_f\right]$, where m is the number of recorded signals (in this case $m = 6$) and N_f is the number of discrete output frequencies. This matrix contains the auto spectral density functions along the diagonal and the cross spectral density function as off diagonal terms and is here estimated through the built-in MATLAB function cpsd [32] from the filtered accelerometer signals. Once an estimate of the output PSD matrix $\hat{\boldsymbol{G}}_{yy}$ is obtained at discrete frequencies ω_i, the singular value decomposition (SVD) of this Hermitian matrix is performed as follows:

$$\hat{\boldsymbol{G}}_{yy}(\omega_i) = \boldsymbol{U}_i \boldsymbol{S}_i \boldsymbol{U}_i^H, \quad (1)$$

where $\boldsymbol{U}_i$ is the unitary (complex orthogonal) matrix in which each column collects the singular vectors, $\boldsymbol{S}_i$ is a diagonal matrix collecting the scalar singular values, i.e., the eigenvalues of the matrix $\boldsymbol{G}_{yy}(\omega)$ in descending order and $\boldsymbol{U}_i^H$ is the conjugate transpose matrix of $\boldsymbol{U}_i$. The SVD is performed through the built-in MATLAB function svd [32] from the estimated PSD matrix and provides six singular values and corresponding singular vectors.

The determination of natural frequencies and mode shapes from the PSD matrix assumes that near the resonant frequency of the structure (where there is only one dominant mode) the singular vectors represent estimates of the mode shapes, and the corresponding singular values represent estimates of corresponding natural frequencies. Some representative examples of singular values of the PSD matrix and the resulting identification of the natural frequencies by peak picking technique is illustrated in Figure 8.

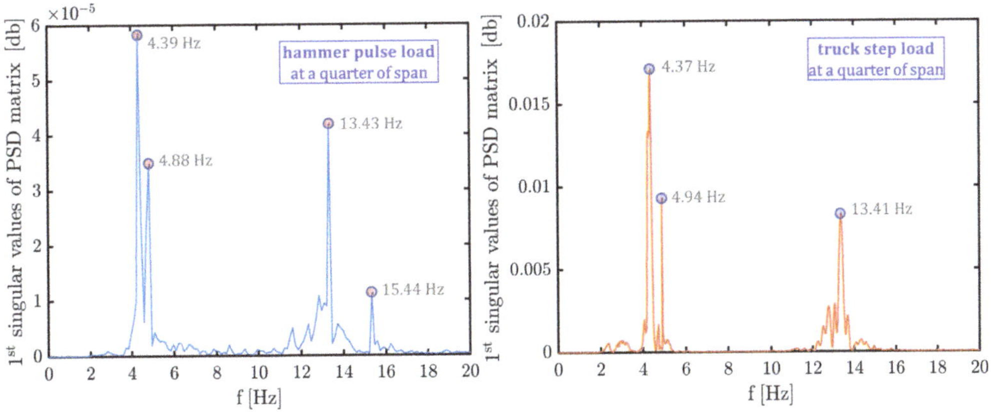

Figure 8. Identification of natural frequencies based on peak picking from the first singular values of the PSD matrix.

The list of natural frequencies identified from OMA is reported in Table 1 for each excitation configuration considered in the experimental campaign, while the first four mode shapes are illustrated in Figure 9. These results are relevant to the dynamic tests performed in the roadway from Palermo to Messina; however, similar results (not shown here for brevity) are obtained in the twin roadway from Messina to Palermo. It is worth noting that the mode shapes are identified by the modal coordinate corresponding to each of the six accelerometers. Since these six accelerometers are placed on the two sides of the bridge deck, it is not easy to distinguish between longitudinal flexural modes (mainly involving longitudinal PC girders) and transversal flexural mode (mainly involving transverse diaphragms) of the bridge deck. To capture this detail, additional accelerometers along the width of the deck (e.g., at $W/4$, $W/2$ and $3/4$ W) should have been placed. Despite the relatively low number of signals, it is possible to foresee, based on the results in Figure 9, that the first and third vibration modes are of flexural type, while the second and fourth modes are of torsional type. These conclusions can be inferred by the critical analysis of the modal displacements: in the first and third modes the modal displacements related to the two sensor lines (i.e., S1-S2-S3 and S4-S5-S6) have the same sign, which indicates a flexural (longitudinal or transversal) deformation of the bridge deck, whereas in the second and fourth mode, they have opposite signs, which denotes a torsional deformation of the deck. The obtained results are useful for the development of the numerical FE model of the bridge deck, which is described in the following subsection.

In addition to natural frequencies and mode shapes, dynamic tests are useful to identify the inherent damping ratio. Experimental measurements of damping ratios, typically achieved through the logarithmic decrement method or through the half-power bandwidth method [33], are useful to detect signs of damage of an existing structure. In this work, the second method is used to identify the damping ratio corresponding to the first and second modes of vibration. To this aim, the peak of the response amplitude (i.e., the resonant amplitude) u_0 is first identified from the frequency response curve (FRC), then the half-power bandwidth is computed from the frequency points at which the amplitude is $u_0/\sqrt{2}$ on either side of the resonant frequency f_n, namely f_1 and f_2. These two frequencies define an interval of frequencies, called half-power bandwidth, that is twice the damping ratio ζ, see Figure 10. Thus, the damping ratio can be estimated as:

$$\frac{f_2 - f_1}{f_n} = 2\zeta. \tag{2}$$

The right part of Figure 10 shows the application of Equation (2) for the determination of the first-mode damping ratio based on the acceleration frequency response curves of the S1 signal, which results in $\zeta_1 \approx 2.4\%$. Similar results are obtained for the other five signals. Experimental damping ratios (average values out of the six measures from the six accelerometers) of the first two modes identified for the two roadways separately are listed in Table 2. It is noted that average values of the first two damping ratios are approximately 2.5%, which are reasonable values for PC structural elements in elastic regime. These results suggest that there is no irreversible damage occurred or ongoing in the bridge deck, at least in the considered operating conditions.

Table 1. Identified natural frequencies from OMA related to different excitation configurations before static load tests.

Excitation Configuration	f_1 [Hz]	f_2 [Hz]	f_3 [Hz]	f_4 [Hz]
hammer pulse load at midspan	4.45	4.88	13.48	-
hammer pulse load at a quarter of span	4.39	4.88	13.43	15.44
truck load step at midspan	4.50	4.87	13.18	-
truck load step at a quarter of span	4.37	4.94	13.41	-

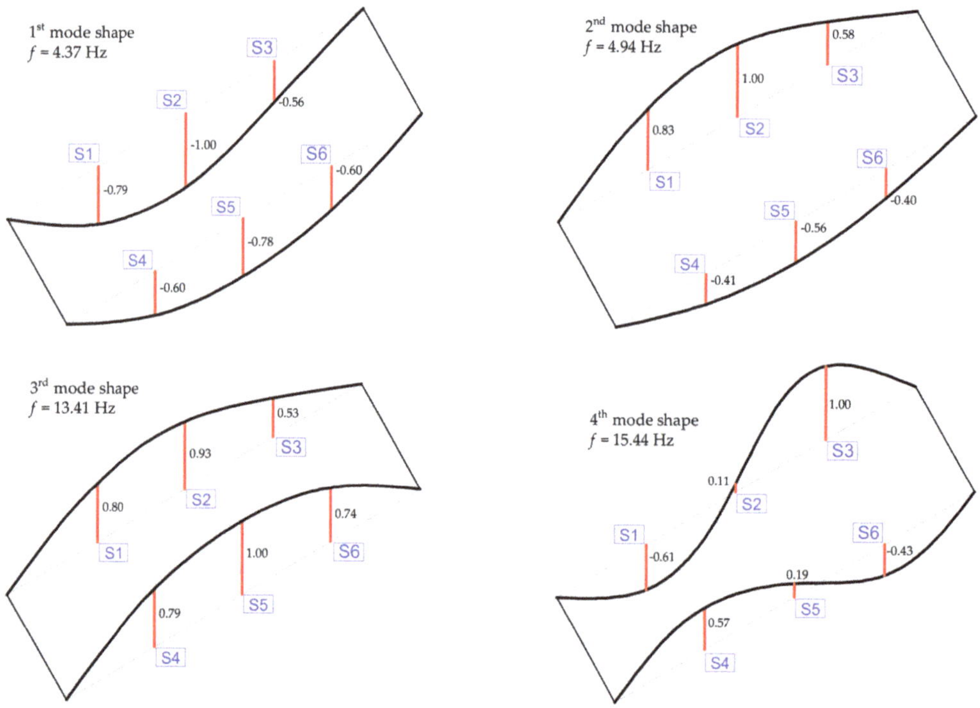

Figure 9. First four mode shapes identified from OMA before static load tests.

Table 2. Experimental damping ratios (average values) for the first two modes of vibration.

Mode Shape	Average Damping Ratio ζ Messina to Palermo Roadway	Average Damping Ratio ζ Palermo to Messina Roadway
mode 1	2.51%	2.10%
mode 2	2.39%	2.61%

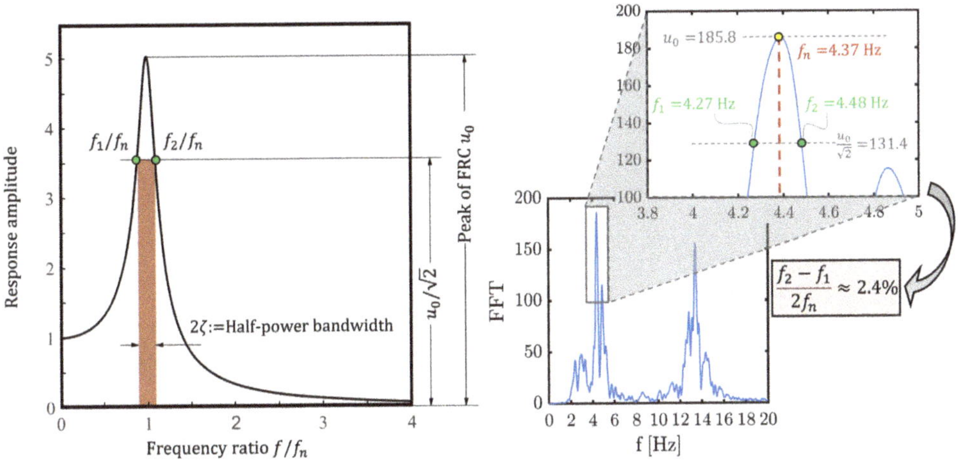

Figure 10. Identification of modal damping via half-power bandwidth: qualitative representation (**left**, adapted from Chopra [33]) and application to experimental frequency response curve for detecting the first mode damping ratio (**right**).

4.2. Development and Preliminary Calibration of the Numerical FE Model

A simplified numerical FE model of the bridge deck is realized via a mesh of 1D beam elements representing the longitudinal PC girders and the transverse diaphragms, each incorporating a collaborating portion of RC slab whose effective width b_{eff} is calculated according to prescriptions from NTC2018 [10] (in a similar fashion to what reported in Eurocode 4 for composite steel-concrete structures [34]).

A sketch of the FE model realized with the structural analysis software SAP2000 [35] is illustrated in Figure 11. Only one span (length 29.00 m) is simulated in the model to reproduce the structural behavior of the bridge deck. Each node has six degrees of freedom. Pinned restraints are assumed at the abutment and at the intermediate pier. The exact geometry of the bridge deck is reproduced by also accounting for variation (widening) of the cross section at the two terminals of the girders, near the supports. Sectional properties (moments of inertia, torsional stiffness, etc.) are directly calculated by the Section Designer tool [35] that is integrated within SAP2000, based on geometrical input characteristics. A diaphragm constraint is assigned to all nodes of the bridge deck to simulate the membrane stiffening effect provided by RC slab and transverse diaphragms. Mass distribution is computed from the load analysis of the bridge deck, including self-weight of structural elements and superimposed dead loads (road pavement and guardrails).

Modal analysis (eigenvector analysis) is performed to determine the natural frequencies and mode shapes of the bridge deck. The first four modes of vibration along with corresponding frequencies are depicted in Figure 12. It is observed that the first (dominant) mode of vibration is, as reasonably expected, relevant to a longitudinal flexural mode of the bridge deck and has a frequency equal to 4.01 Hz. This frequency is only slightly different (8% lower) from the first natural frequency identified from OMA (4.37 Hz). The first natural frequency obtained by the FE model matches well with the theoretical first mode (flexural) frequency of a simply supported continuous beam with length $L = 29.00$ m and cross-sectional properties of the PC longitudinal girders (i.e., area $A = 1.13$ m^2 and moment of inertia $J = 0.4836$ m^4). This first frequency can be analytically calculated as follows:

$$f_{analytical} = \frac{\pi}{2}\sqrt{\frac{EJ}{\mu L^4}} = 4.01 \text{ Hz} \tag{3}$$

where $E = 34.660$ GPa is the elastic modulus for concrete grade C35/45 (calculated in accordance with Eurocode 2 [36] expressions based on hypothetical 28-day concrete characteristic compressive strength equal to $f_{ck} = 37.5$ MPa) and $\mu = G_{tot}/g$ is the linear mass density calculated as the ratio between the total permanent load acting on a single PC girder $G_{tot} = 35.65$ kN/m (from load analysis) and the acceleration of gravity g. It is worth noting that the analytical frequency in Equation (3) is perfectly identical to the first frequency detected from the numerical FE model of the bridge deck, which is reasonable considering that this first mode is related to the flexural longitudinal response of the bridge deck, thus uniquely involving the flexural behavior of the PC longitudinal girders.

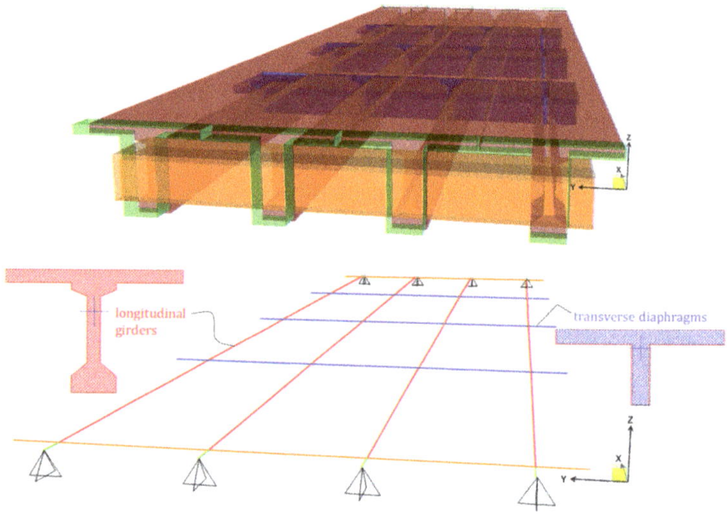

Figure 11. FE model of the bridge deck with 1D beam elements: extruded and standard view.

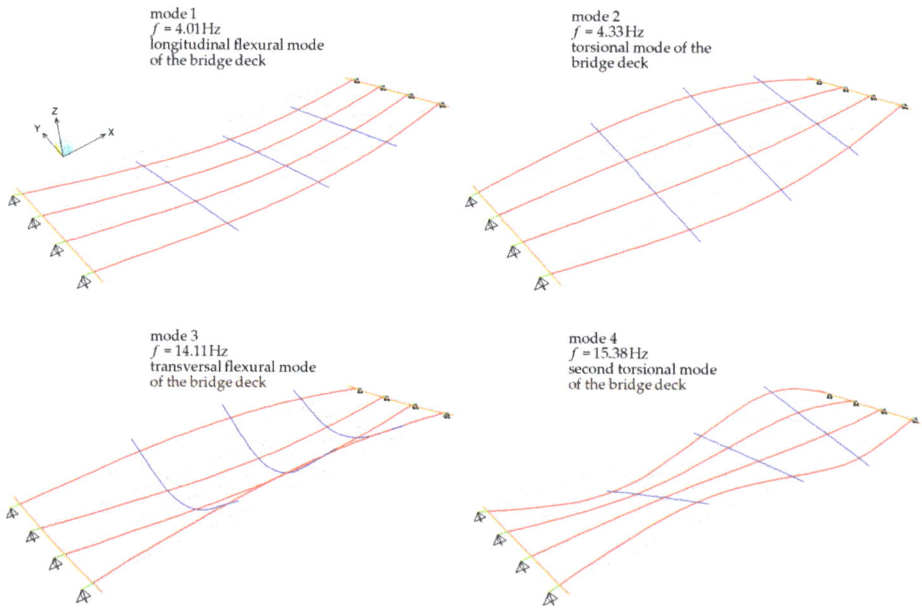

Figure 12. First four mode shapes obtained from FE model of the bridge deck.

A stiffer behavior of the bridge deck is identified from OMA (first two frequencies equal to 4.37 Hz and 4.94 Hz) compared to that obtained from the numerical FE model (first two frequencies equal to 4.01 Hz and 4.33 Hz), which may be due to the stiffening effect of the RC slab in the actual configuration of the bridge that is only approximately included in the simplified FE model, or to slightly different support conditions related to possible wear in the neoprene bearings. Apart from these minor differences, the first four mode shapes from the FE model are qualitatively consistent with the experimental one, cf. Figure 12 with Figure 9, and the comparison between experimental and numerical natural frequencies is also reasonably acceptable (relative errors are listed in Table 3). It can be concluded that the simplified FE model developed is accurate enough to capture the dynamic characteristics of the bridge deck. Further validation of the model is performed in the next subsection to simulate static load tests.

Table 3. Comparison between experimental and numerical natural frequencies of the bridge deck.

Mode Number	Type of Vibration (Exp = FE Model)	Frequency (Exp) [Hz]	Frequency (FE Model) [Hz]	Relative Error [%]
1	longitudinal flexural	4.37	4.01	8.23
2	1st torsional	4.94	4.33	12.35
3	transversal flexural	13.41	14.11	5.22
4	2nd torsional	15.44	15.41	0.19

4.3. Static Load Tests

Static load tests are performed on the Longano viaduct to investigate the bridge structural behavior when subjected to service load conditions. The aim of these tests is to verify whether the bridge remains in elastic phase (without any cracking and damage sign) under the maximum allowed serviceability loads prescribed by the NTC2018 [10], by checking that the residual deflection measured after the load tests does not exceed 15% of the maximum deflection measured during the tests [10]. Code-conforming traffic loads on the bridge are computed based on the NTC2018 prescriptions [10] by dividing the deck in three equivalent lanes and applying uniformly distributed loads combined with tandem concentrated loads simulating heavy vehicle wheels. The worst loading position in longitudinal direction is first identified via girder analysis, and subsequently deck analysis is performed to evaluate the transversal distribution of loads.

For preliminary design of static load tests, a complete finite element model (FEM) is realized. Load schemes adopted for the static load tests are then designed to reproduce the same stress levels on the bridge deck (bending moments and shear forces) as those obtained under the characteristic combination of loads (with unitary combination coefficient for variable actions) prescribed by the NTC2018 [10].

In the static tests, the loads are applied to the bridge deck by means of six heavy trucks identified by a unique ID (from 1 to 6), having either three or four axles. The truck loads are applied gradually in subsequent phases of increasing magnitude to check any possible sign of cracking or other damage-related phenomena. The layout of the actual truck loads relevant to the three phases is depicted in Figure 13 for the representative roadway from Palermo to Messina. However, a similar layout of truck loads is applied to the other roadway in the opposite direction (from Messina to Palermo).

Some measurement points of displacement are located on the two sides of the deck. More specifically, measurements of deflections are carried out via surveyor's levels located in eight measurement points of displacement P1-P8, namely near the two end supports (P1 and P5), at the midspan (P3, P7) and at the two quarters of the bridge span (P2, P4, P6, P8). The sensitivity of the considered instrumentation is of the order of two hundredths of mm, which represents a reasonable threshold considering the deflection values usually measured in bridges (including those reported in this experimental campaign).

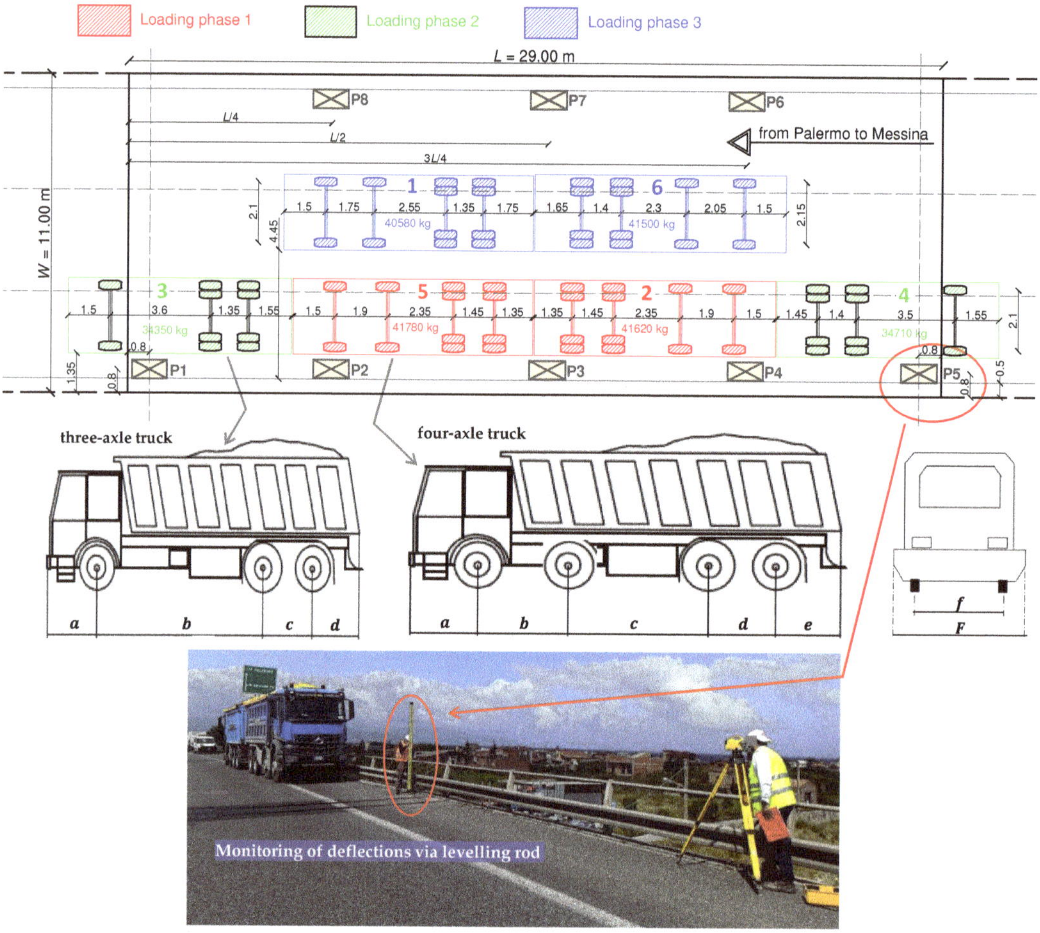

Figure 13. Plan of static load tests on the Longano viaduct deck with location of eight measurement points of displacement (P1–P8) and actual arrangement of three-axle and four-axle heavy trucks in three subsequent loading phases.

Geometrical details of the trucks sketched in Figure 13 are listed in Table 4. Each truck is filled with coarse gravel such that the final gross mass reaches up to more than 34 t and 40 t for the three-axle and four-axle trucks, respectively. Some photographs taken during the various loading phases of the tests (pertinent to the roadway from Palermo to Messina) are reported in Figure 14. Between subsequent loading phases, a certain waiting time is expected to allow the stabilization of the deflection value. The time of loading phase 1 and 2 is of around 10 min, while that of phase 3 is of around 30 min to ensure stabilization of deflection. During the loading phases, the deck is monitored from below with a movable basket to control that all girders (including prestressing strands) do no exhibit any damage sign. After the three load steps, the unloading phase is performed by removing the truck loads following the opposite order of the loading phases.

The measured deflections on the eight measurement points are reported in Figure 15, for each loading phase and for each traffic direction (roadway). It can be noted that the trend of the static deflections in the two roadways is substantially similar. Minor differences are observed when comparing the results from phase 2, which may be due to a slightly different arrangement of the trucks in the two roadways. The first loadings phases (especially phase 1) cause a positive displacement (i.e., raising of the deck) in the side opposite to the loading area (see measured deflections for points P6–P8) as the eccentric truck loads induce a torsional response of the bridge deck. The maximum deflections measured in the midspan of the deck (P3) are approximately 17 mm for both the roadways. It can be observed that the residual deflection measured upon unloading are negligible and are largely below the 15% limit threshold prescribed by the NTC2018 [10]. These results lead to the conclusion that the corrosion of the prestressing strands (documented above in Figure 3) does not affect the structural behavior of the bridge under service loads, i.e., does not produce decompression phenomena under service loads. This seems to reasonably indicate that the level of initial prestressing force that was considered in the design stage is large enough to compensate the reduction of resistant area of the prestressing strands induced by corrosion.

Table 4. Characteristics of the heavy trucks used for the load tests on the Longano viaduct.

Truck ID	N. of Axles	*a* [m]	*b* [m]	*c* [m]	*d* [m]	*e* [m]	*f* [m]	*F* [m]	Gross Mass [kg]
1	4	1.50	1.75	2.55	1.35	1.75	2.10	2.60	40,580
2	4	1.50	1.90	2.35	1.45	1.35	2.10	2.60	41,620
3	3	1.50	3.60	1.35	1.55	-	2.10	2.60	34,350
4	3	1.55	3.50	1.40	1.45	-	2.10	2.60	34,710
5	4	1.50	1.90	2.35	1.45	1.35	2.10	2.60	41,780
6	4	1.50	2.05	2.30	1.40	1.65	2.15	2.65	41,500

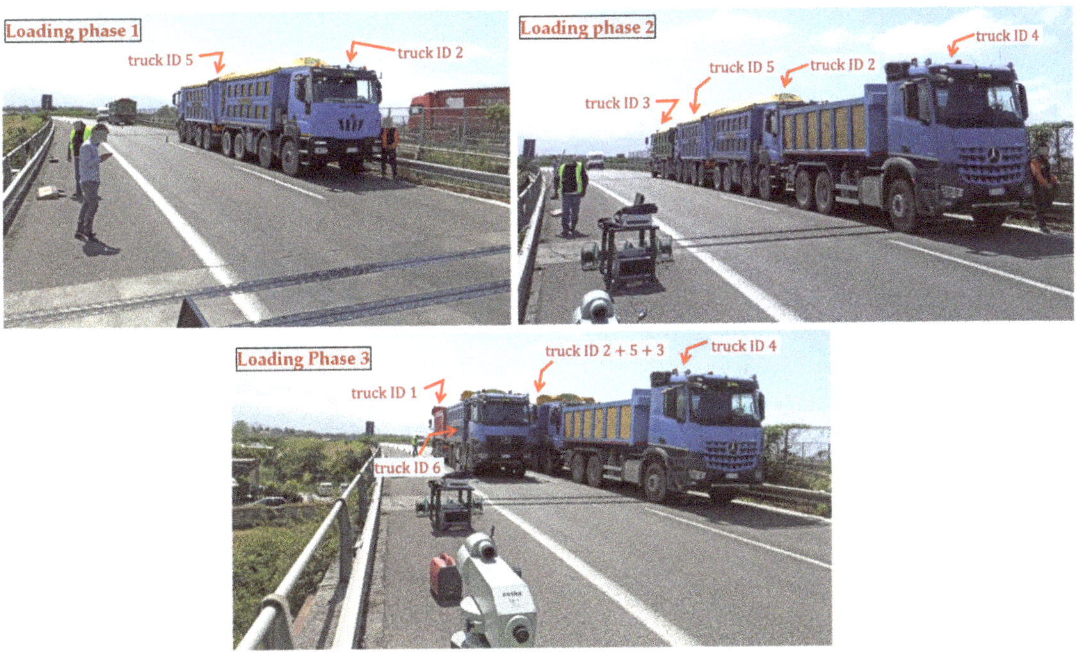

Figure 14. Photographs of static load tests in three subsequent phases for the roadway from Palermo to Messina.

To support the above conclusions, additional dynamic tests are performed after the removal of truck loads to check whether the natural frequencies of the bridge deck are altered by the static load tests. Pertinent OMA results for the same excitation configurations described before (i.e., hammer pulse and truck load step) are listed in Table 5 for the Palermo to Messina roadway. Comparing the natural frequencies before (Table 1) and after (Table 5) the static load tests, minimal differences are found, which cannot be certainly ascribed to a potential accumulation of damage of the Longano viaduct. It can be inferred that the bridge deck remains in elastic regime under the maximum allowed serviceability loads prescribed by the NTC2018 [10].

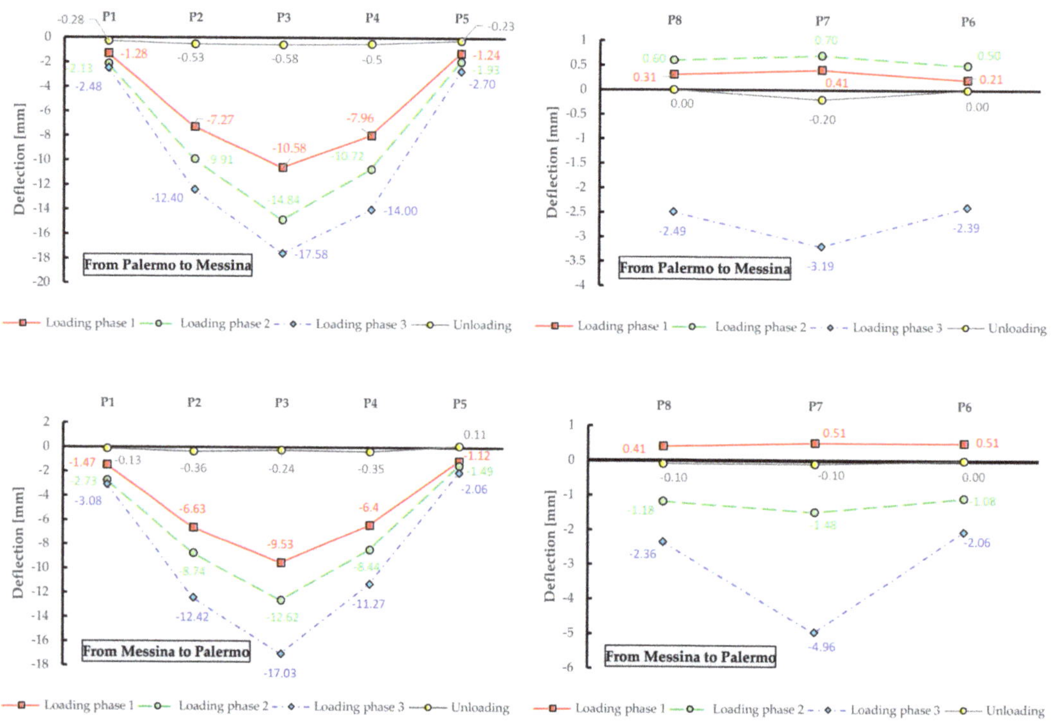

Figure 15. Experimental deflections for each measurement point obtained from static load tests on Longano bridge deck.

Table 5. List of identified natural frequencies from OMA after static load tests (to compare with results from Table 1).

Excitation Configuration	f_1 [Hz]	f_2 [Hz]	f_3 [Hz]	f_4 [Hz]
hammer pulse load at midspan	4.39	4.87	13.48	-
hammer pulse load at a quarter of span	4.41	4.82	13.43	15.38
truck load step at midspan	4.34	4.85	-	-
truck load step at a quarter of span	4.37	4.88	13.42	-

4.4. Further Validation of the FE Model against Load Test Results

The numerical FE model of the bridge deck developed for simulating the dynamic tests is further validated against the load test results. To reproduce the actual load conditions of the tests, the FE model include self-weight of the structure, superimposed dead loads (road surface and parapets) and truck loads with appropriate distributions. For computational simplicity, it is assumed that the gross weight of each truck is equally distributed on the wheels.

Using 1D beams elements, loads can only be applied to the beam axis, whereas truck loads are transferred on the RC slab through wheel imprints that are generally not aligned with the beam axis. Consequently, to calculate equivalent loads on the longitudinal girders generated by the trucks located on the bridge deck, a simplified scheme for a reasonable transversal distribution is adopted as shown in Figure 16. Here, the reactions of the supports in the deck analysis are applied as concentrated loads (in opposite direction) to the longitudinal girders of the FE model.

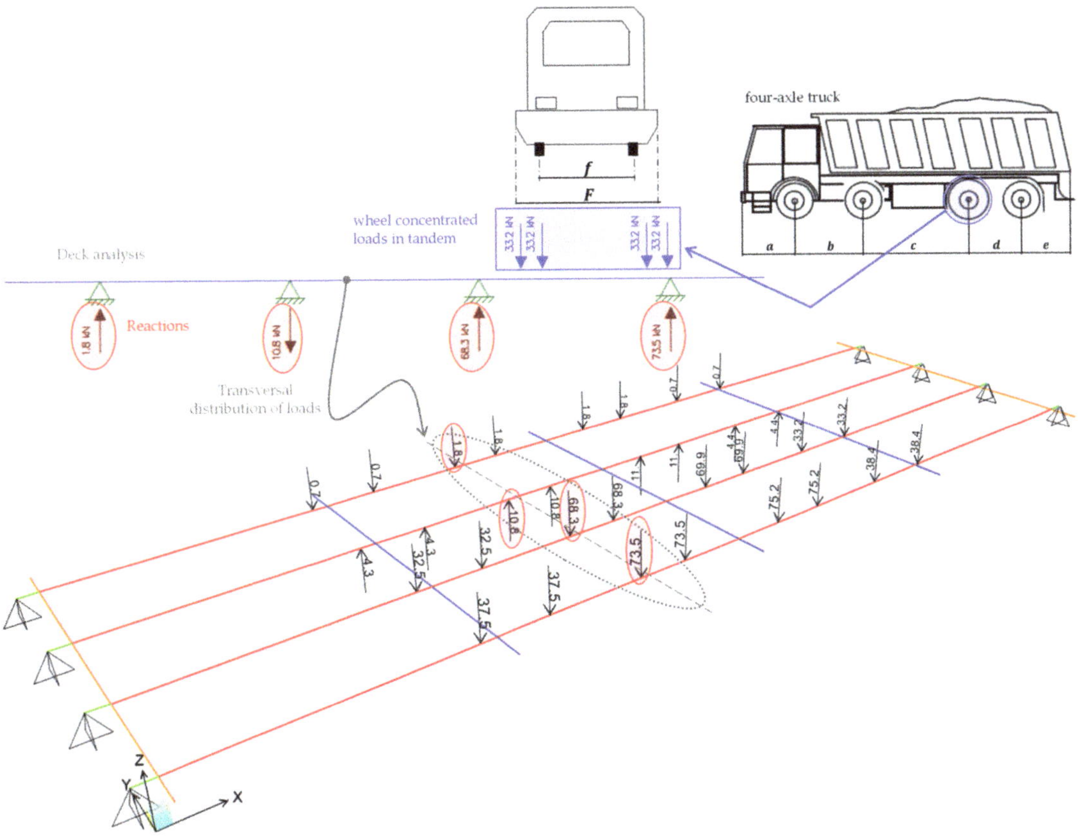

Figure 16. Simplified calculation of load distribution in FE model, representative example simulating loading phase 1.

Once the truck loads are defined and implemented in the FE model, static linear analysis is performed to compute the deflections of the bridge deck. For comparative purposes, particular attention is paid to the deflections of those nodes closest to the eight locations of the measurement points considered in the experimental campaign. The comparison of the experimental (exp) and numerical (FEM) deflections is illustrated in Figure 17 for both the roadways (from Palermo to Messina and vice versa) and for the eight monitored points.

There is a reasonable agreement between the FE results and the experimental displacements, with average relative errors in the order of 10–15%. The highest discrepancies (30–40%) are observed in the measurement line P6-P7-P8, which is opposite to the loaded portion of the bridge deck. This may be ascribed to the simplified transversal load distribution adopted. It is also found that the FE model generally provides slightly larger deflections than the static load tests, which can be justified by the additional stiffening contribution of the RC slab that is only approximately incorporated in the resisting section of the longitudinal girders. Finally, some deviations between numerical and experimental results are noted (although not clearly shown in Figure 17) near the supports: it is expected that some minor displacements occur due to deformability of abutment and intermediate pier in the experimental case that are not captured in the FE model using pinned restraints that prevent translations in the three directions. Apart from these critical considerations, it can be concluded the analysis results further validate the numerical FE model. In the next section, the validated FE model is used to investigate the ultimate limit state behavior of the bridge deck accounting for different corrosion scenarios.

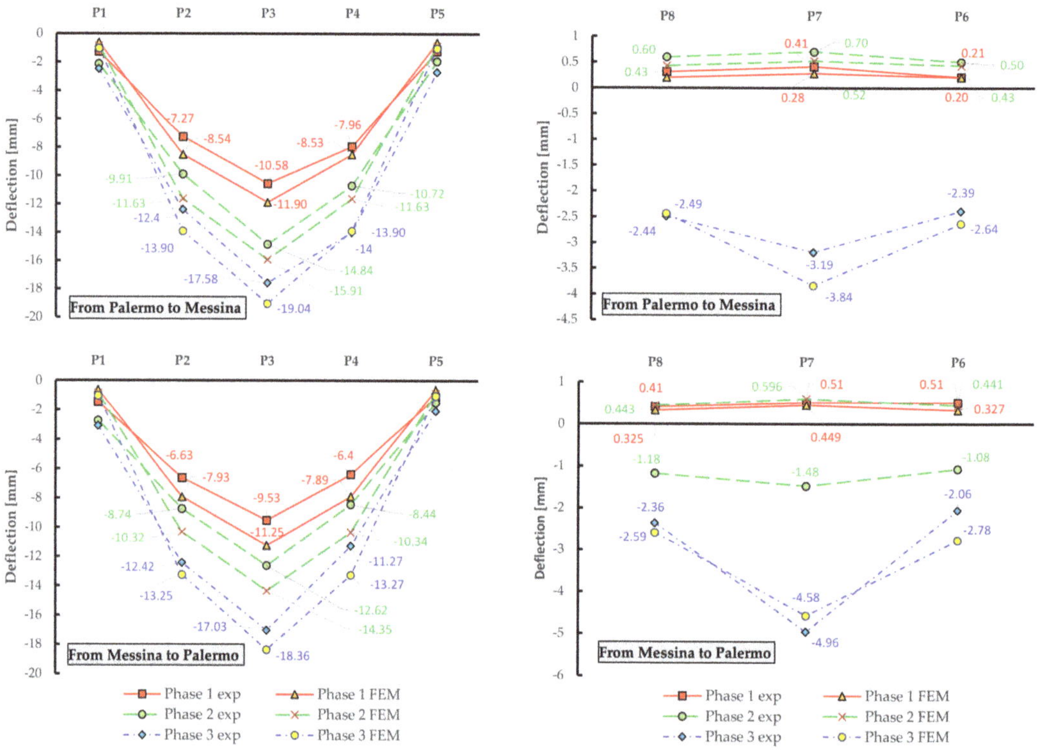

Figure 17. Comparison between experimental (exp) and numerical (FEM) displacements of the Longano bridge deck.

5. Nonlinear Static Analysis Accounting for Different Corrosion Scenarios

Based on the visual inspection, the bridge is subject to widespread corrosion phenomena concentrated only in the prestressing strands (cf. again Figure 3). Although it has been experimentally and numerically demonstrated that the corrosion does not affect the structural behavior of the bridge under service loads, it is reasonable to expect that the load-bearing capacity of the bridge deck at the ultimate limit states is seriously influenced by the corrosion-induced reduction of the cross-sectional area of the prestressing strands. Unfortunately, no experimental measurements of the actual corrosion rates are available so far, but they are planned for the near future. However, it is of utmost importance to gain insights, or at least to obtain preliminary estimates, on how the corrosion of steel strands affects the load-bearing capacity of the bridge. To this aim, this section is devoted to investigating, from a numerical point of view, the ultimate limit-state behavior of the bridge deck accounting for different (hypothetical) corrosion scenarios of prestressing strands. The results from this numerical investigation will be useful to quantify the risk level of the infrastructure associated with different extents of corrosion (whose actual values will be identified through subsequent experimental findings).

Numerical static nonlinear analysis under incremental loads is performed to quantify the post-elastic structural response focusing on the reduction of the load-bearing capacity of the bridge deck induced by corrosion. It is worth noting that in the literature many models were developed to describe the effect of corrosion on the strength and ductility of RC structures [37,38]. Generally, the reduction of mechanical properties is more severe for pitting corrosion than for uniform corrosion, especially in presence of permanent cracking. Without any experimental results concerning the actual corrosion state of the bridge deck and considering the absence of cracking, here, some simplified hypotheses must be declared; in particular, a status of uniform corrosion of the prestressing strands is assumed and the indirect degradation of the mechanical properties of surrounding concrete close to corroded steel is neglected. For comparative purposes, six different nonlinear FE models of the bridge deck are realized, each characterized by a different corrosion rate (CR), or mass loss in percentage, namely 0% (uncorroded bridge configuration), 5%, 10%, 20%, and 30%. In each FE model, the distribution of vertical loads is consistent with the static load tests. Assuming as initial condition the scheme of the loading phase 3 (cf. Figure 13), these loads are increased monotonically up to the collapse of the structure. The choice of this initial loading configuration (loading phase 3 related to the static load tests previously described) is motivated by the eccentricity of the loads in the deck that generate simultaneous flexural effects on the girders and torsional effects of the deck.

To perform nonlinear static analysis, material nonlinearity is incorporated in the numerical FE model. A concentrated plasticity approach based on plastic hinges located in some specific portions of the beam elements is adopted. Preliminary calibration of the moment-curvature $M-\chi$ relationships of some representative sections of the longitudinal girders, namely section A-A, B-B and C-C depicted in Figure 2b), is made. These three sections are characterized by three different positions of the prestressing strands and, thus, by different yielding and ultimate moment values and moment-curvature relationships. Assuming an ultimate tensile strength of the prestressing strands of longitudinal girders equal to $f_{pk}=1700$ MPa, the resulting characteristic yielding stress is $f_{p(0.1)k}=1544$ MPa (a hardening ratio equal to 1.1 is adopted). Initial stress (allowable stress) is then calculated as $\sigma_{pi}=0.90f_{p(0.1)k}=1360$ MPa according to EC2 [39,40]. Since the Longano viaduct was built in 1970, stress losses in the prestressing strands play an important role in the definition of the sectional capacity and cannot be ignored in the analysis. The stress losses due to concrete shrinkage, creep, and steel relaxation are computed and combined in accordance with Eurocode 2 expressions [39], assuming class 1 for ordinary wires and strands [40]. The final value of combined stress losses is equal to 345.45 MPa, thus leading to an effective stress for the strands equal to 1016.55 MPa. This value is used to compute the $M-\chi$ relationships for the calibration of the plastic hinges.

The nonlinear (concentrated-plasticity) FE model of the bridge deck is shown in Figure 18 along with the corresponding moment-curvature laws for the three representative sections A-A, B-B, and C-C, and for the six corrosion scenarios analyzed in this study. It can be observed that the ultimate moment decreases with increasing values of CR, while the ultimate curvature increases with CR because the failure tends to be more ductile, as it is achieved with a higher strain level in the corroded prestressing strands (having reduced resistant sections due to corrosion) than the uncorroded case. From the $M-\chi$ relationships, normalized moment-rotation curves are constructed and implemented in SAP2000 [35] based on a plastic hinge length equal to the section height. In a separate model, plastic hinges are also calibrated and implemented for the transverse diaphragms, besides those in longitudinal PC girders; however, it has been found that the yielding moment of these transversal elements, even in the most extreme loading and corrosion scenarios, is never exceeded, which implies that transverse diaphragms can be assumed as linear elastic elements for computational simplicity.

Displacement-controlled nonlinear static analysis is carried out in SAP2000 [35], by monitoring the displacement of the midspan node (on the roadway side that is mostly loaded) under loads of increasing amplitudes, as sketched in Figure 19. As reasonably expected, the development of the plastic hinges under monotonically increasing loads indicates that the most critical sections are those located near the mid span.

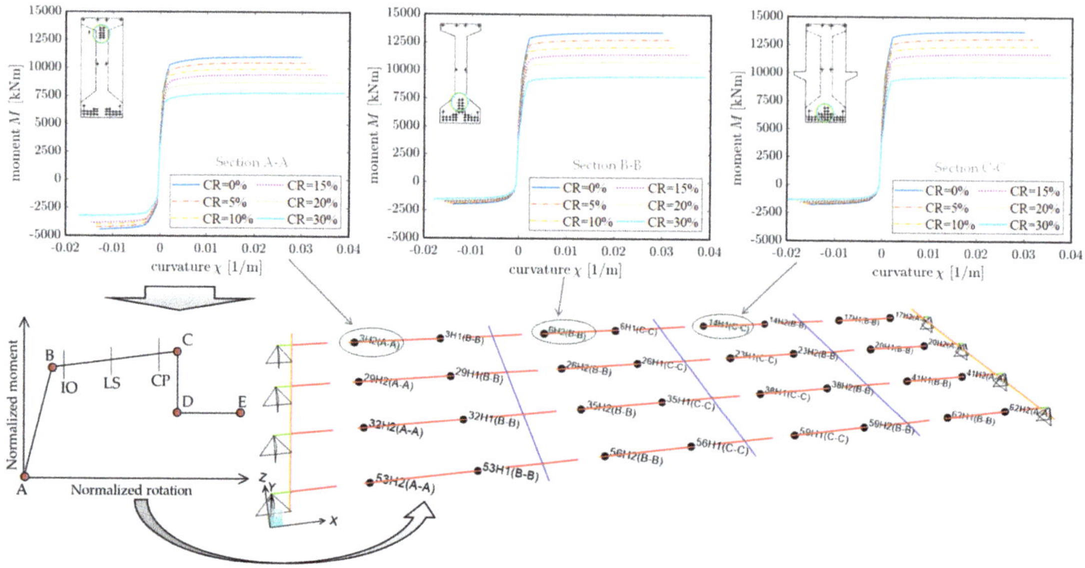

Figure 18. FE model of the corroded bridge deck with material nonlinearity simulated through a set of plastic hinges in longitudinal girders (concentrated plasticity approach) with properly calibrated moment-curvature relationships.

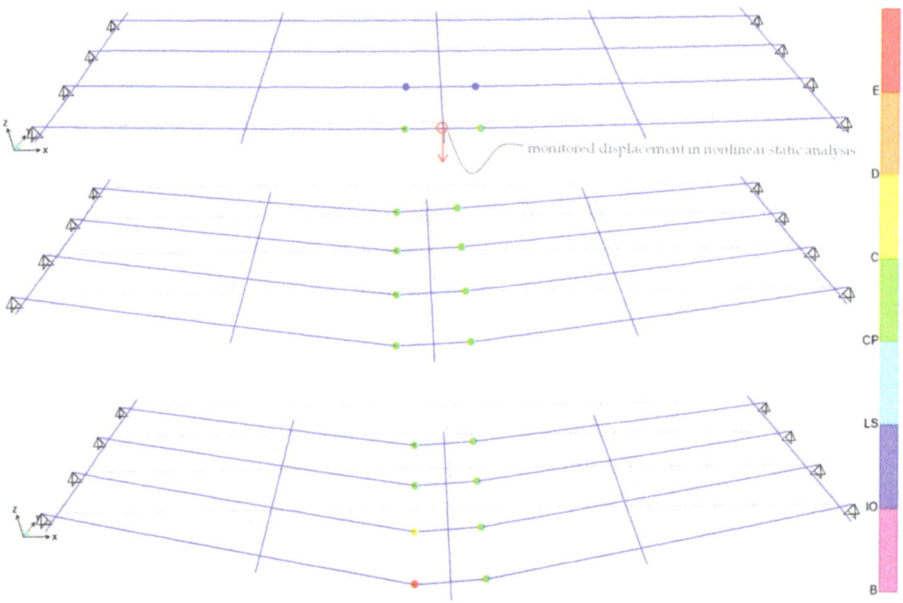

Figure 19. Deformed shape and plastic hinge development corresponding to three incremental loading steps (of increasing magnitude) during the nonlinear static analysis.

The pushover curves in terms of base reactions (sum of reactions of abutment supports and intermediate pier supports) versus monitored displacement are shown in Figure 20. It can be observed that the pushover curve starts from a base reaction $R_0 = 4305$ kN for all the corrosion scenarios, which corresponds to the end of loading phase 3, i.e., the reaction computed under the maximum allowed serviceability loads. Further load increase (up to yielding load 8000–10,000 kN, depending on CR) is allowed by the bridge deck while remaining in elastic regime, after which the behavior becomes nonlinear. The corrosion evidently affects (reduces) the load-bearing capacity of the bridge deck.

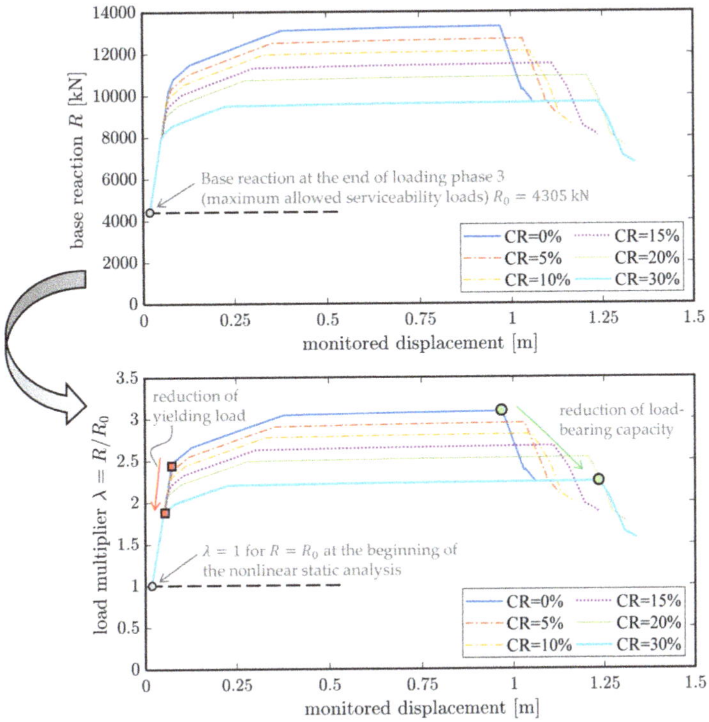

Figure 20. Results from nonlinear static analysis of the bridge deck: pushover curve (**top**) and corresponding load multiplier λ computed by normalizing the base reaction with respect to the value obtained at the end of loading phase 3 (**bottom**).

For the generic step s of the nonlinear static analysis, the base reaction can be expressed as $R(s) = \lambda(s) \cdot R_0$, where $\lambda(s)$ represents a dimensionless load multiplier at the considered step s that quantifies the increase of the load beyond the elastic threshold at the end of loading phase 3. In addition to the pushover curve in dimensional form, the bottom part of Figure 20 reports the trend of such load multiplier λ versus the monitored displacement for the six considered corrosion scenarios. It can be easily observed that both the yielding load (i.e., the base reaction corresponding to the formation of the first plastic hinge) and the load-bearing capacity of the bridge deck (peak of the pushover curve) decrease with increasing CR values.

To summarize the main results, Figure 21 illustrates the trend of three characteristic load multiplies versus the CR value, namely the yielding load, the load-bearing capacity (maximum load) and the ultimate load calculated as the 85% of the maximum load. These diagrams are useful to preliminarily estimate the reduction of the load-bearing capacity of the bridge deck depending on the corrosion rate. The yielding load multiplier in the uncorroded case is 2.37, which means that the first plastic hinge formation occurs at a load level approximately two times and a half higher than that simulated in the loading phase 3 for the on-site static tests. This is a relatively high value considering the time of construction of the bridge, which confirms that the bridge is rather safe under the service loads prescribed by the current regulations NTC2018 [10]. Moreover, the corrosion leads to a considerable drop of the three load multipliers, following a linear trend with almost comparable slope. In particular, the following linear regression formula well describes the decreasing trend of the load-bearing capacity of the bridge deck:

$$F_{\text{corroded}} = F_{\text{uncorroded}} - 0.028 \cdot \text{CR}[\%] \quad (4)$$

which is valid under the simplified assumption of uniform corrosion for all the prestressing strands and within the limitations of corrosion levels considered in this numerical study (i.e., up to 30%). This formula can be applied to predict a first estimate of the corrosion-induced reduction of the load-bearing capacity of existing PC bridge decks sharing similar features to the Longano viaduct here analyzed as case study. This is particularly important when a large infrastructure network is analyzed, in which generally many bridges are similar from the perspective of (aggressive) environmental conditions, static scheme and age of construction. Therefore, it is expected that the degradation of material parameters is comparable for a series of bridge structures belonging to the same infrastructure network. As an example, for a reasonable corrosion rate equal to 10%, Equation (4) predicts a considerable reduction of the load-bearing capacity of almost 30%. Although this is only a preliminary estimate that is acceptable under the strict assumption of uniform corrosion, in most cases it is difficult to obtain a punctual and precise evaluation of the mass loss in percentage of each prestressing strand. Indeed, many practical difficulties exist when performing experimental measurements, including (among others) costs and feasibility considerations for finding a convenient access to girders and other structural elements. Considering such circumstances, the proposed experimental-numerical approach can be used as a preliminary assessment tool to identify critical parts of a large infrastructure network prior to performing widespread and expensive material test campaigns.

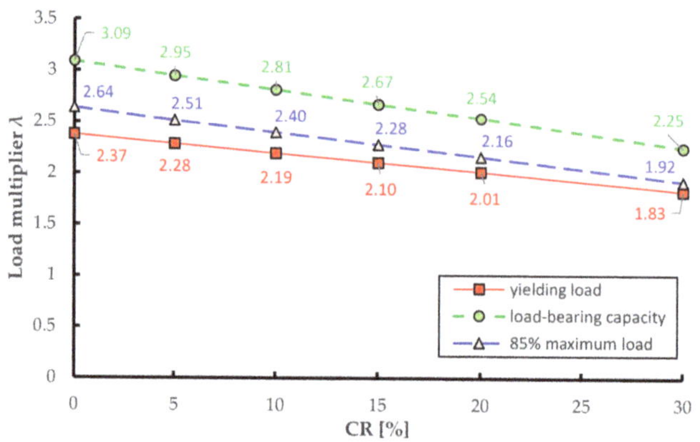

Figure 21. Reduction of load multipliers obtained from nonlinear static analysis of the bridge deck for increasing corrosion scenarios.

6. Conclusions

This paper has presented a systematic approach for the preliminary assessment of the load-bearing capacity of existing PC bridge decks by combining experimental measurements and numerical analyses. The methodology proves to be particularly useful for existing PC bridges exhibiting corrosion in the prestressing strands in order to ascertain whether and to what extent these phenomena affect the structural behavior of the bridge in both serviceability and ultimate conditions. The main steps of the procedure, here applied to the real case study of the PC bridge deck of the Longano viaduct (southern Italy), are summarized as follows:

1. Dynamic identification tests are preliminary performed based on OMA from vibration data collected in situ. These tests are performed in free vibration mode and do not cause any damage to the bridge deck. In addition to providing natural frequencies and mode shapes of the bridge deck, these measurements also serve to obtain estimates of the damping ratios (for instance, by the half-power bandwidth method), which are useful indicators of the potential ongoing damage of the bridge deck.
2. The results of the dynamic tests from step 1 are then used to develop and calibrate a numerical FE model. For simplified estimates of the structural response, 1D beam elements can be used for the structural analysis of the bridge deck.
3. Static load tests are performed in situ to investigate the structural behavior (deflections) of the bridge deck under service loads. These tests are designed and performed in different loading phases of increasing amplitude to avoid sudden damage of structural members in case of advanced material deterioration state. The test results are used to study whether the bridge behavior remains in elastic regime under the maximum allowed (code-conforming) serviceability loads, i.e., by checking that the residual deformation upon unloading does not exceed a critical threshold, which may be an indicator of irreversible damage in some part of the bridge. These test results, if possible, should be followed by a second series of dynamic tests to compare the natural frequencies after the application of loads with those identified in the previous step 1.
4. The results of the static load tests from step 3 are then used to further verify the validity of the FE model developed in step 2, by reproducing the actual load conditions of the tests and comparing numerical deflections with experimental measurements.
5. In addition to the response under service loads, numerical static nonlinear analysis with the previously validated FE model is performed to investigate the influence of different corrosion scenarios on the resulting structural behavior of the bridge deck at ultimate limit states. This numerical analysis is useful to quantify the variation of the load-bearing capacity depending on some hypothetical corrosion rates of the prestressing strands. To this aim, a regression formula has been proposed based on the numerical results found in this study that can be utilized to obtain preliminary estimates of the corrosion-induced degradation of the bridge structural performance for other viaducts sharing similar features to those of the Longano viaduct here analyzed as case study.

In the authors' opinion, the proposed experimental-numerical framework can provide a rapid overview of the bridge structural health in both serviceability and ultimate conditions without performing detailed and extensive tests on material and structural components (which would be certainly needed for deepening the level of investigation). In this context, the methodology can represent a convenient assessment tool to rapidly identify critical portions of a large infrastructure network prior to performing detailed analyses to establish a list of intervention priorities in a timely and reasonable way. Future research developments concern the validation of the findings from this study with additional material test results, and the comparison of the numerical outcomes with alternative time-dependent approaches available in the literature.

Author Contributions: Conceptualization, D.D.D., D.M. and A.R.; methodology, D.D.D., D.M. and A.R.; software, D.D.D., D.M. and A.R.; validation, D.D.D., D.M. and A.R.; formal analysis, D.D.D., D.M. and A.R.; investigation, D.D.D., D.M. and A.R.; data curation, D.D.D., D.M. and A.R.; writing—original draft preparation, D.D.D.; writing—review and editing, D.M. and A.R.; supervision, A.R.; funding acquisition, A.R. All authors have read and agreed to the published version of the manuscript.

Funding: This research received external funding by the Consorzio Autostrade Siciliane (C.A.S.).

Institutional Review Board Statement: Not applicable.

Informed Consent Statement: Not applicable.

Data Availability Statement: Data sharing not applicable.

Acknowledgments: Special thanks are due to the Italian Superior Council of Public Works (CC.SS.LL.PP.) and the Network of University Laboratories of Seismic Engineering (RELUIS). The results were achieved in the national agreement for implementing the agreement pursuant to art. 15 law 7 August 1990, No. 241 between the Superior Council of Public Works and RELUIS.

Conflicts of Interest: The authors declare no conflict of interest.

References

1. Pinto, P.E.; Franchin, P. Issues in the upgrade of Italian highway structures. *J. Earthq. Eng.* **2010**, *14*, 1221–1252. [CrossRef]
2. Borzi, B.; Ceresa, P.; Franchin, P.; Noto, F.; Calvi, G.M.; Pinto, P.E. Seismic vulnerability of the Italian roadway bridge stock. *Earthq. Spectra* **2015**, *231*, 2137–2161. [CrossRef]
3. Colajanni, P.; Recupero, A.; Ricciardi, G.; Spinella, N. Failure by corrosion in PC bridges: A case history of a viaduct in Italy. *Int. J. Struct. Integr.* **2016**, *7*, 181–193. [CrossRef]
4. Morgese, M.; Ansari, F.; Domaneschi, M.; Cimellaro, G.P. Post-collapse analysis of Morandi's Polcevera viaduct in Genoa Italy. *J. Civ. Struct. Health Monit.* **2020**, *10*, 69–85. [CrossRef]
5. Di Prisco, M. Critical infrastructures in Italy: State of the art, case studies, rational approaches to select the intervention priorities. In Proceedings of the Fib Symposium 2019: Concrete-Innovations in Materials, Design and Structures, Krakow, Poland, 27–29 May 2019.
6. Herbrand, M.; Adam, V.; Classen, M.; Kueres, D.; Hegger, J. Strengthening of existing bridge structures for shear and bending with carbon textile-reinforced mortar. *Materials* **2017**, *10*, 1099. [CrossRef]
7. Moslehi Tabar, A.; De Domenico, D.; Dindari, H. Seismic Rehabilitation of Steel Arch Bridges Using Nonlinear Viscous Dampers: Application to a Case Study. *Pract. Period. Struct. Des. Constr.* **2021**, *26*, 04021012. [CrossRef]
8. Italian Ministry of Public Works. *Controllo Delle Condizioni di Stabilità Delle Opere D'arte Stradali;* Circolare 19/07/1967 n. 6736/61A1; Italian Ministry of Public Works: Roma, Italy, 1967. (In Italian)
9. High Council of Public Works, Italian Ministry of Infrastructures and Transportations. *Linee Guida Per La Classificazione E Gestione Del Rischio, La Valutazione Della Sicurezza Ed il Monitoraggio Dei Ponti Esistenti;* High Council of Public Works, Italian Ministry of Infrastructures and Transportations: Roma, Italy, 2020. (In Italian)
10. Ministry of Infrastructures and Transportations. Aggiornamento delle «Norme tecniche per le costruzioni». Decreto 17 Gennaio 2018. In *Supplemento Ordinario Alla "Gazzetta Ufficiale n. 42 Del;* Ministry of Infrastructures and Transportations: Roma, Italy, 2018. (In Italian)
11. Rashetnia, R.; Ghasemzadeh, F.; Hallaji, M.; Pour-Ghaz, M. Quantifying prestressing force loss due to corrosion from dynamic structural response. *J. Sound Vib.* **2018**, *433*, 129–137. [CrossRef]
12. Recupero, A.; Spinella, N. Experimental tests on corroded prestressed concrete beams subjected to transverse load. *Struct. Concr.* **2019**, *20*, 2220–2229. [CrossRef]
13. Imperatore, S.; Rinaldi, Z.; Drago, C. Degradation relationships for the mechanical properties of corroded steel rebars. *Constr. Build. Mater.* **2017**, *148*, 219–230. [CrossRef]
14. Tondolo, F. Bond behaviour with reinforcement corrosion. *Constr. Build. Mater.* **2015**, *93*, 926–932. [CrossRef]
15. Li, F.; Yuan, Y. Effects of corrosion on bond behavior between steel strand and concrete. *Constr. Build. Mater.* **2013**, *38*, 413–422. [CrossRef]
16. Cairns, J.; Plizzari, G.A.; Du, Y.; Law, D.W.; Franzoni, C. Mechanical properties of corrosion-damaged reinforcement. *ACI Mater. J.* **2005**, *102*, 256.
17. Meda, A.; Mostosi, S.; Rinaldi, Z.; Riva, P. Experimental evaluation of the corrosion influence on the cyclic behaviour of RC columns. *Eng. Struct.* **2014**, *76*, 112–123. [CrossRef]
18. Bae, J.H.; Hwang, H.H.; Park, S.Y. Structural Safety Evaluation of Precast, Prestressed Concrete Deck Slabs Cast Using 120-MPa High-Performance Concrete with a Reinforced Joint. *Materials* **2019**, *12*, 3040. [CrossRef]
19. Tu, B.; Fang, Z.; Dong, Y.; Frangopol, D.M. Time-variant reliability analysis of widened deteriorating prestressed concrete bridges considering shrinkage and creep. *Eng. Struct.* **2017**, *153*, 1–16. [CrossRef]
20. Cai, C.S.; Shahawy, M. Predicted and measured performance of prestressed concrete bridges. *J. Bridge Eng.* **2004**, *9*, 4–13. [CrossRef]
21. Bagge, N.; Plos, M.; Popescu, C. A multi-level strategy for successively improved structural analysis of existing concrete bridges: Examination using a prestressed concrete bridge tested to failure. *Struct. Infrastruct. Eng.* **2019**, *15*, 27–53. [CrossRef]
22. Biondini, F.; Camnasio, E.; Palermo, A. Lifetime seismic performance of concrete bridges exposed to corrosion. *Struct. Infrastruct. Eng.* **2014**, *10*, 880–900. [CrossRef]
23. Tu, B.; Dong, Y.; Fang, Z. Time-dependent reliability and redundancy of corroded prestressed concrete bridges at material, component, and system levels. *J. Bridge Eng.* **2019**, *24*, 04019085. [CrossRef]
24. Cai, C.S.; Shahawy, M. Understanding capacity rating of bridges from load tests. *Pract. Period. Struct. Des. Constr.* **2003**, *8*, 209–216. [CrossRef]

25. Hernandez, E.S.; Myers, J.J. Strength evaluation of prestressed concrete bridges by load testing. Life Cycle Analysis and Assessment in Civil Engineering: Towards an Integrated Vision. In Proceedings of the Sixth International Symposium on Life-Cycle Civil Engineering (IALCCE 2018), Ghent, Belgium, 28–31 October 2018; CRC Press: Boca Raton, FL, USA; Volume 5, p. 169.
26. Yoon, Y.G.; Lee, J.Y.; Choi, H.; Oh, T.K. A Study on the Detection of Internal Defect Types for Duct Depth of Prestressed Concrete Structures Using Electromagnetic and Elastic Waves. *Materials* **2021**, *14*, 3931. [CrossRef] [PubMed]
27. Cosenza, E.; Losanno, D. Assessment of existing reinforced-concrete bridges under road-traffic loads according to the new Italian guidelines. *Struct. Concr.* **2021**, 1–14. [CrossRef]
28. European Committee for Standardization. *Basis of Structural Design, European Standard EN 1990:2002*; European Committee for Standardization: Brussels, Belgium, 2002.
29. Brincker, R.; Ventura, C. *Introduction to Operational Modal Analysis*; John Wiley & Sons: Hoboken, NJ, USA, 2015.
30. Ewins, D.J. *Modal Testing: Theory, Practice and Application*; John Wiley & Sons: Hoboken, NJ, USA, 2009.
31. Brincker, R.; Zhang, L.; Andersen, P. Modal identification of output-only systems using frequency domain decomposition. *Smart Mater. Struct.* **2001**, *10*, 441. [CrossRef]
32. MATLAB R2018a. Available online: https://it.mathworks.com/help/matlab (accessed on 5 August 2021).
33. Chopra, A.K. *Dynamics of Structures*, 4th ed.; Pearson Prentice Hall: Hoboken, NJ, USA, 2012.
34. European Committee for Standardization. *Design of Composite Steel and Concrete Structures, Part 1-1: General Rules and Rules for Buildings, European Standard EN 1994-1-1:2004*; European Committee for Standardization: Brussels, Belgium, 2004.
35. Computers and Structures, Inc. Integrated Solution for Structural Analysis and Design. In *CSi Analysis Reference Manual for SAP2000®*; SAP2000 Version 21; Computers and Structures, Inc.: Berkeley, CA, USA, 2017.
36. European Committee for Standardization. *Design of Concrete Structures, Part. 1-1: General Rules and Rules for Buildings, European Standard EN 1992-1-1: 2005*; European Committee for Standardization: Brussels, Belgium, 2005.
37. Di Sarno, L.; Pugliese, F. Critical review of models for the assessment of the degradation of reinforced concrete structures exposed to corrosion. In Proceedings of the Conference SECED 2019, Earthquake Risk and Engineering towards a Resilient World, Greenwich, London, UK, 9–10 September 2019. [CrossRef]
38. Coronelli, D.; Gambarova, P. Structural assessment of corroded reinforced concrete beams: Modeling guidelines. *J. Struct. Eng.* **2004**, *130*, 1214–1224. [CrossRef]
39. Toniolo, G.; Di Prisco, M. *Reinforced Concrete Design to Eurocode 2*; Springer: Berlin, Germany, 2017.
40. Beeby, A.W.; Narayanan, R. *Designer's Guide to Eurocode 2: Design of Concrete Structures: Designers' Guide to EN 1992-1-1 and EN 1992-1-2, Eurocode 2: Design of Concrete Structures, Design of Concrete Structures, General Rules and Rules for Buildings and Structural Fire Design*; Thomas Telford Limited: London, UK, 2010.

materials

MDPI

Article

Evaluation of Smeared Constitutive Laws for Tensile Concrete to Predict the Cracking of RC Beams under Torsion with Smeared Truss Model

Mafalda Teixeira and Luís Bernardo *

Centre of Materials and Building Technologies (C-MADE), Department of Civil Engineering and Architecture, University of Beira Interior, 6201-001 Covilhã, Portugal; mafalda.m.teixeira@ubi.pt

* Correspondence: lfb@ubi.pt

Abstract: In this study, the generalized softened variable angle truss-model (GSVATM) is used to predict the response of reinforced concrete (RC) beams under torsion at the early loading stages, namely the transition from the uncracked to the cracked stage. Being a 3-dimensional smeared truss model, the GSVATM must incorporate smeared constitutive laws for the materials, namely for the tensile concrete. Different smeared constitutive laws for tensile concrete can be found in the literature, which could lead to different predictions for the torsional response of RC beams at the earlier stages. Hence, the GSVATM is used to check several smeared constitutive laws for tensile concrete proposed in previous studies. The studied parameters are the cracking torque and the corresponding twist. The predictions of these parameters from the GSVATM are compared with the experimental results from several reported tests on RC beams under torsion. From the obtained results and the performed comparative analyses, one of the checked smeared constitutive laws for tensile concrete was found to lead to good predictions for the cracking torque of the RC beams regardless of the cross-section type (plain or hollow). Such a result could be useful to help with choosing the best constitutive laws to be incorporated into the smeared truss models to predict the response of RC beams under torsion.

Keywords: RC beams; torsion; generalized softened variable angle truss-model (GSVATM); tensile concrete; smeared constitutive law; cracking torque; cracking twist

Citation: Teixeira, M.; Bernardo, L. Evaluation of Smeared Constitutive Laws for Tensile Concrete to Predict the Cracking of RC Beams under Torsion with Smeared Truss Model. *Materials* **2021**, *14*, 1260. https://doi.org/10.3390/ma14051260

Academic Editor: Lizhi Sun

Received: 16 February 2021
Accepted: 4 March 2021
Published: 7 March 2021

1. Introduction

In the second half of the last century, the Space Truss Analogy (STA) was successively refined in order to better predict the response of structural concrete beams under torsion. Nowadays, modern truss-based models can be considered reliable, comprehensive and unified analytical models. They are able to simulate the complex 3-dimensional features of the torsional phenomenon, including the nonlinear behavior and the interaction between the material components of the beam in all loading stages. Models based on the STA constitute the basis models for most codes of practice to establish the design procedures for torsion and still continue to be improved and extended [1,2].

A STA-based model assumes that a reinforced concrete (RC) beam under torsion behaves like a cracked thin tube, where the external torque is resisted through a transversal circulatory shear flow. The tube is modeled with a spatial truss, which includes longitudinal and transverse steel reinforcement under tension interacting with inclined concrete struts under compression. The model satisfies the three Navier's principles of the mechanics of materials, namely, stress equilibrium, strain compatibility and constitutive laws.

Among the STA-based models that have been developed, one of the most commonly used and extended is the Variable-Angle Truss Model (VATM), which was originally proposed by Hsu and Mo in 1985 [3]. This model incorporated for the first time smeared constitutive laws, or smeared stress (σ)—strain (ε) relationships, for both tensile steel reinforcement embedded in concrete and compressive concrete. Such constitutive laws are established from controlled experimental tests on RC panels under in-plane shear, in order to account for, on average (considering an area sufficiently wide to include several cracks), the effect of the biaxial stress state in the principal direction of stresses, the effect of cracking, the interaction between the material components, and both the softening and stiffening effects. The Universal Panel Tester at the University of Houston is one of the testing devices which has most contributed to the establishment of smeared constitutive laws for smeared truss models [4].

Despite being a nonlinear model with an incremental and iterative calculation procedure, the VATM is relatively simple to implement, having access to programming languages in a computer. The model allows us to calculate the full response of RC beams under torsion, namely the torque (M_T)–twist (θ) curve. The predictions from the VATM showed good agreement with experimental results, namely when predicting the response of RC beams under torsion at the ultimate stage [3,5,6]. When compared with more complex models also proposed for the RC beam under torsion, which sometimes involve large computational effort (for instance [7–10]), the VATM is recognized as a simpler and more reliable model for predicting the torsional strength of RC beams under torsion, which is one of the most important key parameters for design. It should also be mentioned that smeared approaches, such as the VATM, constitute an alternative approach to local ones in which the local fracture properties are directly accounted for, such as in the numerical models from [11,12]. In smeared approaches, smeared constitutive laws for the materials are incorporated into the model. Such models have been shown to be reliable, on average, for modeling the global behavior of structural elements, such as for the RC beams under torsion.

The VATM has been extended for prestress concrete (PC) beams [13] and also for axially restrained RC beams [14,15]. The VATM was also improved in order to reliably predict the response of RC beam under torsion for the low loading stages, namely the transition between the uncracked stage and the cracked stage. This was achieved by incorporating into the model the contribution of the tensile concrete (neglected in the VATM) through an additional smeared σ—ε constitutive law in the perpendicular direction to the concrete struts. The new model, called generalized softened variable angle truss-model (GSVATM), was proposed in 2015 for RC solid beams under torsion [16]. The predictions from the GSVATM showed good agreement with experimental results for all loading stages. The GSVATM was recently extended for PC beams [17], hollow RC beams [18] and RC flanged beams [19]. A unified version of the model was also recently proposed [2].

The predictions from any smeared truss model, such as the VATM or the GSVATM, strongly depend on the smeared σ—ε relationships for the materials. This important aspect was previously demonstrated by Bernardo et al. in 2012 [20] for the prediction of the torsional strength and corresponding twist for the RC beams under torsion. The study aimed to find the most reliable smeared σ—ε relationships for the materials, among the several ones found in the literature, to be incorporated into the VATM to better predict the ultimate response of RC beams under torsion. The best constitutive laws found in [20], for both the concrete in compression and steel reinforcement in tension, were posteriorly incorporated in the GSVATM [16]. Bernardo et al. in 2012 [20] did not include in their study the prediction of the key parameters for the low loading stages because, as referred to before, the predictions from the VATM were shown to be in good agreement with the experimental results only for the ultimate stage. This is mainly because the model assumes that the member has been fully cracked since the beginning of loading, which is not true.

For design, it is also important to reliably predict the behavior for the low loading levels. The current codes of practice compel us to check the structural members for both the serviceability and ultimate limit states. For the first one, it is important that the cracking torque is known. As previously referred to, the GSVATM is able to predict the full response of the RC beams under torsion, including the transition between the uncracked stage and the cracked stage. The prediction of such a transition zone highly depends on the smeared constitutive law for the tensile concrete. As for the other constitutive laws referred to (for concrete in compression and steel reinforcement in tension), different proposals of smeared constitutive laws for tensile concrete can be found in the literature. To the best of the authors' knowledge, no previous study was found with the aim of checking such constitutive laws in smeared truss models, in order to evaluate which features allow the model to give the best predictions for the low loading stages. Usually, researchers working with smeared truss models use their own smeared constitutive laws or choose them based on the proposals from other studies.

In this study, the GSVATM is used to check some proposed smeared constitutive laws for tensile concrete found in the literature. The GSVATM was the chosen model because, as previously stated, it is able to predict the full response of the RC beams under torsion for all loading stages. In addition, this model was proposed by the corresponding author [16] and has also been successfully used in previous studies [2,17–19,21]. The chosen parameters to be studied are the cracking torque and the corresponding twist. The theoretical predictions of such parameters are compared with the experimental results from several reported tests on RC beams under torsion. Only RC beams with rectangular sections are studied because they constitute the current solution used in practice. In addition, the number of reported experimental results in the literature for such beams is much higher than for other typologies such as PC beams or beams with a flanged cross-section.

2. The Generalized Softened Variable Angle Truss-Model

For the sake of the readers of this article, a brief description of the GSVATM is presented. The GSVATM was initially proposed for RC plain beams under torsion [16]. Recently, the model was extended and unified for RC hollow beams under torsion [18]. Details about the assumptions of the model, the derivation of the equations and the justification of the calculation solution procedure can be found [16,18].

According to the GSVATM, a cracked RC thin beam element under a vertical shear force V, which induces a shear flow q in the cross-section, is modeled with a smeared plain truss analogy, as illustrated in Figure 1. The behavior of the RC thin beam is governed by Equations (1) to (5). The smeared plain truss incorporates inclined concrete struts (with compressive force C) with an angle α to the longitudinal axis, and perpendicular concrete ties (with tensile force T). The corresponding stress fields are denoted by σ_2^c and σ_1^c, respectively. The meanings of the parameters are (see Figure 1): R is the resultant force, β is the angle of R to the force C, γ is the angle of R to the longitudinal axis, t_c is the width of the cross-section and d_v is the distance between centers of the longitudinal bars.

$$R = \sqrt{C^2 + T^2} \tag{1}$$

$$\beta = \arctan(T/C) \tag{2}$$

$$\gamma = \alpha + \beta \tag{3}$$

$$C = \sigma_2^c t_c d_v \cos\alpha \tag{4}$$

$$T = \sigma_1^c t_c d_v \sin\alpha \tag{5}$$

An equivalent cracked RC hollow beam under a torque M_T, as illustrated in Figure 2, is modeled as the union of four thin beam elements as in Figure 1. Each thin beam constitutes a wall of the RC hollow beam. As a result of this union, the torque M_T induces a circulatory shear flow q and the beam can be modeled with a smeared spatial truss analogy. The center line of the circulatory shear flow q coincides with the center line of the walls. The behavior of the RC hollow beam is governed by equilibrium equations, Equations (6) to (8), and compatibility equations, Equations (9) to (12). If $\gamma = \alpha + \beta > 90°$, Equation (7) must be multiplied by (-1). The previous equations account for the strain gradient along the walls' thickness due to the bidirectional opposite curvatures induced by bending (Figure 3).

$$M_T = \frac{2AR\sin\gamma}{d_v} \tag{6}$$

$$t_c = \frac{A_{sl} f_{sl}}{\sigma_2^c p} \frac{\cos \beta}{\cos \alpha \cos \gamma} \text{ for } \gamma = \alpha + \beta \leq 90^\circ \tag{7}$$

$$\alpha = \arctan\left(\frac{\sqrt{F^2 (\tan \beta)^2 + F (\tan \beta)^4 + F + (\tan \beta)^2}}{F (\tan \beta)^2 + 1} \right) \text{ with } F = \frac{A_{st} f_{st} p}{A_{sl} f_{sl} s} \tag{8}$$

$$\varepsilon_{st} = \left(\frac{A_o^2 \sigma_2^c \sin \gamma}{p M_T \cos \beta \tan \alpha \sin \alpha} - \frac{1}{2} \right) \varepsilon_{2s}^c \tag{9}$$

$$\varepsilon_{sl} = \left(\frac{A_o^2 \sigma_2^c \sin \gamma}{p M_T \cos \beta \cot \alpha \sin \alpha} - \frac{1}{2} \right) \varepsilon_{2s}^c \tag{10}$$

$$\theta = \frac{\varepsilon_{2s}^c}{2 t_c \sin \alpha \cos \alpha} \tag{11}$$

$$\varepsilon_{1s}^c = 2\varepsilon_1^c = 2\varepsilon_{sl} + 2\varepsilon_{st} + \varepsilon_{2s}^c \tag{12}$$

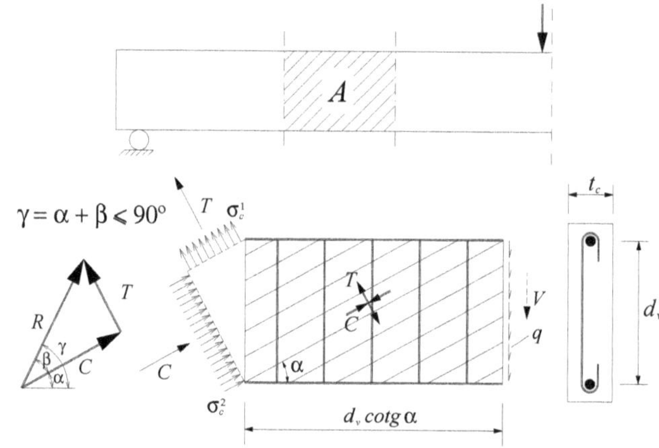

Figure 1. Reinforced concrete (RC) thin beam element [18].

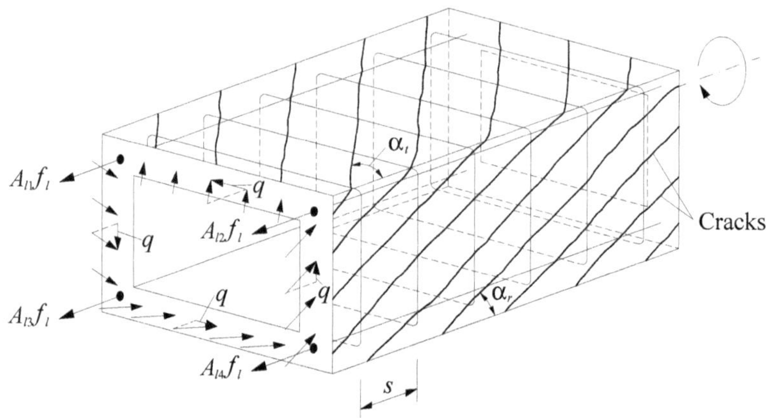

Figure 2. RC hollow beam element [18].

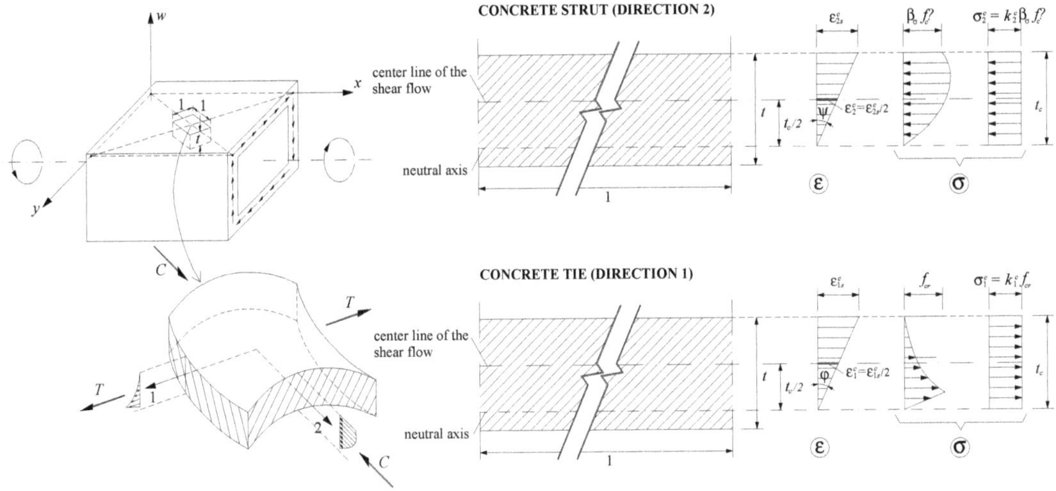

Figure 3. Curvatures and strain gradient in the walls [18].

In the previous equations (see Figures 2 and 3), t_c is the effective thickness of the concrete strut and tie in the walls, $A = (x - t_c)(y - t_c)$ and $p = 2(x - t_c) + 2(y - t_c)$ are the area enclosed and the perimeter of the center line of the shear flow q (with x the minor and y the major outer dimension of the beam's cross-section), respectively, A_{sl} is the total area of the longitudinal reinforcement, A_{st} is the area of one rebar of the transverse reinforcement, s is the longitudinal spacing of the transverse reinforcement, f_{sl} and f_{st} are the tensile stresses in the longitudinal and transverse reinforcement, respectively, ε_{sl} and ε_{st} are the tensile strains in the longitudinal and transverse reinforcement, respectively, ε_{2s}^{c} and ε_{1s}^{c} are the strains at the outer fiber of the concrete strut and concrete tie, respectively, ε_1^c is the average strain in the concrete tie, and θ is the twist per unit length.

As referred to in the introduction section, the GSVATM incorporates smeared σ—ε relationships to model the behavior of the compressive concrete in the struts, the tensile concrete in the ties and the tensile longitudinal and transverse steel reinforcement (rebars embedded in concrete). For the RC beams under torsion, some suitable smeared σ—ε relationships were previously found by Bernardo et al. in 2012 [20] and are also used in this study. For the compressive concrete, the smeared σ—ε relationship proposed by Belarbi and Hsu in 1995 [22] (Equations (13) and (14)) with softening factor $\beta_* = \beta_\sigma = \beta_\varepsilon$, for both the peak stress and corresponding strain, proposed by Zhang and Hsu in 1998 [23] (Equations (15) to (18)) are used. For the steel reinforcement in tension, the smeared σ—ε relationship proposed by Belarbi and Hsu in 1994 [24] (Equations (19) to (21)) is used.

The meaning of the parameters are: f_c' is the average uniaxial concrete compressive strength, ε_o is the strain corresponding to f_c', ε_2^c is the average strain in the concrete strut (Figure 3), ρ_l and ρ_t are the longitudinal ($\rho_l = A_{sl}/A_c$, with $A_c = xy$) and transverse ($\rho_t = A_{st}u/A_c s$, with $u = 2x + 2y$) reinforcement ratios, respectively, f_{ly} and f_{ty} are the yielding stresses for the longitudinal and transverse reinforcement, respectively, f_{cr} is the tensile concrete strength, ε_{cr} is the strain corresponding to f_{cr}, f_s and ε_s are the stress and strain in the steel reinforcement (longitudinal or transversal), respectively, E_s is the Young's Modulus for steel reinforcement, f_y is the yielding stress of steel reinforcement (longitudinal or transversal) and ρ is the reinforcement ratio (longitudinal or transversal).

$$\sigma_2^c = \beta_\sigma f_c' \left[2\left(\frac{\varepsilon_2^c}{\beta_\varepsilon \varepsilon_o}\right) - \left(\frac{\varepsilon_2^c}{\beta_\varepsilon \varepsilon_o}\right)^2 \right] \text{ if } \varepsilon_2^c \leq \beta_\varepsilon \varepsilon_o \tag{13}$$

$$\sigma_2^c = \beta_\sigma f_c' \left[1 - \left(\frac{\varepsilon_2^c - \beta_\varepsilon \varepsilon_o}{2\varepsilon_o - \beta_\varepsilon \varepsilon_o}\right)^2\right] \text{ if } \varepsilon_2^c > \beta_\varepsilon \varepsilon_o \tag{14}$$

$$\beta_* = \beta_\sigma = \beta_\varepsilon = \frac{R(f_c')}{\sqrt{1 + \frac{400\varepsilon_1^c}{\eta'}}} \tag{15}$$

$$\eta = \frac{\rho_l f_{ly}}{\rho_t f_{ty}} \tag{16}$$

$$\begin{cases} \eta \leq 1 \Rightarrow \eta' = \eta \\ \eta > 1 \Rightarrow \eta' = 1/\eta \end{cases} \tag{17}$$

$$R(f_c') = \frac{5.8}{\sqrt{f_c'(\text{MPa})}} \leq 0.9 \tag{18}$$

$$f_s = \frac{0.975 E_s \varepsilon_s}{\left[1 + \left(\frac{1,1 E_s \varepsilon_s}{f_y}\right)^m\right]^{\frac{1}{m}}} + 0.025 E_s \varepsilon_s \tag{19}$$

$$m = \frac{1}{9B - 0.2} \leq 25 \tag{20}$$

$$B = \frac{1}{\rho}\left(\frac{f_{cr}}{f_y}\right)^{1.5} \tag{21}$$

For the tensile concrete, the smeared σ—ε relationships checked in this study are presented in more detail in the Section 3. However, in order to present the equations for some correction coefficients and also the flowchart with the calculation procedure for the GSVATM, the following general and common form of the equations are written ($F(\ldots)$ stands for "function of ... "):

$$\sigma_1^c = E_c \varepsilon_1^c \text{ if } \varepsilon_1^c \leq \varepsilon_{cr} \tag{22}$$

$$\sigma_1^c = F(f_{cr}; \varepsilon_1^c) \text{ if } \varepsilon_1^c > \varepsilon_{cr} \tag{23}$$

In Equation (22), which models the linear–elastic stage before cracking, E_c is the Young's Modulus for the concrete. Based on the proposals from previous studies [9,25,26], and in order to unify the GSVATM for both the RC plain and the hollow beams under torsion, and also to improve the predictions from the model for the low loading stages, in 2019 Bernardo [18] presented a set of equations (Equations (24) to (29)) to compute the parameters ε_{cr} and E_c, accounting for the correction coefficients μ and λ. These equations apply for all smeared σ—ε relationships for tensile concrete presented in the Section 3 and checked in this study.

$$\varepsilon_{cr} = 0.00008\mu \tag{24}$$

$$E_c = 3875\lambda\sqrt{f_c' \text{ (MPa)}} \tag{25}$$

$$\mu = \lambda = 1.45 \text{ (RC solid beams)} \tag{26}$$

$$\mu = \lambda = 0.93 \text{ (RC thin-walled hollow beams)} \tag{27}$$

$$\mu = \lambda = 1.20 \left(\text{RC thick-walled hollow beams and } f_c' \leq 48 \text{ MPa}\right) \tag{28}$$

$$\mu = \lambda = 1.129 \left(\text{RC thick-walled hollow beams and } f_c' > 48 \text{ MPa}\right) \tag{29}$$

The classification of the RC hollow beams into "thin wall" or "thick wall" [26] is done during the calculation procedure of the GSVATM. The RC hollow beam is firstly calculated as an equivalent RC plain beam until both the cracking torque $M_{Tcr,plain}$ and the corresponding value for the effective wall's thickness $t_{c,cr,plain}$ are computed. Then, the following classification applies (with t being the real thickness of the wall of the RC hollow beam):

- if $t \leq 0.91 t_{c,cr,plain}$ the RC hollow beam has a "thin wall";

- if $t > 0.91t_{c,cr,plain}$ the RC hollow beam has a "thick wall".

Then, the beam is recalculated considering the real cross-section (hollow).

For the RC beams under torsion, average stresses σ_2^c (Equation (30)) and σ_1^c (Equation (31)) are computed for the concrete strut and tie, respectively, accounting for the section type through the correction coefficient η (Equations (32) to (35)). This simplification is assumed because the real stress diagrams along the effective wall's thickness t_c are not uniform due to the strain gradient (Figure 3). The coefficients k_2^c and k_1^c are computed from the numerical integration of the smeared σ—ε relationships.

$$\sigma_2^c = \eta k_2^c \beta_\sigma f_c' \tag{30}$$

$$\sigma_1^c = \eta k_1^c f_{cr} \tag{31}$$

$$\eta = 1 \text{ (RC solid beams)} \tag{32}$$

$$\eta = 0.033\sqrt{f_c' \text{ (MPa)}} + 0.73 \text{ (RC thin-walled hollow beams)} \tag{33}$$

$$\eta = 0.0938\sqrt{f_c' \text{ (MPa)}} + 0.43 \left(\text{RC thick-walled hollow beams and } f_c' \leq 48 \text{ MPa}\right) \tag{34}$$

$$\eta = \frac{8.45}{\sqrt{f_c' \text{ (MPa)}}} + 0.17 \left(\text{RC thick-walled hollow beams and } f_c' > 48 \text{ MPa}\right) \tag{35}$$

To solve the nonlinear procedure of the GSVATM, an algorithm incorporating a trial-and-error technique was implemented using the programming language Delphi (see flowchart in Figure 4) [16,18]. For each iteration, the input parameter $\varepsilon_{2s}^c = 2\varepsilon_2^c$ (strain at the outer fiber of the concrete strut) is incremented in order to compute each solution point to draw the theoretical M_T—θ curve. The calculation procedure ends when the assumed failure strains for the materials is reached, either for concrete in compression (ε_{cu}) or for steel reinforcement in tension (ε_{su}). In this study, European code Eurocode 2 was used to define the conventional failure strains for the materials.

3. Smeared Constitutive Laws for Tensile Concrete

This section presents eight smeared σ—ε relationships for tensile concrete proposed in previous studies (laws *l*1 to *l*8), so that they can be implemented in the GSVATM and checked (Section 4). In a previous study, it was showed that these relationships are suitable to be implemented in smeared truss models, such as the GSVATM, to account for the contribution of the tensile concrete [27].

Some of the presented smeared σ—ε relationships for tensile concrete were proposed based on the experimental results from concrete panels under shear. In such cases, the average stress σ_1^c in the tensile concrete after cracking ($\varepsilon_1^c > \varepsilon_{cr}$) is usually obtained from the equilibrium of the stress fields applied to the panels by separating the average stresses in both the tensile steel reinforcement and the tensile concrete. The other smeared σ—ε relationships for tensile concrete were proposed by refining the previous ones in order to improve the predictions of the used smeared models.

For all presented smeared σ—ε relationships for tensile concrete, two equations are written. The first one aims to model the tensile behavior of the concrete before cracking and is equal for all smeared constitutive laws:

$$\sigma_1^c = E_c \varepsilon_1^c \text{ if } \varepsilon_1^c \leq \varepsilon_{cr} \tag{36}$$

The second equation aims to model the tensile behavior of the concrete after cracking, and accounts for the tension softening (the influence of the cracks) and the tension stiffening (the retention of concrete tensile stress due to the interaction with steel reinforcement).

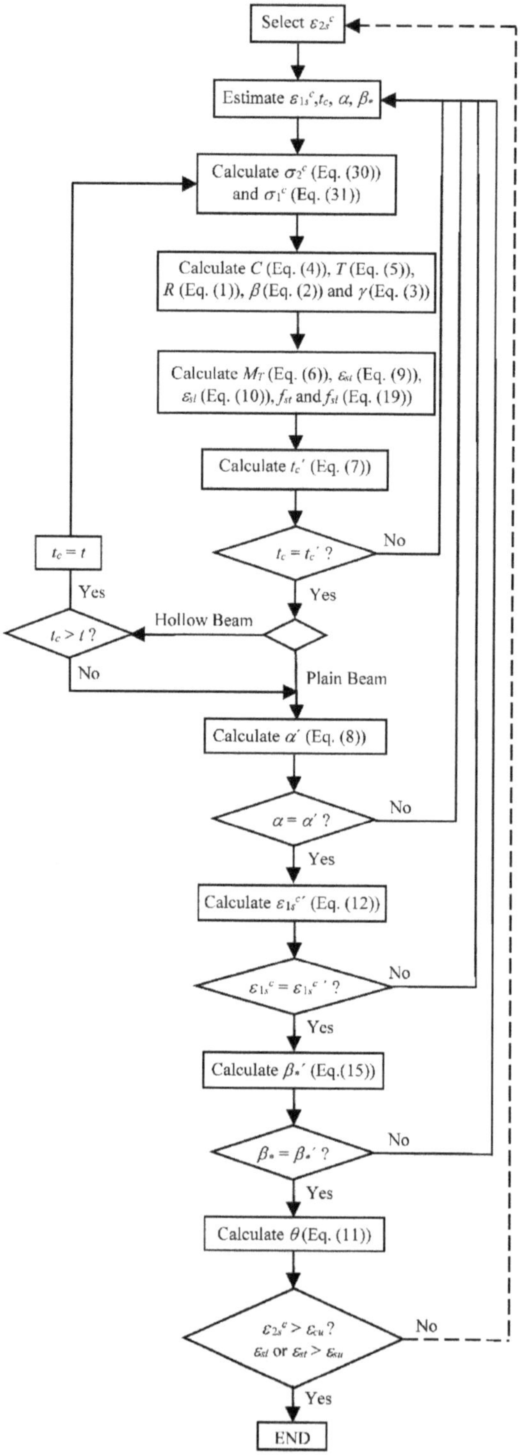

Figure 4. Flowchart.

As presented in Section 2, parameters ε_{cr} and E_c are computed according to Equations (24) and (25), which apply for all the presented smeared σ—ε relationships. Further, for all presented equations, the symbology was adapted to the same one used in the previous section.

3.1. *Law l1—Cervenka in 1985*

In 1985, Cervenka proposed a smeared model for cracked RC panels. In this model, the author implemented the following equation for the descending branch of the smeared σ—ε relationships for tensile concrete [28]:

$$\sigma_1^c = f_{cr}\left[1 - \left(\frac{\varepsilon_1^c}{c}\right)^{k_2}\right] \text{ if } \varepsilon_1^c > \varepsilon_{cr} \tag{37}$$

Parameter c is the average tensile strain (ε_1^c) for which the principal tensile stress can be considered null. The author observed that c ranges between 0.004 and 0.005. For this study, the average value (0.0045) was considered. The exponent k_2 is related with the curvature shape of the descending branch of the σ—ε curve after the peak tensile stress. Cervenka proposed to consider $k_2 = 0.5$.

3.2. *Law l2—Vecchio and Collins in 1986*

In 1986, based on several experimental results from RC panels under shear performed at the University of Toronto, Vecchio and Collins proposed the smeared model called Modified Compression Field Theory. For this model, the following postpeak smeared σ—ε relationship for tensile concrete was proposed [29]:

$$\sigma_1^c = \frac{f_{cr}}{1 + \sqrt{200\varepsilon_1^c}} \text{ if } \varepsilon_1^c > \varepsilon_{cr} \tag{38}$$

3.3. *Law l3—Hsu in 1991*

In 1991, Hsu [30] proposed an efficient algorithm for his softened truss model theory to analyze the nonlinear behavior of concrete membrane elements. For this model, a refined version of the postpeak smeared σ—ε relationship for tensile concrete from Vecchio and Collins in 1986 [29] was proposed:

$$\sigma_1^c = \frac{f_{cr}}{1 + \sqrt{\frac{\varepsilon_1^c - \varepsilon_{cr}}{0.005}}} \text{ if } \varepsilon_1^c > \varepsilon_{cr} \tag{39}$$

3.4. *Law l4—Belarbi and Hsu in 1994*

Based on experimental studies on RC panels under shear performed at the University of Houston, Belarbi and Hsu in 1994 [24] proposed Equation (40) for the descending branch of the smeared constitutive law for tensile concrete.

$$\sigma_1^c = f_{cr}\left(\frac{\varepsilon_{cr}}{\varepsilon_1^c}\right)^{0.4} \text{ if } \varepsilon_1^c > \varepsilon_{cr} \tag{40}$$

3.5. *Law l5—Collins and Colaborators in 1996*

In 1996, Collins et al. [31] proposed a postpeak smeared constitutive law for tensile concrete slightly different from the one proposed by Vecchio and Collins in 1986 [29]:

$$\sigma_1^c = \frac{f_{cr}}{1 + \sqrt{500\varepsilon_1^c}} \text{ if } \varepsilon_1^c > \varepsilon_{cr} \tag{41}$$

3.6. *Law l6—Vecchio in 2000*

The Disturbed Stress Field Model for RC was proposed by Vecchio in 2000 [32]. For this model, the author proposed a somewhat more complicated postpeak smeared constitutive law for tensile concrete, in order to account more precisely for the tension stiffening. The author proposed two equations, with a maximum condition, to also account indirectly for the level of reinforcement ratio (Equations (42) to (45)). When a low (high) reinforcement ratio exists, tension softening (stiffening) is more relevant.

$$\sigma_1^c = \max\left(f_{c1}^a; f_{c1}^b\right) \text{ if } \varepsilon_1^c > \varepsilon_{cr} \tag{42}$$

$$f_{c1}^a = f_{cr}\left(1 - \frac{\varepsilon_1^c - \varepsilon_{cr}}{\varepsilon_{ts} - \varepsilon_{cr}}\right) \tag{43}$$

$$f_{c1}^b = \frac{f_{cr}}{1 + \sqrt{c_t \varepsilon_1^c}} \tag{44}$$

$$\varepsilon_{ts} = 2.0\frac{G_f}{f_{cr} L_r} \tag{45}$$

Parameter ε_{ts} represents the terminal strain, which depends on the fracture energy (G_f), assumed to be constant and equal to 75 N/m by Vecchio, and also on half of the distance between cracks (L_r). Parameter c_t can be simply considered equal to 200 for small members or for members incorporating steel reinforcement grids with very small spacing, and 500 for large members. For this study, L_r was infered from the experimental data of the used reference beams (Section 4).

3.7. *Law l7—Bentz in 2005*

In 2005, Bentz proposed Equations (46) and (47) for the smeared postpeak tension stiffening relationship of tensile concrete [33].

$$\sigma_1^c = \frac{f_{cr}}{1 + \sqrt{3.6 M \varepsilon_1^c}} \text{ if } \varepsilon_1^c > \varepsilon_{cr} \tag{46}$$

$$M = \frac{A_c}{\sum \phi \pi} \tag{47}$$

Parameter M (in "mm" units) accounts for the effective tensile concrete area around the rebars (A_c) and for the rebars' diameter (ϕ). For this study, A_c was computed considering the effective thickness of the concrete tie (t_c), which is computed from the GSVATM.

3.8. *Law l8—Stramandinoli and Rovere in 2008*

In 2008, for the nonlinear analysis of reinforced concrete members, Stramandinoli and Rovere proposed equations for the postpeak smeared constitutive law for tensile concrete [34] (Equations (48) to (50)). The law accounts directly for the longitudinal reinforcement ratio ρ.

$$\sigma_1^c = f_{cr} e^{-\alpha\left(\frac{\varepsilon_1^c}{\varepsilon_{cr}}\right)} \text{ if } \varepsilon_1^c > \varepsilon_{cr} \tag{48}$$

$$\alpha = 0.017 + 0.255(n\rho) - 0.106(n\rho)^2 + 0.016(n\rho)^3 \tag{49}$$

$$n = \frac{E_s}{E_c} \tag{50}$$

3.9. *Comparison between the Smeared Constitutive Laws*

For comparison, Figure 5 illustrates the smeared σ—ε curves for tensile concrete for each of the proposals presented in the previous subsections. The curves were computed considering the same arbitrary and typical cross-section with current materials.

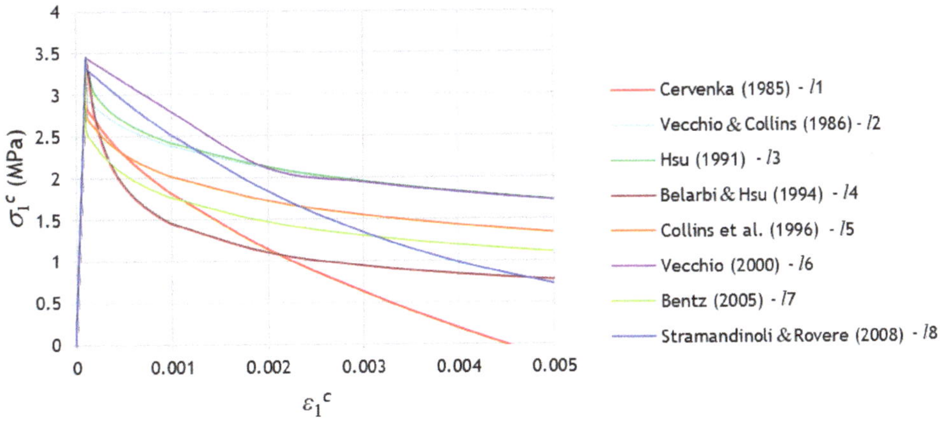

Figure 5. Comparison between the smeared constitutive laws.

After the peak stress, namely for the descending branch, Figure 5 shows high variability between the σ—ε curves. In spite of the peak stress coincides for all the curves, it should be noted that the referred variability will influence the calculation of the cracking torque and corresponding twist with the GSVATM. This is because, as previously referred, the tensile stress σ_1^c computed from Equation (31) represents an average stress since the real stress diagram along the effective tie's thickness is not uniform due to the strain gradient (Figure 3). The representative concrete tensile stress in the GSVATM (σ_1^c) does not coincide with the maximum tensile stress. Hence, the strain ε_1^c corresponding to the effective cracking torque in the M_T—θ curve computed with the GSVATM does not coincide with the strain ε_{cr} corresponding to the peak stress in the smeared σ—ε curves for tensile concrete. This is illustrated in Figure 6, where an example of M_T—θ and corresponding σ—ε curves for tensile concrete, computed with the GSVATM, are presented. The highlighted point in the curves (with marker " ●") corresponds to the effective cracking torque, which is reached for a strain $\varepsilon_1^c > \varepsilon_{cr}$, i.e., in the descending branch of the smeared σ—ε curve for tensile concrete. This explains why different smeared σ—ε curves for tensile concrete incorporated in the GSVATM will lead to different coordinates for the cracking torque (cracking torque and corresponding twist).

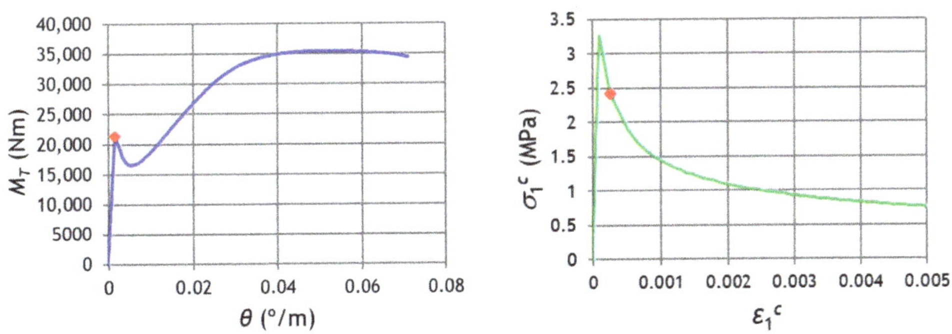

(a) (b)

Figure 6. Location of the effective cracking point.

4. Comparison with Experimental Results

For this study, the experimental results of 103 RC beams tested under torsion were collected from the literature. Both RC beams with plain and hollow rectangular cross section were considered. These beams were selected based on criteria related to minimum requirements from codes of practice (for instance, the beams should incorporate a minimum torsional reinforcement, the spacing between rebars should not exceed the maximum allowed, etc.) in order to ensure a typical behavior under torsion. A detailed discussion on such applied criteria can be found in [21]. For the RC plain beams, the data were collected from the following studies: Hsu in 1968 [35], McMullen and Rangan in 1978 [36], Koutchkali and Belarbi in 2001 [37], Fang and Shiau in 2004 [38], and Peng and Wong in 2011 [39]. For RC hollow beams, the following studies were consulted: Hsu in 1968 [35], Lampert and Thürlimann in 1969 [40], Leonhardt and Schelling in 1974 [41], Bernardo and Lopes in 2009 [42], and Jeng in 2015 [26].

Table A1 in Appendix A summarizes the main properties for each reference beam. In Table 1, "P" and "H" stand for "plain" and "hollow" cross-section, respectively. For all the reference beams from Table A1, the experimental values of the cracking torque (M_{Tcr}^{exp}) and corresponding twist (θ_{cr}^{exp}) were obtained from the data or graphs given by the authors [26,35–42]. Such values are presented for each reference beam in Tables A2–A4 (see Appendix A).

Table 1. Comparative analysis.

Cross-Section		P		H		P + H	
Constitutive law		$\frac{M_{Tcr}^{exp}}{M_{Tcr}^{thli}}$	$\frac{\theta_{cr}^{exp}}{\theta_{cr}^{thli}}$	$\frac{M_{Tcr}^{exp}}{M_{Tcr}^{thli}}$	$\frac{\theta_{cr}^{exp}}{\theta_{cr}^{thli}}$	$\frac{M_{Tcr}^{exp}}{M_{Tcr}^{thli}}$	$\frac{\theta_{cr}^{exp}}{\theta_{cr}^{thli}}$
*l*1—Cervenka (1985) [28]	$\overline{x} =$	1.02	1.16	1.29	1.71	1.05	1.23
	$cv(\%) =$	12.13	25.19	21.47	35.22	21.29	38.54
*l*2—Vecchio and Collins (1986) [29]	$\overline{x} =$	0.96	1.11	1.24	1.62	0.99	1.16
	$cv(\%) =$	12.41	25.68	21.47	35.61	21.73	39.13
*l*3—Hsu (1991) [30]	$\overline{x} =$	0.91	1.04	1.18	1.55	0.94	1.10
	$cv(\%) =$	12.38	25.83	21.66	36.11	22.13	40.05
*l*4—Belarbi and Hsu (1994) [24]	$\overline{x} =$	1.00	1.16	1.03	1.42	1.01	1.19
	$cv(\%) =$	11.35	24.10	32.17	46.05	21.45	39.02
*l*5—Collins et al. (1996) [31]	$\overline{x} =$	1.04	1.20	1.33	1.74	1.07	1.24
	$cv(\%) =$	12.31	25.29	21.49	35.97	21.36	39.23
*l*6—Vecchio (2000) [32]	$\overline{x} =$	0.96	1.08	1.18	1.50	1.00	1.15
	$cv(\%) =$	11.82	26.20	19.22	37.88	18.56	37.43
*l*7—Bentz (2005) [33]	$\overline{x} =$	0.94	1.08	1.26	1.68	0.99	1.16
	$cv(\%) =$	12.24	25.91	21.23	34.90	22.79	40.35
*l*8—Stramandinoli and Rovere (2008) [34]	$\overline{x} =$	0.86	0.98	1.13	1.46	0.89	1.06
	$cv(\%) =$	12.84	26.75	22.11	36.09	22.72	39.36

The torsional response of all the reference beams was computed using the GSVATM, for each of the smeared σ—ε relationships for the tensile concrete presented in Section 3 (laws *l*1 to *l*8). From the obtained theoretical M_T—θ curves, the theoretical coordinates of the cracking point, i.e., the cracking torque (M_{Tcr}^{thli}, with $i = 1$ to 8) and corresponding twists, i.e., the cracking twists (θ_{cr}^{thli}, with $i = 1$ to 8), were obtained. Such values are also presented for each reference beam in Tables A2–A4 (see Appendix A). In addition, the ratios between the experimental to the theoretical values are also presented for each reference beam ($M_{Tcr}^{exp}/M_{Tcr}^{thli}$ and $\theta_{cr}^{exp}/\theta_{cr}^{thli}$, with $i = 1$ to 8).

Figure 7 presents, as an example, a graph with the experimental and theoretical M_T—θ curves, computed for each smeared constitutive law for tensile concrete, for reference beam N-20-20 [38]. Figure 7 confirms that the coordinates of the cracking point, namely the cracking torque, as well as the postcracking response, highly depends on the used smeared constitutive law for the tensile concrete. The influence of the used smeared constitutive law is residual at the ultimate stage, namely for the maximum torque.

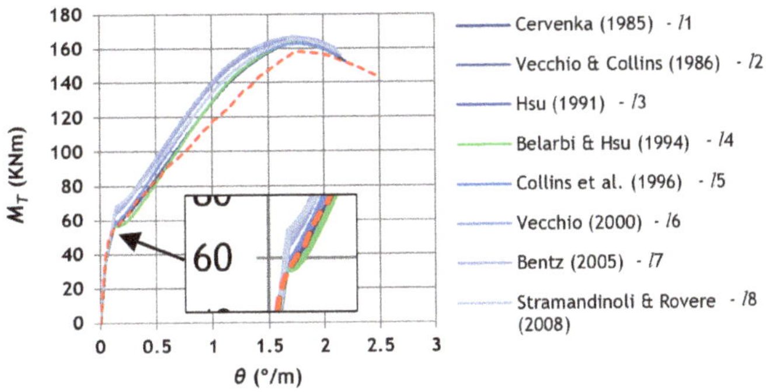

Figure 7. Example of M_T—θ curves for reference beam N-20-20.

Table 1 summarizes and compares the results from Tables A2–A4 (Appendix A) for the cracking torque (M_{Tcr}^{exp}) and corresponding twist (θ_{cr}^{exp}). For this, the following statistical parameters were computed for each ratio $M_{Tcr}^{exp}/M_{Tcr}^{thli}$ and $\theta_{cr}^{exp}/\theta_{cr}^{thli}$ ($i = 1$ to 8): the average value ($\overline{x}$) and the coefficient of variation ($cv(\%) = 100 \times s/\overline{x}$, with s being the sample standard deviation). Table 1 also presents separately the results for plain (P) and hollow (H) beams. This is because some studies showed that noticeable differences exist between the response of plain and hollow beams under torsion for the low loading stages, namely for the transition between the uncracked and the cracked stage [26].

Table 1 shows that, for the RC plain beams, the smeared constitutive laws *l*1, *l*2, *l*4, *l*5 and *l*6 allow us to predict the cracking torque M_{Tcr} (with $0.95 < \overline{x} < 1.05$) very well and with a very acceptable degree of dispersion ($cv < 13\%$). Among those models, the smeared constitutive law *l*4 from Belarbi and Hsu (1994) [24] is the best (with $\overline{x} = 1.00$ and $cv = 11.35\%$). For the RC hollow beams, this constitutive law gives the better average value $\overline{x} = 1.03$, although the degree of dispersion is high ($cv = 32.17\%$). The higher difficulty of reliably predicting the cracking torque for the RC hollow beams, when compared with the RC plain beams, was also observed and discussed in previous studies [18,26,27]. In particular, the RC hollow beams are more sensitive to the high variability of concrete tensile strength, which highly influences the cracking torque. When all beams are considered together, the smeared constitutive laws *l*2, *l*4, *l*6 and *l*7 give the best results with $\overline{x} \approx 1.00$, although the degree of dispersion is higher ($cv < 23\%$) due to the influence of the results for the RC hollow beams. In general, it can be stated that the smeared constitutive law *l*4 from Belarbi and Hsu (1994) [24] allows us to best predict the cracking torque, regardless of the cross-section type. This constitutive law has been widely used by authors in previous studies [9,16–19,23,26]. The results from Table 1 confirm the validity of such studies having chosen this smeared constitutive law for tensile concrete.

Regarding the twist corresponding to the cracking torque (θ_{cr}), Table 1 shows that, in general, there is a higher difficulty in obtaining a good prediction of this parameter. The constitutive laws *l*3 and *l*8 give the best average values for both the RC plain beams ($0.95 < \overline{x} < 1.05$) and also for all the RC beams together ($\overline{x} \leq 1.10$). However, the dispersion of these results is high ($cv > 25\%$). The results are the worst for the RC hollow beams, which was also reported in previous studies [17,18,25,27]. One possible explanation for this is that the experimental twists are very small until the end of the uncracked stage. Hence, experimental limitations related to the accurate measurement of the twists at this stage are expected. However, since the cracking twist is not very important for design, the previously reported worst results can also be considered not very important.

Figure 8 presents, for each smeared constitutive law (*l*1 to *l*8), scatter graphs showing the experimental versus the theoretical values for the cracking torque. Similar graphs are not presented for the cracking twist because of the high dispersion of the results previously reported. In the graphs, different markers were used to distinguish the results regarding the cross-section type, namely " ■" for RC plain beams and " □" for RC hollow beams.

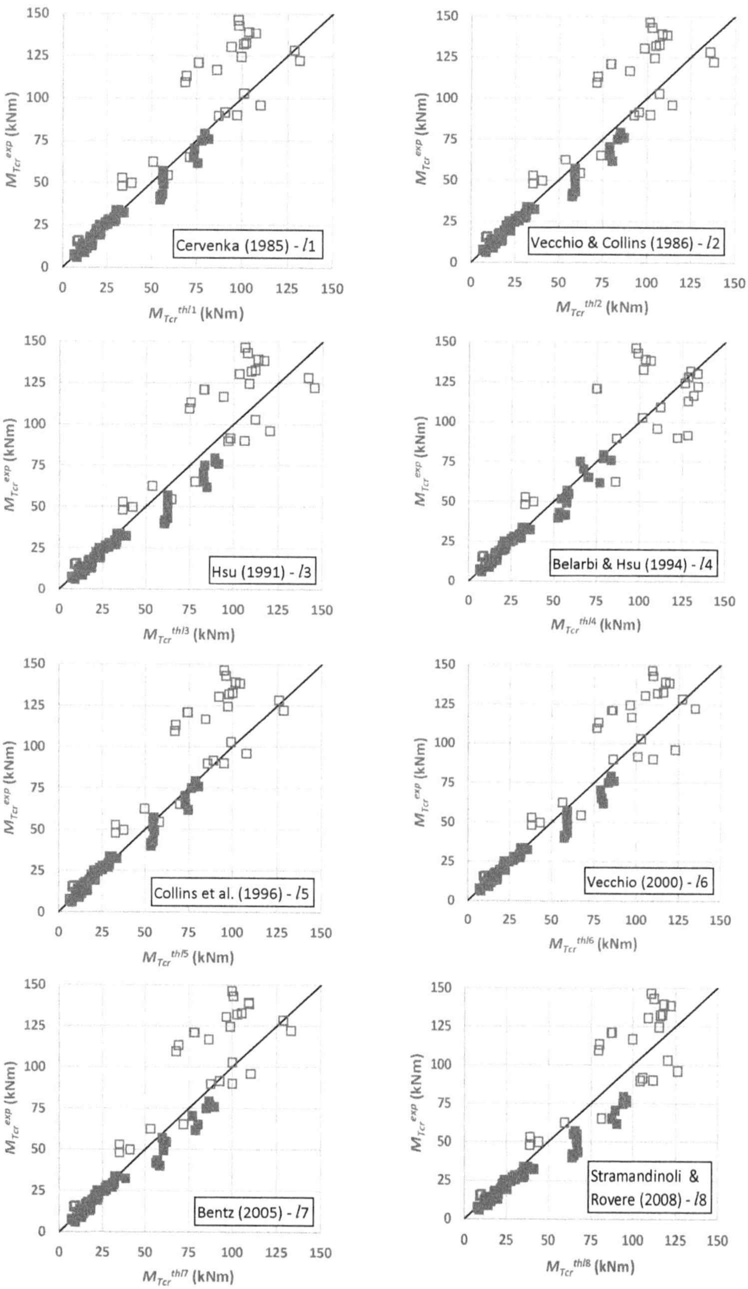

Figure 8. Experimental versos theoretical cracking torque.

Figure 8 visually confirms the observations previously stated from Table 1, namely the higher dispersion of the results for the RC hollow beams.

5. Conclusions

In this study, the GSVATM was used to check some proposed smeared constitutive laws for tensile concrete found in the literature in order to predict the response of the RC beams under torsion for the low loading stage; namely the transition from the uncracked stage to the cracked stage. As referred to in the introduction section, the smeared model GSVATM is simpler than the other, more complex models for the RC beams under torsion. In addition, it was also validated in several previous studies. Hence, the GSVATM was considered to be sufficiently simpler and reliable to evaluate the smeared constitutive laws for tensile concrete. From the obtained results, the following conclusions can be drawn:

(1) The different proposals for the smeared constitutive law for tensile concrete analyzed in this study lead to high differences in the shape of the postpeak descending branch of the corresponding smeared σ—ε curves;
(2) The obtained results confirm that the predicted response of the RC beams under torsion, for the transition from the uncracked stage to the cracked stage highly depends on the smeared constitutive law for tensile concrete incorporated into the model;
(3) The predictions for the cracking torque of the RC plain beams are better than the same ones for the RC hollow beams for which higher variability of the results is observed, as also reported in previous studies;
(4) Regardless of the used smeared constitutive law for tensile concrete, the cracking twist is not very well predicted. Namely, higher variability of the results is observed, as also reported in previous studies;
(5) Among the studied smeared constitutive laws for tensile concrete, the one proposed by Belarbi and Hsu in 1994 allows us to reliably predict the cracking torque of the RC beams under torsion, regardless of the cross-section type (plain or hollow). This result confirms the validity of several previous studies having incorporated this constitutive law in the used smeared truss models.

Finally, the authors consider that the results obtained in this study, using the smeared model GSVATM as reference model, can be extrapolated and could be useful to other smeared models for the RC beams under torsion. It must also be pointed out that additional solutions of experiments on the different failure mechanisms and related suitable approaches for the identification process for the parameters of relations of concrete are greatly needed and should be further studied, namely for the cracking of the RC beams under torsion.

Author Contributions: Conceptualization, L.B.; methodology, L.B.; validation, L.B.; formal analysis, M.T.; investigation, M.T.; data curation, M.T.; writing—original draft preparation, L.B.; writing—review and editing, L.B. and M.T.; supervision, L.B.; All authors have read and agreed to the published version of the manuscript.

Funding: No funding supported this work.

Institutional Review Board Statement: Not applicable.

Informed Consent Statement: Not applicable.

Data Availability Statement: Data is contained within the article.

Conflicts of Interest: The authors declare no conflict of interest.

Appendix A

Table A1. Properties of the reference beams.

Beam		x cm	y cm	t cm	x_1 cm	y_1 cm	A_{st}/s cm^2/m	A_{sl} cm^2	ρ_t %	ρ_l %	f_{ty} MPa	f_{ly} MPa	f'_c MPa	ε_o %
B3 [35]	P	25.4	38.1	-	21.6	34.3	10.16	11.36	1.17	1.17	320	328	28.1	0.20
B4 [35]	P	25.4	38.1	-	21.6	34.3	14.01	15.48	1.62	1.60	323	320	29.2	0.20
B5 [35]	P	25.4	38.1	-	21.6	34.3	18.47	20.39	2.13	2.11	321	332	30.6	0.20
B6 [35]	P	25.4	38.1	-	21.6	34.3	22.58	25.81	2.61	2.67	323	332	28.8	0.20
B7 [35]	P	25.4	38.1	-	21.6	34.3	10.16	5.16	1.17	0.53	319	320	26.0	0.19
B8 [35]	P	25.4	38.1	-	21.6	34.3	22.58	5.16	2.61	0.53	320	322	26.8	0.19
B9 [35]	P	25.4	38.1	-	21.6	34.3	4.66	11.36	0.54	1.17	343	319	28.8	0.20
C4 [35]	P	25.4	25.4	-	21.6	21.6	13.11	11.36	1.76	1.76	328	337	27.2	0.20
C5 [35]	P	25.4	25.4	-	21.6	21.6	17.67	15.48	2.37	2.40	329	328	27.2	0.20
C6 [35]	P	25.4	25.4	-	21.6	21.6	23.91	20.39	3.20	3.16	328	316	27.6	0.20
G3 [35]	P	25.4	50.8	-	21.6	47.0	8.29	11.36	0.88	0.88	328	339	26.8	0.19
G4 [35]	P	25.4	50.8	-	21.6	47.0	11.29	15.48	1.20	1.20	321	326	28.3	0.20
G5 [35]	P	25.4	50.8	-	21.6	47.0	15.05	20.39	1.60	1.58	328	331	26.9	0.19
G7 [35]	P	25.4	50.8	-	21.6	47.0	8.84	12.00	0.94	0.93	323	319	31.0	0.20
G8 [35]	P	25.4	50.8	-	21.6	47.0	12.32	17.03	1.31	1.32	329	322	28.3	0.20
I3 [35]	P	25.4	38.1	-	21.6	34.3	10.16	11.36	1.17	1.17	334	343	44.8	0.23
I4 [35]	P	25.4	38.1	-	21.6	34.3	14.01	15.48	1.62	1.60	326	315	45.0	0.23
I5 [35]	P	25.4	38.1	-	21.6	34.3	18.47	20.39	2.13	2.11	326	310	45.0	0.23
I6 [35]	P	25.4	38.1	-	21.6	34.3	22.58	25.81	2.61	2.67	329	326	45.8	0.23
J1 [35]	P	25.4	38.1	-	21.6	34.3	4.66	5.16	0.54	0.53	346	328	14.3	0.16
J2 [35]	P	25.4	38.1	-	21.6	34.3	7.21	8.00	0.83	0.83	341	320	14.6	0.16
J3 [35]	P	25.4	38.1	-	21.6	34.3	10.16	11.36	1.17	1.17	337	389	16.9	0.17
J4 [35]	P	25.4	38.1	-	21.6	34.3	14.01	15.48	1.62	1.60	332	324	16.8	0.17
K2 [35]	P	15.2	49.5	-	11.4	45.7	6.77	7.74	1.03	1.03	338	336	30.6	0.20
K3 [35]	P	15.2	49.5	-	11.4	45.7	10.42	12.00	1.58	1.59	321	316	29.0	0.20
K4 [35]	P	15.2	49.5	-	11.4	45.7	15.05	17.03	2.28	2.26	340	344	28.6	0.20
M1 [35]	P	25.4	38.1	-	21.6	34.3	4.76	8.00	0.55	0.83	353	326	29.9	0.20
M2 [35]	P	25.4	38.1	-	21.6	34.3	6.77	11.36	0.78	1.17	357	329	30.6	0.20
M3 [35]	P	25.4	38.1	-	21.6	34.3	9.24	15.48	1.07	1.60	326	322	26.8	0.29
M4 [35]	P	25.4	38.1	-	21.6	34.3	12.33	20.39	1.42	2.11	327	319	26.6	0.19
M5 [35]	P	25.4	38.1	-	21.6	34.3	15.63	25.81	1.81	2.67	331	335	28.0	0.20
M6 [35]	P	25.4	38.1	-	21.6	34.3	15.63	30.58	1.81	3.16	341	318	29.4	0.20
N1 [35]	P	15.2	30.5	-	13.0	28.3	3.50	2.84	0.62	0.61	341	352	29.5	0.20
N1a [35]	P	15.2	30.5	-	13.0	28.3	3.50	2.84	0.62	0.61	345	346	28.7	0.20
N2 [35]	P	15.2	30.5	-	13.0	28.3	6.35	5.16	1.13	1.11	338	331	30.4	0.20
N2a [35]	P	15.2	30.5	-	13.0	28.3	6.21	1.61	1.10	1.11	361	333	28.4	0.20
N3 [35]	P	15.2	30.5	-	13.0	28.3	5.08	4.26	0.90	0.92	352	352	27.3	0.20
N4 [35]	P	15.2	30.5	-	13.0	28.3	7.98	6.58	1.42	1.42	356	341	27.3	0.20
A2 [36]	P	25.4	25.4	-	22.2	22.2	7.82	5.16	1.08	0.80	285	380	38.2	0.22
A3 [36]	P	25.4	25.4	-	21.9	21.9	8.94	8.00	1.22	1.24	360	352	39.4	0.22
A4 [36]	P	25.4	25.4	-	21.9	21.9	12.42	11.36	1.69	1.76	360	351	39.2	0.22
B3 [36]	P	17.8	35.6	-	14.3	32.1	8.60	8.00	1.26	1.27	360	352	38.6	0.22
B4 [36]	P	17.8	35.6	-	14.3	32.1	11.76	11.36	1.73	1.80	360	351	38.5	0.22
B5UR1 [37]	P	20.3	30.5	-	16.5	26.7	6.56	5.16	0.92	0.83	373	386	39.6	0.20
B9UR1 [37]	P	20.3	30.5	-	16.5	26.7	6.56	5.16	0.92	0.83	373	386	75.0	0.27
B12UR1 [37]	P	20.3	30.5	-	16.5	26.7	6.56	5.16	0.92	0.83	399	386	80.6	0.27
B14UR1 [37]	P	20.3	30.5	-	16.5	26.7	6.56	5.16	0.92	0.83	386	386	93.9	0.29
B12UR2 [37]	P	20.3	30.5	-	16.5	26.7	6.95	5.16	0.97	0.83	386	386	76.2	0.27
B12UR3 [37]	P	20.3	30.5	-	16.5	26.7	7.46	6.58	1.04	1.06	386	380	72.9	0.26
B12UR4 [37]	P	20.3	30.5	-	16.5	26.7	7.88	7.74	1.10	1.25	386	373	75.9	0.27
B12UR5 [37]	P	20.3	30.5	-	16.5	26.7	10.13	8.00	1.41	1.29	386	380	76.7	0.27
H-06-12 [38]	P	35.0	50.0	-	30.0	45.0	7.10	20.65	0.61	1.18	440	410	78.5	0.27
H-07-10 [38]	P	35.0	50.0	-	30.0	45.0	7.89	17.03	0.68	0.97	420	500	68.4	0.26
H-07-16 [38]	P	35.0	50.0	-	30.0	45.0	7.89	28.39	0.68	1.62	420	500	68.4	0.26
H-12-12 [38]	P	35.0	50.0	-	30.0	45.0	14.19	20.65	1.22	1.18	440	410	78.5	0.27
H-12-16 [38]	P	35.0	50.0	-	30.0	45.0	14.19	28.39	1.22	1.62	440	520	78.5	0.27
H-14-10 [38]	P	35.0	50.0	-	30.0	45.0	16.13	17.03	1.38	0.97	360	500	68.4	0.26
H-20-20 [38]	P	35.0	50.0	-	30.0	45.0	23.46	34.06	2.01	1.95	440	560	78.5	0.27
N-06-06 [38]	P	35.0	50.0	-	30.0	45.0	7.10	12.00	0.61	0.69	440	440	35.5	0.21
N-06-12 [38]	P	35.0	50.0	-	30.0	45.0	7.10	20.65	0.61	1.18	440	410	35.5	0.21
N-07-10 [38]	P	35.0	50.0	-	30.0	45.0	7.89	17.03	0.68	0.97	420	500	33.5	0.21
N-07-16 [38]	P	35.0	50.0	-	30.0	45.0	7.89	28.39	0.68	1.62	420	500	33.5	0.21
N-12-12 [38]	P	35.0	50.0	-	30.0	45.0	14.19	20.65	1.22	1.18	440	410	35.5	0.21

Table A1. *Cont.*

Beam		x cm	y cm	t cm	x_1 cm	y_1 cm	A_{st}/s cm^2/m	A_{sl} cm^2	ρ_t %	ρ_l %	f_{ty} MPa	f_{ly} MPa	f'_c MPa	ε_o %
N-12-16 [38]	P	35.0	50.0	-	30.0	45.0	14.19	28.39	1.22	1.62	440	520	35.5	0.21
N-14-10 [38]	P	35.0	50.0	-	30.0	45.0	16.13	17.03	1.38	0.97	360	500	33.5	0.21
N-20-20 [38]	P	35.0	50.0	-	30.0	45.0	23.46	34.06	2.01	1.95	440	560	35.5	0.21
SW12-1 [39]	P	15.0	120.0	-	10.0	115.0	3.93	11.31	0.55	1.26	459	480	44.2	0.23
SW10-1 [39]	P	15.0	100.0	-	10.0	95.0	3.93	9.05	0.55	1.21	459	499	29.5	0.20
SW10-2 [39]	P	15.0	100.0	-	9.8	94.8	7.54	9.05	1.05	1.21	480	480	44.2	0.23
SW10-3 [39]	P	15.0	100.0	-	9.8	94.8	11.31	9.05	1.58	1.21	499	499	29.5	0.20
SW10-4 [39]	P	15.0	100.0	-	9.4	94.4	16.08	16.08	2.23	2.14	497	497	33.8	0.21
SW8-1 [39]	P	15.0	80.0	-	10.2	75.2	4.02	7.07	0.57	1.18	433	459	29.5	0.20
SW8-2 [39]	P	15.0	80.0	-	9.8	74.8	11.31	7.07	1.59	1.18	499	459	29.5	0.20
D3 [35]	H	25.4	38.1	6.4	21.6	34.3	10.16	11.36	1.17	1.17	333	341	28.4	0.20
D4 [35]	H	25.4	38.1	6.4	21.6	34.3	14.01	15.48	1.62	1.60	333	330	30.6	0.20
T0 [40]	H	50.0	50.0	8.0	43.0	43.0	10.28	32.16	0.71	1.29	357	345	45.1	0.23
T1 [40]	H	50.0	50.0	8.0	45.4	45.4	10.28	18.10	0.75	0.72	357	357	35.3	0.21
T2 [40]	H	50.0	50.0	8.0	43.0	43.0	10.28	18.10	0.71	0.72	357	357	35.3	0.21
T5 [40]	H	80.0	40.0	8.0	73.0	33.0	10.28	10.00	0.68	0.31	513	529	47.1	0.23
VH1 [41]	H	32.4	32.4	8.0	30.4	30.4	2.88	3.46	0.33	0.33	447	447	17.2	0.17
VH2 [41]	H	32.4	32.4	8.0	30.4	30.4	5.76	6.91	0.67	0.66	447	447	17.2	0.17
A2 [42]	H	60.0	60.0	10.7	53.8	53.1	6.28	13.95	0.37	0.39	696	672	47.3	0.23
A3 [42]	H	60.0	60.0	10.9	53.5	53.5	8.27	18.10	0.49	0.50	715	672	46.2	0.23
A4 [42]	H	60.0	60.0	10.4	52.0	52.5	11.22	23.75	0.65	0.66	715	724	54.8	0.24
A5 [42]	H	60.0	60.0	10.4	52.8	52.8	14.14	30.66	0.83	0.85	672	724	53.1	0.24
B2 [42]	H	60.0	60.0	10.8	53.3	53.4	6.70	14.58	0.40	0.41	696	672	69.8	0.26
B3 [42]	H	60.0	60.0	10.9	53.5	53.7	11.22	23.75	0.67	0.66	715	724	77.8	0.27
B4 [42]	H	60.0	60.0	11.2	52.3	53.6	15.08	32.17	0.89	0.89	672	724	79.8	0.27
B5 [42]	H	60.0	60.0	11.7	51.8	51.8	18.85	40.21	1.09	1.12	672	724	76.4	0.27
C2 [42]	H	60.0	60.0	10.0	53.2	53.3	6.28	13.95	0.37	0.39	696	672	94.8	0.28
C3 [42]	H	60.0	60.0	10.3	54.5	54.0	10.47	23.75	0.63	0.66	715	724	91.6	0.28
C4 [42]	H	60.0	60.0	10.3	54.6	54.5	14.14	30.66	0.86	0.85	672	724	91.4	0.28
C5 [42]	H	60.0	60.0	10.4	54.0	54.3	17.40	36.69	1.05	1.02	672	724	96.7	0.28
C6 [42]	H	60.0	60.0	10.4	53.3	52.9	22.62	48.25	1.34	1.34	672	724	87.5	0.28
A095c [26]	H	49.7	71.1	14.5	43.7	65.1	9.93	13.16	0.61	0.37	381	371	35.1	0.21
A120a [26]	H	50.2	71.9	18.4	44.2	65.9	7.59	20.00	0.46	0.55	380	464	27.6	0.20
B065b [26]	H	50.3	71.0	9.2	44.3	65.0	9.93	50.97	0.61	1.43	380	452	39.2	0.22
B080a [26]	H	50.0	72.1	11.2	44.0	66.1	12.90	28.39	0.79	0.79	392	454	46.5	0.23
B110a [26]	H	49.8	71.0	15.5	43.8	65.0	8.60	20.00	0.53	0.57	369	453	48.1	0.23
C065a [26]	H	49.5	78.1	8.5	43.5	72.1	9.93	20.00	0.59	0.52	376	338	78.8	0.27
C100a [26]	H	49.9	72.3	12.7	43.9	66.3	12.90	28.39	0.79	0.79	447	466	90.6	0.28
D075a [26]	H	49.8	73.4	8.7	43.8	67.4	12.90	28.39	0.79	0.78	381	469	94.9	0.29
D090a [26]	H	50.1	72.2	10.5	44.1	66.2	12.90	28.39	0.79	0.79	447	466	105.7	0.30

Table A2. Cracking torques and corresponding twists (smeared constitutive laws *l*1 to *l*3).

Beam	M_{Tcr}^{exp} kNm	θ_{cr}^{exp} °/m	M_{Tcr}^{thl1} kNm	$\frac{M_{Tcr}^{exp}}{M_{Tcr}^{thl1}}$	θ_{cr}^{thl1} °/m	$\frac{\theta_{cr}^{exp}}{\theta_{cr}^{thl1}}$	M_{Tcr}^{thl2} kNm	$\frac{M_{Tcr}^{exp}}{M_{Tcr}^{thl2}}$	θ_{cr}^{thl2} °/m	$\frac{\theta_{cr}^{exp}}{\theta_{cr}^{thl2}}$	M_{Tcr}^{thl3} kNm	$\frac{M_{Tcr}^{exp}}{M_{Tcr}^{thl3}}$	θ_{cr}^{thl3} °/m	$\frac{\theta_{cr}^{exp}}{\theta_{cr}^{thl3}}$
B3 [35]	20.1	0.12	20.9	0.96	0.10	1.21	22.0	0.91	0.11	1.15	23.2	0.87	0.11	1.09
B4 [35]	21.9	0.12	21.0	1.05	0.10	1.20	22.0	0.99	0.10	1.15	23.1	0.95	0.11	1.09
B5 [35]	22.6	0.14	21.4	1.05	0.10	1.42	22.2	1.02	0.10	1.36	23.4	0.97	0.11	1.30
B6 [35]	25.0	0.16	20.6	1.21	0.09	1.75	21.7	1.15	0.10	1.67	22.8	1.09	0.10	1.58
B7 [35]	20.2	0.11	20.0	1.01	0.10	1.07	21.0	0.96	0.11	1.02	22.1	0.91	0.11	0.97
B8 [35]	21.8	0.13	20.3	1.07	0.10	1.28	21.3	1.02	0.10	1.22	22.3	0.98	0.11	1.17
B9 [35]	19.6	0.11	20.8	0.94	0.10	1.04	22.0	0.89	0.11	0.99	23.2	0.85	0.11	0.94
C4 [35]	11.9	0.13	11.3	1.05	0.11	1.18	11.8	1.01	0.12	1.12	12.4	0.96	0.13	1.07
C5 [35]	14.0	0.17	11.2	1.25	0.11	1.51	11.9	1.17	0.12	1.41	12.5	1.12	0.12	1.35
C6 [35]	13.9	0.17	11.3	1.23	0.11	1.61	11.5	1.20	0.11	1.57	12.0	1.15	0.11	1.51
G3 [35]	27.1	0.10	29.5	0.92	0.09	1.05	31.0	0.87	0.10	1.00	32.7	0.83	0.10	0.95
G4 [35]	28.7	0.12	30.1	0.95	0.09	1.29	31.6	0.91	0.10	1.23	33.4	0.86	0.10	1.16
G5 [35]	29.5	0.11	29.2	1.01	0.09	1.30	30.7	0.96	0.09	1.24	32.3	0.91	0.10	1.17
G7 [35]	33.6	0.13	31.7	1.06	0.09	1.45	33.3	1.01	0.10	1.38	35.1	0.96	0.10	1.31
G8 [35]	33.6	0.12	30.1	1.12	0.09	1.37	31.6	1.06	0.09	1.30	33.4	1.01	0.10	1.23
I3 [35]	25.5	0.11	25.5	1.00	0.11	0.97	27.2	0.94	0.12	0.91	28.7	0.89	0.12	0.86
I4 [35]	28.0	0.12	25.7	1.09	0.11	1.15	27.3	1.03	0.11	1.08	28.8	0.97	0.12	1.02
I5 [35]	28.1	0.15	26.0	1.08	0.11	1.43	27.4	1.02	0.11	1.36	28.9	0.97	0.12	1.29
I6 [35]	27.5	0.13	26.2	1.05	0.10	1.22	27.7	0.99	0.11	1.15	29.2	0.94	0.12	1.09

Table A2. *Cont.*

Beam	M_{Tcr}^{exp} kNm	θ_{cr}^{exp} °/m	M_{Tcr}^{thl1} kNm	$\frac{M_{Tcr}^{exp}}{M_{Tcr}^{thl1}}$	θ_{cr}^{thl1} °/m	$\frac{\theta_{cr}^{exp}}{\theta_{cr}^{thl1}}$	M_{Tcr}^{thl2} kNm	$\frac{M_{Tcr}^{exp}}{M_{Tcr}^{thl2}}$	θ_{cr}^{thl2} °/m	$\frac{\theta_{cr}^{exp}}{\theta_{cr}^{thl2}}$	M_{Tcr}^{thl3} kNm	$\frac{M_{Tcr}^{exp}}{M_{Tcr}^{thl3}}$	θ_{cr}^{thl3} °/m	$\frac{\theta_{cr}^{exp}}{\theta_{cr}^{thl3}}$
J1 [35]	14.0	0.09	15.3	0.92	0.09	0.92	15.9	0.88	0.10	0.88	16.5	0.85	0.10	0.85
J2 [35]	17.1	0.12	15.1	1.13	0.09	1.38	15.7	1.09	0.09	1.32	16.5	1.03	0.10	1.26
J3 [35]	16.9	0.10	15.9	1.06	0.09	1.08	16.6	1.02	0.09	1.03	17.5	0.97	0.10	0.98
J4 [35]	18.0	0.11	15.6	1.15	0.09	1.28	16.3	1.10	0.09	1.23	17.1	1.05	0.09	1.17
K2 [35]	12.2	0.18	12.0	1.02	0.14	1.31	12.3	0.99	0.14	1.27	13.1	0.93	0.15	1.20
K3 [35]	12.4	0.19	11.5	1.08	0.13	1.49	11.9	1.05	0.13	1.45	12.6	0.98	0.14	1.37
K4 [35]	13.1	0.21	11.1	1.19	0.12	1.71	11.6	1.13	0.13	1.63	12.4	1.06	0.13	1.53
M1 [35]	19.2	0.11	21.2	0.90	0.10	1.03	22.5	0.85	0.11	0.97	23.6	0.81	0.12	0.92
M2 [35]	20.6	0.11	21.5	0.96	0.10	1.08	22.5	0.92	0.11	1.03	23.8	0.86	0.11	0.97
M3 [35]	20.7	0.12	20.0	1.03	0.10	1.24	21.0	0.98	0.10	1.18	22.2	0.93	0.11	1.12
M4 [35]	20.7	0.13	19.9	1.04	0.10	1.41	20.8	0.99	0.10	1.34	22.0	0.94	0.11	1.27
M5 [35]	21.7	0.12	20.2	1.07	0.09	1.30	21.3	1.02	0.10	1.23	22.4	0.97	0.10	1.17
M6 [35]	22.7	0.15	20.7	1.10	0.09	1.57	21.7	1.05	0.10	1.49	22.8	0.99	0.10	1.42
N1 [35]	7.6	0.13	6.7	1.14	0.16	0.81	7.0	1.08	0.17	0.77	7.4	1.02	0.18	0.73
N1a [35]	7.0	0.11	6.6	1.06	0.16	0.69	6.9	1.01	0.17	0.66	7.3	0.96	0.17	0.62
N2 [35]	7.4	0.22	6.8	1.10	0.15	1.40	7.1	1.05	0.16	1.34	7.5	0.99	0.17	1.27
N2a [35]	7.5	0.21	6.6	1.14	0.15	1.37	6.9	1.09	0.16	1.31	7.3	1.03	0.17	1.24
N3 [35]	7.4	0.21	6.4	1.15	0.15	1.39	6.7	1.10	0.16	1.33	7.1	1.04	0.17	1.25
N4 [35]	7.6	0.21	6.4	1.19	0.15	1.43	6.7	1.13	0.16	1.36	7.1	1.07	0.16	1.29
A2 [36]	11.3	0.12	13.4	0.84	0.13	0.94	14.0	0.81	0.13	0.89	14.8	0.76	0.14	0.84
A3 [36]	12.2	0.12	13.5	0.90	0.13	0.98	14.3	0.85	0.13	0.92	15.1	0.81	0.14	0.88
A4 [36]	12.5	0.15	13.5	0.93	0.12	1.20	14.3	0.88	0.13	1.14	15.1	0.83	0.14	1.08
B3 [36]	8.8	0.15	12.0	0.73	0.14	1.06	12.6	0.70	0.14	1.01	13.3	0.66	0.15	0.96
B4 [36]	10.2	0.15	12.0	0.85	0.13	1.09	12.6	0.81	0.14	1.04	13.3	0.77	0.15	0.98
B5UR1 [37]	11.6	0.09	12.4	0.94	0.14	0.63	13.1	0.89	0.14	0.60	13.8	0.84	0.15	0.57
B9UR1 [37]	13.0	0.13	16.2	0.80	0.15	0.91	17.3	0.75	0.16	0.86	18.2	0.71	0.17	0.81
B12UR1 [37]	16.2	0.09	16.7	0.97	0.15	0.61	17.8	0.91	0.16	0.57	18.8	0.86	0.17	0.55
B14UR1 [37]	19.3	0.12	17.7	1.09	0.15	0.78	18.9	1.02	0.16	0.73	19.9	0.97	0.17	0.69
B12UR2 [37]	17.8	0.11	16.4	1.09	0.15	0.75	17.4	1.02	0.16	0.71	18.4	0.97	0.17	0.67
B12UR3 [37]	16.0	0.10	16.1	1.00	0.15	0.70	17.3	0.93	0.16	0.65	18.0	0.89	0.16	0.62
B12UR4 [37]	16.9	0.14	16.4	1.03	0.15	0.96	17.5	0.96	0.16	0.89	18.4	0.92	0.16	0.85
B12UR5 [37]	13.6	0.04	16.7	0.81	0.15	0.24	17.8	0.76	0.16	0.23	18.6	0.73	0.16	0.22
H-06-12 [38]	75.0	0.09	78.0	0.96	0.09	1.00	83.5	0.90	0.09	0.93	83.5	0.90	0.09	0.93
H-07-10 [38]	70.5	0.09	73.9	0.95	0.09	1.08	79.0	0.89	0.09	1.01	82.9	0.85	0.10	0.96
H-07-16 [38]	65.3	0.09	73.5	0.89	0.08	1.03	79.2	0.82	0.09	0.95	83.1	0.79	0.09	0.91
H-12-12 [38]	77.1	0.07	79.7	0.97	0.09	0.85	85.0	0.91	0.09	0.80	89.5	0.86	0.10	0.76
H-12-16 [38]	79.3	0.09	79.7	1.00	0.09	1.06	85.3	0.93	0.09	0.99	89.4	0.89	0.10	0.95
H-14-10 [38]	61.8	0.09	75.7	0.82	0.09	1.00	80.6	0.77	0.09	0.94	84.9	0.73	0.10	0.89
H-20-20 [38]	76.0	0.09	81.8	0.93	0.09	1.05	87.1	0.87	0.09	0.99	91.6	0.83	0.10	0.94
N-06-06 [38]	43.2	0.08	56.1	0.77	0.08	1.02	59.3	0.73	0.08	0.96	62.5	0.69	0.09	0.91
N-06-12 [38]	51.8	0.11	56.1	0.92	0.08	1.47	59.1	0.88	0.08	1.39	62.4	0.83	0.09	1.32
N-07-10 [38]	41.6	0.11	54.7	0.76	0.08	1.40	57.6	0.72	0.08	1.33	60.7	0.68	0.09	1.26
N-07-16 [38]	40.0	0.11	54.7	0.73	0.08	1.43	57.5	0.70	0.08	1.36	60.6	0.66	0.08	1.29
N-12-12 [38]	49.3	0.09	56.7	0.87	0.08	1.18	59.2	0.83	0.08	1.13	62.5	0.79	0.09	1.07
N-12-16 [38]	57.1	0.12	56.3	1.02	0.08	1.58	59.2	0.96	0.08	1.50	62.5	0.91	0.08	1.42
N-14-10 [38]	41.8	0.12	55.2	0.76	0.08	1.56	57.9	0.72	0.08	1.49	61.1	0.68	0.09	1.41
N-20-20 [38]	55.0	0.13	56.6	0.97	0.08	1.68	58.9	0.93	0.08	1.61	62.4	0.88	0.08	1.52
SW12-1 [39]	32.3	0.15	34.6	0.93	0.13	1.16	36.4	0.89	0.14	1.11	38.8	0.83	0.15	1.03
SW10-1 [39]	24.6	0.13	23.2	1.06	0.13	1.00	24.7	1.00	0.14	0.93	25.8	0.95	0.14	0.90
SW10-2 [39]	29.6	0.20	29.1	1.02	0.14	1.47	31.1	0.95	0.15	1.37	32.0	0.92	0.15	1.36
SW10-3 [39]	26.6	0.15	23.9	1.11	0.13	1.16	25.3	1.05	0.14	1.10	26.7	1.00	0.15	1.04
SW10-4 [39]	27.7	0.16	25.7	1.08	0.13	1.24	27.1	1.02	0.14	1.17	28.8	0.96	0.15	1.10
SW8-1 [39]	19.7	0.16	18.6	1.06	0.14	1.13	20.0	0.98	0.15	1.03	21.0	0.94	0.16	0.99
SW8-2 [39]	22.5	0.14	18.9	1.19	0.14	1.06	20.1	1.12	0.15	0.98	21.2	1.06	0.15	0.94
D3 [35]	15.2	0.08	8.0	1.89	0.05	1.49	8.5	1.79	0.06	1.40	8.8	1.73	0.06	1.36
D4 [35]	15.8	0.12	8.9	1.78	0.06	2.05	9.3	1.70	0.06	1.96	9.7	1.63	0.06	1.88
T0 [40]	49.8	0.06	38.8	1.28	0.03	1.92	40.8	1.22	0.03	1.81	42.5	1.17	0.04	1.75
T1 [40]	48.0	0.04	33.7	1.43	0.03	1.35	35.3	1.36	0.03	1.27	36.8	1.31	0.03	1.23
T2 [40]	52.8	0.10	33.7	1.57	0.03	3.07	35.3	1.49	0.03	2.90	36.8	1.44	0.03	2.81
T5 [40]	62.5	0.06	50.8	1.23	0.03	2.07	53.7	1.16	0.03	1.89	53.6	1.17	0.03	2.10
VH1 [41]	12.0	0.12	9.8	1.22	0.07	1.65	10.4	1.15	0.08	1.55	10.9	1.10	0.08	1.48
VH2 [41]	11.5	0.07	10.4	1.10	0.08	0.90	11.1	1.03	0.08	0.85	11.7	0.99	0.09	0.81
A2 [42]	109.5	0.06	68.6	1.60	0.03	2.44	71.6	1.53	0.03	2.34	74.8	1.46	0.03	2.25
A3 [42]	113.3	0.06	69.2	1.64	0.03	2.18	72.3	1.57	0.03	2.09	75.5	1.50	0.03	2.02
A4 [42]	120.9	0.06	75.9	1.59	0.03	2.42	79.8	1.51	0.03	2.25	83.4	1.45	0.03	2.17
A5 [42]	120.9	0.04	76.1	1.59	0.03	1.66	79.4	1.52	0.03	1.59	82.9	1.46	0.03	1.53

Table A2. *Cont.*

Beam	M_{Tcr}^{exp} kNm	θ_{cr}^{exp} °/m	M_{Tcr}^{thl1} kNm	$\frac{M_{Tcr}^{exp}}{M_{Tcr}^{thl1}}$	θ_{cr}^{thl1} °/m	$\frac{\theta_{cr}^{exp}}{\theta_{cr}^{thl1}}$	M_{Tcr}^{thl2} kNm	$\frac{M_{Tcr}^{exp}}{M_{Tcr}^{thl2}}$	θ_{cr}^{thl2} °/m	$\frac{\theta_{cr}^{exp}}{\theta_{cr}^{thl2}}$	M_{Tcr}^{thl3} kNm	$\frac{M_{Tcr}^{exp}}{M_{Tcr}^{thl3}}$	θ_{cr}^{thl3} °/m	$\frac{\theta_{cr}^{exp}}{\theta_{cr}^{thl3}}$
B2 [42]	116.7	0.04	86.3	1.35	0.03	1.69	90.2	1.29	0.03	1.60	94.4	1.24	0.03	1.53
B3 [42]	130.5	0.05	94.2	1.39	0.03	1.74	98.4	1.33	0.03	1.65	103.1	1.26	0.03	1.57
B4 [42]	142.9	0.07	98.3	1.45	0.03	2.73	102.7	1.39	0.03	2.58	107.7	1.33	0.03	2.47
B5 [42]	146.3	0.06	98.0	1.49	0.03	2.44	101.4	1.44	0.03	2.38	106.3	1.38	0.03	2.28
C2 [42]	124.5	0.05	99.6	1.25	0.03	1.81	104.1	1.20	0.03	1.70	108.6	1.15	0.03	1.67
C3 [42]	131.9	0.06	100.8	1.31	0.03	2.35	104.7	1.26	0.03	2.29	109.9	1.20	0.03	2.18
C4 [42]	132.6	0.05	102.1	1.30	0.03	1.92	106.8	1.24	0.03	1.82	112.0	1.18	0.03	1.73
C5 [42]	138.3	0.05	107.4	1.29	0.03	1.91	111.3	1.24	0.03	1.85	116.9	1.18	0.03	1.76
C6 [42]	139.1	0.05	103.7	1.34	0.03	2.02	108.3	1.28	0.03	1.91	113.5	1.23	0.03	1.83
A095c [26]	102.9	0.03	101.0	1.02	0.04	0.82	106.9	0.96	0.04	0.77	112.0	0.92	0.05	0.74
A120a [26]	89.8	0.05	87.4	1.03	0.04	1.14	92.8	0.97	0.04	1.07	97.1	0.92	0.05	1.03
B065b [26]	54.4	0.03	59.1	0.92	0.03	1.23	62.1	0.88	0.03	1.16	64.7	0.84	0.03	1.12
B080a [26]	65.2	0.03	71.2	0.92	0.03	1.24	74.2	0.88	0.03	1.19	78.0	0.84	0.03	1.12
B110a [26]	128.3	0.04	128.6	1.00	0.04	0.99	135.8	0.94	0.04	0.93	141.8	0.90	0.05	0.90
C065a [26]	91.7	0.03	90.9	1.01	0.03	1.06	95.5	0.96	0.03	0.98	98.4	0.93	0.03	1.03
C100a [26]	122.2	0.03	131.6	0.93	0.03	0.85	137.8	0.89	0.04	0.81	145.3	0.84	0.04	0.76
D075a [26]	90.1	0.03	97.3	0.93	0.03	0.99	101.8	0.88	0.03	0.94	106.0	0.85	0.03	0.92
D090a [26]	96.1	0.03	110.0	0.87	0.03	1.08	114.3	0.84	0.03	1.05	120.2	0.80	0.03	0.99

Table A3. Cracking torques and corresponding twists (smeared constitutive laws *l*4 to *l*6).

Beam	M_{Tcr}^{exp} kNm	θ_{cr}^{exp} °/m	M_{Tcr}^{thl4} kNm	$\frac{M_{Tcr}^{exp}}{M_{Tcr}^{thl4}}$	θ_{cr}^{thl4} °/m	$\frac{\theta_{cr}^{exp}}{\theta_{cr}^{thl4}}$	M_{Tcr}^{thl5} kNm	$\frac{M_{Tcr}^{exp}}{M_{Tcr}^{thl5}}$	θ_{cr}^{thl5} °/m	$\frac{\theta_{cr}^{exp}}{\theta_{cr}^{thl5}}$	M_{Tcr}^{thl6} kNm	$\frac{M_{Tcr}^{exp}}{M_{Tcr}^{thl6}}$	θ_{cr}^{thl6} °/m	$\frac{\theta_{cr}^{exp}}{\theta_{cr}^{thl6}}$
B3 [35]	20.1	0.12	21.3	0.94	0.10	1.19	20.4	0.99	0.10	1.24	22.0	0.91	0.11	1.15
B4 [35]	21.9	0.12	21.5	1.02	0.10	1.17	20.4	1.07	0.10	1.24	22.0	0.99	0.10	1.15
B5 [35]	22.6	0.14	22.0	1.02	0.10	1.37	20.9	1.08	0.09	1.45	22.2	1.02	0.10	1.36
B6 [35]	25.0	0.16	20.9	1.19	0.09	1.73	20.1	1.24	0.09	1.80	21.7	1.15	0.10	1.67
B7 [35]	20.2	0.11	20.7	0.98	0.10	1.04	19.5	1.04	0.10	1.10	21.0	0.96	0.11	1.02
B8 [35]	21.8	0.13	20.8	1.05	0.10	1.25	19.7	1.10	0.10	1.32	21.3	1.02	0.10	1.22
B9 [35]	19.6	0.11	20.0	0.98	0.10	1.09	20.3	0.97	0.10	1.07	22.0	0.89	0.11	0.99
C4 [35]	11.9	0.13	11.6	1.02	0.12	1.15	11.0	1.08	0.11	1.21	11.8	1.01	0.12	1.12
C5 [35]	14.0	0.17	11.6	1.21	0.11	1.46	11.0	1.27	0.11	1.53	11.9	1.17	0.12	1.41
C6 [35]	13.9	0.17	11.4	1.21	0.11	1.58	10.9	1.27	0.10	1.65	12.0	1.16	0.11	1.51
G3 [35]	27.1	0.10	30.5	0.89	0.09	1.01	28.7	0.94	0.09	1.08	31.0	0.87	0.10	1.00
G4 [35]	28.7	0.12	31.0	0.93	0.09	1.25	29.4	0.98	0.09	1.32	31.6	0.91	0.10	1.23
G5 [35]	29.5	0.11	29.8	0.99	0.09	1.27	28.5	1.03	0.09	1.33	30.7	0.96	0.09	1.24
G7 [35]	33.6	0.13	32.8	1.02	0.10	1.40	30.9	1.09	0.09	1.49	33.3	1.01	0.10	1.38
G8 [35]	33.6	0.12	31.2	1.08	0.09	1.32	29.3	1.15	0.09	1.40	31.5	1.07	0.09	1.30
I3 [35]	25.5	0.11	26.3	0.97	0.11	0.94	25.1	1.02	0.11	0.99	27.2	0.94	0.12	0.91
I4 [35]	28.0	0.12	26.5	1.06	0.11	1.11	25.2	1.11	0.11	1.17	27.3	1.03	0.11	1.08
I5 [35]	28.1	0.15	26.8	1.05	0.11	1.39	25.4	1.11	0.10	1.46	27.4	1.02	0.11	1.36
I6 [35]	27.5	0.13	26.9	1.02	0.11	1.19	25.7	1.07	0.10	1.24	27.7	0.99	0.11	1.15
J1 [35]	14.0	0.09	15.2	0.92	0.09	0.92	14.8	0.95	0.09	0.95	15.9	0.88	0.10	0.88
J2 [35]	17.1	0.12	15.4	1.10	0.09	1.35	14.3	1.19	0.09	1.45	15.7	1.09	0.09	1.32
J3 [35]	16.9	0.10	16.5	1.03	0.09	1.04	15.4	1.10	0.09	1.11	16.6	1.02	0.09	1.03
J4 [35]	18.0	0.11	15.8	1.13	0.09	1.26	15.1	1.19	0.08	1.33	16.3	1.10	0.09	1.23
K2 [35]	12.2	0.18	12.7	0.96	0.14	1.24	11.8	1.04	0.13	1.33	12.7	0.96	0.14	1.24
K3 [35]	12.4	0.19	12.3	1.01	0.14	1.40	11.0	1.13	0.12	1.56	12.2	1.02	0.14	1.42
K4 [35]	13.1	0.21	12.1	1.08	0.13	1.56	10.8	1.21	0.12	1.75	12.0	1.09	0.13	1.58
M1 [35]	19.2	0.11	19.7	0.97	0.10	1.11	20.7	0.93	0.10	1.05	22.5	0.85	0.11	0.97
M2 [35]	20.6	0.11	21.5	0.96	0.10	1.08	20.9	0.98	0.10	1.10	22.5	0.92	0.11	1.03
M3 [35]	20.7	0.12	20.7	1.00	0.10	1.19	19.5	1.06	0.10	1.27	21.0	0.98	0.10	1.18
M4 [35]	20.7	0.13	20.1	1.03	0.10	1.39	19.3	1.07	0.09	1.45	20.8	0.99	0.10	1.34
M5 [35]	21.7	0.12	20.6	1.05	0.10	1.27	19.6	1.11	0.09	1.34	21.3	1.02	0.10	1.23
M6 [35]	22.7	0.15	21.0	1.08	0.09	1.54	20.0	1.13	0.09	1.62	21.7	1.05	0.10	1.49
N1 [35]	7.6	0.13	6.9	1.10	0.16	0.79	6.5	1.17	0.15	0.83	7.0	1.08	0.17	0.77
N1a [35]	7.0	0.11	6.4	1.11	0.15	0.72	6.4	1.10	0.15	0.71	6.9	1.01	0.17	0.66
N2 [35]	7.4	0.22	7.0	1.06	0.16	1.36	6.5	1.14	0.15	1.45	7.1	1.05	0.16	1.34
N2a [35]	7.5	0.21	6.7	1.12	0.16	1.35	6.4	1.17	0.15	1.41	6.9	1.09	0.16	1.31
N3 [35]	7.4	0.21	6.7	1.11	0.16	1.34	6.2	1.19	0.15	1.43	6.7	1.10	0.16	1.33
N4 [35]	7.6	0.21	6.6	1.16	0.15	1.39	6.2	1.22	0.14	1.46	6.7	1.13	0.16	1.36

Table A3. *Cont.*

Beam	M_{Tcr}^{exp} kNm	θ_{cr}^{exp} °/m	M_{Tcr}^{thl4} kNm	$\frac{M_{Tcr}^{exp}}{M_{Tcr}^{thl4}}$	θ_{cr}^{thl4} °/m	$\frac{\theta_{cr}^{exp}}{\theta_{cr}^{thl4}}$	M_{Tcr}^{thl5} kNm	$\frac{M_{Tcr}^{exp}}{M_{Tcr}^{thl5}}$	θ_{cr}^{thl5} °/m	$\frac{\theta_{cr}^{exp}}{\theta_{cr}^{thl5}}$	M_{Tcr}^{thl6} kNm	$\frac{M_{Tcr}^{exp}}{M_{Tcr}^{thl6}}$	θ_{cr}^{thl6} °/m	$\frac{\theta_{cr}^{exp}}{\theta_{cr}^{thl6}}$
A2 [36]	11.3	0.12	13.6	0.83	0.13	0.92	13.1	0.86	0.13	0.96	14.0	0.81	0.13	0.89
A3 [36]	12.2	0.12	13.9	0.87	0.13	0.95	13.2	0.92	0.12	1.00	14.3	0.85	0.13	0.92
A4 [36]	12.5	0.15	14.0	0.90	0.13	1.16	13.2	0.95	0.12	1.22	14.3	0.88	0.13	1.14
B3 [36]	8.8	0.15	12.2	0.72	0.14	1.05	11.6	0.76	0.13	1.10	12.6	0.70	0.14	1.01
B4 [36]	10.2	0.15	12.3	0.83	0.14	1.07	11.7	0.87	0.13	1.12	12.6	0.81	0.14	1.04
B5UR1 [37]	11.6	0.09	12.4	0.93	0.14	0.63	12.0	0.97	0.13	0.65	13.1	0.89	0.14	0.60
B9UR1 [37]	13.0	0.13	16.4	0.79	0.15	0.90	16.0	0.81	0.14	0.92	17.3	0.75	0.16	0.86
B12UR1 [37]	16.2	0.09	16.9	0.96	0.15	0.61	16.5	0.98	0.15	0.62	17.8	0.91	0.16	0.57
B14UR1 [37]	19.3	0.12	17.9	1.08	0.15	0.77	17.4	1.11	0.15	0.79	18.9	1.02	0.16	0.73
B12UR2 [37]	17.8	0.11	16.6	1.07	0.15	0.74	16.1	1.10	0.15	0.76	17.4	1.02	0.16	0.71
B12UR3 [37]	16.0	0.10	15.9	1.01	0.14	0.71	15.8	1.01	0.14	0.71	17.3	0.93	0.16	0.65
B12UR4 [37]	16.9	0.14	16.1	1.05	0.14	0.97	16.1	1.05	0.14	0.97	17.5	0.96	0.16	0.89
B12UR5 [37]	13.6	0.04	16.8	0.81	0.15	0.24	16.4	0.83	0.14	0.25	17.8	0.76	0.16	0.23
H-06-12 [38]	75.0	0.09	65.2	1.15	0.07	1.19	76.5	0.98	0.08	1.02	83.5	0.90	0.09	0.93
H-07-10 [38]	70.5	0.09	67.2	1.05	0.08	1.18	72.8	0.97	0.08	1.09	79.0	0.89	0.09	1.01
H-07-16 [38]	65.3	0.09	69.8	0.93	0.08	1.08	73.0	0.89	0.08	1.03	79.2	0.82	0.09	0.95
H-12-12 [38]	77.1	0.07	78.9	0.98	0.09	0.86	78.6	0.98	0.09	0.87	85.0	0.91	0.09	0.80
H-12-16 [38]	79.3	0.09	79.2	1.00	0.09	1.07	78.6	1.01	0.09	1.08	85.3	0.93	0.09	0.99
H-14-10 [38]	61.8	0.09	76.9	0.80	0.09	0.98	74.7	0.83	0.09	1.01	80.6	0.77	0.09	0.94
H-20-20 [38]	76.0	0.09	83.5	0.91	0.09	1.03	80.6	0.94	0.09	1.07	87.1	0.87	0.09	0.99
N-06-06 [38]	43.2	0.08	53.3	0.81	0.07	1.07	54.7	0.79	0.08	1.04	59.3	0.73	0.08	0.96
N-06-12 [38]	51.8	0.11	54.5	0.95	0.08	1.51	54.7	0.95	0.08	1.51	59.1	0.88	0.08	1.39
N-07-10 [38]	41.6	0.11	55.2	0.75	0.08	1.39	53.5	0.78	0.08	1.44	57.6	0.72	0.08	1.33
N-07-16 [38]	40.0	0.11	52.5	0.76	0.07	1.49	53.4	0.75	0.07	1.46	57.5	0.70	0.08	1.36
N-12-12 [38]	49.3	0.09	57.5	0.86	0.08	1.16	55.4	0.89	0.08	1.21	59.2	0.83	0.08	1.13
N-12-16 [38]	57.1	0.12	57.6	0.99	0.08	1.54	55.0	1.04	0.07	1.61	59.2	0.96	0.08	1.50
N-14-10 [38]	41.8	0.12	56.8	0.74	0.08	1.52	53.7	0.78	0.08	1.60	57.9	0.72	0.08	1.49
N-20-20 [38]	55.0	0.13	58.5	0.94	0.08	1.62	54.6	1.01	0.07	1.74	58.9	0.93	0.08	1.61
SW12-1 [39]	32.3	0.15	36.1	0.89	0.13	1.14	33.8	0.96	0.13	1.20	35.7	0.91	0.16	0.98
SW10-1 [39]	24.6	0.13	24.5	1.00	0.14	0.95	22.7	1.08	0.13	1.03	24.7	1.00	0.19	0.70
SW10-2 [39]	29.6	0.20	30.1	0.98	0.14	1.47	28.7	1.03	0.13	1.49	31.5	0.94	0.15	1.33
SW10-3 [39]	26.6	0.15	25.1	1.06	0.14	1.12	23.5	1.13	0.13	1.18	27.6	0.96	0.15	1.01
SW10-4 [39]	27.7	0.16	27.0	1.02	0.14	1.19	25.2	1.10	0.13	1.26	29.4	0.94	0.15	1.08
SW8-1 [39]	19.7	0.16	19.2	1.03	0.14	1.14	18.5	1.07	0.14	1.13	19.4	1.02	0.19	0.84
SW8-2 [39]	22.5	0.14	19.8	1.14	0.14	1.02	18.7	1.20	0.14	1.06	21.7	1.04	0.16	0.92
D3 [35]	15.2	0.08	7.9	1.92	0.05	1.55	7.9	1.92	0.05	1.50	9.3	1.62	0.06	1.27
D4 [35]	15.8	0.12	8.7	1.83	0.06	2.14	8.7	1.82	0.06	2.10	10.2	1.55	0.07	1.79
T0 [40]	49.8	0.06	38.1	1.31	0.03	2.09	37.9	1.32	0.03	1.99	43.2	1.15	0.03	1.79
T1 [40]	48.0	0.04	33.0	1.45	0.03	1.46	32.8	1.46	0.03	1.39	37.9	1.27	0.04	1.22
T2 [40]	52.8	0.10	33.0	1.60	0.03	3.32	32.8	1.61	0.03	3.16	37.9	1.39	0.04	2.78
T5 [40]	62.5	0.06	86.3	0.72	0.07	0.91	49.9	1.25	0.03	2.08	56.5	1.11	0.03	1.92
VH1 [41]	12.0	0.12	9.7	1.24	0.07	1.66	9.7	1.24	0.07	1.66	9.9	1.21	0.10	1.23
VH2 [41]	11.5	0.07	10.2	1.12	0.08	0.90	10.3	1.12	0.08	0.91	11.0	1.05	0.10	0.73
A2 [42]	109.5	0.06	112.0	0.98	0.06	1.13	66.9	1.64	0.03	2.54	76.6	1.43	0.03	2.27
A3 [42]	113.3	0.06	128.4	0.88	0.04	1.37	67.5	1.68	0.03	2.27	77.7	1.46	0.03	1.98
A4 [42]	120.9	0.06	75.2	1.61	0.02	2.60	74.5	1.62	0.03	2.44	86.2	1.40	0.03	2.14
A5 [42]	120.9	0.04	74.7	1.62	0.02	1.82	74.1	1.63	0.03	1.72	85.6	1.41	0.03	1.50
B2 [42]	116.7	0.04	131.8	0.89	0.04	1.25	84.5	1.38	0.03	1.70	97.5	1.20	0.03	1.49
B3 [42]	130.5	0.05	134.1	0.97	0.03	1.31	92.3	1.41	0.03	1.76	105.3	1.24	0.03	1.57
B4 [42]	142.9	0.07	99.0	1.44	0.03	2.76	96.4	1.48	0.03	2.76	109.9	1.30	0.03	2.46
B5 [42]	146.3	0.06	97.8	1.50	0.03	2.54	95.2	1.54	0.03	2.54	109.4	1.34	0.03	2.21
C2 [42]	124.5	0.05	126.5	0.98	0.03	1.49	97.4	1.28	0.03	1.83	96.3	1.29	0.04	1.38
C3 [42]	131.9	0.06	129.7	1.02	0.03	1.94	98.1	1.34	0.03	2.47	112.4	1.17	0.03	2.18
C4 [42]	132.6	0.05	102.1	1.30	0.03	2.02	100.1	1.32	0.03	1.95	116.0	1.14	0.03	1.69
C5 [42]	138.3	0.05	106.4	1.30	0.02	2.06	104.2	1.33	0.03	2.00	119.5	1.16	0.03	1.75
C6 [42]	139.1	0.05	103.4	1.34	0.03	2.10	101.7	1.37	0.03	2.04	117.0	1.19	0.03	1.78
A095c [26]	102.9	0.03	101.6	1.01	0.04	0.80	99.2	1.04	0.04	0.83	102.9	1.00	0.05	0.64
A120a [26]	89.8	0.05	86.8	1.03	0.04	1.15	85.7	1.05	0.04	1.18	86.4	1.04	0.04	1.11
B065b [26]	54.4	0.03	58.0	0.94	0.03	1.33	58.0	0.94	0.03	1.24	67.4	0.81	0.03	1.09
B080a [26]	65.2	0.03	70.3	0.93	0.02	1.32	69.9	0.93	0.03	1.26	79.7	0.82	0.03	1.13
B110a [26]	128.3	0.04	128.5	1.00	0.04	0.96	125.9	1.02	0.04	1.01	127.1	1.01	0.05	0.80
C065a [26]	91.7	0.03	128.2	0.72	0.03	0.83	89.0	1.03	0.03	1.07	101.0	0.91	0.03	0.99
C100a [26]	122.2	0.03	134.3	0.91	0.03	0.85	128.6	0.95	0.03	0.87	134.8	0.91	0.04	0.67
D075a [26]	90.1	0.03	121.9	0.74	0.03	0.84	95.3	0.95	0.03	1.00	109.8	0.82	0.03	0.90
D090a [26]	96.1	0.03	110.3	0.87	0.03	1.14	107.8	0.89	0.03	1.10	122.8	0.78	0.03	0.98

Table A4. Cracking torques and corresponding twists (smeared constitutive laws *l7* to *l8*).

Beam	M_{Tcr}^{exp} kNm	θ_{cr}^{exp} °/m	M_{Tcr}^{thl7} kNm	$\frac{M_{Tcr}^{exp}}{M_{Tcr}^{thl7}}$	θ_{cr}^{thl7} °/m	$\frac{\theta_{cr}^{exp}}{\theta_{cr}^{thl7}}$	M_{Tcr}^{thl8} kNm	$\frac{M_{Tcr}^{exp}}{M_{Tcr}^{thl8}}$	θ_{cr}^{thl8} kNm	$\frac{\theta_{cr}^{exp}}{\theta_{cr}^{thl8}}$
B3 [35]	20.1	0.12	21.5	0.94	0.11	1.18	24.6	0.82	0.12	1.03
B4 [35]	21.9	0.12	22.1	0.99	0.10	1.14	24.3	0.90	0.11	1.04
B5 [35]	22.6	0.14	22.8	0.99	0.10	1.33	24.7	0.91	0.11	1.23
B6 [35]	25.0	0.16	21.9	1.14	0.10	1.65	23.7	1.05	0.11	1.53
B7 [35]	20.2	0.11	20.2	1.00	0.10	1.06	23.9	0.84	0.12	0.90
B8 [35]	21.8	0.13	20.6	1.06	0.10	1.26	24.2	0.90	0.12	1.07
B9 [35]	19.6	0.11	21.9	0.90	0.11	0.99	24.7	0.79	0.12	0.88
C4 [35]	11.9	0.13	11.9	0.99	0.12	1.11	13.1	0.90	0.13	1.01
C5 [35]	14.0	0.17	11.8	1.18	0.12	1.42	12.8	1.10	0.13	1.32
C6 [35]	13.9	0.17	11.8	1.18	0.11	1.54	12.8	1.09	0.12	1.42
G3 [35]	27.1	0.10	30.5	0.89	0.09	1.02	35.0	0.77	0.11	0.88
G4 [35]	28.7	0.12	31.6	0.91	0.10	1.23	35.5	0.81	0.11	1.09
G5 [35]	29.5	0.11	30.9	0.95	0.09	1.23	34.2	0.86	0.10	1.11
G7 [35]	33.6	0.13	33.1	1.01	0.10	1.38	37.5	0.90	0.11	1.22
G8 [35]	33.6	0.12	32.1	1.05	0.10	1.28	35.3	0.95	0.11	1.16
I3 [35]	25.5	0.11	27.0	0.94	0.12	0.92	30.4	0.84	0.13	0.81
I4 [35]	28.0	0.12	27.4	1.02	0.11	1.08	30.3	0.92	0.13	0.97
I5 [35]	28.1	0.15	27.8	1.01	0.11	1.34	30.2	0.93	0.12	1.23
I6 [35]	27.5	0.13	28.3	0.97	0.11	1.13	29.9	0.92	0.12	1.07
J1 [35]	14.0	0.09	15.3	0.92	0.09	0.92	18.1	0.77	0.11	0.77
J2 [35]	17.1	0.12	15.3	1.11	0.09	1.36	17.6	0.97	0.11	1.18
J3 [35]	16.9	0.10	16.5	1.03	0.09	1.04	18.6	0.91	0.10	0.92
J4 [35]	18.0	0.11	16.4	1.09	0.09	1.22	18.1	0.99	0.10	1.10
K2 [35]	12.2	0.18	12.8	0.95	0.15	1.22	14.0	0.87	0.16	1.12
K3 [35]	12.4	0.19	12.4	1.01	0.14	1.40	13.1	0.95	0.15	1.32
K4 [35]	13.1	0.21	12.2	1.07	0.13	1.55	12.5	1.05	0.14	1.51
M1 [35]	19.2	0.11	21.9	0.88	0.11	0.99	25.4	0.76	0.12	0.86
M2 [35]	20.6	0.11	22.4	0.92	0.11	1.03	25.3	0.81	0.12	0.92
M3 [35]	20.7	0.12	21.1	0.98	0.10	1.17	23.4	0.88	0.11	1.06
M4 [35]	20.7	0.13	21.2	0.98	0.10	1.32	22.9	0.90	0.11	1.22
M5 [35]	21.7	0.12	22.0	0.99	0.10	1.19	23.1	0.94	0.11	1.13
M6 [35]	22.7	0.15	22.6	1.01	0.10	1.43	23.4	0.97	0.11	1.38
N1 [35]	7.6	0.13	7.1	1.07	0.17	0.77	8.0	0.95	0.19	0.68
N1a [35]	7.0	0.11	6.9	1.01	0.16	0.66	7.9	0.89	0.19	0.58
N2 [35]	7.4	0.22	7.3	1.02	0.17	1.30	8.0	0.93	0.18	1.19
N2a [35]	7.5	0.21	7.0	1.08	0.16	1.29	7.6	0.98	0.18	1.18
N3 [35]	7.4	0.21	7.0	1.06	0.17	1.28	7.6	0.97	0.18	1.18
N4 [35]	7.6	0.21	7.0	1.08	0.16	1.30	7.5	1.01	0.17	1.22
A2 [36]	11.3	0.12	13.8	0.82	0.13	0.91	15.9	0.71	0.15	0.79
A3 [36]	12.2	0.12	14.3	0.86	0.13	0.93	16.0	0.76	0.15	0.83
A4 [36]	12.5	0.15	14.4	0.87	0.13	1.13	15.8	0.79	0.15	1.03
B3 [36]	8.8	0.15	13.0	0.68	0.15	0.99	14.2	0.62	0.16	0.90
B4 [36]	10.2	0.15	13.0	0.78	0.15	1.01	14.0	0.73	0.16	0.93
B5UR1 [37]	11.6	0.09	13.0	0.89	0.14	0.60	14.8	0.79	0.16	0.53
B9UR1 [37]	13.0	0.13	17.2	0.76	0.16	0.86	19.7	0.66	0.18	0.75
B12UR1 [37]	16.2	0.09	17.8	0.91	0.16	0.58	20.2	0.80	0.18	0.51
B14UR1 [37]	19.3	0.12	18.8	1.02	0.16	0.73	21.5	0.90	0.18	0.64
B12UR2 [37]	17.8	0.11	17.4	1.03	0.16	0.71	19.9	0.89	0.18	0.62
B12UR3 [37]	16.0	0.10	17.5	0.91	0.16	0.64	19.5	0.82	0.18	0.58
B12UR4 [37]	16.9	0.14	18.0	0.94	0.16	0.87	19.7	0.86	0.18	0.80
B12UR5 [37]	13.6	0.04	18.0	0.75	0.16	0.22	20.0	0.68	0.18	0.20
H-06-12 [38]	75.0	0.09	85.0	0.88	0.09	0.92	94.2	0.80	0.10	0.83
H-07-10 [38]	70.5	0.09	77.1	0.91	0.09	1.03	89.5	0.79	0.10	0.89
H-07-16 [38]	65.3	0.09	80.3	0.81	0.09	0.94	87.5	0.75	0.10	0.86
H-12-12 [38]	77.1	0.07	87.0	0.89	0.10	0.78	96.3	0.80	0.11	0.71
H-12-16 [38]	79.3	0.09	86.5	0.92	0.09	0.98	94.6	0.84	0.10	0.89
H-14-10 [38]	61.8	0.09	78.8	0.78	0.09	0.96	90.5	0.68	0.10	0.83

Table A4. *Cont.*

Beam	M_{Tcr}^{exp} kNm	θ_{cr}^{exp} °/m	M_{Tcr}^{thl7} kNm	$\frac{M_{Tcr}^{exp}}{M_{Tcr}^{thl7}}$	θ_{cr}^{thl7} °/m	$\frac{\theta_{cr}^{exp}}{\theta_{cr}^{thl7}}$	M_{Tcr}^{thl8} kNm	$\frac{M_{Tcr}^{exp}}{M_{Tcr}^{thl8}}$	θ_{cr}^{thl8} kNm	$\frac{\theta_{cr}^{exp}}{\theta_{cr}^{thl8}}$
H-20-20 [38]	76.0	0.09	89.7	0.85	0.10	0.96	96.2	0.79	0.10	0.89
N-06-06 [38]	43.2	0.08	57.1	0.76	0.08	1.00	67.4	0.64	0.09	0.85
N-06-12 [38]	51.8	0.11	60.5	0.86	0.08	1.36	66.5	0.78	0.09	1.24
N-07-10 [38]	41.6	0.11	56.3	0.74	0.08	1.36	64.9	0.64	0.09	1.18
N-07-16 [38]	40.0	0.11	58.4	0.68	0.08	1.33	64.0	0.63	0.09	1.22
N-12-12 [38]	49.3	0.09	60.5	0.81	0.08	1.10	66.8	0.74	0.09	1.00
N-12-16 [38]	57.1	0.12	59.9	0.95	0.08	1.48	65.8	0.87	0.09	1.35
N-14-10 [38]	41.8	0.12	56.5	0.74	0.08	1.52	64.1	0.65	0.09	1.34
N-20-20 [38]	55.0	0.13	61.2	0.90	0.08	1.55	64.7	0.85	0.09	1.47
SW12-1 [39]	32.3	0.15	38.7	0.84	0.15	1.04	41.3	0.78	0.16	0.97
SW10-1 [39]	24.6	0.13	26.1	0.94	0.15	0.87	27.8	0.88	0.16	0.82
SW10-2 [39]	29.6	0.20	32.1	0.92	0.15	1.34	34.6	0.86	0.16	1.24
SW10-3 [39]	26.6	0.15	26.6	1.00	0.15	1.04	28.7	0.93	0.16	0.96
SW10-4 [39]	27.7	0.16	28.9	0.96	0.15	1.09	30.3	0.92	0.15	1.04
SW8-1 [39]	19.7	0.16	21.1	0.93	0.16	0.97	21.9	0.90	0.16	0.98
SW8-2 [39]	22.5	0.14	21.1	1.07	0.15	0.94	22.7	0.99	0.17	0.87
D3 [35]	15.2	0.08	8.4	1.81	0.06	1.42	9.1	1.67	0.06	1.32
D4 [35]	15.8	0.12	9.4	1.69	0.06	1.94	10.0	1.58	0.06	1.82
T0 [40]	49.8	0.06	41.2	1.21	0.03	1.78	44.1	1.13	0.04	1.68
T1 [40]	48.0	0.04	34.9	1.38	0.03	1.29	38.8	1.24	0.04	1.15
T2 [40]	52.8	0.10	34.9	1.51	0.03	2.93	38.8	1.36	0.04	2.63
T5 [40]	62.5	0.06	53.1	1.18	0.03	1.92	59.5	1.05	0.04	1.71
VH1 [41]	12.0	0.12	10.1	1.19	0.08	1.60	11.8	1.02	0.09	1.37
VH2 [41]	11.5	0.07	11.4	1.01	0.09	0.82	12.4	0.93	0.09	0.76
A2 [42]	109.5	0.06	67.7	1.62	0.03	2.49	79.5	1.38	0.03	2.09
A3 [42]	113.3	0.06	69.1	1.64	0.03	2.17	80.0	1.42	0.03	1.87
A4 [42]	120.9	0.06	77.9	1.55	0.03	2.34	87.7	1.38	0.03	2.05
A5 [42]	120.9	0.04	78.4	1.54	0.03	1.61	87.0	1.39	0.03	1.44
B2 [42]	116.7	0.04	86.4	1.35	0.03	1.69	99.8	1.17	0.03	1.44
B3 [42]	130.5	0.05	96.6	1.35	0.03	1.66	108.8	1.20	0.03	1.47
B4 [42]	142.9	0.07	100.7	1.42	0.03	2.60	112.4	1.27	0.03	2.36
B5 [42]	146.3	0.06	100.0	1.46	0.03	2.42	110.6	1.32	0.03	2.17
C2 [42]	124.5	0.05	99.0	1.26	0.03	1.79	115.2	1.08	0.03	1.56
C3 [42]	131.9	0.06	102.9	1.28	0.03	2.33	116.0	1.14	0.03	2.04
C4 [42]	132.6	0.05	105.4	1.26	0.03	1.86	117.1	1.13	0.03	1.66
C5 [42]	138.3	0.05	109.5	1.26	0.03	1.88	122.1	1.13	0.03	1.67
C6 [42]	139.1	0.05	109.5	1.27	0.03	1.87	118.1	1.18	0.03	1.75
A095c [26]	102.9	0.03	100.2	1.03	0.04	0.82	120.2	0.86	0.05	0.68
A120a [26]	89.8	0.05	87.6	1.03	0.04	1.15	104.2	0.86	0.05	0.95
B065b [26]	54.4	0.03	62.0	0.88	0.03	1.16	66.8	0.81	0.03	1.07
B080a [26]	65.2	0.03	72.0	0.91	0.03	1.22	81.3	0.80	0.03	1.09
B110a [26]	128.3	0.04	128.6	1.00	0.04	0.99	152.2	0.84	0.05	0.82
C065a [26]	91.7	0.03	92.8	0.99	0.03	1.01	105.7	0.87	0.03	0.88
C100a [26]	122.2	0.03	133.1	0.92	0.03	0.84	153.1	0.80	0.04	0.72
D075a [26]	90.1	0.03	100.1	0.90	0.03	0.95	111.8	0.81	0.03	0.85
D090a [26]	96.1	0.03	110.6	0.87	0.03	1.09	126.0	0.76	0.03	0.94

References

1. Hsu, T.T.C.; Mo, Y.L. *Unified Theory of Concrete Structures*; John Wiley Sons Ltd.: Chichester, UK, 2010; p. 500.
2. Bernardo, L.F.; Andrade, J.M. A unified softened truss model for RC and PC beams under torsion. *J. Build. Eng.* **2020**, *32*, 101467. [CrossRef]
3. Hsu, T.T.C.; Mo, Y.L. Softening of concrete in torsional members—Theory and tests. *J. Am. Concr. Inst.* **1985**, *82*, 290–303.
4. Hsu, T.; Belarbi, A.; Pang, X. A Universal panel tester. *J. Test. Eval.* **1995**, *23*, 41. [CrossRef]
5. Bernardo, L.F.A.; Lopes, S.M.R. Behavior of concrete beams under torsion—NSC plain and hollow beams. *Mater. Struct.* **2008**, *41*, 1143–1167. [CrossRef]
6. Bernardo, L.F.A.; Lopes, S.M.R. Theoretical behavior of HSC sections under torsion. *Eng. Struct.* **2011**, *33*, 3702–3714. [CrossRef]

7. Bairan Garcia, J.M.; Mari Bernat, A.R. Coupled model for the non-linear analysis of anisotropic sections subjected to general 3D loading. Part 1: Theoretical formulation. *Comput. Struct.* **2006**, *84*, 2254–2263. [CrossRef]
8. Bairan Garcia, J.M.; Mari Bernat, A.R. Coupled model for the nonlinear analysis of sections made of anisotropic materials, subjected to general 3D loading. Part 2: Implementation and validation. *Comput. Struct.* **2006**, *84*, 2264–2276. [CrossRef]
9. Jeng, C.-H.; Hsu, T.T. A softened membrane model for torsion in reinforced concrete members. *Eng. Struct.* **2009**, *31*, 1944–1954. [CrossRef]
10. Mostofinejad, D.; Behzad, T.S. Nonlinear modeling of RC beams subjected to torsion using smeared crack model. *Procedia Eng.* **2011**, *14*, 1447–1454. [CrossRef]
11. Valikhani, A.; Jahromi, A.J.; Mantawy, I.M.; Azizinamini, A. Numerical modelling of concrete-to-UHPC bond strength. *Materials* **2020**, *13*, 1379. [CrossRef]
12. Sucharda, O. Identification of fracture mechanic properties of concrete and analysis of shear capacity of reinforced concrete beams without transverse reinforcement. *Materials* **2020**, *13*, 2788. [CrossRef]
13. Hsu, T.T.C.; Mo, Y.L. Softening of concrete in torsional members—Prestressed concrete. *J. Am. Concr. Inst.* **1985**, *82*, 603–615.
14. Bernardo, L.F.; Taborda, C.S.; Andrade, J.M. Ultimate torsional behaviour of axially restrained RC beams. *Comput. Concr.* **2015**, *16*, 67–97. [CrossRef]
15. Taborda, C.S.B.; Bernardo, L.F.B.; Gama, J.M.R. Effective torsional strength of axially restricted RC beams. *Struct. Eng. Mech.* **2018**, *67*, 465–479.
16. Bernardo, L.F.A.; Andrade, J.M.A.; Nunes, N.C.G. Generalized softened variable angle truss-model for reinforced concrete beams under torsion. *Mater. Struct.* **2014**, *48*, 2169–2193. [CrossRef]
17. Bernardo, L.F.A.; Taborda, C.S.B.; Andrade, J.M.A. Generalized softened variable angle truss model for PC Beams under torsion. *Int. J. Concr. Struct. Mater.* **2018**, *12*, 62. [CrossRef]
18. Bernardo, L. Generalized softened variable angle truss model for RC hollow beams under torsion. *Materials* **2019**, *12*, 2209. [CrossRef] [PubMed]
19. Bernardo, L. Modeling the full behavior of reinforced concrete flanged beams under torsion. *Appl. Sci.* **2019**, *9*, 2730. [CrossRef]
20. Bernardo, L.; Andrade, J.; Lopes, S. Softened truss model for reinforced NSC and HSC beams under torsion: A comparative study. *Eng. Struct.* **2012**, *42*, 278–296. [CrossRef]
21. Teixeira, M.; Bernardo, L. Ductility of RC beams under torsion. *Eng. Struct.* **2018**, *168*, 759–769. [CrossRef]
22. Belarbi, A.; Hsu, T.T.C. Constitutive laws of softened concrete in biaxial tension—Compression. *Struct. J. Am. Concr. Inst.* **1995**, *92*, 562–573.
23. Zhang, L.X.; Hsu, T.C. Behaviour and analysis of 100 MPa concrete membrane elements. *J. Struct. Eng.* **1998**, *124*, 24–34. [CrossRef]
24. Belarbi, A.; Hsu, T.C. Constitutive laws of concrete in tension and reinforcing bars stiffened by concrete. *Struct. J. Am. Concr. Inst.* **1994**, *91*, 465–474.
25. Bernardo, L.F.A.; Andrade, J.M.A.; Oliveira, L.A.P. Reinforced and prestressed concrete hollow beams under torsion. *J. Civ. Eng. Manag.* **2013**, *19*, S141–S152. [CrossRef]
26. Jeng, C.H. Unified softened membrane model for torsion in hollow and solid reinforced concrete members: Modeling precracking and postcracking behavior. *J. Struct. Eng.* **2015**, *141*. [CrossRef]
27. Nobre, S.S. Evaluation of the Constitutive Law for Tensile Concrete for the Cracking of Reinforced Concrete Beams under Torsion. Master's Thesis, University of Beira Interior, Covilhã, Portugal, 2014; p. 150. (In Portuguese).
28. Cervenka, V. Constitutive model for cracked reinforced concrete. *J. Am. Concr. Inst.* **1985**, *82*, 877–882.
29. Vecchio, F.J.; Collins, M.P. The modified compression—Field theory for reinforced concrete elements subjected to shear. *J. Am. Concr. Inst.* **1986**, *83*, 219–231.
30. Hsu, T.T.C. Nonlinear analysis of concrete membrane elements. *Struct. J. Am. Concr. Inst.* **1991**, *88*, 552–561.
31. Collins, M.P.; Mitchell, D.; Adebar, P.; Vecchio, F.J. A General shear design method. *Am. Concr. Inst. Struct. J.* **1996**, *93*, 36–45.
32. Vecchio, F.J. Disturbed Stress field model for reinforced concrete: Formulation. *J. Struct. Eng.* **2000**, *126*, 1070–1077. [CrossRef]
33. Bentz, E.C. Explaining the riddle of tension stiffening models for shear panel experiments. *J. Struct. Eng.* **2005**, *131*, 1422–1425. [CrossRef]
34. Stramandinoli, R.S.B.; Rovere, H.L. An efficient tension-stiffening model for nonlinear analysis of reinforced concrete members. *Eng. Struct.* **2008**, *30*, 2069–2080. [CrossRef]
35. Hsu, T.T.C. *Torsion of Structural Concrete-Behavior of Reinforced Concrete Rectangular Members*; Torsion of Structural Concrete SP-18; American Concrete Institute: Detroit, MI, USA, 1968; pp. 261–306.
36. McMullen, A.E.; Rangan, B.V. Pure torsion in rectangular sections—A re-examination. *J. Am. Concr. Inst.* **1978**, *75*, 511–519.
37. Koutchoukali, N.E.; Belarbi, A. Torsion of high-strength reinforced concrete beams and minimum reinforcement requirement. *Am. Concr. Inst. Struct. J.* **2001**, *98*, 462–469.
38. Fang, I.K.; Shiau, J.K. Torsional behavior of normal and high-strength concrete beams. *Am. Concr. Inst. Struct. J.* **2004**, *101*, 304–313.
39. Peng, X.-N.; Wong, Y.-L. Behavior of reinforced concrete walls subjected to monotonic pure torsion—An experimental study. *Eng. Struct.* **2011**, *33*, 2495–2508. [CrossRef]

40. Lampert, P.; Thürlimann, B. Essais de Poutre en Béton Armé sous Torsion Simple et Flexion Combinées (Torsionsversuche und Torsions-Biege-Versuche an Stahlbetonbalken). *Comitée Européen du Béton B.I.* **1969**, *71*, 177–207.
41. Leonhardt, F.; Schelling, G. *Torsionsversuche na Stahlbetonbalken*; Bulletin No. 239; Deutscher Ausschuss für Stahlbeton: Berlin, Germany, 1974; 122p.
42. Bernardo, L.F.A.; Lopes, S.M.R. Torsion in HSC hollow beams: Strength and ductility analysis. *Am. Concr. Inst. Struct. J.* **2009**, *106*, 39–48.

materials

MDPI

Article

Prediction of Compressive Strength of Fly Ash Based Concrete Using Individual and Ensemble Algorithm

Ayaz Ahmad [1], Furqan Farooq [1,2,*], Pawel Niewiadomski [2], Krzysztof Ostrowski [3], Arslan Akbar [4,*], Fahid Aslam [5] and Rayed Alyousef [5]

1 Department of Civil Engineering, Abbottabad Campus, COMSATS University Islamabad, Islamabad 22060, Pakistan; ayazahmad@cuiatd.edu.pk
2 Faculty of Civil Engineering, Wroclaw University of Science and Technology, Wybrzeże Wyspiańskiego 27, 50-370 Wroclaw, Poland; pawel.niewiadomski@pwr.edu.pl
3 Faculty of Civil Engineering, Cracow University of Technology, 24 Warszawska Str., 31-155 Cracow, Poland; krzysztof.ostrowski.1@pk.edu.pl
4 Department of Architecture and Civil Engineering, City University of Hong Kong, Kowloon, Hong Kong
5 Department of Civil Engineering, College of Engineering in Al-Kharj, Prince Sattam Bin Abdulaziz University, Al-Kharj 11942, Saudi Arabia; f.aslam@psau.edu.sa (F.A.); r.alyousef@psau.edu.sa (R.A.)
* Correspondence: furqan@cuiatd.edu.pk (F.F.); aakbar4-c@my.cityu.edu.hk (A.A.)

Abstract: Machine learning techniques are widely used algorithms for predicting the mechanical properties of concrete. This study is based on the comparison of algorithms between individuals and ensemble approaches, such as bagging. Optimization for bagging is done by making 20 sub-models to depict the accurate one. Variables like cement content, fine and coarse aggregate, water, binder-to-water ratio, fly-ash, and superplasticizer are used for modeling. Model performance is evaluated by various statistical indicators like mean absolute error (MAE), mean square error (MSE), and root mean square error (RMSE). Individual algorithms show a moderate bias result. However, the ensemble model gives a better result with R^2 = 0.911 compared to the decision tree (DT) and gene expression programming (GEP). K-fold cross-validation confirms the model's accuracy and is done by R^2, MAE, MSE, and RMSE. Statistical checks reveal that the decision tree with ensemble provides 25%, 121%, and 49% enhancement for errors like MAE, MSE, and RMSE between the target and outcome response.

Keywords: concrete compressive strength; fly ash waste; ensemble modeling; decision tree; DT-bagging regression; cross-validation python

Citation: Ahmad, A.; Farooq, F.; Niewiadomski, P.; Ostrowski, K.; Akbar, A.; Aslam, F.; Alyousef, R. Prediction of Compressive Strength of Fly Ash Based Concrete Using Individual and Ensemble Algorithm. *Materials* **2021**, *14*, 794. https://doi.org/10.3390/ma14040794

Academic Editor: Dario De Domenico
Received: 28 November 2020
Accepted: 1 February 2021
Published: 8 February 2021

1. Introduction

Carbon dioxide produced from the cement industry has a malignant adamant effect on environmental conditions [1]. Its utilization and excessive use in modern construction around the world produces greenhouse gases (GHG) [2]. Moreover, countless amounts of gases are emitted during the production of cement due to the burning of natural resources and fossil fuels [3]. Annually, 4 billion tons of Portland cement (PC) is being produced and approximately one ton of cement generates one ton of CO_2 gas [4]. This huge amount of carbon dioxide is a serious threat to the environment. The report shows that a 1.6% increment (3.4% to 5%) of global CO_2 gas discharge was observed from the year 2000 to 2006. The cement industry contributes 18% of industrial greenhouse gases (GHG to the environment. This is due to the direct process-related activity, energy-related combustion, and remaining use of electricity, which is termed as indirect energy [5]. To overcome the above-mentioned issue, a process of replacing the cement material with an alternative binder is of great research interest [6].

The supplementary cementation materials (SCMs) can be used for many purposes, especially in the concrete industry. Their utilization in concrete gives a benignant effect by reducing the percentage of CO_2 gas emitted. SCMs used in the cement industry can be industrial and agricultural waste products, which includes olive oil, bagasse ash, sugarcane, rice husk ash, palm oil fuel ash, etc. However, commonly adopted and used in the construction industry are silica fume, fly ash, and ground granulated blast furnace slag [7–9]. Their utilization in concrete reduces the malignant effect on the environment [10]. The replacement of cement in concrete with the waste material helps both in the utilization of the wastes and fulfills the increasing demand for the concrete. What is more, it has been observed that the use of waste materials as pozzolanic in high-strength concrete improves its strength and durability. This alternately helps minimize the impending environmental degradation [11].

Concrete is stated as the second-highest used material after water in the world [12]. This is due to the intensive use of concrete in the construction industries and the field of civil engineering. Concrete requires a comprehensive technique to produce. It is a mixture of multiple materials like coarse aggregate, fine aggregate, water, binder, admixtures, and supplementary raw materials [13]. The concrete matrix is the random distribution of the previously mentioned variables [14]. The extensive use of it can be seen as a building material around the globe. For the effective evaluation of the performance of concrete according to the advanced design technologies, its mechanical properties must be examined [14]. One of its supreme mechanical properties is its compressive strength, which is alternately the sign of structural safety throughout life [15]. This remarkable property of concrete can be affected by numerous factors, like particle size, water-to-cement ratio, waste composition, and use of chemicals. However, casting concrete by using the proper techniques in the laboratory and conducting experimental tests to find the mechanical properties after the setting is quite a time-consuming task [14]. Moreover, using the previously mentioned technique in the recent and modern period of life is quite uneconomical. Thus, the modern methodologies of machine learning techniques can be adapted to predict the desired result in advance [16]. The prediction of variables can be done from regressions and machine learning models. These algorithm-based techniques give a precise relation and predict the accurate model by the use of input variables [17].

Machine learning approaches are raising trends in the domain of civil engineering. They are extensively used in forecasting the mechanical properties of concrete [18–21]. These techniques use extensive data to build a precise model. Their prediction accuracy is dependent upon the data sample used in experimental work during casting of the specimen or upon the literature study. Researchers use these algorithms for the prediction of the mechanical properties of concrete. Javed et al. [22] predict the compressive strength of sugarcane bagasse ash (SCBA) concrete using gene expression programming (GEP). The author used the experimental test for calibration and validation of the model. Similarly, Aslam et al. [23] predict the compressive strength of high-strength concrete (HSC) by employing GEP. The author used 357 data points and reported an adamant relationship between the target and predicted values. Hosein et al. [24] forecast the mechanical properties of recycled concrete (REC) by using an artificial neural network (ANN). Correspondingly, Getahun et al. [25] forecast the strength of concrete incorporating waste materials using ANN. The author concluded that ANN gives adamant relation with fewer errors. Similarly, Qing et al. [26] predict the diffusion capability of chloride in reinforced concrete structures with ANN. The result indicates better prediction by employing an individual algorithm based on 653 data samples. Farooq et al. [15] predict the compressive nature of HSC by developing two models with random forest (RF) and GEP. RF gives a robust performance with precise correlation with strong predicted values. That machine learning algorithm is not limited to predict only the compressive or tensile nature of concrete but can be used to forecast any response in any engineering or data sciences domain. In turn, Ahmad et al. [27] employ supervised machine learning (ML) algorithms to predict energy in the distinct buildings. Similarly, Wang et al. [28] predict the COVID-19 response by employing different ML-based algorithms. Cladera et al. [29] predict the response of a structural beam with and without stirrups by using ANN. The author achieved a better response from modeled than empirical relations. Similarly, Onyari et al. [30] reveal robust performance by utilizing ANN to predict the flexural and compressive strength of modified mortar. Previously mentioned examples show the overwhelming response of individual algorithms.

Recently, application of ensemble modeling is perceived as a chance for enhancement of the model's overall efficiency. It can be achieved due to taking a weak leaner to build strong, predictive learners than individual learners [31]. Feng et al. [32] use ensemble algorithm techniques for the prediction of failure mode classification and bearing capacity of reinforced concrete (RC) structural element (column). Both models give robust performance. However, bearing capacity is characterized by better correlation than failure mode classification. Bui et al. [33] employed a modified firefly algorithm with ANN on high performance concrete (HPC) and reported better performance of the model. Moreover, Salamai et al. [34] report good accuracy of R^2 = 0.9867 by using the RF algorithm. In turn, Cai et al. [35] use various supervised machine ensemble algorithms for the prediction of chloride penetration in the RC structure situated in a marine environment. Ensemble models outclass individual algorithms to predict chloride penetration in RC. Hacer et al. [36] present the comparative assessment of bagging as the ensemble approach for high-performance concrete mix slump flow. Ensemble models with bagging were found to be superior with regard to standalone approaches. Halil et al. [37] predict the strength of HPC by employing three ensemble modeling approaches. The author used the decision tree as a base learner for other models and found that the hybrid model outperforms with the output result of R^2 = 0.9368 among the several proposed models. Kermani et al. [38] represents the performance of five soft, computing base learners for predicting concrete corrosion in sewers. The author used both tree-based and network-based learners and reported that RF ensemble learners give a better result with R^2 = 0.872. These ensemble approaches give an enhanced effect with robust performance of the overall models.

Taking the above into consideration, it may seem that ensemble learning models have more favorable features and give better results than individual learning models. The difference between individual and ensemble model is illustrated in Figure 1.

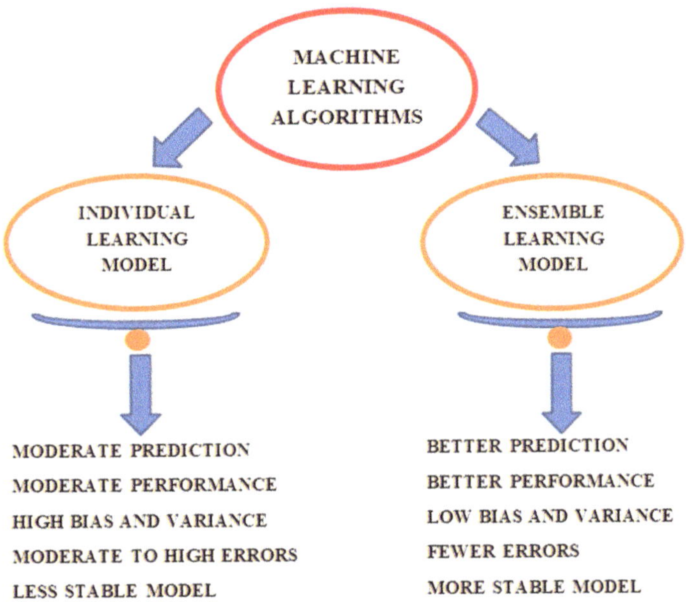

Figure 1. Comparison between individual and ensemble approaches.

2. Research Significance

The aim of this study is to use the machine-learning algorithm with ensemble modeling using Anaconda Python to predict the compressive strength of fly-ash-based concrete using different algorithms. A decision tree with a bagging algorithm is used and optimization is done by making 20 sub-models to give a strong outcome. A comparison is made with the individual, ensemble algorithms, and with gene expression programming to give the best model. Moreover, K-fold cross-validation and a statistical check are applied to evaluate the model performance.

3. Data Description

The efficiency of the model is completely dependent upon the variables and the number of data samples used. The parameters used in models preparation in order to predict the strength of concrete were taken from published literature [39] and are summarized in Appendix A. Eight variables concerning composition of the concrete mixture and including cement, fine and coarse aggregate, superplasticizer, water, waste material, age, and a water-to-binder ratio were taken into analysis. The overall distribution in terms of the relative frequency distribution is illustrated in Figure 2. The range of variables of each parameter used in the study, with a minimum and maximum value, is illustrated in Figure 3. Statistical descriptive analysis for the variables in terms of strength is listed in Table 1.

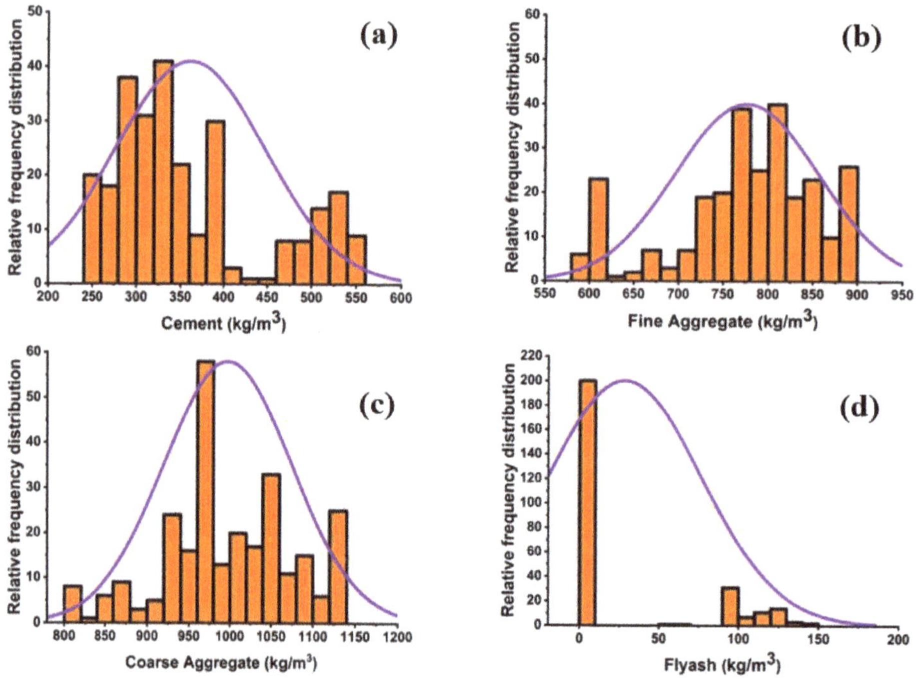

Figure 2. *Cont.*

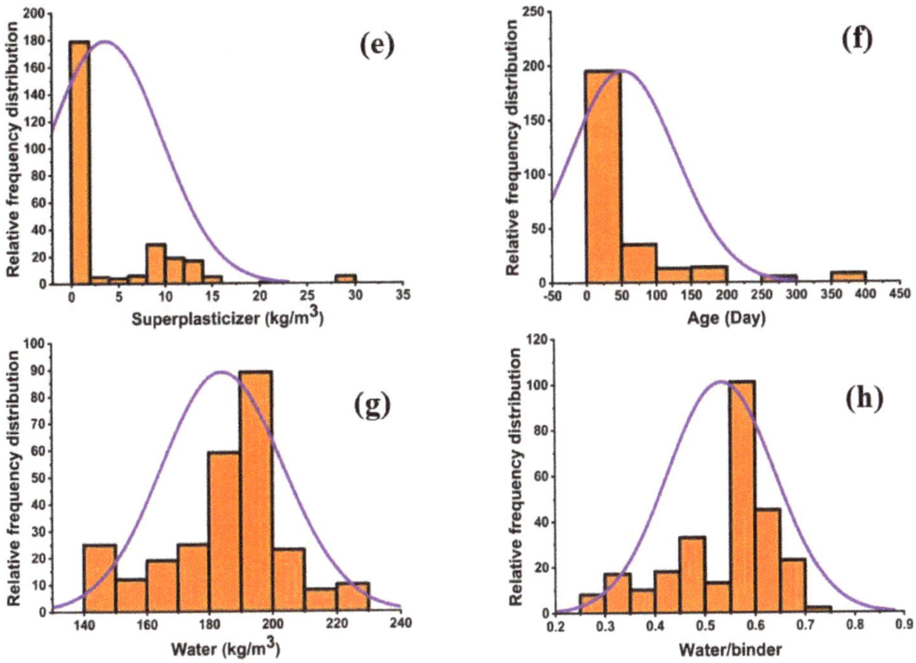

Figure 2. Relative frequency distribution of variables, (**a**) cement, (**b**) fine aggregate, (**c**) coarse aggregate, (**d**) fly ash, (**e**) superplasticizer, (**f**) age, (**g**) water, and (**h**) water-to-binder ratio.

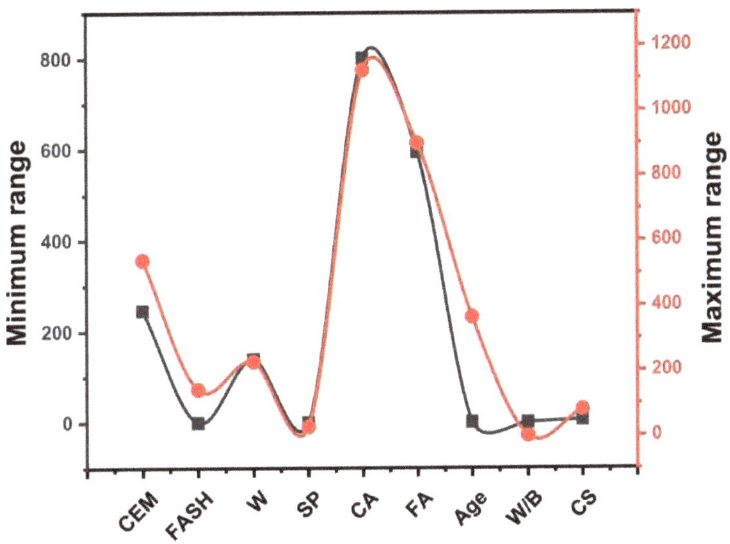

Figure 3. Parameters range of variables with minimum and maximum values.

Table 1. Statistical measures on variables.

Statistics	Cem *	FASH *	W *	SP *	CA *	FA *	Age *	W/B *
Mean	361.39	28.15	184.15	3.68	996.90	775.93	53.31	0.53
Standard Error	5.20	2.94	1.17	0.36	4.70	4.86	4.63	0.01
Median	336.25	0.00	189.00	0.00	987.80	781.95	28.00	0.58
Mode	349.00	0.00	192.00	0.00	1125.00	613.00	28.00	0.58
Standard Deviation	85.49	48.35	19.29	5.95	77.26	79.92	76.06	0.11
Sample Variance	7309.14	2337.79	372.16	35.39	5969.32	6387.59	5784.50	0.01
Kurtosis	−0.50	−0.44	0.29	3.52	−0.19	−0.07	7.01	−0.04
Skewness	0.83	1.20	−0.38	1.77	−0.26	−0.67	2.62	−0.92
Range	293.20	142.00	88.00	28.20	324.00	305.80	364.00	0.43
Minimum	246.80	0.00	140.00	0.00	801.00	594.00	1.00	0.27
Maximum	540.00	142.00	228.00	28.20	1125.00	899.80	365.00	0.70
Sum	97,574.60	7601.70	49,720.30	993.40	269,163.90	209,502.40	14,394.00	143.89
Count	270.00	270.00	270.00	270.00	270.00	270.00	270.00	270.00

* CEM = Cement (kg/m^3), FASH = Fly ash (kg/m^3), W = Water (kg/m^3), SP = Super plasticizer (kg/m^3), CA = Coarse aggregate (kg/m^3), FA = Fine aggregate (kg/m^3), and W/B = Water to binder.

4. Methodology

Individual and ensemble model techniques used to predict the properties in a limited time that are of great interest. The accuracy level between the actual and prediction level is typically obtained from the R^2 value (ranges from 0–0.99). A high R^2 value indicates the satisfactory results of the selected technique. This study uses three approaches to predict the compressive strength of concrete with waste material. A decision tree with ensemble algorithms such as bagging with a learning rate of 0.9 and gene expression programming is used. These techniques are selected due to their popularity among other algorithms. The overall machine learning model methodology in the form of a diagram is illustrated in Figure 4.

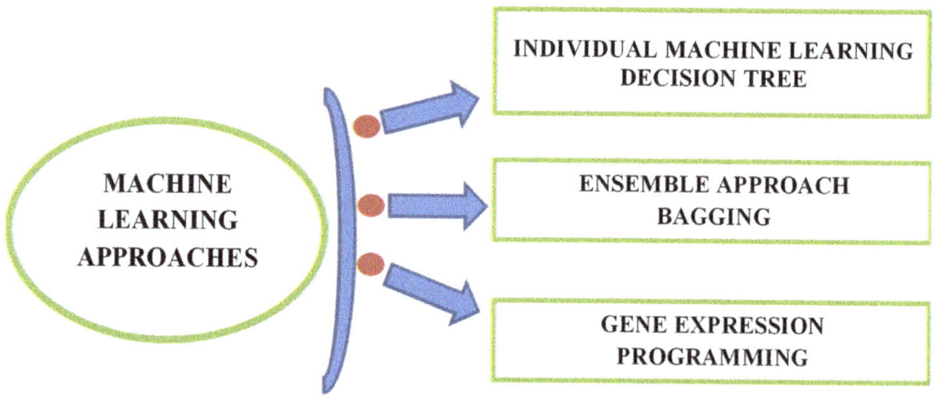

Figure 4. Machine learning model methodology.

4.1. Decision Tree

The decision tree is one of the supervised learning techniques used for categorizing regression problems but is also commonly used for classification problems [40]. There are classes inside the tree. However, if there is no class, then the regression technique can predict the outcome by independent variables [37]. A decision tree is a tree-structured classifier in which the inner nodes reflect the attribute of a database. Branches indicate the conclusion rules, and every leaf node constitutes the outcome. The decision tree consists of two nodes known as a decision node and a leaf node. Decision nodes have multiple branches with the capability to make any decision, while leaf nodes do not have branches and are considered as the output of the decisions. It is known as a decision tree because it has a similar nature to a tree that starts with the root node and distributes in the number of branches, and reflects a tree-like structure [41]. The decision tree splits the data samples at various points. The executed algorithm finds the error between the target and predicted value at every divided point. The errors are calculated at every divided point, and the variable with the least value for the fitness function is selected as a split point, and the same procedure is repeated again.

4.2. Ensemble Bagging Approach

The ensemble technique is the concept of machine learning used to train numerous models by applying a similar learning algorithm [42]. The ensemble involves a substantial group of methods known as multi-classifiers. The group of hundreds or thousands of learners with a common intent are joined together to fix the problem. Bagging is a parallel type ensemble method that explains the variance of the prediction model by producing supplementary data in the stage of training. This production is from irregular sampling including substituting from the real set of data. Some of the observations can be repeated by sampling with replacement in every new training data set. In bagging, every component has an equal chance to appear in the new dataset. The force of prediction cannot be enhanced by increasing the size of the training set. The variance can also be reduced narrowly by tuning the forecast to an anticipated outcome. All these numbers of sets of the given data are normally used to train other numbers of models. This ensemble of different models uses the average of all the predictions from the other various models. In regression, the prediction may be the mean or average of the predictions taken from the different models [43]. The decision tree with bagging is tuned with 20 sub-models to obtain the optimized value that gives an adamant output result.

4.3. Gene Expression Programming

Gene expression programming (GEP) is a computer programming-based algorithm used to develop different models [44]. GEP, which is initially introduced by Ferreira [45], is considered to be a natural development of genetic programming (GP). Multiple numbers of genetic operators that are being used in genetic algorithms (GAs) can also be used in GEP with the help of a few recommended changes. There are five main components of GEP, namely, function set, terminal set, fitness function, control variables, and termination condition. GEP works as a fixed length of character twine to explain the problems, which are next defined as tree-like structures with different dimensions. This type of tree is known as the GEP expression tree (ETs). Selection of individual chromosomes takes place and then they are copied into the next generation, as per the fitness by roulette wheel sampling with elitism [23]. This ensures the durability and replication of the best individual to the next generation. Fluctuation in the population is shown by applying one or more genetic operators (mutation, crossover, or rotation) on the given chromosomes. Among the number of advantages of GEP, the formation of genetic diversity is remarkably simplified because of the working of genetic operators at the chromosome level. This multi-genic approach of GEP permits the natural selection of other complicated and complex programs composed of numerous subprograms. GEP genes along with a function set and terminal set play a vital role during the process [46].

4.4. K-Fold Cross-Validation and Statistical Measures

The model performance in terms of bias and variance is checked by employing K-fold cross-validation. The data is divided into 10 stratified groups, which randomly distribute the data into a training set and test set. This process takes one part of the overall data into the test sample and the remaining into the training set, as illustrated in Figure 5. The model's overall efficiency by cross-validation is then tested by taking an average of 10 rounds by various errors. Similarly, the model evaluation is also done by using statistical indicators [23]. Three types of the indicator are used in our current study, which is listed below (Equations (1)–(3)).

$$\text{MAE} = \frac{1}{n}\sum_{i=1}^{n}|x_i - x| \tag{1}$$

$$\text{MSE} = \frac{1}{n}\sum_{i=1}^{n}\left(y_{pred} - y_{ref}\right)^2 \tag{2}$$

$$\text{RMSE} = \sqrt{\sum\frac{\left(y_{pred} - y_{ref}\right)^2}{n}} \tag{3}$$

where:

- n = Total number of data samples,
- x, y_{ref} = reference values in the data sample,
- x_i, y_{pred} = predicted values from models.

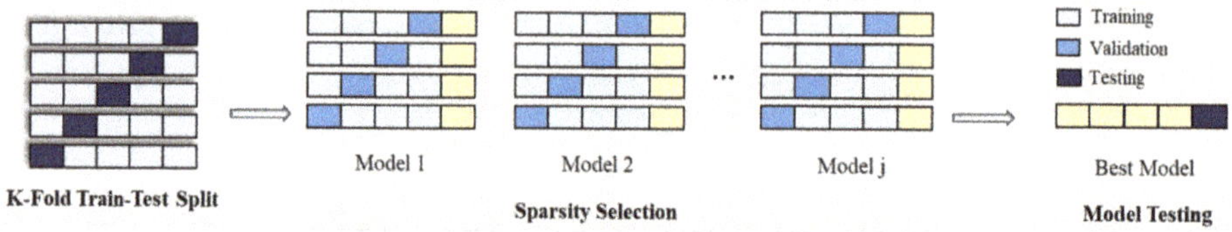

Figure 5. K-fold cross-validation algorithm [46].

5. Model Result

5.1. Decision Tree/Ensemble Model

The prediction of concrete strength by employing a decision tree yields an adamantly strong relationship between targets to output strength, as depicted in Figure 6. It can be seen that the individual model gives a better response with less variance, as illustrated in Figure 6a. However, the decision tree with bagging gives precise performance than an individual one, as illustrated in Figure 6d. This is due to an increase in model efficiency as it takes several data to train the best model by using weak base learners [47]. The ensemble model is optimized by making 20 sub-models, as depicted in Figure 6c. The zero number shows the individual model, which is made by using the decision approach and shows R^2 = 0.812. After the ensemble approach, there is a significant enhancement in the overall response of the model. Every model shows a surpass effect by giving an average score of about R^2 = 0.904 within 20 models. However, the 12th sub-model gives a prime result with R^2 = 0.911, as depicted in Figure 6c. Moreover, the model comparison in terms of errors is depicted in Figure 6b,e. Decision tree (DT) with bagging enhances the model accuracy by giving fewer errors. The test data shows that there is a 20.10% prediction capacity of average errors by bagging than in the individual model. Besides, DT shows the minimum and maximum error of 0 and 21.97 MPa, respectively. Similarly, DT with an ensemble model shows the minimum and maximum error of 0.11, and 12.77 MPa, respectively. The detailed result is shown in Table 2.

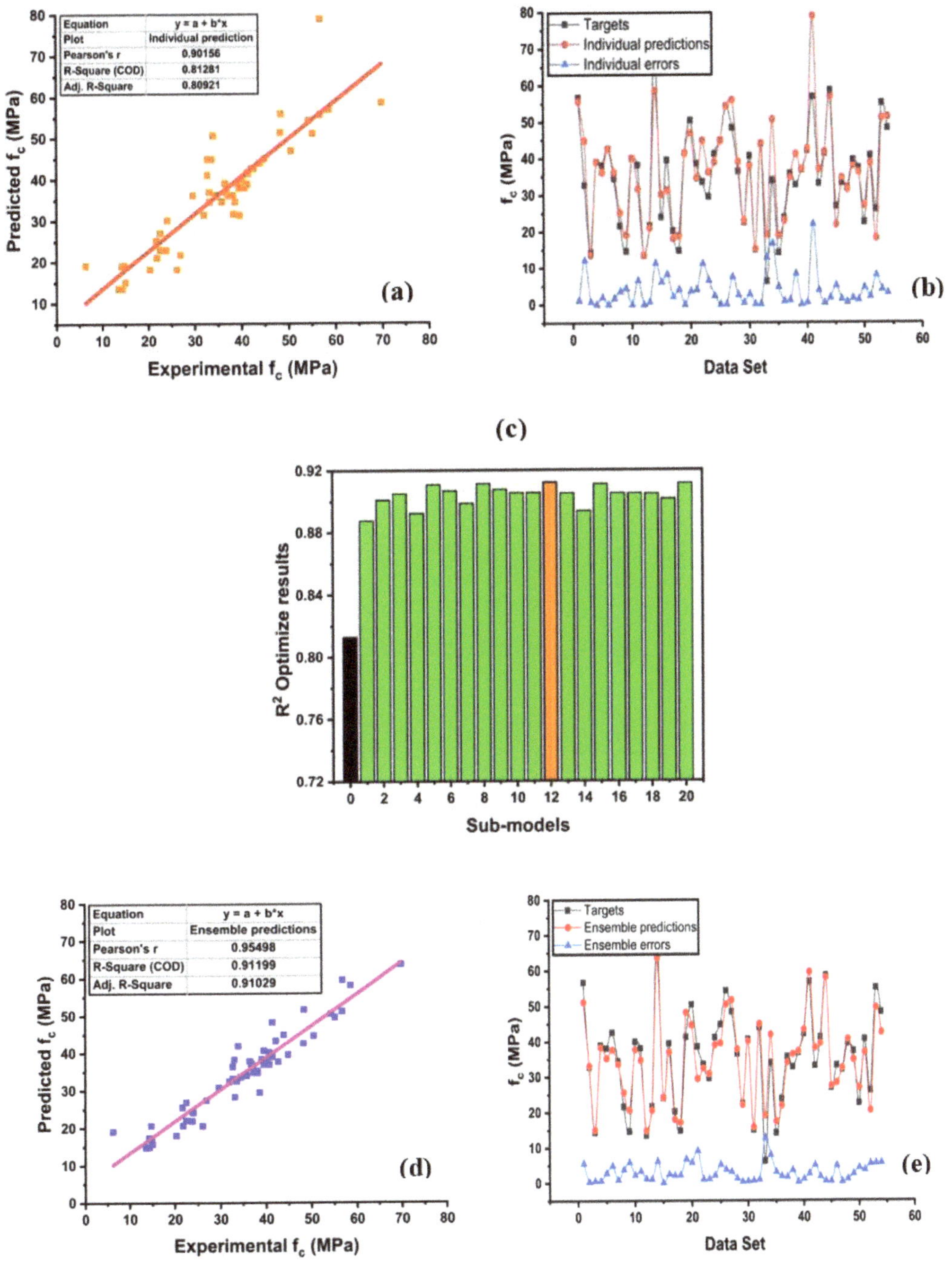

Figure 6. Decision tree (DT) with the ensemble model. (**a**) Predicted regression model with DT. (**b**) Model errors between targets and predictions from the DT technique. (**c**) Optimize model of ensemble. (**d**) Predicted regression model with DT-bagging. (**e**) Model Errors between targets and predictions from the DT-bagging technique.

Table 2. Evaluation of models.

Data Points	Decision Tree (DT) Targets	DT Predictions	Ensemble Prediction	Gene Expression Programming (GEP) Targets	GEP Predictions	DT Errors	Ensemble Errors	GEP Errors
1	56.74	55.64	51.14	26.74	27.66	1.10	5.60	0.92
2	32.72	44.87	33.11	37.44	37.21	12.15	0.39	0.23
3	14.31	13.52	14.94	51.04	49.48	0.79	0.63	1.56
4	39.06	39.05	38.35	18.13	21.68	0.01	0.71	3.55
5	38.11	36.15	35.22	51.33	49.31	1.96	2.89	2.02
6	42.64	42.64	37.67	37.91	39.76	0.00	4.97	1.85
7	34.49	36.15	33.57	25.10	36.42	1.66	0.92	11.32
8	21.65	25.18	25.55	74.17	77.61	3.53	3.90	3.44
9	14.7	19.11	20.62	37.27	40.69	4.41	5.92	3.42
10	40.06	40.06	37.74	15.05	14.58	0.00	2.32	0.47
11	38.21	31.65	34.74	23.52	22.76	6.56	3.47	0.76
12	13.52	13.52	14.79	41.89	42.56	0.00	1.27	0.67
13	21.78	21.02	20.63	48.79	46.44	0.76	1.15	2.35
14	69.84	58.52	63.61	40.68	40.59	11.32	6.23	0.09
15	24	30.14	24.11	32.92	34.99	6.14	0.11	2.07
16	39.58	31.35	37.01	25.18	26.87	8.23	2.57	1.69
17	20.28	18.13	18.00	59.20	85.40	2.15	2.28	26.20
18	14.84	18.91	17.15	33.94	32.67	4.07	2.31	1.27
19	41.37	41.37	48.22	53.30	49.35	0.00	6.85	3.95
20	50.51	46.9	44.65	42.22	46.77	3.61	5.86	4.55
21	38.6	34.57	29.43	30.96	18.90	4.03	9.17	12.06
22	33.61	44.87	32.51	21.75	25.93	11.26	1.10	4.18
23	29.59	36.15	30.80	12.54	9.95	6.56	1.21	2.59
24	41.24	38.89	39.02	31.18	37.50	2.35	2.22	6.32
25	44.86	44.87	39.55	14.20	16.26	0.01	5.31	2.06
26	54.32	54.28	50.46	33.80	36.88	0.04	3.86	3.08
27	48.4	55.94	51.65	30.14	29.77	7.54	3.25	0.37
28	36.45	39	37.80	31.88	36.23	2.55	1.35	4.35
29	22.5	22.95	22.03	30.12	36.52	0.45	0.47	6.40
30	40.66	37.91	40.12	32.72	32.57	2.75	0.54	0.15
31	14.99	15.05	15.71	30.85	41.47	0.06	0.72	10.62
32	43.89	43.94	44.89	43.70	45.88	0.05	1.00	2.18
33	6.27	19.11	19.05	24.50	25.95	12.84	12.78	1.45
34	33.94	50.6	41.89	39.29	41.35	16.66	7.95	2.06
35	14.2	18.91	17.38	32.07	35.21	4.71	3.18	3.14
36	23.8	22.95	21.86	9.01	13.37	0.85	1.94	4.36
37	35.76	34.68	33.99	22.50	18.93	1.08	1.77	3.57
38	32.72	41.05	36.38	14.50	16.89	8.33	3.66	2.39
39	36.8	36.8	37.17	39.06	37.75	0.00	0.37	1.31
40	42.13	42.62	43.28	42.42	43.95	0.49	1.15	1.53
41	56.83	78.8	59.50	42.13	38.72	21.97	2.67	3.41
42	33.08	36.94	38.20	42.03	48.82	3.86	5.12	6.79
43	41.3	41.64	39.45	43.89	46.77	0.34	1.85	2.88
44	58.61	56.85	58.01	40.06	31.79	1.76	0.60	8.27
45	26.85	21.75	27.39	48.28	47.22	5.10	0.54	1.06
46	33.21	34.57	28.27	37.42	37.11	1.36	4.94	0.31
47	31.97	31.45	32.39	39.49	41.69	0.52	0.42	2.20
48	39.49	37.91	40.61	36.94	38.73	1.58	1.12	1.79
49	37.33	36.15	34.77	25.22	27.85	1.18	2.56	2.63
50	22.53	27.04	26.86	55.64	55.64	4.51	4.33	0.00
51	40.68	38.63	36.96	60.95	60.16	2.05	3.72	0.79
52	26.06	18.13	20.51	37.33	40.76	7.93	5.55	3.43
53	55.16	51.04	49.54	25.45	29.65	4.12	5.62	4.20
54	48.28	51.33	42.55	-	-	3.05	5.73	-

5.2. Gene Expression Programming

The performance of the model by GEP yielded a robust relationship between targets and predicted, as illustrated in Figure 7. It can be seen that R^2 by employing GEP is close to 1. Moreover, Figure 7b represents the error distribution of the testing set with fewer errors. Similarly, the predicted value shows a lower error to target values with a minimum, maximum, and average value of 0.00 MPa, 26.20 MPa, and 3.48 MPa, respectively. Table 2 presents detailed results from the models.

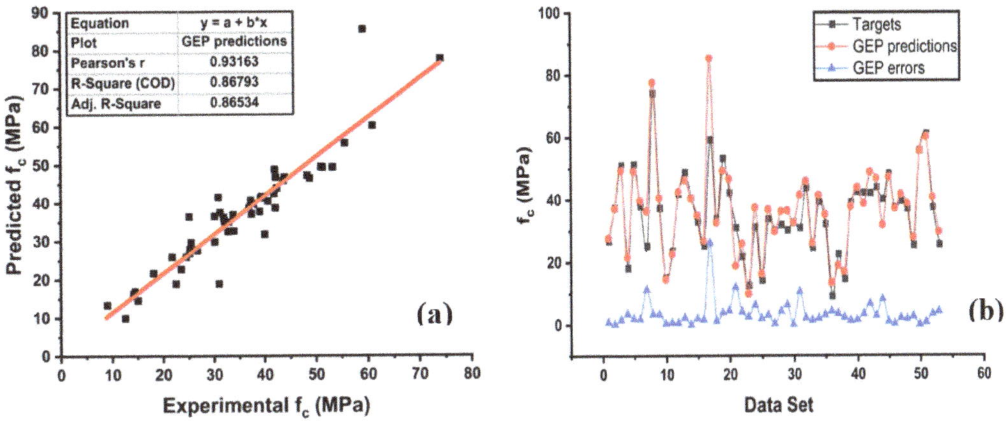

Figure 7. Gene expression programming (GEP) model: (**a**) Predicted regression model. (**b**) Model Errors between targets and predictions from the GEP model.

5.3. Evaluation of the Model by K-Fold and Statistical Checks

Cross-validation is a statistical practice used to evaluate or estimate the actual performance of the machine learning models. It is necessary to know the performance of the selected models. For this purpose, a validation technique is required to find the accuracy level of the model's data. Shuffling of the data set randomly and splitting a dataset into k-groups is required for the k-fold validation test. In the described study, data of experimental samples are equally divided into 10 subsets. It uses nine out of ten subsets, while the only subset is utilized for the validation of the model. The same approach of this process is then repeated 10 times for obtaining the average accuracy of these 10 repetitions. It is clarified widely that the 10-fold cross-validation method well represents the conclusion and accuracy of the model performance [48].

Bias and a variance decrease for the test set can be checked by employing K-fold cross-validation. The results of cross-validation are evaluated by a correlation coefficient (R^2), a mean absolute error (MAE), a mean square error (MSE), and a root mean square error (RMSE), as illustrated in Figure 8. The ensemble model shows fewer errors and better R^2 as compared to GEP. The average R^2 for ensemble modeling is 0.905 with a maximum and minimum values of 0.84 and 0.96, as depicted in Figure 8a. Whereas the GEP model shows an average R^2 = 0.873 of ten folds with 0.76 and 0.95 for a minimum and maximum correlation, respectively, as shown in Figure 8b. Each model shows fewer errors for validation. The validation indicator result shows that ensemble means values of MAE, MSE, and RMSE come to be 6.43 MPa, 6.66 MPa, and 2.55 MPa, respectively. Similarly, the GEP model shows the same trend by showing fewer errors. The GEP model shows mean values of 7.30 MPa, 9.60 MPa, and 3.06 MPa for MAE, MSE, and RMSE, respectively (see Figure 8b). Table 3 represents the validation results of both models.

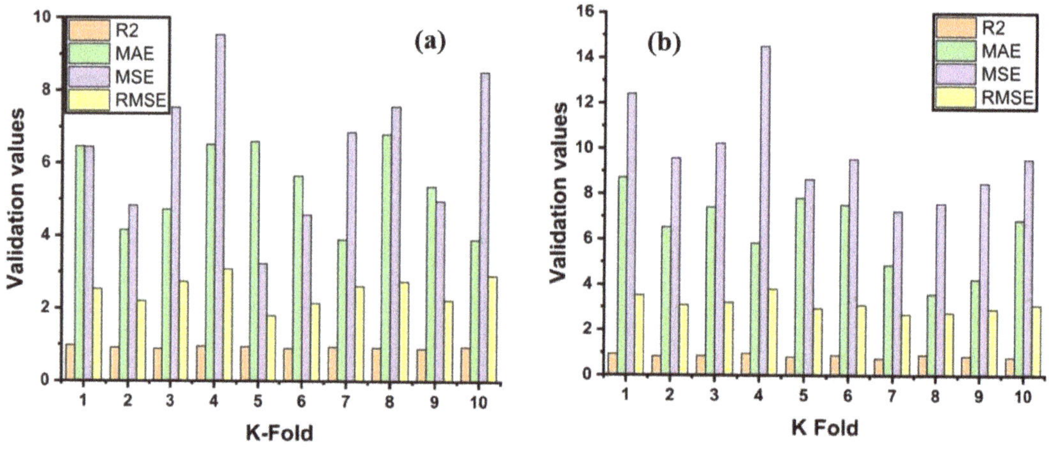

Figure 8. Statistical indicators from K-Fold Cross-validation; (**a**) Ensemble model; (**b**) GEP model.

Table 3. Result of K-Fold Cross-validation.

K Fold	Ensemble Model				GEP Model			
	R^2	MAE	MSE	RMSE	R^2	MAE	MSE	RMSE
1	0.96	8.46	4.45	2.10	0.86	10.71	13.57	3.68
2	0.91	5.17	7.44	2.72	0.94	7.45	7.97	2.82
3	0.84	3.73	8.54	2.92	0.89	6.18	11.24	3.35
4	0.90	9.52	5.84	2.41	0.95	5.84	14.51	3.80
5	0.94	6.81	6.44	2.53	0.93	7.81	9.64	3.10
6	0.90	5.65	5.88	2.42	0.86	7.51	6.51	2.55
7	0.85	7.91	6.87	2.62	0.81	8.47	7.25	2.69
8	0.88	5.81	9.85	3.13	0.76	6.58	7.58	2.75
9	0.95	6.37	4.97	2.22	0.84	5.64	9.47	3.07
10	0.92	4.92	6.35	2.51	0.89	6.84	8.35	2.88

Statistical check is also applied to evaluate the model with regard to the testing results. The statistical check is an indicator that shows the model response towards prediction, as shown in Table 4. It can be seen that models depict bottom-most errors. However, the ensemble model shows a 25% error reduction for MAE as compared to the individual and GEP. Similarly, the bagging approach indicates the robust performance of the model. Moreover, MSE and RMSE for strong learners show 121% and 49% enhancement in the predictions by showing reduced errors between the target and predicted outcomes, as shown in Table 4.

Table 4. Statistical checks.

Statistics	MAE (MPa)	MSE (MPa)	RMSE (MPa)
DT	3.896	36.01	6.00
DT-BAG	3.113	16.28	4.03
GEP	3.478	29.91	5.46

Moreover, permutation feature importance via python is conducted to check the influence of variables on strength, as depicted in Figure 9. These variables have a vital influence on the prediction of compressive strength of concrete. The concrete age, cement, and water-to-cement ratio have a significant influence on model analysis. Whereas water, filler material (fly ash), superplasticizer, fine aggregate, and coarse aggregate have moderate influences in making the model. Thus, it can be concluded that every parameter is crucial in the forecasting of the strength properties. However, cement, age, and the water-to-cement ratio should be given more importance while casting of specimens.

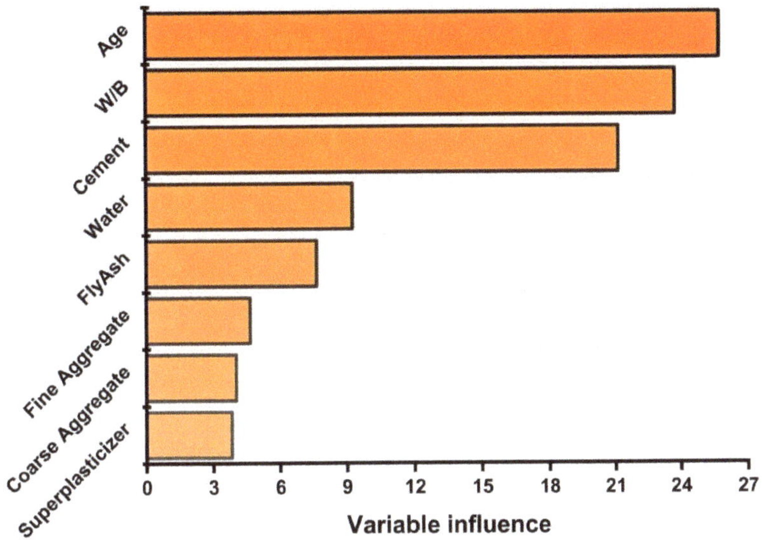

Figure 9. Variable influence on compressive strength of fly ash-based concrete.

5.4. Limitation and Future Work

Despite the fact that, in the work, a thorough analysis based on a large number of data points was conducted and an extensive machine learning algorithm with evaluation was implemented, the limitations of work should be mentioned. Described in the paper selection, an approach can be enhanced by using other appropriate methods. A clear limitation of work is the number of data points equal to 270. The study is also limited to predict only one result from various mechanical properties of concrete. Tensile strength, durability, corrosion, toughness, and abrasion behavior of concrete is not considered in this work. Other algorithm-based techniques, like artificial neural network (ANN), support vector machine (SVM), gradient boosting, and AdaBoost may also be applied to the same dataset for a better understanding. However, this research work does not only focus on algorithm-based techniques but also involves the programming-based GEP, which indicated the wide scope of this work.

Since concrete is the most widely used material after water on this earth, it is further recommended that other properties of this material should be incorporated except for its compressive strength. Machine learning techniques should also be used to predict the environmental effects on concrete properties. To achieve high accuracy in the actual and predicted results, the multi-stage genetic programming approach may also be used. It is also recommended that models can be run for the concrete modified with different fibers as: jute fibers, glass fibers, polypropylene fibers, nylon fibers, and steel fibers.

6. Conclusions

This study describes the supervised machine learning approaches with ensemble modeling and gene expression programming to predict concrete strength. The following points are drawn from the analysis:

1. A decision tree with ensemble modeling gives a robust performance compared to a decision tree individually and with gene expression programming. The correlation coefficient of R^2 = 0.911 is reported for DT with bagging.

2. Optimization of the model for the decision tree with bagging is done by making twenty sub-models. Magnificent enhancement is observed from the twelve, which shows $R^2 = 0.911$ as compared to the individual model with $R^2 = 0.812$.
3. Validation score is conducted by different indicators. Both models (DT with bagging and GEP) show better anticipation for testing results.
4. Statistical analysis checks reveal that the decision tree with bagging shows enhancement in model accuracy by minimizing the error difference between targeted and predicted values.

To summarize, all applied algorithms show a significant effect on the model's quality by predicting the target response more accurately. As described in the paper, machine learning approaches can save experimental time and predict the outcome by gathering extensive data from laboratory and published papers. It can help the scientific society to predict the properties and responses in the coming month or year.

Author Contributions: A.A. (Ayaz Ahmad)—conceptualization, modeling, and write up. F.F.—visualization and review. P.N.—writing, validation and supervision. K.O.—funding and review. A.A. (Arslan Akbar)—review, editing and visualization. F.A. and R.A.—data acquisition. All authors have read and agreed to the published version of the manuscript.

Funding: The APC was funded by Wroclaw University of Science and Technology.

Data Availability Statement: The data presented in this article is available within the article.

Conflicts of Interest: The authors declared no conflict of interest.

Appendix A

Table A1. Experimental variable data.

S. No.	Cement	Fly Ash	Water	Super Plasticizer	Coarse Aggregate	Fine Aggregate	Days	W/C	Strength
1	540	0	162	2.5	1040	676	28	0.3	79.99
2	540	0	162	2.5	1055	676	28	0.3	61.89
3	475	0	228	0	932	594	28	0.48	39.29
4	380	0	228	0	932	670	90	0.6	52.91
5	475	0	228	0	932	594	180	0.48	42.62
6	380	0	228	0	932	670	365	0.6	52.52
7	380	0	228	0	932	670	270	0.6	53.3
8	475	0	228	0	932	594	7	0.48	38.6
9	475	0	228	0	932	594	270	0.48	42.13
10	475	0	228	0	932	594	90	0.48	42.23
11	380	0	228	0	932	670	180	0.6	53.1
12	349	0	192	0	1047	806.9	3	0.55	15.05
13	475	0	228	0	932	594	365	0.48	41.93
14	310	0	192	0	971	850.6	3	0.62	9.87
15	485	0	146	0	1120	800	28	0.3	71.99
16	531.3	0	141.8	28.2	852.1	893.7	3	0.27	41.3
17	531.3	0	141.8	28.2	852.1	893.7	7	0.27	46.9
18	531.3	0	141.8	28.2	852.1	893.7	28	0.27	56.4
19	531.3	0	141.8	28.2	852.1	893.7	56	0.27	58.8
20	531.3	0	141.8	28.2	852.1	893.7	91	0.27	59.2
21	290.4	96.2	168.1	9.4	961.2	865	3	0.58	22.5
22	290.4	96.2	168.1	9.4	961.2	865	14	0.58	34.67
23	290.4	96.2	168.1	9.4	961.2	865	28	0.58	34.74
24	290.4	96.2	168.1	9.4	961.2	865	56	0.58	45.08
25	290.4	96.2	168.1	9.4	961.2	865	100	0.58	48.97
26	277.1	97.4	160.6	11.8	973.9	875.6	3	0.58	23.14
27	277.1	97.4	160.6	11.8	973.9	875.6	14	0.58	41.89
28	277.1	97.4	160.6	11.8	973.9	875.6	28	0.58	48.28

Table A1. *Cont.*

S. No.	Cement	Fly Ash	Water	Super Plasticizer	Coarse Aggregate	Fine Aggregate	Days	W/C	Strength
29	277.1	97.4	160.6	11.8	973.9	875.6	56	0.58	51.04
30	277.1	97.4	160.6	11.8	973.9	875.6	100	0.58	55.64
31	295.7	95.6	171.5	8.9	955.1	859.2	3	0.58	22.95
32	295.7	95.6	171.5	8.9	955.1	859.2	14	0.58	35.23
33	295.7	95.6	171.5	8.9	955.1	859.2	28	0.58	39.94
34	295.7	95.6	171.5	8.9	955.1	859.2	56	0.58	48.72
35	295.7	95.6	171.5	8.9	955.1	859.2	100	0.58	52.04
36	251.8	99.9	146.1	12.4	1006	899.8	3	0.58	21.02
37	251.8	99.9	146.1	12.4	1006	899.8	14	0.58	33.36
38	251.8	99.9	146.1	12.4	1006	899.8	28	0.58	33.94
39	251.8	99.9	146.1	12.4	1006	899.8	56	0.58	44.14
40	251.8	99.9	146.1	12.4	1006	899.8	100	0.58	45.37
41	249.1	98.8	158.1	12.8	987.8	889	3	0.63	15.36
42	249.1	98.8	158.1	12.8	987.8	889	14	0.63	28.68
43	249.1	98.8	158.1	12.8	987.8	889	28	0.63	30.85
44	249.1	98.8	158.1	12.8	987.8	889	56	0.63	42.03
45	249.1	98.8	158.1	12.8	987.8	889	100	0.63	51.06
46	252.3	98.8	146.3	14.2	987.8	889	3	0.58	21.78
47	252.3	98.8	146.3	14.2	987.8	889	14	0.58	42.29
48	252.3	98.8	146.3	14.2	987.8	889	28	0.58	50.6
49	252.3	98.8	146.3	14.2	987.8	889	56	0.58	55.83
50	252.3	98.8	146.3	14.2	987.8	889	100	0.58	60.95
51	246.8	125.1	143.3	12	1086.8	800.9	3	0.58	23.52
52	246.8	125.1	143.3	12	1086.8	800.9	14	0.58	42.22
53	246.8	125.1	143.3	12	1086.8	800.9	28	0.58	52.5
54	246.8	125.1	143.3	12	1086.8	800.9	56	0.58	60.32
55	246.8	125.1	143.3	12	1086.8	800.9	100	0.58	66.42
56	275.1	121.4	159.5	9.9	1053.6	777.5	3	0.58	23.8
57	275.1	121.4	159.5	9.9	1053.6	777.5	14	0.58	38.77
58	275.1	121.4	159.5	9.9	1053.6	777.5	28	0.58	51.33
59	275.1	121.4	159.5	9.9	1053.6	777.5	56	0.58	56.85
60	275.1	121.4	159.5	9.9	1053.6	777.5	100	0.58	58.61
61	297.2	117.5	174.8	9.5	1022.8	753.5	3	0.59	21.91
62	297.2	117.5	174.8	9.5	1022.8	753.5	14	0.59	36.99
63	297.2	117.5	174.8	9.5	1022.8	753.5	28	0.59	47.4
64	297.2	117.5	174.8	9.5	1022.8	753.5	56	0.59	51.96
65	297.2	117.5	174.8	9.5	1022.8	753.5	100	0.59	56.74
66	376	0	214.6	0	1003.5	762.4	3	0.57	16.28
67	376	0	214.6	0	1003.5	762.4	14	0.57	25.62
68	376	0	214.6	0	1003.5	762.4	28	0.57	31.97
69	376	0	214.6	0	1003.5	762.4	56	0.57	36.3
70	376	0	214.6	0	1003.5	762.4	100	0.57	43.06
71	500	0	140	4	966	853	28	0.28	67.57
72	475	59	142	1.9	1098	641	28	0.3	57.23
73	505	60	195	0	1030	630	28	0.39	64.02
74	451	0	165	11.3	1030	745	28	0.37	78.8
75	516	0	162	8.2	801	802	28	0.31	41.37
76	520	0	170	5.2	855	855	28	0.33	60.28
77	528	0	185	6.9	920	720	28	0.35	56.83
78	520	0	175	5.2	870	805	28	0.34	51.02
79	385	136	158	20	903	768	28	0.41	55.55
80	500.1	0	200	3	1124.4	613.2	28	0.4	44.13
81	405	0	175	0	1120	695	28	0.43	52.3
82	516	0	162	8.3	801	802	28	0.31	41.37
83	475	0	162	9.5	1044	662	28	0.34	58.52

Table A1. *Cont.*

S. No.	Cement	Fly Ash	Water	Super Plasticizer	Coarse Aggregate	Fine Aggregate	Days	W/C	Strength
84	500	0	151	9	1033	655	28	0.3	69.84
85	436	0	218	0	838.4	719.7	28	0.5	23.85
86	289	0	192	0	913.2	895.3	90	0.66	32.07
87	289	0	192	0	913.2	895.3	3	0.66	11.65
88	393	0	192	0	940.6	785.6	3	0.49	19.2
89	393	0	192	0	940.6	785.6	90	0.49	48.85
90	393	0	192	0	940.6	785.6	28	0.49	39.6
91	480	0	192	0	936.2	712.2	28	0.4	43.94
92	480	0	192	0	936.2	712.2	7	0.4	34.57
93	480	0	192	0	936.2	712.2	90	0.4	54.32
94	480	0	192	0	936.2	712.2	3	0.4	24.4
95	333	0	192	0	931.2	842.6	3	0.58	15.62
96	289	0	192	0	913.2	895.3	7	0.66	14.6
97	333	0	192	0	931.2	842.6	28	0.58	31.97
98	333	0	192	0	931.2	842.6	7	0.58	23.4
99	289	0	192	0	913.2	895.3	28	0.66	25.57
100	333	0	192	0	931.2	842.6	90	0.58	41.68
101	393	0	192	0	940.6	785.6	7	0.49	27.74
102	397	0	185.7	0	1040.6	734.3	28	0.47	33.08
103	382.5	0	185.7	0	1047.8	739.3	7	0.49	24.07
104	295.8	0	185.7	0	1091.4	769.3	7	0.63	14.84
105	397	0	185.7	0	1040.6	734.3	7	0.47	25.45
106	381.4	0	185.7	0	1104.6	784.3	28	0.49	22.49
107	295.8	0	185.7	0	1091.4	769.3	28	0.63	25.22
108	339.2	0	185.7	0	1069.2	754.3	7	0.55	21.18
109	381.4	0	185.7	0	1104.6	784.3	7	0.49	14.54
110	339.2	0	185.7	0	1069.2	754.3	28	0.55	31.9
111	382.5	0	185.7	0	1047.8	739.3	28	0.49	37.44
112	339	0	197	0	968	781	3	0.58	13.22
113	339	0	197	0	968	781	7	0.58	20.97
114	339	0	197	0	968	781	14	0.58	27.04
115	339	0	197	0	968	781	28	0.58	32.04
116	339	0	197	0	968	781	90	0.58	35.17
117	339	0	197	0	968	781	180	0.58	36.45
118	339	0	197	0	968	781	365	0.58	38.89
119	277	0	191	0	968	856	14	0.69	21.26
120	277	0	191	0	968	856	28	0.69	25.97
121	277	0	191	0	968	856	3	0.69	11.36
122	277	0	191	0	968	856	90	0.69	31.25
123	277	0	191	0	968	856	180	0.69	32.33
124	277	0	191	0	968	856	360	0.69	33.7
125	307	0	193	0	968	812	180	0.63	34.49
126	307	0	193	0	968	812	365	0.63	36.15
127	307	0	193	0	968	812	3	0.63	12.54
128	307	0	193	0	968	812	28	0.63	27.53
129	307	0	193	0	968	812	90	0.63	32.92
130	325	0	184	0	1063	783	7	0.57	17.54
131	325	0	184	0	1063	783	28	0.57	30.57
132	275	0	183	0	1088	808	7	0.67	14.2
133	275	0	183	0	1088	808	28	0.67	24.5
134	300	0	184	0	1075	795	7	0.61	15.58
135	300	0	184	0	1075	795	28	0.61	26.85
136	375	0	186	0	1038	758	7	0.5	26.06
137	375	0	186	0	1038	758	28	0.5	38.21

Table A1. *Cont.*

S. No.	Cement	Fly Ash	Water	Super Plasticizer	Coarse Aggregate	Fine Aggregate	Days	W/C	Strength
138	400	0	187	0	1025	745	28	0.47	43.7
139	400	0	187	0	1025	745	7	0.47	30.14
140	350	0	186	0	1050	770	7	0.53	20.28
141	350	0	186	0	1050	770	28	0.53	34.29
142	310	0	192	0	1012	830	3	0.62	11.85
143	310	0	192	0	1012	830	7	0.62	17.24
144	310	0	192	0	1012	830	28	0.62	27.83
145	310	0	192	0	1012	830	90	0.62	35.76
146	310	0	192	0	1012	830	120	0.62	38.7
147	331	0	192	0	1025	821	3	0.58	14.31
148	331	0	192	0	1025	821	7	0.58	17.44
149	331	0	192	0	1025	821	28	0.58	31.74
150	331	0	192	0	1025	821	90	0.58	37.91
151	331	0	192	0	1025	821	120	0.58	39.38
152	349	0	192	0	1056	809	3	0.55	15.87
153	349	0	192	0	1056	809	7	0.55	9.01
154	349	0	192	0	1056	809	28	0.55	33.61
155	349	0	192	0	1056	809	90	0.55	40.66
156	349	0	192	0	1056	809	120	0.55	40.86
157	296	0	186	0	1090	769	7	0.63	18.91
158	296	0	186	0	1090	769	28	0.63	25.18
159	297	0	186	0	1040	734	7	0.63	30.96
160	480	0	192	0	936	721	28	0.4	43.89
161	480	0	192	0	936	721	90	0.4	54.28
162	397	0	186	0	1040	734	28	0.47	36.94
163	281	0	186	0	1104	774	7	0.66	14.5
164	281	0	185	0	1104	774	28	0.66	22.44
165	500	0	200	0	1125	613	1	0.4	12.64
166	500	0	200	0	1125	613	3	0.4	26.06
167	500	0	200	0	1125	613	7	0.4	33.21
168	500	0	200	0	1125	613	14	0.4	36.94
169	500	0	200	0	1125	613	28	0.4	44.09
170	540	0	173	0	1125	613	7	0.32	52.61
171	540	0	173	0	1125	613	14	0.32	59.76
172	540	0	173	0	1125	613	28	0.32	67.31
173	540	0	173	0	1125	613	90	0.32	69.66
174	540	0	173	0	1125	613	180	0.32	71.62
175	540	0	173	0	1125	613	270	0.32	74.17
176	350	0	203	0	974	775	7	0.58	18.13
177	350	0	203	0	974	775	14	0.58	22.53
178	350	0	203	0	974	775	28	0.58	27.34
179	350	0	203	0	974	775	56	0.58	29.98
180	350	0	203	0	974	775	90	0.58	31.35
181	350	0	203	0	974	775	180	0.58	32.72
182	385	0	186	0	966	763	1	0.48	6.27
183	385	0	186	0	966	763	3	0.48	14.7
184	385	0	186	0	966	763	7	0.48	23.22
185	385	0	186	0	966	763	14	0.48	27.92
186	385	0	186	0	966	763	28	0.48	31.35
187	331	0	192	0	978	825	180	0.58	39
188	331	0	192	0	978	825	360	0.58	41.24
189	349	0	192	0	1047	806	3	0.55	14.99
190	331	0	192	0	978	825	3	0.58	13.52
191	382	0	186	0	1047	739	7	0.49	24
192	382	0	186	0	1047	739	28	0.49	37.42

Table A1. *Cont.*

S. No.	Cement	Fly Ash	Water	Super Plasticizer	Coarse Aggregate	Fine Aggregate	Days	W/C	Strength
193	382	0	186	0	1111	784	7	0.49	11.47
194	281	0	186	0	1104	774	28	0.66	22.44
195	339	0	185	0	1069	754	7	0.55	21.16
196	339	0	185	0	1069	754	28	0.55	31.84
197	295	0	185	0	1069	769	7	0.63	14.8
198	295	0	185	0	1069	769	28	0.63	25.18
199	296	0	192	0	1085	765	7	0.65	14.2
200	296	0	192	0	1085	765	28	0.65	21.65
201	296	0	192	0	1085	765	90	0.65	29.39
202	331	0	192	0	879	825	3	0.58	13.52
203	331	0	192	0	978	825	7	0.58	16.26
204	331	0	192	0	978	825	28	0.58	31.45
205	331	0	192	0	978	825	90	0.58	37.23
206	349	0	192	0	1047	806	7	0.55	18.13
207	349	0	192	0	1047	806	28	0.55	32.72
208	349	0	192	0	1047	806	90	0.55	39.49
209	349	0	192	0	1047	806	180	0.55	41.05
210	349	0	192	0	1047	806	360	0.55	42.13
211	302	0	203	0	974	817	14	0.67	18.13
212	302	0	203	0	974	817	180	0.67	26.74
213	525	0	189	0	1125	613	180	0.36	61.92
214	500	0	200	0	1125	613	90	0.4	47.22
215	500	0	200	0	1125	613	180	0.4	51.04
216	500	0	200	0	1125	613	270	0.4	55.16
217	540	0	173	0	1125	613	3	0.32	41.64
218	339	0	185	0	1060	754	28	0.55	31.65
219	393	0	192	0	940	758	3	0.49	19.11
220	393	0	192	0	940	758	28	0.49	39.58
221	393	0	192	0	940	758	90	0.49	48.79
222	382	0	185	0	1047	739	7	0.48	24
223	382	0	185	0	1047	739	28	0.48	37.42
224	310	0	192	0	970	850	7	0.62	14.99
225	310	0	192	0	970	850	28	0.62	27.92
226	310	0	192	0	970	850	90	0.62	34.68
227	310	0	192	0	970	850	180	0.62	37.33
228	310	0	192	0	970	850	360	0.62	38.11
229	525	0	189	0	1125	613	3	0.36	33.8
230	525	0	189	0	1125	613	7	0.36	42.42
231	525	0	189	0	1125	613	14	0.36	48.4
232	525	0	189	0	1125	613	28	0.36	55.94
233	525	0	189	0	1125	613	90	0.36	58.78
234	525	0	189	0	1125	613	270	0.36	67.11
235	322	0	203	0	974	800	14	0.63	20.77
236	322	0	203	0	974	800	28	0.63	25.18
237	322	0	203	0	974	800	180	0.63	29.59
238	302	0	203	0	974	817	28	0.67	21.75
239	397	0	185	0	1040	734	28	0.47	39.09
240	480	0	192	0	936	721	3	0.4	24.39
241	522	0	146	0	896	896	7	0.28	50.51
242	522	0	146	0	896	896	28	0.28	74.99
243	374	0	190	7	1013	730	28	0.51	39.05
244	305	100	196	10	959	705	28	0.64	30.12
245	298	107	186	6	879	815	28	0.62	42.64
246	318	126	210	6	861	737	28	0.66	40.06
247	356	142	193	11	801	778	28	0.54	40.87

Table A1. *Cont.*

S. No.	Cement	Fly Ash	Water	Super Plasticizer	Coarse Aggregate	Fine Aggregate	Days	W/C	Strength
248	314	113	170	10	925	783	28	0.54	38.46
249	321	128	182	11	870	780	28	0.57	37.26
250	298	107	210	11	880	744	28	0.7	31.87
251	322	116	196	10	818	813	28	0.61	31.18
252	313	113	178	8	1002	689	28	0.57	36.8
253	326	138	199	11	801	792	28	0.61	40.68
254	336	0	182	3	986	817	28	0.54	44.86
255	298	107	164	13	953	784	28	0.55	35.86
256	313	0	178	8	1000	822	28	0.57	25.1
257	313.3	113	178.5	8	1001.9	688.7	28	0.57	36.8
258	326.5	137.9	199	10.8	801.1	792.5	28	0.61	38.63
259	336.5	0	181.9	3.4	985.8	816.8	28	0.54	44.87
260	298.1	107.5	163.6	12.8	953.2	784	28	0.55	35.87
261	312.7	0	178.1	8	999.7	822.2	28	0.57	25.1
262	374.3	0	190.2	6.7	1013.2	730.4	28	0.51	39.06
263	304.8	99.6	196	9.8	959.4	705.2	28	0.64	30.12
264	298.1	107	186.4	6.1	879	815.2	28	0.63	42.64
265	317.9	126.5	209.7	5.7	860.5	736.6	28	0.66	40.06
266	355.9	141.6	193.3	11	801.4	778.4	28	0.54	40.87
267	313.8	112.6	169.9	10.1	925.3	782.9	28	0.54	38.46
268	321.4	127.9	182.5	11.5	870.1	779.7	28	0.57	37.27
269	298.2	107	209.7	11.1	879.6	744.2	28	0.7	31.88
270	322.2	115.6	196	10.4	817.9	813.4	28	0.61	31.18

References

1. Zhang, N.; Duan, H.; Miller, T.R.; Tam, V.W.Y.; Liu, G.; Zuo, J. Mitigation of carbon dioxide by accelerated sequestration in concrete debris. *Renew. Sustain. Energy Rev.* **2020**, *117*, 109495. [CrossRef]
2. Monkman, S.; MacDonald, M. On carbon dioxide utilization as a means to improve the sustainability of ready-mixed concrete. *J. Clean. Prod.* **2017**, *167*, 365–375. [CrossRef]
3. Qian, X.; Wang, J.; Fang, Y.; Wang, L. Carbon dioxide as an admixture for better performance of OPC-based concrete. *J. CO_2 Util.* **2018**, *25*, 31–38. [CrossRef]
4. Akbar, A.; Liew, K.M. Assessing recycling potential of carbon fiber reinforced plastic waste in production of eco-efficient cement-based materials. *J. Clean. Prod.* **2020**, *274*, 123001. [CrossRef]
5. Jahanzaib Khalil, M.; Aslam, M.; Ahmad, S. Utilization of sugarcane bagasse ash as cement replacement for the production of sustainable concrete—A review. *Constr. Build. Mater.* **2020**, 121371. [CrossRef]
6. Wang, L.; Chen, L.; Tsang, D.C.W.; Guo, B.; Yang, J.; Shen, Z.; Hou, D.; Ok, Y.S.; Poon, C.S. Biochar as green additives in cement-based composites with carbon dioxide curing. *J. Clean. Prod.* **2020**, *258*, 120678. [CrossRef]
7. Kang, S.; Lloyd, Z.; Kim, T.; Ley, M.T. Predicting the compressive strength of fly ash concrete with the Particle Model. *Cem. Concr. Res.* **2020**, *137*, 106218. [CrossRef]
8. Farooq, F.; Akbar, A.; Khushnood, R.A.; Muhammad, W.L.B.; Rehman, S.K.U.; Javed, M.F. Experimental investigation of hybrid carbon nanotubes and graphite nanoplatelets on rheology, shrinkage, mechanical, and microstructure of SCCM. *Materials* **2020**, *13*, 230. [CrossRef] [PubMed]
9. Akbar, A.; Farooq, F.; Shafique, M.; Aslam, F.; Alyousef, R.; Alabduljabbar, H. Sugarcane bagasse ash-based engineered geopolymer mortar incorporating propylene fibers. *J. Build. Eng.* **2021**, *33*, 101492. [CrossRef]
10. Ghouleh, Z.; Shao, Y. Turning municipal solid waste incineration into a cleaner cement production. *J. Clean. Prod.* **2018**, *195*, 268–279. [CrossRef]
11. Lee, M.G.; Kang, D.; Jo, H.; Park, J. Carbon dioxide utilization with carbonation using industrial waste-desulfurization gypsum and waste concrete. *J. Mater. Cycles Waste Manag.* **2016**, *18*, 407–412. [CrossRef]
12. Possan, E.; Thomaz, W.A.; Aleandri, G.A.; Felix, E.F.; dos Santos, A.C.P. CO2 uptake potential due to concrete carbonation: A case study. *Case Stud. Constr. Mater.* **2017**, *6*, 147–161. [CrossRef]
13. Farooq, F.; Ahmed, W.; Akbar, A.; Aslam, F.; Alyousef, R. Predictive modelling for sustainable high-performance concrete from industrial wastes: A comparison and optimization of models using ensemble learners. *J. Clean. Prod.* **2021**, *292*, 126032. [CrossRef]
14. Bin Khairul anuar, M.A.R.; Ngamkhanong, C.; Wu, Y.; Kaewunruen, S. Recycled Aggregates Concrete Compressive Strength Prediction Using Artificial Neural Networks (ANNs). *Infrastructures* **2021**, *6*, 17. [CrossRef]

15. Farooq, F.; Amin, M.N.; Khan, K.; Sadiq, M.R.; Javed, M.F.; Aslam, F.; Alyousef, R. A comparative study of random forest and genetic engineering programming for the prediction of compressive strength of high strength concrete (HSC). *Appl. Sci.* **2020**, *10*, 7330. [CrossRef]
16. Javed, M.F.; Farooq, F.; Memon, S.A.; Akbar, A.; Khan, M.A.; Aslam, F.; Alyousef, R.; Alabduljabbar, H.; Rehman, S.K.U.; Ur Rehman, S.K.; et al. New Prediction Model for the Ultimate Axial Capacity of Concrete-Filled Steel Tubes: An Evolutionary Approach. *Crystals* **2020**, *10*, 741. [CrossRef]
17. Iqbal, M.F.; Liu, Q.F.; Azim, I.; Zhu, X.; Yang, J.; Javed, M.F.; Rauf, M. Prediction of mechanical properties of green concrete incorporating waste foundry sand based on gene expression programming. *J. Hazard. Mater.* **2020**, *384*, 121322. [CrossRef]
18. De Domenico, D.; Ricciardi, G. Shear strength of RC beams with stirrups using an improved Eurocode 2 truss model with two variable-inclination compression struts. *Eng. Struct.* **2019**, *198*. [CrossRef]
19. Sadowski, Ł.; Nikoo, M.; Nikoo, M. Concrete compressive strength prediction using the imperialist competitive algorithm. *Comput. Concr.* **2018**, *22*, 355–363. [CrossRef]
20. Czarnecki, S.; Shariq, M.; Nikoo, M.; Sadowski, Ł. An intelligent model for the prediction of the compressive strength of cementitious composites with ground granulated blast furnace slag based on ultrasonic pulse velocity measurements. *Measurement* **2021**, 108951. [CrossRef]
21. Sadowski, Ł.; Piechówka-Mielnik, M.; Widziszowski, T.; Gardynik, A.; Mackiewicz, S. Hybrid ultrasonic-neural prediction of the compressive strength of environmentally friendly concrete screeds with high volume of waste quartz mineral dust. *J. Clean. Prod.* **2019**, *212*, 727–740. [CrossRef]
22. Javed, M.F.; Amin, M.N.; Shah, M.I.; Khan, K.; Iftikhar, B.; Farooq, F.; Aslam, F.; Alyousef, R.; Alabduljabbar, H. Applications of gene expression programming and regression techniques for estimating compressive strength of bagasse ash based concrete. *Crystals* **2020**, *10*, 737. [CrossRef]
23. Aslam, F.; Farooq, F.; Amin, M.N.; Khan, K.; Waheed, A.; Akbar, A.; Javed, M.F.; Alyousef, R.; Alabdulijabbar, H. Applications of Gene Expression Programming for Estimating Compressive Strength of High-Strength Concrete. *Adv. Civ. Eng.* **2020**. [CrossRef]
24. Naderpour, H.; Rafiean, A.H.; Fakharian, P. Compressive strength prediction of environmentally friendly concrete using artificial neural networks. *J. Build. Eng.* **2018**, *16*, 213–219. [CrossRef]
25. Getahun, M.A.; Shitote, S.M.; Abiero Gariy, Z.C. Artificial neural network based modelling approach for strength prediction of concrete incorporating agricultural and construction wastes. *Constr. Build. Mater.* **2018**, *190*, 517–525. [CrossRef]
26. Liu, Q.; Iqbal, M.F.; Yang, J.; Lu, X.; Zhang, P.; Rauf, M. Prediction of chloride diffusivity in concrete using artificial neural network: Modelling and performance evaluation. *Constr. Build. Mater.* **2020**, *268*, 121082. [CrossRef]
27. Ahmad, T.; Chen, H.; Huang, R.; Yabin, G.; Wang, J.; Shair, J.; Azeem Akram, H.M.; Hassnain Mohsan, S.A.; Kazim, M. Supervised based machine learning models for short, medium and long-term energy prediction in distinct building environment. *Energy* **2018**, *158*, 17–32. [CrossRef]
28. Wang, P.; Zheng, X.; Li, J.; Zhu, B. Prediction of epidemic trends in COVID-19 with logistic model and machine learning technics. *Chaos Solitons Fractals* **2020**, *139*, 110058. [CrossRef]
29. Cladera, A.; Marí, A.R. Shear design procedure for reinforced normal and high-strength concrete beams using artificial neural networks. Part I: Beams without stirrups. *Eng. Struct.* **2004**, *26*, 917–926. [CrossRef]
30. Onyari, E.K.; Ikotun, B.D. Prediction of compressive and flexural strengths of a modified zeolite additive mortar using artificial neural network. *Constr. Build. Mater.* **2018**, *187*, 1232–1241. [CrossRef]
31. Feng, D.C.; Liu, Z.T.; Wang, X.D.; Chen, Y.; Chang, J.Q.; Wei, D.F.; Jiang, Z.M. Machine learning-based compressive strength prediction for concrete: An adaptive boosting approach. *Constr. Build. Mater.* **2020**, *230*, 117000. [CrossRef]
32. Feng, D.C.; Liu, Z.T.; Wang, X.D.; Jiang, Z.M.; Liang, S.X. Failure mode classification and bearing capacity prediction for reinforced concrete columns based on ensemble machine learning algorithm. *Adv. Eng. Inform.* **2020**, *45*. [CrossRef]
33. Bui, D.K.; Nguyen, T.; Chou, J.S.; Nguyen-Xuan, H.; Ngo, T.D. A modified firefly algorithm-artificial neural network expert system for predicting compressive and tensile strength of high-performance concrete. *Constr. Build. Mater.* **2018**, *180*, 320–333. [CrossRef]
34. Salami, B.A.; Rahman, S.M.; Oyehan, T.A.; Maslehuddin, M.; Al Dulaijan, S.U. Ensemble machine learning model for corrosion initiation time estimation of embedded steel reinforced self-compacting concrete. *Meas. J. Int. Meas. Confed.* **2020**, *165*, 108141. [CrossRef]
35. Cai, R.; Han, T.; Liao, W.; Huang, J.; Li, D.; Kumar, A.; Ma, H. Prediction of surface chloride concentration of marine concrete using ensemble machine learning. *Cem. Concr. Res.* **2020**, *136*, 106164. [CrossRef]
36. Aydogmus, H.Y.; Erdal, H.I.; Karakurt, O.; Namli, E.; Turkan, Y.S.; Erdal, H. A comparative assessment of bagging ensemble models for modeling concrete slump flow. *Comput. Concr.* **2015**, *16*, 741–757. [CrossRef]
37. Erdal, H.I. Two-level and hybrid ensembles of decision trees for high performance concrete compressive strength prediction. *Eng. Appl. Artif. Intell.* **2013**, *26*, 1689–1697. [CrossRef]
38. Zounemat-Kermani, M.; Stephan, D.; Barjenbruch, M.; Hinkelmann, R. Ensemble data mining modeling in corrosion of concrete sewer: A comparative study of network-based (MLPNN & RBFNN) and tree-based (RF, CHAID, & CART) models. *Adv. Eng. Inform.* **2020**, *43*, 101030. [CrossRef]
39. Lichman, M. *UCI Machine Learning Repository*; University of California, School of Information and Computer Science: Irvine, CA, USA, 2013; Available online: http://archive.ics.uci.edu/ml (accessed on 27 November 2020).

40. Karbassi, A.; Mohebi, B.; Rezaee, S.; Lestuzzi, P. Damage prediction for regular reinforced concrete buildings using the decision tree algorithm. *Comput. Struct.* **2014**, *130*, 46–56. [CrossRef]
41. Ben Chaabene, W.; Flah, M.; Nehdi, M.L. Machine learning prediction of mechanical properties of concrete: Critical review. *Constr. Build. Mater.* **2020**, *260*, 119889. [CrossRef]
42. Ribeiro, M.H.D.M.; dos Santos Coelho, L. Ensemble approach based on bagging, boosting and stacking for short-term prediction in agribusiness time series. *Appl. Soft Comput. J.* **2020**, *86*, 105837. [CrossRef]
43. Dou, J.; Yunus, A.P.; Bui, D.T.; Merghadi, A.; Sahana, M.; Zhu, Z.; Chen, C.W.; Han, Z.; Pham, B.T. Improved landslide assessment using support vector machine with bagging, boosting, and stacking ensemble machine learning framework in a mountainous watershed, Japan. *Landslides* **2020**, *17*, 641–658. [CrossRef]
44. Kaboli, S.H.A.; Fallahpour, A.; Selvaraj, J.; Rahim, N.A. Long-term electrical energy consumption formulating and forecasting via optimized gene expression programming. *Energy* **2017**, *126*, 144–164. [CrossRef]
45. Ferreira, C. Gene Expression Programming in Problem Solving. In *Soft Computing and Industry*; Springer: London, UK, 2002; pp. 635–653.
46. Shah, M.I.; Javed, M.F.; Abunama, T. Proposed formulation of surface water quality and modelling using gene expression, machine learning, and regression techniques. *Environ. Sci. Pollut. Res.* **2020**, 1–19. [CrossRef] [PubMed]
47. Khan, M.A.; Memon, S.A.; Farooq, F.; Javed, M.F.; Aslam, F.; Alyousef, R. Compressive Strength of Fly-Ash-Based Geopolymer Concrete by Gene Expression Programming and Random Forest. *Adv. Civ. Eng.* **2021**, *2021*, 1–17. [CrossRef]
48. Kohavi, R. A Study of Cross-Validation and Bootstrap for Accuracy Estimation and Model Selection. In Proceedings of the International Joint Conference of Artificial Intelligence, Montreal, QC, Canada, 20–25 August 1995.

materials

MDPI

Article

Long-Term Concrete Shrinkage Influence on the Performance of Reinforced Concrete Structures

Alinda Dey [1,*], Akshay Vijay Vastrad [1], Mattia Francesco Bado [1,2], Aleksandr Sokolov [3] and Gintaris Kaklauskas [1]

1 Department of Reinforced Concrete Structures and Geotechnics, Vilnius Tech University (VGTU), Saulatekio al. 11, 10221 Vilnius, Lithuania; akshay-vijay.vastrad@stud.vgtu.lt (A.V.V.); mattia.francesco.bado1@upc.edu (M.F.B.); gintaris.kaklauskas@vgtu.lt (G.K.)
2 Department of Civil and Environmental Engineering, Technical University of Catalonia (UPC), c/Jordi Girona 1-3, 08034 Barcelona, Spain
3 Laboratory of Innovative Building Structures, Vilnius Tech University (VGTU), Saulatekio al. 11, 10221 Vilnius, Lithuania; aleksandr.sokolov@vgtu.lt
* Correspondence: alinda.dey@vgtu.lt

Abstract: The contribution of concrete to the tensile stiffness (tension stiffening) of a reinforced concrete (RC) member is a key governing factor for structural serviceability analyses. However, among the current tension stiffening models, few consider the effect brought forth by concrete shrinkage, and none studies take account of the effect for very long-term shrinkage. The present work intends to tackle this exact issue by testing multiple RC tensile elements (with different bar diameters and reinforcement ratios) after a five-year shrinking time period. The experimental deformative and tension stiffening responses were subjected to a mathematical process of shrinkage removal aimed at assessing its effect on the former. The results showed shrinkage distinctly lowered the cracking load of the RC members and caused an apparent tension stiffening reduction. Furthermore, both of these effects were exacerbated in the members with higher reinforcement ratios. The experimental and shrinkage-free behaviors of the RC elements were finally compared to the values predicted by the CEB-fib Model Code 2010 and the Euro Code 2. Interestingly, as a consequence of the long-term shrinkage, the codes expressed a smaller relative error when compared to the shrinkage-free curves versus the experimental ones.

Keywords: concrete; reinforced concrete; shrinkage; tension stiffening; concrete cracking

Citation: Dey, A.; Vastrad, A.V.; Bado, M.F.; Sokolov, A.; Kaklauskas, G. Long-Term Concrete Shrinkage Influence on the Performance of Reinforced Concrete Structures. *Materials* **2021**, *14*, 254. https://doi.org/10.3390/ma14020254

Received: 11 December 2020
Accepted: 4 January 2021
Published: 6 January 2021

1. Introduction

1.1. Theoretical Background

The continuous concrete consumption growth across the globe (from buildings to bridges, from dams to power plants, etc.) has led to an increased awareness over its behavior at both earlier and later stages of its structural life-cycle. Despite such, its most common application, reinforced concrete (RC), is characterized by a complex nonlinear behavior that has still not been wholly captured. Amongst its most relevant issues, its time-dependent behavior (concrete shrinkage and creep) is still abundantly in need of investigation. This research is of crucial importance for a proper design of RC structures, as shrinkage and creep may critically affect their stress distribution when subject to a load, thus altering its deformation behavior [1,2]. Indeed, the negligence of these phenomena can potentially lead to harsh structural consequences such as member shortening, excessive deflection, and early cracking [3,4].

Generally, we refer to concrete shrinkage as the reduction of concrete volume due to the evaporation of moisture stored in its gel pores in unsaturated air environments and with no applied stress [5]. Whilst the largest amount of compressive deformations occurs in the short term (due to the withdrawal of water from capillary pores caused by the hydration of the previously unhydrated cement [6,7]), shrinkage keeps occurring along all the service life of the structure, but at a considerably reduced rate [8]. Shrinkage in concrete is influenced by several factors such as temperature, humidity, time, mix design, material characteristics, curing processes, and specimen geometry. Indeed, when dealing with larger structure sizes (skyscraper columns and long-span bridge beams amongst others), shrinkage becomes a predominant structural deformation parameter [9,10]. The observation and monitoring of this phenomenon was recently achieved with Fiber Bragg Grating (FBG) [11] and the cutting edge monitoring technology represented by distributed optical fiber sensors (DOFS) [12].

The shrinkage restraint of such a phenomenon induced by embedded elements (reinforcement bars in the case of RC structures) causes the compression of the latter and the rise of tensile stresses in the concrete surrounding it. This is aggravated by the nonuniform shrinkage occurring inside a member due to its nonuniform moisture distribution. The tensile stresses of the concrete, however, are relieved by the presence of the tensile creep phenomenon [13]. The latter can be defined as the slow and gradual deformation of the material under the continuous influence of mechanical stresses (such as the one induced by concrete shrinkage), leading, in our case, to the relaxation of the concrete. The impact of creep on shrinkage strain is dependent upon the member size, and for small sections (as the ones presented in the current article), it can be assumed null [14].

The consequences that restrained shrinkage brings to the table should be properly taken in consideration when designing an RC structure. First, among the shrinkage-restraint issues is the reduction of the tensile capacity of concrete [15] and thus the lowering of the cracking load of the structure. Secondly, if the tensile strength of concrete is to be surpassed, the development of cracks will be initiated, thus facilitating the corrosion of the reinforcement if the former is to occur in a harsh and chemically aggressive environment. This obviously would negatively influence the durability and serviceability of the structure [16–18]. Finally, it has been observed [19–21] that shrinkage negatively affects the tension stiffening potential pool of the structure.

Tension stiffening is representative of the contribution of the concrete to the stiffness and tensile strength of an RC member. Indeed, in both the elastic and postcracking phases, the concrete surrounding the rebar (the latter being situated between consecutive cracks in the second phase) relieves the latter of a certain amount of tensile stress. This phenomenon, occurring through the bond stress present between concrete and steel [22], allows an RC member to carry an additional load and its bare rebar counterpart [23,24] (thus named as tension stiffening). The latter is measured by subtracting the bare steel response from the measured member response (assuming they share the same origin).

Tension stiffening is indicated as a governing parameter for the crack resistance (in the elastic phase [25]) and deformations (later reported in [26,27]) of an RC structure. Indeed, the incorporation of the tension stiffening effect in the structural analysis of RC members makes their behavioral predictions more realistic [19] and compatible with other analyses such as the layered beam section one and the smeared finite element one [28,29].

Abundant research on the topic both in the previous century [30] and in the current one [31,32] has derived multiple tension stiffening relationships based on different assumptions and testing methodologies [33–35]. Worth of mention are the ones presented in the Euro Code 2 [36] (henceforth referred to as EC2) and the CEB-fib Model Code 2010 [37] (henceforth referred to as MC2010). It should be noted that the deformative response of the former has been found to be particularly stiff, especially for small reinforcement ratios [38], whereas the latter also overestimates the member stiffness only at advance loading stages [39]. However, shrinkage and creep accompany most of the abovementioned proposed tension stiffening relationships and thus were unaccounted for in their formulation, except for Bischoff [2] and Kaklauskas et al. [40] who derived shrinkage-free tension stiffening laws.

As foreshadowed earlier, neglecting the effects of shrinkage in an RC member response leads to a perceived reduction in the cracking strength of concrete and to a perceived influence of the reinforcement ratio on tension stiffening. According to Bischoff [2], these issues grow proportionally with the reinforcement ratio. Indeed, for the same amount of shrinkage, the apparent loss of tension stiffening becomes worse, as the reinforcing percentage increases (becoming unneglectable beyond 1%). Instead, as observed by other authors [35], once the shrinkage effect is removed, the tension stiffening appears to be independent from the reinforcement ratio. For all the above reasons, it is crucial to assess tension stiffening independently from the effect of shrinkage.

It should be mentioned that in most of the aforementioned experimental investigations, the shrinkage effect was studied for short-term shrinkage only (just few days). As a matter of fact, no experimental campaign studying the effect of long-term shrinkage on the tension stiffening of RC members has been reported. The present article, instead, sets itself this goal precisely.

In order to achieve it, an experimental campaign was designed and performed encompassing 14 RC tensile elements (RC ties) that cured during a time of 1947 days (5.3 years). It should be mentioned that RC ties are often used to illustrate cracking, deformation, and bond behavior of RC structures due to their simplicity and reasonably good representation of the internal distribution of forces and strains in the tensile zones of RC structures [40] (such as the one of an RC beam). After having reported the experimental load–strain curves of RC ties, the shrinkage influence on their mechanical response was mathematically extrapolated and both experimental and shrinkage-free results were finally compared with the tension stiffening predictions of the EC2 and the MC2010. This should help assess the performance of the codes when long-term concrete behavior with a magnitude of 5.3 years is concerned.

The following chapter will elucidate the well-established steps that need to be undertaken in order to remove the effect of shrinkage from tension stiffening readings.

1.2. Theoretical Background on the Assessment of the Shrinkage-Free Tension Stiffening Phenomenon

Considering an RC tie subjected to an external tensile load P, the latter is shared among the internal forces of concrete N_c and steel N_s, as in Equation (1) (being these two defined in Equations (2) and (3), respectively):

$$P = N_c + N_s \tag{1}$$

$$N_c = \sigma_{ct} A_c \tag{2}$$

$$N_s = A_s E_s \varepsilon_s \tag{3}$$

where σ_{ct} is the average tensile stress in the concrete, A_c and A_s are the cross-sectional areas of concrete and steel, respectively, E_s is the modulus of elasticity of steel, and ε_s is the average steel strain. Thus, σ_{ct} can be derived as per Equation (4):

$$\sigma_{ct} = \frac{P - \varepsilon_s E_s A_s}{A_c} \tag{4}$$

The two key parameters for the derivation of tension stiffening relations are the average tensile stress in the concrete σ_{ct} and the average member strain ε_m (extracted from an RC tie test). In the past, although the predicted deformative behavior of RC structures usually ignored shrinkage strains, this has the potential to yield flawed assessments of their stress–strain behavior, crack resistance, and carrying capacity. Therefore, the present article makes the removal of the effect of shrinkage one of its priorities and the process with which this is achieved is detailed in the following.

Bischoff [2] addresses the issue with a unique method on the grounds of which the present work is based. The model has three origins (presented graphically in Figure 1):

(1) "Assumed origin O" which is the starting point of an RC tie test whenever the concrete shrinkage effects are ignored, which is how the majority of tests have been up until the present);
(2) "Experimental origin O_{exp}" which identifies the starting point of the RC tie test (external load $P = 0$) except this time acknowledging the compressive effect of shrinkage on the member strains (henceforth referred to as $\varepsilon_{m,sh}$) as in Figure 1;
(3) "Shrinkage-free origin O*" which identifies the true origin of the RC tie test. This time, shrinkage elimination does not apply only to the deformative response but also to the applied load. Indeed, the initial aftermath of the application of an external load P is simply the compensation of the abovementioned shrinkage-induced compression, henceforth referred to as P_{sh} (as in Figure 1).

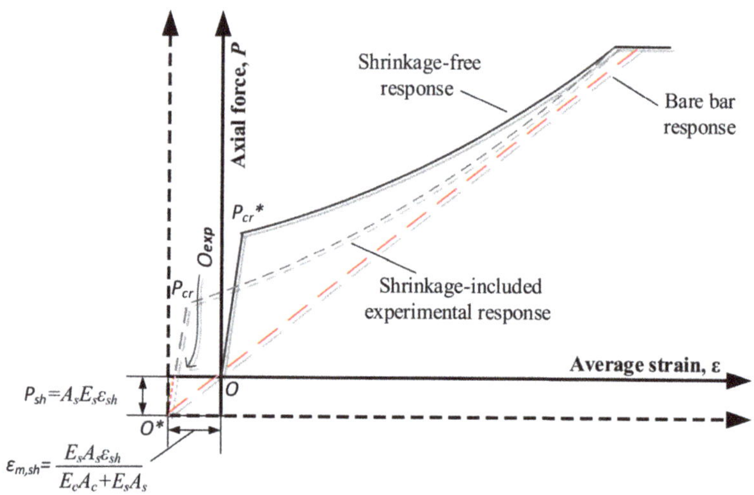

Figure 1. Shrinkage effect on the tension stiffening response of reinforced concrete (RC) ties.

Here, P_{sh} is the fictitious compressive force (defined in Equation (5)) introduced to represent the effect of the free shrinkage strain (ε_{sh}) of the concrete on the behavior of RC members occurring in the former prior to the loading stage:

$$P_{sh} = A_s E_s \varepsilon_{sh} \tag{5}$$

In this study, the free concrete shrinkage strains ε_{sh} of the members were calculated as per the codal provisions, in particular according to the EC2 and the MC2010. On the latter two, it was observed [38] that the EC2 estimates higher free shrinkage strains than the MC2010 for higher specimen sections whilst, for smaller ones (as is the case for the present experimental campaign), the two are quite close. Furthermore, in both standards, the total shrinkage is divided into two components, namely autogenous and drying shrinkage, the second of which is particularly sensible to the environmental humidity in which the drying occurs. Since the specimens of the present study were kept in dry condition for a long time (1947 days), the amount of consequent shrinkage strain was significant.

The free shrinkage also causes an initial member shortening $\varepsilon_{m,sh}$ (expressed in Equation (6)) of which the inclusion would lead to an offset between the experimental response and the bare rebar response, as visible in Figure 1:

$$\varepsilon_{m,sh} = \frac{E_s A_s \varepsilon_{sh}}{E_s A_s + E_c A_c}. \tag{6}$$

Therefore, from a graphical point of view, in order to eliminate the shrinkage effect from the tension stiffening behavior of RC elements, the origin of the load–displacement diagram can simply be shifted downwards in order to coincide with the shrinkage-free origin O*. Instead, from a mathematical point of view, in order to obtain the shrinkage free load P^*, P_{sh} can simply be subtracted from the experimental tensile force applied on the member P_{exp} (Equation (7)):

$$P^* = P_{exp} - P_{sh} \tag{7}$$

The same procedure can be applied for the calculation of the shrinkage-free average member strain $\varepsilon_m{}^*$ as per Equation (8):

$$\varepsilon_s{}^* = \varepsilon_{exp} - \varepsilon_{m,sh} \tag{8}$$

Figure 1 also graphically displays the shrinkage-induced apparent reduction in the cracking load of the RC member ($P_{cr} < P_{cr}{}^*$), where P_{cr} and $P_{cr}{}^*$ are the cracking loads of the same RC member overlooking the shrinkage effect and eliminating the calculated shrinkage from the experimentally obtained load–displacement response, respectively.

2. Experimental Campaign

The experimental campaign, which was the topic of the present article, consisted in the testing of 14 RC ties (visible in Figure 2a), which varied in both geometry and mechanical characteristics.

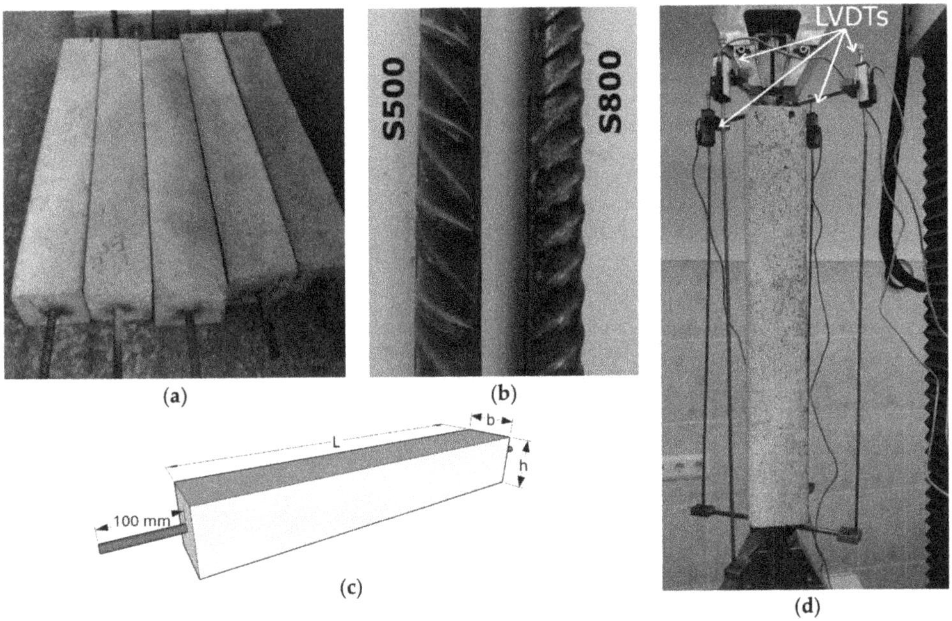

Figure 2. Photographs of the experimental test displaying the RC ties (**a**) and the rib pattern difference between S500 and S800 rebars (**b**). (**c**) An illustration of the RC ties to clarify their dimensional characteristics. (**d**) The RC tie clamping and testing by means of a universal testing machine (UTM).

Table 1 details the concrete prism dimensions of the members (in combination with Figure 2c) and the characteristics of the embedded deformed steel rebars (diameter $\varnothing_s$ and resulting reinforcement ratio ρ_s).

Table 1. Geometrical and mechanical characteristics of the specimens.

Specimen Assigned Code	Number of Samples	h × b × L	Φ_s	ρ_s	f_{cm}	E_c	f_{sy}	E_s
		mm	mm	%	MPa	MPa	MPa	MPa
T_12_500	4	100 × 100 × 655	12	1.13	47.6	32,857	500	198,900
T_10_500	2	65 × 65 × 650	10	1.86				208,400
T_12_800	3	100 × 100 × 655	12	1.13			800	190,000
T_10_800	5	65 × 65 × 650	10	1.86				198,700

In Table 1, every member was assigned a code, where the first numerical digit was indicative of the embedded rebar diameter whilst the second represented its steel grade. The rebars were positioned longitudinally along the centroid axis of the RC tie, and in order to ensure a proper clamping during the test, their lengths were designed in such a manner that they extended beyond the concrete prism of 100 mm on both extremities (as in Figure 2c).

The members were produced with a single concrete batch using CEM II\A-LL 42.5 N cement and the PowerFlow 3100 superplasticizer. They were dried in an environment with an average temperature of 18.4 °C and an average humidity of 48.1%.

The concrete compressive strength f_{cm} was established in accordance with BS EN 12,390 and tested on three 150 mm cubes. The modulus of elasticity E_c and the tensile strength of the concrete f_{ct} were determined according to the EC2 equation displayed as Equations (7)–(10), respectively, while Equations (9) and (10) were used for considering the time factor, as the specimens were kept for a long time (more than 28 days) before testing:

$$E_C = 22000\left(\frac{f_{cm}}{10}\right)^{0.3} \tag{9}$$

$$f_{ct}(t) = (\beta_{cc})^{\alpha} \times f_{ct}\ \alpha = \frac{2}{3}\ \text{for}\ t > 28\ days \tag{10}$$

$$f_{ct} = 0.3[f_{cm} - 8]^{(2/3)} \tag{11}$$

$$\beta_{cc}(t) = exp\left[0.2\left(1 - \frac{28}{t}\right)\right]^{0.5} \tag{12}$$

The steel rebars varied both in diameter (Ø10 and Ø12) and in steel grade (S500 and S800). The contrast in the rib pattern of the different grade bars is shown in Figure 2b. The reinforcement yielding strength f_{sy} and the modulus of elasticity E_s were obtained in accordance with ISO 6892-1:2009 (specified in the standard BS EN 10025).

The specimens were subjected to uniaxial and monotonic tension loads until yielding at a rate of 0.2 mm/min by means of a universal testing machine (UTM). The average member strain was determined by means of four linear variable displacement transformers (LVDTs) fixed along the longitudinal edges of the former as represented in Figure 2d. On a side note, for future RC tie tests, an improved measurement system will be put in use. The latter would include two more LVDTs directly clamped on the rebar ends in order to avoid any measurement alteration in case of bar slippage in the gripping clamps and rigidly constraining one end whilst leaving the opposite one free and monitored by an LVDT mounted upside down in order to avoid any misalignment between the LVDT rod and the extension direction. Finally, in order to discern the location and the width of each crack appearing on the surface of the specimen, digital image correlation (DIC) monitoring was parallelly performed. Two IMAGER E-LITE 5 M cameras from LaVision (Göttingen, Germany) were fixed on a tripod stand with a distance of 0.7 m from each other and 3 m from the specimen. The cameras worked at a resolution of 2456 × 2085 pixels and at a 12.2 fps rate.

3. Results and Discussion

3.1. Standard Test Behavior of an RC Tie

The load–average strain diagram of one of the tested members (T_12_500) is illustrated in Figure 3 in order to elucidate the cracking, deformative, and tension stiffening behavior that characterized all RC ties tests.

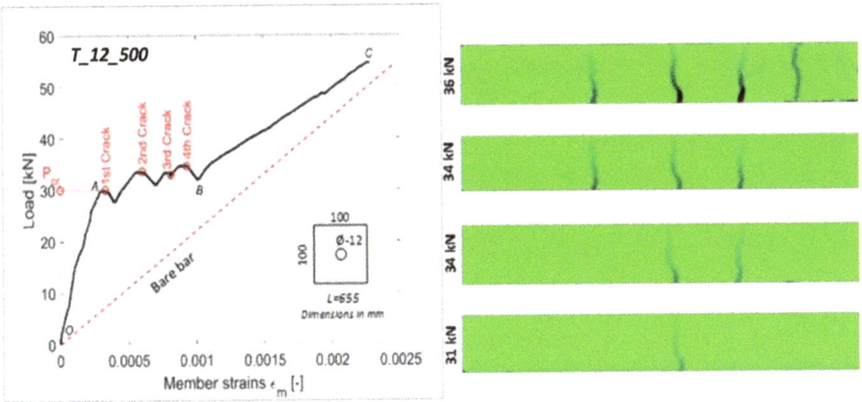

(a) (b)

Figure 3. Failure behavior of a RC tie (study specimen) T_12_500: (**a**) experimental load–displacement diagram; and (**b**) development of crack with the corresponding loads through digital image correlation (DIC) pictures.

The first phase was defined by points O and A in Figure 3a and was defined "elastic phase" to evaluate the elastic behaviors of both constituting materials. It is at this stage that tension stiffening provided the largest contribution. Indeed, as visible in Figure 3a, the difference between the average deformation of the RC tie and the one of the bare steel rebar was largest in the interval segment OA. As described previously, the increasingly larger reinforcement stresses and strains were transferred to the concrete by means of bond stresses present on the surface between the two. This transfer continued uninterruptedly until point A, where the first crack appeared in the cross-section where the concrete stress σ_{ct} first surpassed its tensile strength f_{ct}. This represented the beginning of the "cracking phase" (segment AB). The corresponding load was defined as cracking load P_{cr} which, for the study case member T_12_500, corresponded to 31 kN.

As soon as the crack appeared, the contribution of the concrete drastically decreased, thus reducing the amount of tension stiffening it provided to the member. Consequently, the average strain of the latter plummets was visible in the segment following point A. The latter occurred, whenever an RC tie is loaded with a deformation-controlled regimen, as was the case for the present test. Indeed, as soon as the UTM detected a decrease in the specimen stiffens (due to the crack formation), it suddenly decreased the applied load in order to match the defined deformation speed (0.2 mm/min in the present test). Whenever this adjustment was performed, the profile inverted its trend and increased one more. The same occurred for all the following cracks. It can be noticed that the trend of ε_m progressively approached the one of the bare rebar after the appearance of every crack. This is indicative of the progressive loss of the initial extra stiffness provided by the concrete.

It was noticed for most members the first crack appeared close to the mid-section. As visible in Figure 3a, as the applied tensile stress increased in magnitude, new cracks kept appearing (their respective loads being 34, 34, and 36 kN) whilst older ones were widened. The DIC pictures of the latter with their respective loads are shown in Figure 3b. It should be kept in mind that the loads at which the cracks appeared in the DIC pictures did not necessarily correspond to the loads at which the crack-indicative profile drops appeared in Figure 3a. Indeed, the former only reported the loads at which the cracks appeared on the RC tie surface. The latter was equivalent to the drops shown in Figure 3a, only if the initial width of the crack was sufficiently wide.

Beyond point B, the lengths of the various concrete segments in which the RC tie was subdivided were insufficient for the transferred tensile stresses to reach the concrete maximum capacity, and thus, no new cracks can appeared. Indeed, in Figure 3a, no new crack-indicative drops were noticeable, and the profile acquired a linear trend. Therefore, point B represented the beginning of the third and last behavioral phase known as the "stabilized cracking stage" (section BC).

3.2. *Test Results*

The left column of Figure 3 displays all the load/average deformation graphs of the tested specimens categorized in four classes based on their reinforcement diameters and steel grades (see Table 1).

Expectedly, a behavioral proximity can be discerned among the load–deformation graphs of the specimens pertaining to the same category. As such, for clarity purposes, only one specimen per each category (defined as "study case member" in Figure 4) was subjected to the shrinkage elimination process. The study case members were processed according to the steps detailed in Section 1.2, and the shrinkage-free profiles were reported in the right column of Figure 4 (in yellow). For the latter, the estimated ε_{sh} after 1947 days performed with the EC2 and the MC2010 are reported in Table 2 and categorized per specimen geometry.

Table 2. Free shrinkage predicted by the Euro Code 2 (EC2) and the CEB-fib Model Code 2010 (MC2010).

Specimen Geometry	Time (days)	ε_{sh} According to the EC2	ε_{sh} According to the MC2010
100 × 100 × 655	1947	−0.000715	−0.000699
65 × 65 × 650	1947	−0.000722	−0.000729

The process of shrinkage removal based on the EC2 and the MC2010, free shrinkage strains yielded curves that were practically identical and graphically overlapped as visible in Figure 4. Consequently, in their stead, a single shrinkage-free curve was displayed and simply named shrinkage-free experimental curve. It can be observed in Figure 4 that some of the ascending segments of the initial load–displacement curves were not perfectly linear due to the occurrence of some rebar slips in the gripping clamps. Figure 4 also shows that specimens with lower reinforcement ratios (T_12 specimens) had higher cracking loads (around 42 kN) than those with higher reinforcement ratio ones (T_10 specimens), of which P_{cr} oscillated around 19 kN. This is due to the larger amount of concrete volume present in the former combined with the larger stiffness of its rebar. Figure 4 also displays how overlooking the shrinkage effect caused a 25% underestimation of the P_{cr} for the T_12 specimens and about 38% for the T_10 specimens. This clearly demonstrated the significance of the long-age (5.3 years) shrinkage influence on an RC member response.

Figure 5 compares the shrinkage-free load–deformation curves of the above-defined study case members against the ones predicted by the EC2 and the MC2010.

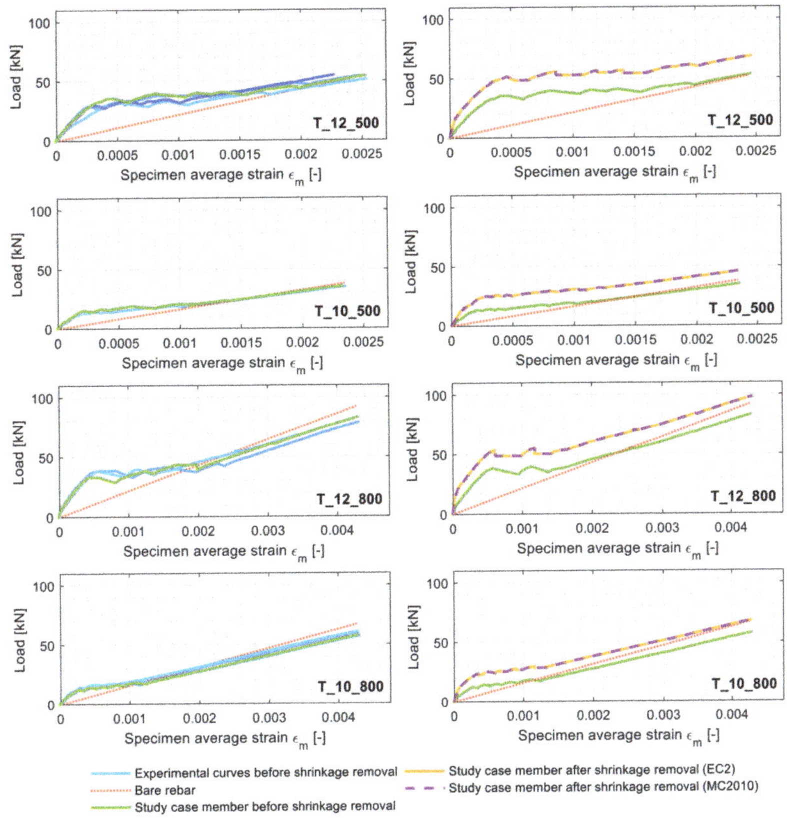

Figure 4. Load–specimen average strain diagrams of all the members (**left column**) and those of the study case members (**right column**) before and after shrinkage elimination.

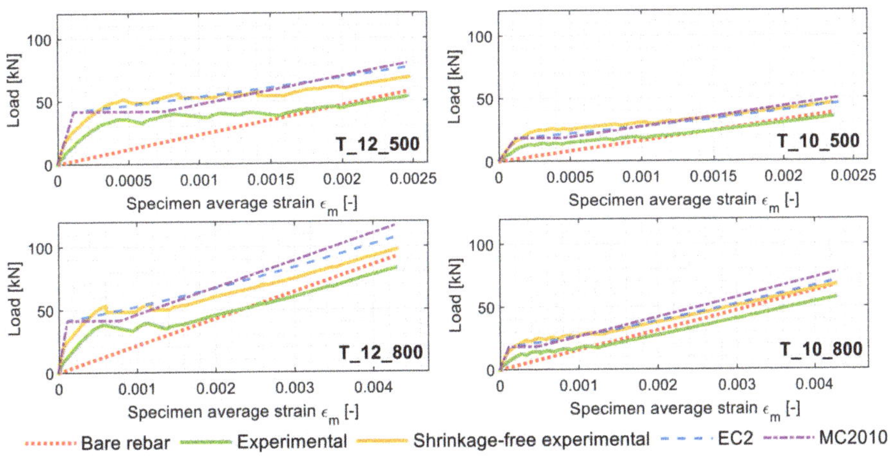

Figure 5. Load–specimen average strain diagrams of the study case members before and after shrinkage elimination along with the EC2 and MC2010 predictions.

In Figure 5, it can be noticed that the experimental load–displacement curves of some members (green lines) display strains exceeding the bare rebar ones at later load levels (particularly for T_12_800 and T_10_800). This might be caused by an imperfect positioning of the LVDTs. Apart from this issue, an unexpected close match between the shrinkage-free load–displacement curves and the codal predictions (EC2 and MC2010) can be discerned for all the members, except T_12_800. The radical difference between the codal predictions and the experimental curves was highly noticeable. The tension stiffening prediction power of the codes for long-term-curing RC members is of interest, considering their equations were built on databases collecting nonshrinkage-free RC structural tests results.

3.3. Experiemental Tension Stiffening and Shrinkrage-Free Tension Stiffenning Relationships

The tension stiffening relationships displayed in this section were derived on the grounds of the above load–deformation curves and the Equations (3)–(12) of Section 1.2. The left column of Figure 6 shows the tension stiffening relationships of the various specimens, named as the fluctuations in the tensile stress of the concrete against the average member strain, divided in the same four categories as the ones of the previous subsection. The right column of Figure 6, instead, displays the shrinkage-free tension-stiffening curves against the experimental ones of their respective study case members.

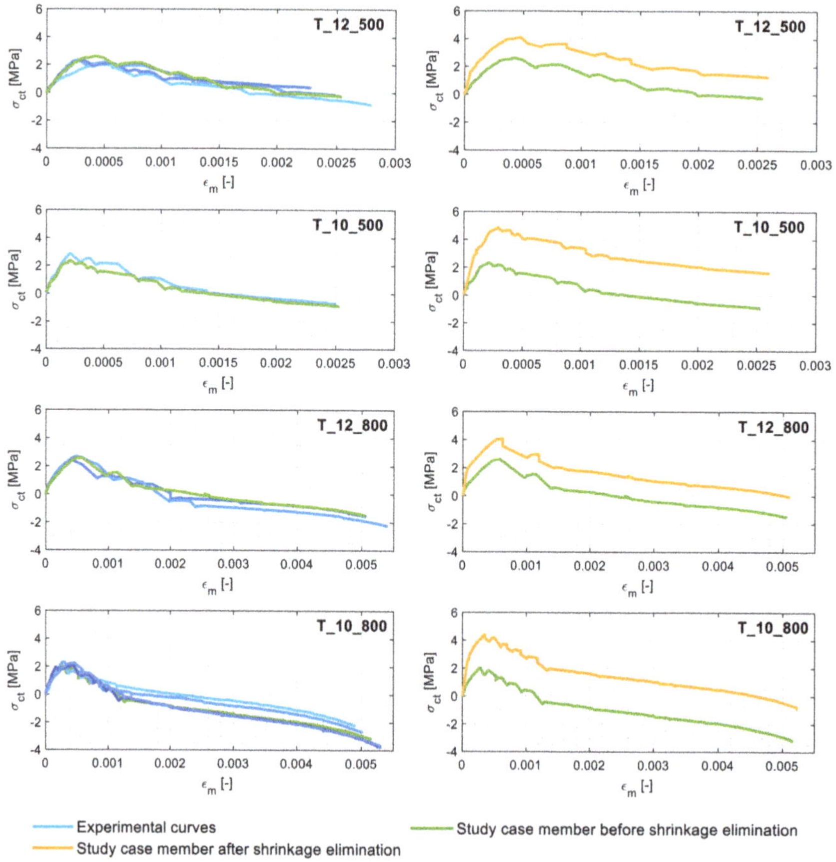

Figure 6. Tension stiffening behaviors of all the members (**left column**) and those of a study case member (**right column**) before and after shrinkage elimination.

Expectedly, the concrete gained its maximum tensile stress just prior to the appearance of the first crack (coinciding with the end of the elastic behavior), beyond which any subsequent crack led to rapid σ_{ct} drops. Once the cracking stabilized, the tension stiffening curve does not welcome any more sudden drops but instead gradually declines due to the slow concrete/rebar bond deterioration until the reinforcement was yielded. Also expectedly, the maximum concrete stresses for all the specimens were found to be similar among all the specimens (in the range of 2–3 MPa), as they were composed of the same concrete.

The right column of Figure 6, instead, confirms how the unprocessed observations of the experimental results (thus ignoring the effect of shrinkage) led to an underestimation of the tension stiffening. In particular, it is evident that, as foreshadowed earlier, the reinforcement ratio had an undeniable influence on the tension stiffening value alteration. Figure 7 displays the tension stiffening curves of the abovementioned study case members before (Figure 7a) and after (Figure 7b) the shrinkage elimination as a function of their reinforcement ratios.

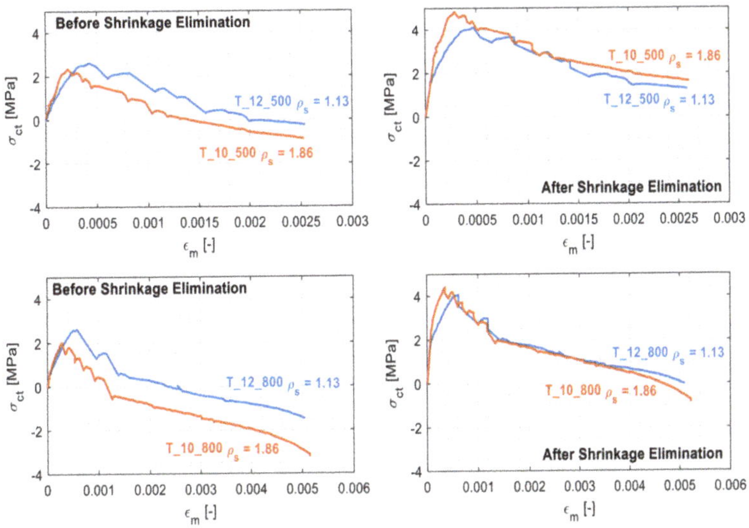

(a) (b)

Figure 7. Comparison of the tension stiffening behaviors of the study case members with different reinforcement ratios before (**a**) and after (**b**) the shrinkage elimination.

The separation present between the curves in Figure 7a is a consequence of the reinforcement ratio-induced tension stiffening alteration. Expectedly, it disappears in Figure 7b as a result of the shrinkage elimination. The apparent tension stiffening reductions were 38% and 80% for the reinforcement ratios of 1.13% and 1.86% respectively. Hence, it can be stated that for similar shrinkage-free ratios, the apparent loss in tension stiffening was more pronounced for members with a higher reinforcement ratio. Furthermore, Figure 7b, different from Figure 7a, also properly reports that the concrete still absorbed a progressively inferior amount of tensile stress σ_{ct} until the end of the test (once again, due to the insurgence of concrete/steel slip and the consequent degradation of their bond). This is in agreement with the shrinkage-free curves of Figure 4. Indeed, the latter do not intersect the bare steel strain curve (always remaining above it) due to the continuous effect of the steel strain relaxation from the part of the concrete. Finally, the largest discrepancy between the curves in Figure 7b was concentrated in the section corresponding to their cracking phases. The reason lies in the aforementioned inferior cracking load of the T10 members, leading to a curve trend inversion in correspondence of smaller deformative values. For this reason, the T10 graphs are shifted leftwards when compared to their T12 counterparts.

Finally, Figure 8 plots simultaneously the tension stiffening curves extracted from the experimental campaign for the study case members, their corresponding shrinkage-free tension stiffening curves and their respective predicted tension stiffening profiles as per the EC2 and MC2010.

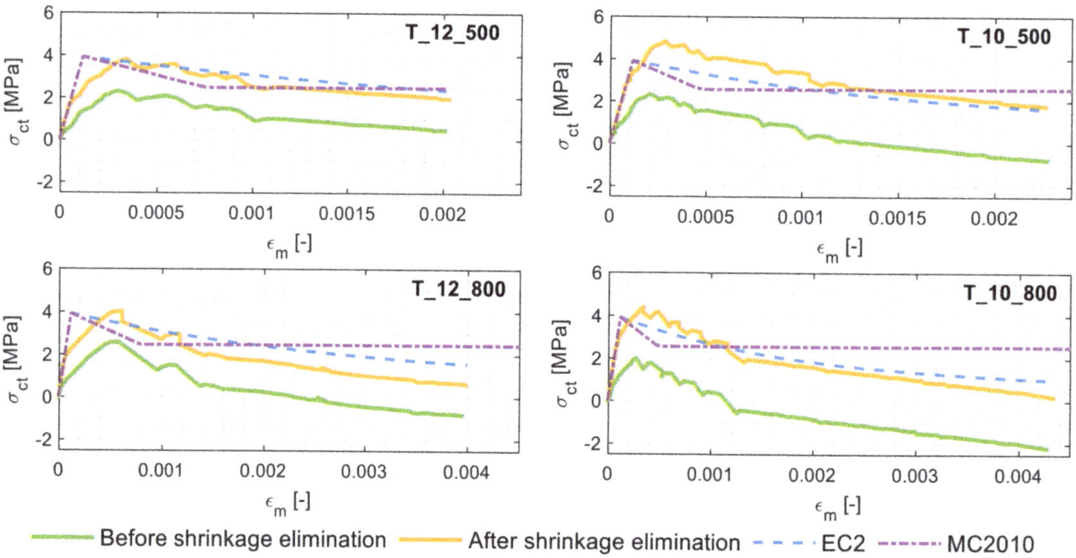

Figure 8. Tension stiffening behaviors of the study case members before and after shrinkage elimination along with the EC2 and MC2010 predictions.

A substantial difference can be noticed between the predictions of the codes and the experimental response of the RC ties (green lines). With respect to the shrinkage-free response (yellow lines), instead, a much closer match can be spotted with the codal predictions. In spite of multiple research articles reporting that both the EC2 and MC2010 generally overestimate the shrinkage-free member stiffness [38,39], the "long age" could be the key explaining factor here. Further research on the topic could confirm the present hypothesis.

Figure 8 further shows that whilst the two codes yielded similar patterns in the ascending branch, they cannot be said for the descending one. Of the two, the EC2 seemed to provide closer predictions to the shrinkage-free tension stiffening curves as it includes a gradual decline in the tensile stress of the concrete in the postcracking stage (different from the MC2010).

4. Statistical Analysis

This chapter statistically compares the experimental tension stiffening contribution of concrete σ_{ct} with and without shrinkage elimination, against their predicted values according to the EC2 and the MC2010. Their differences were quantified through a relative error calculated as $(\sigma_{ct,predicted} - \sigma_{ct,experimental})/\sigma_{ct,predicted}$. In order to normalize the results, otherwise differing in cracking load, final load and deformation ratio, the $\sigma_{ct,experimental}$ values of the tested members were sampled at four specific test instances. The first corresponded to their cracking deformation ε_{cr}, thus coinciding with the beginning of the descending branch of the tension stiffening graph (therefore labeled as 0% strain). The last sampling point coincided with the steel yielding strain of the members (labeled as 100%). Two intermediate levels of 33% and 66% were also introduced in order to provide an indication on the evolution of the relative errors of the readings. Figure 9 displays an example of the evolution of the relative error obtained for each of the four T_12_500 specimens along the abovementioned four stages versus the code predictions with (w) and without (w/o) shrinkage elimination.

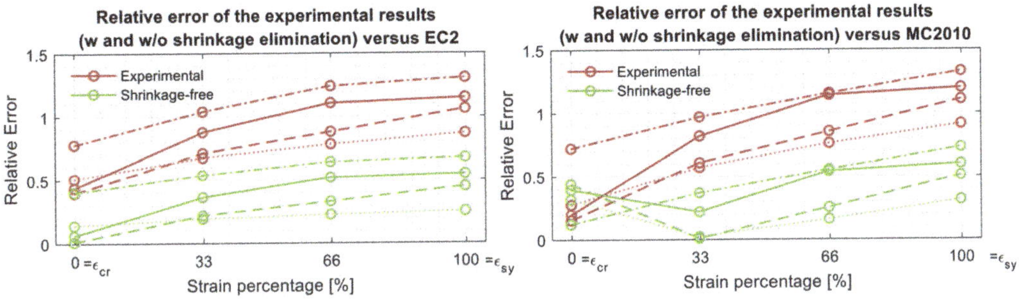

Figure 9. Relative errors of the experimental and shrinkage-free tension stiffening values of member T_12_500 versus the EC2 (**left**) and MC2010 (**right**) predictions.

In Figure 9, the red lines are representative of the relative errors between the experimental outputs and the relative code, whilst the green ones are representative of their shrinkage-free variances. First of all, it is observable that the relative errors of both model predictions were smaller when compared against the tested shrinkage-free results and the experimental results. As a matter of fact, the experimental results exhibited average relative errors of 0.87 and 0.79 with respect to EC2 and MC2010 predictions, respectively, and 0.35 with respect to both models for the shrinkage-free scenario. Furthermore, it is observable the relative error increased proportionally with the strain level. The relative errors of all 14 tested specimens were calculated in a similar way, and their overall mean relative error values and corresponding standard deviation for each strain stage are reported in Table 3.

Table 3. Mean relative error and standard deviation values for all specimens at different strain levels.

Model	Results Processing	Statistical Parameter	Strain Ratio			
			0% (ε_{cr})	33%	66%	100% (ε_{sy})
EC2	Experimental	Mean relative error (Re. Er.)	0.416	1.064	1.561	2.078
		Standard deviation (St. Dev.)	0.156	0.295	0.586	0.976
	Shrinkage-free	Mean Re. Er.	0.231	0.324	0.488	0.698
		St. Dev.	0.112	0.199	0.336	0.493
MC2010	Experimental	Mean Re. Er.	0.470	0.944	1.294	1.644
		St. Dev.	0.209	0.250	0.313	0.445
	Shrinkage-free	Mean Re. Er.	0.331	0.275	0.533	0.884
		St. Dev.	0.211	0.189	0.289	0.376

The normal probabilistic distribution data of Table 3 are further graphically represented in Figure 10. The plots show the tension stiffening prediction errors of the EC2 and the MC2010 versus the experimental and its shrinkage-free counterpart. Furthermore, the mean relative errors are clearly indicated on the x-axis with the vertical dashed lines adjoining the respective peaks. It is clear that the experimental tension stiffening results had overall average relative errors of 1.25 and 1.2 with respect to the EC2 and the MC2010, respectively, whereas the shrinkage-free data were overall characterized by smaller relative errors, namely 0.45 and 0.5. Once again, it is of interest to notice the smaller prediction errors of the codes for the shrinkage-free data by considering the lack of the shrinkage elimination procedure in their constitutive equations. Additionally, a comparison of the model predictive errors concerning the raw test data elevated the MC2010 curve over the

EC2 curve, as it is characterized by a smaller data spread (thus thinner bell curve) despite having a similar average.

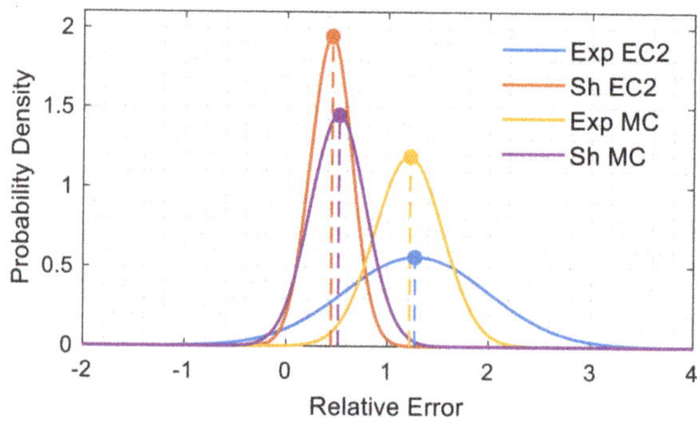

Figure 10. Normal distribution relative error curves between the experimental and shrinkage-free tension stiffening values versus their EC2 and MC2010 predictions.

Interestingly, the situation was reversed, whenever the elimination of shrinkage was included in Figure 10. Indeed, the EC2 exhibited a smaller data spread and a slightly higher accuracy (smaller average error) versus the MC2010.

5. Conclusions

The present article studied the long-age shrinkage effect on the tensile behavior, in particular on tension stiffening, of 14 tested RC tensile elements. Careful consideration of concrete shrinkage and its mechanism elimination is the key factor of this study. The assessment of the experimentally obtained results in accordance with the EC2 and the MC2010 has led to the following conclusions:

(1) The accumulated shrinkage strain during 5.3 years was quite significant and capable of making serious impact on the load–deformative behavior of the member as well as on their tension stiffening behaviors;

(2) The shrinkage effect lowered the apparent RC member cracking load. This underestimation increased with the increase in reinforcement ratio (25% for ϱ = 1.13% and 38% for ϱ = 1.86%);

(3) The shrinkage effect caused an apparent reduction of the tension stiffening mechanism on an average of 40% for a lower reinforcement ratio (ϱ = 1.13%) and about 80% for a higher one (ϱ = 1.86%);

(4) After the process of shrinkage elimination, the tension stiffening behaviors of members with different reinforcement ratios were in good agreement with each other, confirming the influence of the reinforcement ratio on the alteration of the tension stiffening effect;

(5) A statistical analysis on the tension stiffening-predicted power of the EC2 and MC2010 model codes against the experimental and shrinkage-free results showed an overall increase in relative error proportional to the strain level increase;

(6) The predictions of both codes displayed a much smaller relative error (66%) when compared against the shrinkage-free tension stiffening results than against the experimental one;

(7) According to the literature review, the previous point does not occur for short-term shrinkage, thus suggesting the increased accuracy of the model for members that include very long-term shrinkage.

(8) Among the predictions of the two models, the MC2010 one exhibited a slightly closer match to the raw test result, whereas the EC2 predictions were marginally more accurate to the shrinkage-free tension stiffening.

Author Contributions: Conceptualization and design, G.K. and A.S.; execution of laboratory experiments, A.S. with the aid of A.D. and A.V.V.; analysis and postprocess, A.D., A.V.V., and M.F.B.; software, validation, formal analysis, investigation, and article writing, A.D. and M.F.B.; review and editing, M.F.B. and G.K.; funding acquisition, G.K. All authors have read and agreed to the published version of the manuscript.

Funding: The study was performed within the project (No. 09.3.3-LMT-K-712-01-0145) that has received funding from the European Social Fund under grant agreement with the Research Council of Lithuania (LMTLT).

Institutional Review Board Statement: Not applicable.

Informed Consent Statement: Not applicable.

Data Availability Statement: The data presented in this study are available on request from the corresponding author.

Conflicts of Interest: The authors declare no conflict of interest.

References

1. Gribniak, V.; Kaklauskas, G.; Bacinskas, D. Shrinkage in reinforced concretet structures: A computational aspect. *J. Civ. Eng. Manag.* **2008**, *14*, 49–60. [CrossRef]
2. Bischoff, P.H. Effects of shrinkage on tension stiffening and cracking in reinforced concrete. *Can. J. Civ. Eng.* **2001**, *28*, 363–374. [CrossRef]
3. American Concrete Institute. *Control of Cracking in Concrete Structures*; AAI 224R-01; American Concrete Institute: San Bernardino, CA, USA, 2008.
4. Sakata, K. Prediction of Creep and Shrinkage of Concrete. *Jpn. Soc. Civ. Eng.* **1996**, 160–171.
5. Collins, F.; Sanjayan, J.G. Effect of pore size distribution on drying shrinkage of alkali-activated slag concrete. *Cem. Concrete Res.* **2000**, *30*, 1401–1406. [CrossRef]
6. Barcelo, L.; Moranville, M.; Clavaud, B. Autogenous shrinkage of concrete: A balance between autogenous swelling and self-desiccation. *Cem. Concr. Res.* **2005**, *35*, 177–183. [CrossRef]
7. Aili, A.; Vandamme, M.; Torrenti, J.; Masson, B. Is long-term autogenous shrinkage a creep phenomenon induced by capillary effects due to self-desiccation? *Cem. Concr. Res.* **2018**, *108*, 186–200. [CrossRef]
8. Holt, E. Contribution of mixture design to chemical and autogenous shrinkage of concrete at early ages. *Cem. Concr. Res.* **2005**, *35*, 464–472. [CrossRef]
9. Seddik Meddah, M.; Tagnit-Hamou, A. Evaluation of rate of deformation for early-age concrete shrinkage analysis and time zero determination. *J. Mater. Civ. Eng.* **2011**, *23*, 1076–1086. [CrossRef]
10. Jaafari, C.; Bertrand, D.; Guillot, T.; Prudhomme, E.; Tardif, N.; Georgin, J.F.; Delhomme, F.; Trunfio, R.; Chateur, N.; Bruyere, E.; et al. Effect of early age drying shrinkage on the seismic response of RC structures. *Mater. Struct.* **2020**, *53*. [CrossRef]
11. Yazdizadeh, Z.; Marzouk, H.; Hadianfard, M.A. Monitoring of concrete shrinkage and creep using Fiber Bragg Grating sensors. *Constr. Build. Mater.* **2017**, *137*, 505–512. [CrossRef]
12. Davis, M.B.; Hoult, N.A.; Bajaj, S.; Bentz, E.C. Distributed Sensing for Shrinkage and Tension Stiffening Measurement. *ACI Struct. J.* **2017**, *114*. [CrossRef]
13. Han, B.; Xiang, T.; Xie, H. A Bayesian inference framework for predicting the long-term deflection of concrete structures caused by creep and shrinkage. *Eng. Struct.* **2017**, *142*, 46–55. [CrossRef]
14. Bazant, Z.P.; Osman, E.; Thonguthai, W. Practical formulation of shrinkage and creep of concrete. *Mater. Constr.* **1976**, *9*, 395–406. [CrossRef]
15. Nguyen, D.H.; Dao, V.T.N.; Lura, P. Tensile properties of concrete at very early ages. *Constr. Build. Mater.* **2017**, *134*, 563–573. [CrossRef]
16. Liu, J.; Yan, K.Z.; Zhao, X.; Hu, Y. Prediction of autogenous shrinkage of concretes by support vector machine. *Int. J. Pavement Res. Technol.* **2016**, *9*, 169–177. [CrossRef]
17. Goel, R.; Kumar, R.; Paul, D.K. Comparative study of various creep and shrinkage prediction models for concrete. *J. Mater. Civ. Eng.* **2007**, *19*, 249–260. [CrossRef]
18. Shadravan, S.; Ramseyer, C.; Kang, T.H. A long term restrained shrinkage study of concrete slabs on ground. *Eng. Struct.* **2015**, *102*, 258–265. [CrossRef]
19. Gilbert, R.I. Tension stiffening in lightly reinforced concrete slabs. *J. Struct. Eng.* **2007**, *133*, 899–903. [CrossRef]
20. Gilbert, R.I. Shrinkage, cracking and deflection—The serviceability of concrete structures. *Electron. J. Struct. Eng.* **2001**, *1*, 15–37.
21. Kaklauskas, G.; Gribniak, V. Eliminating Shrinkage Effect from Moment Curvature and Tension Stiffening Relationships of Reinforced Concrete Members. *J. Struct. Eng.* **2011**, *137*, 1460–1469. [CrossRef]
22. Bado, M.F.; Casas, J.R.; Kaklauskas, G. Distributed Sensing (DOFS) in Reinforced Concrete members for reinforcement strain monitoring, crack detection and bond-slip calculation. *Eng. Struct.* **2020**. [CrossRef]

23. Fields, K.; Bischoff, P.H. Tension Stiffening and Cracking of High-Strength Reinforced Concrete Tension Members. *ACI Mater. J.* **2004**, *101*, 447–456.
24. Gribniak, V.; Cervenka, V.; Kaklauskas, G. Deflection prediction of reinforced concrete beams by design codes and computer simulation. *Eng. Struct.* **2013**, *56*, 2175–2186. [CrossRef]
25. Al-Saleh, S.A. Comparison of theoretical and experimental shrinkage in concrete. *Constr. Build. Mater.* **2014**, *72*, 326–332. [CrossRef]
26. Khalfallah, S. Tension stiffening bond modelling of cracked flexural reinforced concrete beams. *J. Civ. Eng. Manag.* **2008**, *14*, 131–137. [CrossRef]
27. Borosnyoi, A.; Balazs, G.L. Models for flexural cracking in concrete: The state of the art. *Struct. Concr.* **2005**, *6*, 53–62. [CrossRef]
28. Soltani, M.; An, X.; Maekawa, K. Computational model for post cracking analysis of RC membrane elements based on local stress-strain characteristics. *Eng. Struct.* **2003**, *25*, 993–1007. [CrossRef]
29. Kwak, H.-G.; Kim, D.-Y. Material nonlinear analysis of RC shear walls subject to monotonic loadings. *Eng. Struct.* **2004**, *26*, 1517–1533. [CrossRef]
30. Lash, S.D. Ultimate strength and cracking resistance of lightly reinforced beams. *ACI J.* **1953**, *49*, 573–582.
31. Gilbert, R.I. Deflection Calculation for Reinforced Concrete Structures—Why We Sometimes Get It Wrong. *ACI Struct. J.* **1999**, *96*, 1027–1033.
32. Scanlon, A.; Bischoff, P.H. Shrinkage Restraint and Loading History Effects on Deflections of Flexural Members. *ACI Struct. J.* **2009**, *105*, 498–506.
33. Kaklauskas, G.; Ghaboussi, J. Stress-strain relations for cracked tensile concrete from RC beam tests. *J. Struct. Eng.* **2001**, *127*, 64–73. [CrossRef]
34. Gribniak, V.; Mang, H.A.; Kupliauskas, R. Stochastic Tension-Stiffening Approach for the Solution of Serviceability Problems in Reinforced Concrete: Constitutive Modeling. *Comput. Civ. Infrastruct. Eng.* **2015**, *30*, 684–702. [CrossRef]
35. Kaklauskas, G.; Tamulenas, V.; Bado, M.F.; Bacinskas, D. Shrinkage-free tension stiffening law for various concrete grades. *Constr. Build. Mater.* **2018**, *189*, 736–744. [CrossRef]
36. *Design of Concrete Structures—Part 1-1: General Rules and Rules for Buildings*; Federal Institute of Technology Lausanne-EPFI: Geneva, Switzerland, 2004.
37. Kaklauskas, G.; Tamulenas, V.; Gribniak, V.; Ng, P.L.; Kupliauskas, R.; Structures, S.; Kong, H. Tension-stiffening behaviour of reinforced concrete ties of various strength classes. In Proceedings of the International Conference on Performance-based and Life-cycle Structural Engineering, School of Civil Engineering, The University of Queensland, Brisbane, QLD, Australia, 9–11 December 2015; pp. 582–590.
38. Sokolov, A.; Kaklauskas, G.; Jakubovskis, R.; Juknys, M.; Kliukas, R.; Ng, P.L.; Gribniak, V. Experimental Investigation of Tension Stiffening in RC Ties. *Adv. Mater. Sci. Eng.* **2016**, *2016*, 1–8. [CrossRef]
39. Kaklauskas, G.; Gribniak, V.; Bacinskas, D.; Vainiunas, P. Shrinkage influence on tension stiffening in concrete members. *Eng. Struct.* **2009**, *31*, 1305–1312. [CrossRef]
40. Ruiz, M.F.; Muttoni, A.; Gambarova, P.G. Analytical modeling of the pre- and postyield behavior of bond in reinforced concrete. *J. Struct. Eng.* **2007**, *133*, 1364–1372. [CrossRef]

materials

MDPI

Review

Advances in the Deformation and Failure of Concrete Pavement under Coupling Action of Moisture, Temperature, and Wheel Load

Wanguo Dong [1], Chunlin Liu [2,3], Xueben Bao [2], Tengfei Xiang [2,3] and Depeng Chen [2,3,*]

1 School of Management Science and Engineering, Anhui University of Technology, Ma'anshan 243032, China; wanguo88@ahut.edu.cn

2 School of Architectural and Civil Engineering, Anhui University of Technology, Ma'anshan 243032, China; chunlinliu@ahut.edu.cn (C.L.); xbb2020123@163.com (X.B.); xiangtf@ahut.edu.cn (T.X.)

3 Institute of Green Building Materials, Anhui University of Technology, Ma'anshan 243032, China

* Correspondence: dpchen@ahut.edu.cn; Tel.: +86-158-5550-3077

Received: 7 November 2020; Accepted: 2 December 2020; Published: 4 December 2020

Abstract: The deformation and cracking of concrete will lead to various deterioration processes, which will greatly reduce the durability and service life of the concrete pavement. The relating previous studies and analysis revealed that the coupling action of environmental temperature, moisture, and wheel load will cause cracking and seriously affect the normal service and durability of pavement concrete. This paper presents theoretical and numerical state-of-the-art information in the field of deformation and failure of pavement concrete under coupling action of moisture, temperature, and wheel load and draws some conclusions. (a) Concrete is a typical porous material, moisture and heat transfer theory has obtained enough data to simulate the hygro-thermo properties of concrete, and the relationship between moisture and heat is very clear. (b) There are few studies on concrete pavement or airport pavement considering the coupling action of moisture, temperature, and wheel load. (c) Concrete pavement is subjected to hygro-thermal-mechanical coupling action in service, which has the characteristics of a similar period and its possible fatigue effect. (d) COMSOL software has certain advantages for solving the coupled hygro-thermal-mechanical of concrete.

Keywords: hygro-thermo-mechanical; coupling action; pavement concrete; deformation and failure

1. Introduction

Cement concrete pavement is widely used in the construction of the airport runway and high-grade highway in China. More than 80% of the airport pavement in China adopts cement concrete [1]. For pavement concrete, the exposed surface formed after the construction of the whole cast-in-place pavement is large, and the trend of early shrinkage cracks caused by concrete water loss is more obvious. Temperature cracks, including temperature shrinkage and temperature fatigue cracks caused by the temperature difference between day and night, are also cracks of the airport or highway pavement form. At present, there are more researches on temperature cracks [2–4]. It is found that the cement concrete airport and highway pavement which has not reached the design service life has been found to have various degrees of defects, cracks, and even complete damage. This phenomenon brings huge maintenance pressure and economic burden. Dry shrinkage cracks and warpage cracks are also the main forms of concrete cracks on airport pavement, especially in areas with drought, wind, less rainfall, and large evaporation, the drying shrinkage deformation of concrete is more significant. For the airport pavement during the service period, the diurnal (seasonal) temperature difference and the diurnal (seasonal) coupled with dry-wet changes present quasi-periodic changes, and the effects of temperature and humidity changes also have coupling effects. In addition, airport pavement or highway pavement also bears the action of aircraft or driving wheel load and its dynamic effect, which will make pavement concrete appear fatigue failure. Therefore, the deformation, cracking, and fracture failure of airport and highway pavement are the comprehensive embodiment of temperature, humidity fatigue effect, and wheel load fatigue effect.

This paper will start from the coupling action of moisture, temperature, and wheel load that affects the deformation and failure of concrete pavement in the service life. The basic model of the multi-physical fields of concrete, research on the fatigue effect of concrete pavement, research on the deformation and failure of concrete pavement or airport pavement (except for concrete pavement or airport pavement), and the numerical simulation method of hygro-thermal-mechanical coupling deformation of concrete are discussed. This paper will provide new research ideas for the study of concrete deformation and failure of the pavement.

2. The Basic Model of The Multiple Physical Fields of Concrete

Concrete is a typical porous medium material [5,6], and its internal moisture and heat transfer are similar to the transmission principle and model of general porous media materials. The mechanism of moisture transmission in concrete is the diffusion of steam and liquid water. The transmission of water vapor conforms to the principle of liquid diffusion and is also the main form of concrete wet transmission. According to the actual hole structure size of concrete, the internal water diffusion should be combined with molecular diffusion and Knudsen diffusion. These provide a theoretical basis for the numerical simulation calculation of concrete hygro-thermo coupling deformation. The heat and moisture transfer can be commonly expressed based on Fourier's Law, Fick's law, and the law of conservation of mass, respectively [7–11]. Experiments on temperature and relative humidity were conducted by Haitao Zhao [11], and a coupled model of temperature and relative humidity was proposed. The test results showed that the coupled phenomenon of temperature and relative humidity could be observed and divided into three stages.

2.1. Moisture Diffusion Model

Generally, it is assumed that the moisture transfer in porous cement-based materials is in the form of diffusion, and the humidity gradient is the driving force of humidity (liquid water and steam). According to Fick's law and mass conservation equation, the one-dimensional transmission formula is as follows [12]:

$$\frac{\partial}{\partial x}\left(D_m\frac{\partial M}{\partial x}\right)+Q_m=\frac{\partial M}{\partial t} \tag{1}$$

In the formula, M is the moisture content: D_m is the moisture diffusion coefficient. In the actual calculation, the moisture diffusion coefficient should be modified, that is, $D_{mk} = K_f D_m$, K_f is the Knudsen diffusion influence coefficient; Q_m is the wet source.

2.2. Heat Transfer Model

Fourier's law is usually used to describe heat transfer in cement-based materials. The thermal conductivity is the apparent thermal conductivity which has taken into account the influence of internal convection. The Formula (2) is as follows [12]:

$$\left(-\lambda\frac{\partial T}{\partial x}\right)+Q=\rho c_p\frac{\partial T}{\partial t} \tag{2}$$

In the formula, Q is the heat source, the heat change caused by cement hydration heat release or other non-heat transfer processes; T is temperature; ρ is the apparent density of cement-based materials; c_p is the specific heat of cement-based materials; λ is the nominal thermal conductivity of cement-based materials.

2.3. Microprestress-Consolidation Theory

Consolidation theory [13] and micro prestressing consolidation theory [14] are widely accepted and used theoretical models. Consolidation theory provides a physical mechanism description for the basic creep and aging effect of concrete, and its B3 model [15] has been widely used in the analysis and calculation of compressive creep of mature concrete. At present, the Microprestress-Solidification (MPS) theory is widely adopted to analyze the complex influence of temperature and humidity on concrete creep. However, the MPS theory still has some shortcomings. The MPS theory was improved for concrete creep under a complex environment [16].

Under the action of uniaxial stress, the total strain of concrete can be regarded as the sum of five parts of strain, which can be expressed by Formula (3). The rheological model is shown in Figure 1.

$$\varepsilon = \varepsilon^{i} + \varepsilon^{ev} + \varepsilon^{f} + \varepsilon^{sh} + \varepsilon^{T} \tag{3}$$

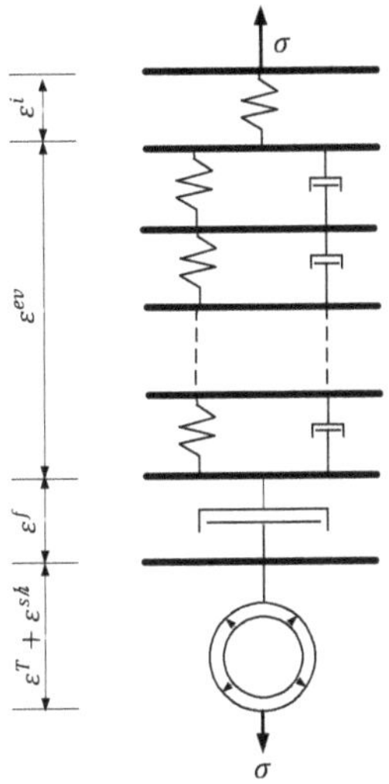

Figure 1. Rheological model.

In the formula, the ε^{i} is instantaneous strain; the ε^{ev} is viscoelastic strain; the ε^{f} is the viscous strain; ε^{sh} is the shrinkage strain caused by the humidity change; ε^{T} is the temperature strain caused by the temperature change.

3. Research on the Fatigue Effect of Concrete Pavement

For the pavement in the area with significant temperature and humidity changes, frequent takeoff and landing of aircraft or airport runway of large military multi-wheel aircraft, heavy load and frequent traffic, the pavement is more prone to deformation and damage. The problem of deformation and cracking of airport runway, highway, and urban road pavement has not been well solved because the influence factors of pavement concrete deformation and failure are not fully considered in the design stage. The above factors include temperature change, dry and wet change, and dynamic and static action of wheel load and its coupling effect. The research on the fatigue effect of airport concrete is mainly as follows: large scale airfield concrete slabs were tested by Jeffery R. Roesler [17] in the laboratory to evaluate the effect of multi-wheel gears on the fatigue resistance of concrete slabs. The test program solved the influence of peak stress ratio, stress range and stress pulse type on the fatigue resistance of concrete slabs. The mechanical properties of pavement concrete under the joint action of corrosion, fatigue, and fiber content were assessed by response surface methodology (RSM) [18]. The RSM model fitted well and indeed effectively revealed the mechanical properties of pavement concrete under the joint action of corrosion, fatigue, and fiber content. Frost damage was a common durability problem for concrete structures in cold and wet regions, and in many cases, the frost damage was coupled with fatigue loadings such as the traffic loads on bridge decks or pavements. To investigate the basic fatigue behavior of concrete materials affected by frost damage, a mesoscale approach based on Rigid Body Spring Method (RBSM) had been developed [19]. Some researchers studied the fatigue failure law of concrete pavement by adding new materials in the concrete pavement [20–23]. To sum up, concrete pavement is subjected to hygro-thermal-mechanical action in service, which has the characteristics of a similar period and its possible fatigue effect.

4. Research on Deformation and Failure of Concrete Pavement or Airport Pavement

In the related research on deformation and failure of airport pavement or cement concrete pavement, many researchers mainly focus on the influence of temperature, fatigue load and freeze-thaw, temperature and humidity changes on the shrinkage, and cracking of pavement concrete [24–32], and there is a little previous literature on the coupling of humidity, heat and wheel load or considering the effect of temperature, humidity and wheel load at the same time [33]. Yinchuan Guo [33] studied micropore deterioration and its mechanism in pavement concrete in seasonally frozen regions, the dynamic deterioration rules of micropores under different coupling levels were discussed at the micro scale. To clarify the influence of the coupling conditions on the evolution of pavement concrete micropores, a fatigue load single field and fatigue load and freeze-thaw double field were designed as the control groups. The detailed programs are shown in Figures 2 and 3. Xiaolong Yang [30] studied the micropore change and its mechanism of pavement concrete in the seasonal freeze-thaw region. The coupling effect tests of fatigue load, freeze-thaw cycle, and dry wet cycle were carried out, and the evolution mechanism of micropore in seasonally frozen regions was proposed.

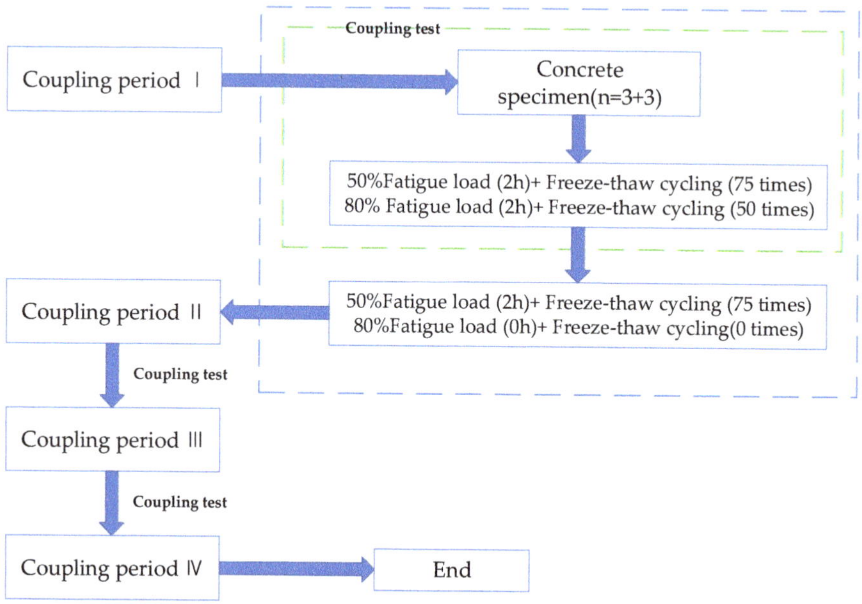

Figure 2. Flow chart of the double-field coupling experiment [30].

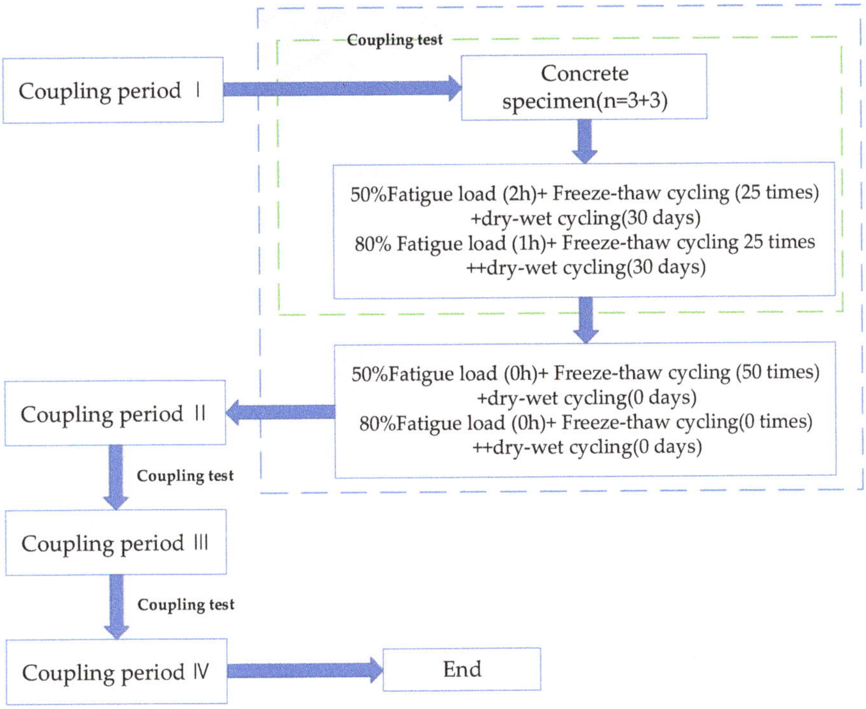

Figure 3. Flow chart of the triple-field coupling experiment [30].

Seongcheol. Choi [29] et al. studied the influence of environmental temperature and humidity on the behavior of continuously reinforced concrete pavement. The drying shrinkage of concrete and the elastic modulus of coarse aggregate played a decisive role in the development of nonstructural stress of concrete pavement. The behavior analysis of continuously reinforced concrete pavement (CRCP) under environmental load was divided into two stages, namely the early stage and the late stage. Two models were proposed, and Figure 4a illustrated the structural model used to analyze the behavior of CRCP at the early stages before transverse cracking. Figure 4b showed the structural model used to analyze the behavior of CRCP at later stages after transverse crack development. Arunw. Dhawale [34] concluded that the stress in the rigid concrete slab was caused by wheel load and slab movement caused by moisture loss and temperature change in the whole slab depth. The warpage stress caused by the difference of moisture content between the top and bottom of the slab was equal to the stress caused by the heaviest load in importance, and the warpage stress response was more important than the wheel load stress. H. Tomas Yu [35] measured the response of concrete pavement surface temperature and wheel load with an instrument panel. For cracking performance, most of the pavement with stable base showed an unbonded response, although the back-calculation results showed that most of the same cross-sections an adhesive response, which may indicate that most slab cracking occurred under the corner loading. Ye. Dan and Mukhopadhyay, Anal. K [26,36] studied the warpage stress and deformation of pavement concrete slab caused by early moisture and heat transfer. It was considered that the damp heat effect had a decisive influence on pavement concrete deformation and cracking under the condition of environmental climate change, and emphasized the importance of strengthening early maintenance. Many researchers [24–27,29,37] had studied the warpage of concrete pavement, which showed that the early warpage deformation behavior of concrete pavement was not only affected by temperature change but also affected by humidity change, drying shrinkage, and temperature conditions during construction. The core viewpoints of relevant studies are shown in Table 1.

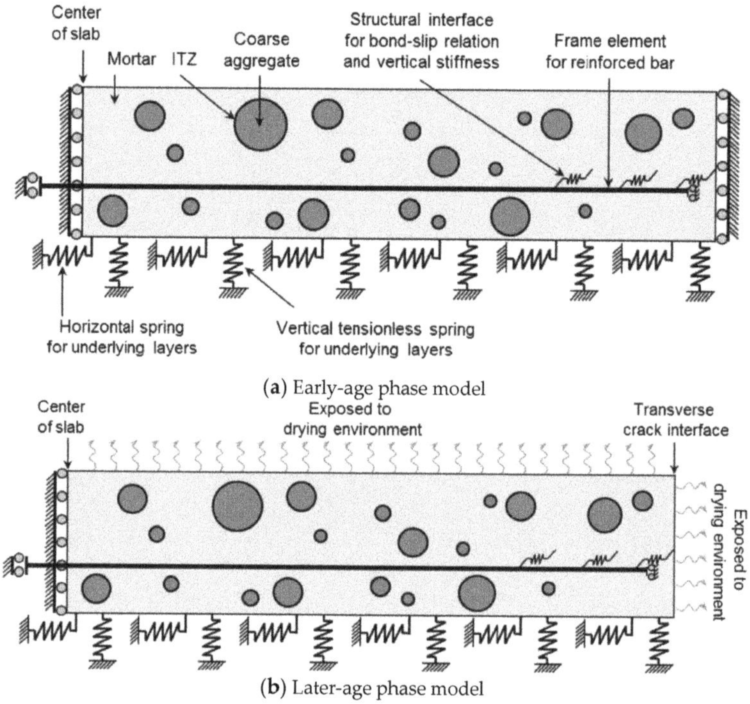

Figure 4. Early-age and Later-age phase model [29].

Many Chinese researchers have carried out research on warpage stress. In recent years, there have been studies considering the joint effect of temperature effect and wheel load stress. Tan Zhiming et al. [38] earlier studied the numerical simulation of temperature field, temperature warpage stress, and load stress of cement concrete pavement under different conditions. Wu Jun et al. [39,40] studied the dynamic performance test of a multi-layer pavement system under impact load and concluded that the impact resistance of rigid pavement was weak and brittle fracture occurred easily under impact load. The impact resistance performance test and numerical simulation of composite pavement structure were carried out. The three-dimensional numerical model of the composite pavement system under impact load was established. Zhao fangran et al. [41] studied the damage of the airport pavement thermal blowing-snow process and considered that the comprehensive effects of thermal shock stress, the thermal expansion force caused by ice crystal vaporization, and freeze-thaw cycle were the fundamental reasons for the damage of the airport concrete pavement. Ling Jianming et al. [42,43] used the finite element software ABAQUS to analyze the mechanical responses of cement concrete airport pavement to multiple-gear military aircraft loadings.

A 3D finite element model was established for analyzing the mechanical responses of cement concrete airport pavement to multiple-gear military aircraft loadings. The test results showed that the maximum tensile stress in the pavement occurred at the bottom of the slab, the tensile stress and the deformation of the pavement under two 8-wheeled gears were smaller than that under one 8-wheeled gear. The elasticity modulus of the slab, the resilience modulus of the base and the reaction modulus of the subgrade all had significant effects on the mechanical responses of the pavement.

Table 1. Research progress abroad.

Number of Cycles	Author	Viewpoints
1	Shadravan. Shideh [24]	Study on the behavior of concrete slabs on the ground in controlled moisture and temperature environment.
2	Maekawa. Koichi [25]	Structural creep deformations were reproduced by using the multi-scale coupled thermo-hygro-mechanical modeling.
3	Ye. Dan [26]	Moisture capacity was induced into the temperature and moisture analysis for curing concrete to solve the coupled and nonlinear heat transfer and moisture transport problems in early-age concrete.
4	Hiller. Jacob E. [27]	A simplified method termed NOLA(Nonlinear Area) and a mechanistic-based rigid pavement analysis program called RadiCAL was proposed to determine fatigue damage levels and critical cracking locations.
5	Simonova. Anna [28]	The scope of the research is to analyze the impact of ambient temperature variations in water, and the thermal state of the subgrade on different types of road structures.
6	Seongcheol. Choi [29]	Mesoscale analysis of continuously reinforced concrete pavement behavior subjected to temperature and moisture variations.

Zheng Fei and Weng Xingzhong [44] analyzed the main influence factors of pavement slab stress using elastic foundation plate theory and proposed the stress calculation method of cement concrete pavement slab under aircraft load. Wang Zhenhui et al. [45] considered that the cumulative damage model of rigid pavement should consider the influence of load stress distribution of pavement slab, and established the cumulative damage optimization model suitable for rigid pavement by using the covering action curve and stress distribution function. Huang Xiaoming et al. [46] studied the temperature warpage stress of continuously reinforced cement concrete pavement and considered that the temperature warpage stress of a single slab in the Winkler foundation model can be calculated by using the warpage stress calculation formula of ordinary cement concrete slab consistent with its size. Li Xinkai et al. [47] analyzed the deformation and stress of cement pavement slab under the action of axial load and temperature. Slab deflection and stress were calculated under axial loads at different slab positions and negative or positive temperature gradient coupling. The calculated results show that the different conditions of axial loads and temperature gradient coupling will change the maximum tension and cause various types of cracks in a slab. Yang Jinzhi [48] analyzed the deformation and crack generation mechanism of concrete pavement surface under the coupling effect of temperature and cyclic load by using finite element software, and obtained the relationship among temperature, load, and pavement displacement. The failure law of pavement under the coupling effect of temperature and the cyclic load was obtained. Some valuable conclusions for concrete pavement construction were obtained. Hu Changbin et al. [49] studied the temperature field and temperature stress of cement concrete pavement in hot and humid areas, and concluded that "with the periodic change of external environmental meteorological conditions, there were many typical distribution shapes of pavement temperature gradient and they changed with the external environment". Combined with relevant research experience and results [6,50–52], the deformation and cracking of pavement concrete during the service period are related to the times of wheel load action and wheel load impact effect, and closely related to the temperature and humidity changes in the actual use environment. Especially for areas where the temperature and humidity changed dramatically in day and night or season, it was very important to study the deformation and cracking behavior and meso mechanical mechanism of concrete pavement under the coupling action of moisture-heat-force.

5. Research on Deformation and Failure of Concrete (Except for Airport Pavement and Highway Pavement) under the Action of Moisture-Heat-Force

The research on the hygro-thermal-mechanical coupling action analysis was previously in the field of rock engineering and had always been a research hotspot [53,54]. The moisture-heat-force multi-field coupling research in the field of rock engineering was basically about the deterioration of concrete performance under high temperature or fire [55–61]. Schrefler et al. [55,56] put forward the mathematical and numerical model of the nonlinear performance of porous multiphase concrete according to the characteristics of porous multiphase concrete and the principle of thermodynamic equilibrium. The early performance of self-drying and high-temperature deformation of concrete was simulated and analyzed. A moisture-heat-force coupling model of concrete at high temperature was proposed, its numerical simulation was realized by a finite element method, and the performance of high-performance concrete walls and columns in the fire was studied. Through moving boundary problems, Beneš and Štefan [59] put forward a mathematical model of hygro-thermal-mechanical analysis for a high-temperature burst of the concrete wall. The thermal stress effect and pore pressure change of burst were considered, and the validity was verified. S. Grasberger and G. Meschke [62] established a 3D coupled thermo-hygro-mechanical model for concrete accounting for moisture and heat transport, cracking and irreversible deformations, and the various interactions between these processes. The effects of drying shrinkage, non-isothermal transmission, and cracking on the drying process of concrete were studied. Bangert et al. [63] considered the mechanical damage and the interaction of moisture heat transfer and established a concrete hygro-thermal-mechanical coupling model based on the theoretical mechanism analysis and experimental results. A two-dimensional simulation of a concrete slab under the condition of moisture thermal-mechanical coupling was carried out. D. Gawin et al. [64,65] regarded concrete as a multiphase porous material and proposed a mathematical model for analyzing the hygro-thermal behavior of high-temperature concrete. Jaroslav Kruis et al. [66] studied and analyzed the effective computer implementation of the coupled analysis of prestressed concrete nuclear reactor based on the hygro-thermal-mechanical coupled analysis. The hydro-thermo-mechanical analysis of reactor vessels based on the finite element method was a very difficult task because of its complexity and numerous unknowns. This contribution involved efficient computer implementation of coupling analysis and was also devoted to the implementation of domain decomposition methods that could utilize parallel computers. Parallel processing could achieve very good acceleration and solve large problems in an acceptable time. The proposed strategy is demonstrated in the coupling analysis of existing reactor vessels. C.T. Davie et al. [67] analyzed the sensitivity of typical prestressed pressure vessels through the hygro-thermal-mechanical fully coupled model of concrete. The results showed that changes to operating procedures only led to minor changes in the behavior of the structure throughout its life cycle but the unplanned thermal excursions could have a greater impact on the concrete structure. Li Rongtao and Li Xikun [57,58] studied the failure process of concrete at high temperature, the constitutive relationship of concrete at high temperature, the chemical hygro-thermal-mechanical coupling process, and the numerical calculation method. Li Zhongyou and Liu Yuanxue [68] studied the evolution process of mechanical damage, which started from the characteristics of energy dissipation in the process of material deformation (failure) and based on the mixture theory. Considering the thermal damage caused by high temperature, the heat-water-force coupling damage model of concrete under high temperature was established. Liu Jiaping et al. [69] studied the early cracking of sidewall concrete. A multi-field (hygro–thermo-chemo-mechanical) coupling model based on Fourier's law, Fick's law, and mass and energy balance equations, was adopted to describe the cement hydration, temperature, and humidity evolution for early-age sidewall concrete. Gasch et al. [70,71] established the hygro-thermal-mechanical coupling model of hardened concrete based on microprestress-consolidation theory, which took into account the factors such as age, creep, shrinkage, thermal expansion, and cracking under the condition of varying temperature and humidity, which had inspirations on this project.

6. Numerical Simulation Method of Hygro-Thermal-Mechanical Coupling Deformation of Concrete

In the research of the numerical simulation method of hygro-thermal-mechanical coupling deformation of concrete, there are four main numerical simulation methods:

6.1. Heat and Mass Transfer in Porous Media of Phenomenological Thermodynamic Method

This method does not involve the internal heat and mass transfer mechanism and specific processes of the porous media, but only considers the relationship and cross effect between various flows and forces of heat and mass transfer. A phenomenological equation can be obtained to describe various flows. Phenomenological methods are difficult to apply in practice due to the influence of test conditions and many material parameters. Due to the lack of effective means and methods for testing moisture distribution, the moisture migration rate is slower than the thermal migration rate. In the process of solving various models, the values of some coefficients are basically based on experimental or empirical data, which affects the versatility of this method.

6.2. Numerical Analysis Method Based on Luikov's Coupled Heat and Mass Transfer Equation

This method considers the heat absorption or exotherm in the process of heat and mass transfer, and establishes a partial differential equation system based on the principles of mass conservation and energy conservation in the process of moisture migration. The method of numerical analysis is used to solve the equations to obtain the analytical solutions of the heat, humidity fields, and their dynamic changes [72–75]. This method is also mainly used for the analysis of heat and mass transfer in porous media. With the development of computer technology, the difficulty of implementing numerical analysis methods is greatly reduced. Concrete is also a typical porous medium [5,6] and the above method can be considered for application in concrete. In the process of solving the temperature and humidity coupling equations, there are different solutions; for example, Bouddour [76] et al. used the asymptotic continuous method of periodic structure to study the heat and humidity transfer in the evaporation and condensation process. However, the research methods involved material did not absorb wet steam, and the hygroscopicity of the material had an effect on the heat transfer process. Lobo and Mikhailov [77,78] used the classic integral transformation method to solve the heat and mass transfer problem in porous media. Because of its complex characteristics and difficult calculation, it could not accurately reflect the distribution of temperature and humidity to a large extent, and can not get the correct results. Chang [79] et al. used decoupling technology to solve the coupled equations of the heat and mass transfer process; however, this appeared powerless when the governing equations and boundary conditions were coupled at the same time. Cheroto [80] et al. used an improved lumped system analysis method to find the approximate solution of the coupling equation. Although it avoided the problems encountered in the calculation of complex eigenvalues and Chang's decoupling technology, its accuracy was not enough. It can not truly reflect the temperature and humidity distribution in porous media.

6.3. Finite Element Analysis Method

The finite element method is the most widely used numerical calculation method in scientific research and has become the main tool for solving scientific and technological problems. At present, many commercial finite element softwares have been developed, such as ANSYS, MSC, ADINA, ABAQUS, etc. In the finite element analysis method, the model should be established first, and then solved by self-programming [57,58,63,81], ANSYS and ABAQUS software are used more in pratice [82–85]. Bernard et al. [82] proposed a three-dimensional multi-scale simulation method for the mechanical properties of cement-based materials. Firstly, the CEMHYD3D model was used to generate ideal three-dimensional representative volume elements of cement-based materials at different scales, and then the mechanical behavior was calculated by using the finite element analysis software ABAQUS. Li Zhaoxia et al. [83–85] studied the damage analysis and state assessment of long-span structures. A series of multi-scale modeling processes, multi-scale damage, and its homogenization algorithms in the structural multi-scale simulation were studied based on the nonlinear finite element software ABAQUS.

6.4. Application of Multiphysics Coupling Analysis Software

The analysis method of COMSOL multi-physical field coupling calculation software [6,70,86] was used to model and simulate engineering problems based on partial differential equations. This method was easy for a couple of different physical fields for simulation. In the author's previous research, the analytical calculation of the temperature and humidity field based on the Luikov humidity–heat coupled transfer equation had been studied carefully. The analytical calculation of temperature and humidity field and finite element analysis of coupled deformation were realized by mixed programming based on the principle of heat and moisture transfer in porous media. The above research can simulate the moisture heat coupling deformation behavior of some concrete within a certain engineering precision range. The differential equation of this model had considered the mutual influence between temperature and humidity, which reflected the coupling of humidity and heat, but it was more difficult to solve. Moreover, through the Laplace transform and the transfer function method, the calculation requirements were stricter, and the selection of the transfer function was more difficult, especially for the study of the internal temperature and humidity distribution of the concrete under the change of environmental temperature and humidity conditions. The author published an academic paper [6] in 2013. Based on the theory of moisture and heat transfer in porous media and microprestress consolidation theory, the coupled model of concrete moisture-heat-force was established, and the numerical solution of the coupling was realized by using COMSOL multi-physical field numerical simulation software and proposed the concrete moisture expansion coefficient. Gasch et al. [70] published a paper in 2016, using COMSOL software to carry out coupled finite element simulation analysis of strong form partial differential equation, which has important reference value for this research.

7. Conclusions and Perspective

The studies of the coupled hygro-thermo-mechanical behavior of concrete pavement have been reviewed in this paper. The various aspects on the deformation and failure of concrete pavement under coupling action of moisture, temperature and wheel load presented in this paper could be summarized and concluded as:

1. Concrete is a typical porous material, moisture and heat transfer theory had been obtained enough data to simulate the hygro-thermo-mechanical properties of concrete, and the relationship between moisture and heat is very clear.
2. There has been some research on the moisture-heat-mechanical coupling of concrete materials, but most of the research focused on the performance of concrete in fire or high temperature. There are few studies on the moisture-heat-mechanical coupling deformation of concrete under normal service conditions.
3. In the related research on deformation and failure of airport pavement or cement concrete pavement, many researchers mainly focused on the influence of temperature, fatigue load and freeze-thaw, and temperature and humidity changes, which lead to the shrinkage and cracking of pavement concrete. There are few kinds of literature considering the coupling action of moisture, temperature, and wheel load at the same time.
4. Concrete pavement is subjected to hygro-thermal-mechanical coupled action in service, which has the characteristics of a similar period and its possible fatigue effect. In the existing moisture-heat-mechanical coupling research, the research on "force" is mostly the load of a certain stress level continuously applied, but not enough attention is paid to the periodicity of wheel load and the change of temperature and humidity.
5. COMSOL software has certain advantages for solving the coupled hygro-thermal-mechanical of concrete.

In the later research, the deformation and failure mechanism of pavement concrete under the coupling action of moisture, temperature, and wheel load need to be studied.

Author Contributions: W.D., C.L., X.B., and T.X. conducted the theoretical analysis and wrote the manuscript. D.C. advised the theoretical research and previous experimental work and revised the manuscript. All authors have read and agreed to the published version of the manuscript.

Funding: This research was funded by the National Natural Science Foundation of China (Grant No. 51978002, 51108002 and 11802001) and the Natural Science Foundation for Higher Education Institutions in the Anhui Province of China (Grant No. KJ2016SD08).

Acknowledgments: The authors gratefully acknowledge the National Natural Science Foundation of China for their financial support of this work. The authors would like to express our gratitude to Anhui University of Technology for providing various research conditions to facilitate this work.

Conflicts of Interest: The authors declare no conflict of interest.

References

1. Cai, Z.; LI, S.; Xu, X.; Wu, C. Comparative study on construction technology of airport road surface and cement concrete pavement. *Highw. Eng.* **2020**, *45*, 167–174. (In Chinese)
2. Li, M.; Xu, W.; Wang, Y.J.; Tian, Q.; Liu, J.P. Shrinkage crack inhibiting of cast in situ tunnel concrete by double regulation on temperature and deformation of concrete at early age. *Constr. Build. Mater.* **2020**, *240*, 117834. [CrossRef]
3. Xin, J.D.; Zhang, G.X.; Liu, Y.; Wang, Z.H.; Wu, Z. Effect of temperature history and restraint degree on cracking behavior of early-age concrete. *Constr. Build. Mater.* **2018**, *192*, 381–390. [CrossRef]
4. Zhu, H.; Hu, Y.; Li, Q.B.; Ma, R. Restrained cracking failure behavior of concrete due to temperature and shrinkage. *Constr. Build. Mater.* **2020**, *244*, 10. [CrossRef]
5. Zeng, Q.; Li, K. Quasi-Liquid Layer on Ice and Its Effect on the Confined Freezing of Porous Materials. *Crystals* **2019**, *9*, 250. [CrossRef]
6. Chen, D.P. Multi-physicalfield coupling simulation of hygro-thermal deformation of concrete. *J. Southeast Univ. Sci.* **2013**, *43*, 582–587. (In Chinese)
7. Bocciarelli, M.; Ranzi, G. Identification of the hygro-thermo-chemical-mechanical model parameters of concrete through inverse analysis. *Constr. Build. Mater.* **2018**, *162*, 202–214. [CrossRef]
8. Isgor, O.B.; Razaqpur, A.G. Finite element modeling of coupled heat transfer, moisture transport and carbonation processes in concrete structures. *Cem. Concr. Compos.* **2004**, *26*, 57–73. [CrossRef]
9. Kim, J.-K.; Lee, C.-S. Prediction of differential drying shrinkage in concrete. *Cem. Concr. Res.* **1998**, *28*, 985–994. [CrossRef]
10. Wan, L.; Wendner, R.; Liang, B.; Cusatis, G. Analysis of the behavior of ultra high performance concrete at early age. *Cem. Concr. Compos.* **2016**, *74*, 120–135. [CrossRef]
11. Zhao, H.T.; Jiang, K.D.; Yang, R.; Tang, Y.M.; Liu, J.P. Experimental and theoretical analysis on coupled effect of hydration, temperature and humidity in early-age cement-based materials. *Int. J. Heat Mass Transf.* **2020**, *146*, 118784. [CrossRef]
12. Chen, D.P. Study on Numerical Simulation of Hygro-Thermal Deformation of Concrete Based on Heat and Moisture Transfer in Porous Medium and Its Application. Ph.D. Thesis, Southeast University, Nanjing, China, October 2007.
13. Bažant, Z.P.; Fellow, A.S.C.E.; Prasannan, S.; Member, S. Solidification Theory for Concrete Creep. I: Formulation. *J. Eng. Mech.* **1989**, *115*, 1691–1703. [CrossRef]
14. Bažant, Z.P.; Hauggaard, A.B.; Baweja, S.; Ulm, F.J. Microprestress-Solidification Theory for Concrete Creep. I: Aging and Drying Effects. *J. Eng. Mech.* **1997**, *123*, 1188–1194. [CrossRef]
15. Baant, Z.P.; Baweja, S. Creep and Shrinkage Prediction Model for Analysis and Design of Concrete Structures: Model B3. *Mater. Struct.* **1995**, *28*, 357–365.
16. Yu, P.; Duan, Y.H.; Fan, Q.X.; Tang, S.W. Improved MPS model for concrete creep under variable humidity and temperature. *Constr. Build. Mater.* **2020**, *243*, 118183. [CrossRef]
17. Roesler, J.R.; Hiller, J.E.; Littleton, P.C. Large-scale airfield concrete slab fatigue tests. *IJCP* **2005**, *1*, 66–87.
18. Long, X.; Cai, L.; Li, W. RSM-based assessment of pavement concrete mechanical properties under joint action of corrosion, fatigue, and fiber content. *Constr. Build. Mater.* **2019**, *197*, 406–420. [CrossRef]

19. Gong, F.; Ueda, T.; Wang, Y.; Zhang, D.; Wang, Z. Mesoscale simulation of fatigue behavior of concrete materials damaged by freeze-thaw cycles. *Constr. Build. Mater.* **2017**, *144*, 702–716. [CrossRef]
20. Alsaif, A.; Garcia, R.; Figueiredo, F.P.; Neocleous, K.; Christofe, A.; Guadagnini, M.; Pilakoutas, K. Fatigue performance of flexible steel fibre reinforced rubberised concrete pavements. *Eng. Struct.* **2019**, *193*, 170–183. [CrossRef]
21. Graeff, A.G.; Pilakoutas, K.; Neocleous, K.; Peres, M.V.N.N. Fatigue resistance and cracking mechanism of concrete pavements reinforced with recycled steel fibres recovered from post-consumer tyres. *Eng. Struct.* **2012**, *45*, 385–395. [CrossRef]
22. Pacheco-Torres, R.; Cerro-Prada, E.; Escolano, F.; Varela, F. Fatigue performance of waste rubber concrete for rigid road pavements. *Constr. Build. Mater.* **2018**, *176*, 539–548. [CrossRef]
23. Xiao, J.; Li, H.; Yang, Z. Fatigue behavior of recycled aggregate concrete under compression and bending cyclic loadings. *Constr. Build. Mater.* **2013**, *38*, 681–688. [CrossRef]
24. Shadravan, S.; Ramseyer, C.; Kang, T.H.K. A long term restrained shrinkage study of concrete slabs on ground. *Eng. Struct.* **2015**, *102*, 258–265. [CrossRef]
25. Maekawa, K.; Chijiwa, N.; Ishida, T. Long-term deformational simulation of PC bridges based on the thermo-hygro model of micro-pores in cementitious composites. *Cem. Concr. Res.* **2011**, *41*, 1310–1319. [CrossRef]
26. Ye, D. *Early-Age Concrete Temperature and Moisture Relative to Curing Effectiveness and Projected Effects on Selected Aspects of Slab Behavior*; Texas A&M University: College Station, TX, USA, 2007.
27. Hiller, J.E.; Roesler, J.R. Simplified Nonlinear Temperature Curling Analysis for Jointed Concrete Pavements. *J. Transp. Eng.* **2010**, *136*, 654–663. [CrossRef]
28. Simonova, A. Evaluation of Functional Safety of Cement Concrete and Composite Pavements Exposed to Weather Conditions. *Procedia.Trans. Res.* **2017**, *20*, 618–623. [CrossRef]
29. Choi, S.; Na, B.-U.; Won, M.C. Mesoscale analysis of continuously reinforced concrete pavement behavior subjected to environmental loading. *Constr. Build. Mater.* **2016**, *112*, 447–456. [CrossRef]
30. Yang, X.; Shen, A.; Guo, Y.; Zhou, S.; He, T. Deterioration mechanism of interface transition zone of concrete pavement under fatigue load and freeze-thaw coupling in cold climatic areas. *Constr. Build. Mater.* **2018**, *160*, 588–597. [CrossRef]
31. Mateos, A.; Harvey, J.; Bolander, J.; Wu, R.; Paniagua, J.; Paniagua, F. Structural response of concrete pavement slabs under hygrothermal actions. *Constr. Build. Mater.* **2020**, *243*, 118261. [CrossRef]
32. Huang, K.; Zollinger, D.G.; Shi, X.; Sun, P. A developed method of analyzing temperature and moisture profiles in rigid pavement slabs. *Constr. Build. Mater.* **2017**, *151*, 782–788. [CrossRef]
33. Guo, Y.; Chen, Z.; Qin, X.; Shen, A.; Zhou, S.; Lyu, Z. Evolution mechanism of microscopic pores in pavement concrete under multi-field coupling. *Constr. Build. Mater.* **2018**, *173*, 381–393. [CrossRef]
34. Dhawale, A.W.; Vishwanath, M. Comparative study of wheel load stress and warping stress on_concrete pavements. *J. Int. Civ. Struct. Environ. Inf. Eng. Res. Dev.* **2014**, *4*, 9–13.
35. Yu, H.; Khazanovich, L.; Darter, M.; Ardani, A. Analysis of Concrete Pavement Responses to Temperature and Wheel Loads Measured from Intrumented Slabs. *J. Trans. Res. Rec. Board* **1998**, *1639*, 94–101. [CrossRef]
36. Mukhopadhyay, A.K.; Ye, D.; Zollinger, D.G. *Moistrue-Related Cracking Effects on Hydrating Concrete Pavement*; Texas Transportation Institute: College Station, TX, USA, 2006.
37. William, G.W.; Shoukry, S.N. 3D Finite Element Analysis of Temperature-Induced Stresses in Dowel Jointed Concrete Pavements. *Int. J. Geomech.* **2001**, *1*, 291–307. [CrossRef]
38. Ming, T.Z.; Kang, Y.Z. Thermal stress in two-layer concrete slab due to restraint of interface. *J. Traffic. Transp. Eng.* **2001**, *1*, 25–28. (In Chinese)
39. Jun, W.U.; Liang, L.I.; Xiu-Li, D.U. Experiment and numerical simulation on impact resistance of composite pavement slab. *J. Archi. Civ. Eng.* **2018**, *35*, 71–78. (In Chinese)
40. Jun, W.U.; Liang, L.I.; Xiuli, D.; Zhongxian, L. Experiment on dynamic properties of multi-layer pavement system under impact load. *J. Tianjin Univ. Sci.* **2017**, *50*, 1321–1328. (In Chinese)
41. Fangran, Z.; Xinchen, L. Study on airport concrete pavement damage law of hot wind snow removal. *Concrete* **2015**, *4*, 74–77. (In Chinese)
42. Zhou, Z.-F.; Ling, J.-M. Finite element model of airport rigid pavement structure based on ABAQUS. *J. Traffic. Transp. Eng.* **2009**, *9*, 39–44. (In Chinese)

43. Ling, J.; Liu, W.; Zhao, H. Mechanical responses of rigid airport pavement to multiple-gear military air cr aft loadings. *J. Chin. Civ. Eng.* **2007**, *40*, 60–65. (In Chinese)
44. Zheng, F.; Weng, X.Z. Calculating methods of stress for cement concrete pavement slab under plane loads. *J. Traffic. Transp. Eng.* **2010**, *10*, 8–15. (In Chinese)
45. Wang, Z.; Cai, L.; Gu, Q.; Wu, A. Airport rigid pavement cumulative damage optimization model considering load stress distribution. *J. Chin. Civ. Eng.* **2011**, *44*, 143–150. (In Chinese)
46. Huang, X.; Bai, T.; Li, C.; Jin, J. Temperature warping stress study on continuously reinforced concrete pavement. *J. Tongji Univ. Nat. Sci.* **2011**, *39*, 1026–1030. (In Chinese)
47. Li, X.K.; Hou, X.S.; Ma, S.L. Deflection and Stress Analysis of a Pavement Slab under Temperature Gradient and Axial Loading Coupling. *J. Chongqing Jianzhu Univ.* **2008**, *30*, 53–57. (In Chinese)
48. Yang, J.; Xu, Y.; Liu, J.; Duan, Y. Failure mechanism analysis of concrete pavement under temperature load coupling effect. *Highw. Transp. Tech.* **2019**, *000*, 171–173. (In Chinese)
49. Chang-Bin, H.U.; Hui-zhen, Z.; Yun, Q. Characteristics of concrete pavement temperature field and temperature stress in hot and humid areas. *J. Fuzhou Univ. Nat. Sci.* **2011**, *39*, 727–737. (In Chinese)
50. Chen, D.P.; Qian, C.X. Numerical simulation of concrete shrinkage based on heat and moisture transfer in porous medium. *J. Southeast Univ.* **2007**, *23*, 75–80.
51. Chen, D.P.; Liu, C.L.; Qian, C.X.; Miao, C.W.; Liu, J.P. Hybrid analytic-FEA method for calculating hygro-thermal deformation of concrete. *Adv. Sci. Lett.* **2011**, *4*, 1711–1716. [CrossRef]
52. Chen, D.P.; Miao, C.W.; Liu, J.P.; Tang, M.S. Advances in multi-scale simulation of hygro-thermo- mechanical deformation behavior of structural concrete. *Int. J. Civ. Eng.* **2015**, *13*, 267–277.
53. Obeid, W.; Mounajed, G.; Alliche, A. Mathematical formulation of thermo-hygro-mechanical coupling problem in non-saturated porous media. *Comput. Methods Appl. Mech. Eng.* **2001**, *190*, 5105–5122. [CrossRef]
54. Tomac, I.; Gutierrez, M. Coupled hydro-thermo-mechanical modeling of hydraulic fracturing in quasi-brittle rocks using BPM-DEM. *J. Rock Mech. Geotech. Eng.* **2017**, *9*, 92–104. [CrossRef]
55. Schrefler, B.A.; Khoury, G.A.; Gawin, D.; Majorana, C.E. Thermo-hydro-mechanical modelling of high performance concrete at high temperatures. *Eng. Comput.* **2002**, *19*, 787–819. [CrossRef]
56. Dal Pont, S.; Durand, S.; Schrefler, B.A. A multiphase thermo-hydro-mechanical model for concrete at high temperatures–Finite element implementation and validation under LOCA load. *Nucl. Eng. Des.* **2007**, *237*, 2137–2150. [CrossRef]
57. Li, X.K.; Li, R.T.; Schrefler, B.A. A coupled chemo-thermo-hygro-mechanical model of concrete at high temperature and failure analysis. *Int. J. Numer. Anal. Methods Geomech.* **2006**, *30*, 635–681. [CrossRef]
58. Li, R.T.; Li, X.K. A Coupled Chemo-Elastoplastic-Damage Constitutive Model for Plain Concrete Subjected to High Temperature. *Int. J. Damage Mech.* **2010**, *19*, 971–1000.
59. Beneš, M.; Štefan, R. Hygro-thermo-mechanical analysis of spalling in concrete walls at high temperatures as a moving boundary problem. *Int. J. Heat Mass Transf.* **2015**, *85*, 110–134. [CrossRef]
60. Xu, C.; Mudunuru, M.K.; Nakshatrala, K.B. Material degradation due to moisture and temperature. Part 1: Mathematical model, analysis, and analytical solutions. *Contin. Mech. Thermodyn.* **2016**, *28*, 1847–1885. (In Chinese) [CrossRef]
61. Melhem, C.; Boussa, H.; Dumontet, H. *Thermo-Hydro-Mechanical Behavior of Concrete at High Temperature*; John Wiley & Sons, Inc.: Hoboken, NJ, USA, 2013.
62. Grasberger, S.; Meschke, G. Thermo-hygro-mechanical degradation of concrete: From coupled 3D material modelling to durability-oriented multifield structural analyses. *Mater. Struct.* **2004**, *37*, 244–256. [CrossRef]
63. Bangert, F.; Grasberger, S.; Kuhl, D.; Meschke, G. Environmentally induced deterioration of concrete: Physical motivation and numerical modeling. *Eng. Fract. Mech.* **2003**, *70*, 891–910. [CrossRef]
64. Gawin, D.; Pesavento, F.; Schrefler, B.A. Towards prediction of the thermal spalling risk through a multi-phase porous media model of concrete. *Comput. Methods Appl. Mech. Eng.* **2006**, *195*, 5707–5729. [CrossRef]
65. Gawin, D.; Pesavento, F.; Schrefler, B.A. Hygro-thermo-chemo-mechanical modelling of concrete at early ages and beyond. Part II: Shrinkage and creep of concrete. *Int. J. Numer. Methods Eng.* **2006**, *67*, 332–363. [CrossRef]
66. Kruis, J.; Koudelka, T.; Krejčí, T. Efficient computer implementation of coupled hydro-thermo-mechanical analysis. *Math. Comput. Simul.* **2010**, *80*, 1578–1588. [CrossRef]
67. Davie, C.T.; Pearce, C.J.; Bićanić, N. Fully coupled, hygro-thermo-mechanical sensitivity analysis of a pre-stressed concrete pressure vessel. *Eng. Struct.* **2014**, *59*, 536–551. [CrossRef]

68. Zhongyou, L.; Yuanxue, L. A Coupled Thermo-Hygro-Mechanical Damage Model for Concrete Subjected to High Temperatures. *Eur. J. Environ. Civ. Eng.* **2012**, *33*, 444–459. (In Chinese)
69. Li, H.; Liu, J.; Wang, Y.; Yao, T.; Tian, Q.; Li, S. Deformation and cracking modeling for early-age sidewall concrete based on the multi-field coupling mechanism. *Constr. Build. Mater.* **2015**, *88*, 84–93. [CrossRef]
70. Gasch, T.; Malm, R.; Ansell, A. A coupled hygro-thermo-mechanical model for concrete subjected to variable environmental conditions. *Int. J. Solids Struct.* **2016**, *91*, 143–156. [CrossRef]
71. Jirásek, M.; Havlásek, P. Microprestress–solidification theory of concrete creep: Reformulation and improvement. *Cem. Concr. Res.* **2014**, *60*, 51–62. [CrossRef]
72. Pandey, R.N.; Pandey, S.K.; Ribeiro, J.W. Complete and satisfactory solutions of luikov equations of heat and moisture transport in a spherical capillary—Porous body. *Int. Commun. Heat Mass Transf.* **2000**, *27*, 975–984. [CrossRef]
73. Pandey, R.N.; Srivastava, S.K.; Mikhailov, M.D. Solutions of Luikov equations of heat and masstransfer in capillary porous bodies through matrixcalculus: A new approach. *Int. J. Heat Mass Transf.* **1999**, *42*, 2649–2660. [CrossRef]
74. Chang, W.J.; Weng, C.I. Analytical solution to coupled heat and moisture diffusion transfer in porous materials. *Int. J. Heat Mass Transf.* **2000**, *43*, 3621–3632. [CrossRef]
75. Gaur, R.C.; Bansal, N.K. Periodic solution of Luikov equations for heat and mass transfer in capillary porous bodies. *Int. J. Energy Res.* **1999**, *23*, 875–885. [CrossRef]
76. Bloch, M.A.F. Heat and mass transfer in wet porous media in presence of evaporation—condensation. *Int. J. Heat Mass Transf.* **1998**, *41*, 2263–2277. [CrossRef]
77. Ribeiro, J.W.; Cotta, R.M.; Mikhailov, M.D. Integral transform solution of Luikov's equations for heat and mass transfer in capillary porous media. *Int. J. Heat Mass Transf.* **1993**, *36*, 4467–4475. [CrossRef]
78. Lobo, P.D.C.; Mikhailov, M.D.; Özisik, M.N. On the complex eigen-values of luikov system of equations. *Dry. Technol.* **1987**, *5*, 273–286. [CrossRef]
79. Chang, W.J.; Chen, T.C.; Weng, C.I. Transient hygrothermal stresses in an infinitely long annular cylinder: Coupling of heat and moisture. *J. Therm. Stress.* **1991**, *14*, 439–454. [CrossRef]
80. Cheroto, S.; Guigon, S.M.S.; Ribeiro, J.W. Lumped- Differential Formulations for Drying in Capillary Porous Media. *Dry Technol.* **1997**, *15*, 811–835. [CrossRef]
81. Li, R.; Li, X. Mathematical model and numerical method for simulation of coupled chemo-thermo-hydro-mechanical process in concrete subjected to fire. *Acta Mech. Solida Sin.* **2006**, *38*, 471–479.
82. Bernard, F.; Kamali-Bernard, S.; Prince, W. 3D multi-scale modelling of mechanical behaviour of sound and leached mortar. *Cem. Concr. Res.* **2008**, *38*, 449–458. [CrossRef]
83. Li, Z.; Wang, Y.; Wu, B.; Wang, C.; Chen, Z. Multi-scale modeling and analyses on structural deterioration and damage in long-span bridges and its application. *Acta Mech. Solida Sin.* **2010**, *31*, 731–756. (In Chinese)
84. Li, Z.X.; Zhou, T.Q.; Chan, T.H.T.; Yu, Y. Multi-scale numerical analysis on dynamic response and local damage in long-span bridges. *Eng. Struct.* **2007**, *29*, 1507–1524. [CrossRef]
85. Li, Z.X.; Chan, T.H.T.; Yu, Y.; Sun, Z.H. Concurrent multi-scale modeling of civil infrastructures for analyses on structural deterioration part I: Modeling methodology and strategy. *Finite Elem. Anal. Des.* **2009**, *45*, 782–794. [CrossRef]
86. Sheng, J. Fully coupled thermo-hydro-mechanical model of saturated porous media and numerical modelling. *Chin. J. Rock Mech. Eng.* **2006**, *25*, 3028–3033. (In Chinese)

Publisher's Note: MDPI stays neutral with regard to jurisdictional claims in published maps and institutional affiliations.

MDPI
St. Alban-Anlage 66
4052 Basel
Switzerland
Tel. +41 61 683 77 34
Fax +41 61 302 89 18
www.mdpi.com

Materials Editorial Office
E-mail: materials@mdpi.com
www.mdpi.com/journal/materials

www.ingramcontent.com/pod-product-compliance
Lightning Source LLC
Chambersburg PA
CBHW061806210926
51850CB00006B/629

* 9 7 8 3 0 3 6 5 4 9 1 2 5 *